AF713442

SPIDERS

Webs, Behavior, and Evolution

SPIDERS

Webs, Behavior, and Evolution

Edited by
William A. Shear

Stanford University Press
Stanford, California 1986

Stanford University Press
Stanford, California
© 1986 by the Board of Trustees of the
Leland Stanford Junior University
Printed in the United States of America

CIP data appear at the end of the book

Preface

AS AN UNDERGRADUATE student at the College of Wooster in Ohio in the late 1950's and early 1960's, I first became interested in spiders through the encouragement of a remarkable teacher, Andrew Weaver. About that time, there could not have been more than a dozen researchers in the United States actively working on all aspects of spider biology. Luckily, beginning students like myself were able to communicate freely with these authorities, who gladly exchanged information and specimens with anyone enthusiastic about their favorite animals.

Ten years later the American Arachnological Society was formed, with an original membership of some 100 scientists, a number of them interested Europeans, and a few from the Latin American countries. Our early meetings were attended by 25 to 30 members, and discussions were informal, though stimulating. By 1976, the AAS was holding triennial meetings, and the 1981 congress at the University of Tennessee was attended by more scientists than the society had charter members only 8 years before. Two symposia and several formal paper presentations were held, and some seasoned participants described the meeting as the best scientific meeting that they had ever attended.

I give this bit of history to draw attention to the phenomenal growth of interest in spiders and other arachnids in the last two or three decades. Researchers in arachnology now find essential the annual volume published by the Centre International de Documentation Arachnologique in Paris, listing the year's publications on arachnids. (The volume for 1985 contained 873 entries.) The AAS publishes a refereed journal, the *Journal of Arachnology*, now in its tenth volume, there are older, similar journals in Britain and Japan, and a French journal has recently begun publication. Research activity has increased to the point that it has become difficult to keep up with the entire field, and we talk of spider physiologists, ecologists, and ethologists, as well as systematists and anatomists. To become thoroughly versed in what is now known about spiders (to say nothing of other arachnids) would be an immense task indeed—and so much remains to be studied. Every year brings more news of startling discoveries of unusual new species, or of strange behavioral vagaries in even the most common spiders.

Like many other biological specialties, arachnology had its beginnings in systematics: in a desire to perceive and describe order in the living world. To recast what has become an old saw, evolution makes sense only in the light of systematics. Right from the start, a strong evolutionary theme unified the study of spiders, as it has the study of other major groups of plants and animals. The theme of this volume, and, we hope, of forthcoming surveys, is that evolution continues to unite and make sense of biology.

This book had its beginnings in one of the symposia held at the 1981 meeting of the AAS at the University of Tennessee. The idea of the symposium was to examine one important aspect of spider behavior—web building and use—from a structural/functional viewpoint, and then, with the results of that examination in hand, to proceed to an evolutionary analysis of spider webs. The topic was broad, and the coverage was by no means comprehensive, but each paper was united to the others by a strong reliance on the evolutionary theme. The symposium was so well received that it was decided to ask the participants to expand their contributions to include reviews and analyses of previous work, as well as new results, so that the papers might be published in book form. You hold the result in your hands.

To cover in depth every aspect of spider biology having to do with webs and silk would require a volume at least five times the size of this one. Our purpose here is to provide an outline for future research, and a framework into which the new results may be placed. We hope that other aspects of spider biology, such as silk chemistry and physics, the physiology of silk production, and the ecological significance of the use of silk traps by spiders, can be treated similarly in future years.

Strangely, spiders are seldom included in general books or surveys of animal behavior. This is remarkable, because so many of the major ideas of behavioral biology can be illustrated by fascinating examples from spiders. Our book should commend itself to ethologists, sociobiologists, and even neuroethologists. They will find much here to interest them.

Nor do I hesitate to say that amateurs in natural history—that rapidly growing and increasingly sophisticated cadre—can read and understand nearly everything in this book. Our approach is straightforward enough, and our story exciting enough, to keep the scientifically inclined layman interested throughout.

A few words about our authors. All are active researchers (as evidenced by the difficulties faced by an editor whose authors have scattered for the field season to Costa Rica, Mexico, New Guinea, the Galápagos, and Africa), yet each of them has a gift for making his or her work accessible to other biologists and even to nonbiologists. Some adopt a highly technical approach, using sophisticated laboratory equipment to isolate single bits of behavior and scrutinize them closely, as Mitchell Masters and his colleagues do in focusing laser beams on single, vibrating silk threads. William Tietjen has found computers useful in monitoring the activities of the social spider *Mallos gregalis.*

Anthony Janetos, Susan Riechert, Rosemary Gillespie, and George Uetz are ecologists. They approach the problems of web-building behavior from the viewpoint of the interaction of spiders with their environment, and with one another. The rest of us are interested in behavioral biology. William Eberhard, Mark Stowe, Yael Lubin, and William Tietjen have conducted most of their research in this area, whereas Frederick Coyle, James Carico, Jonathan Coddington, and I began as taxonomists, and that field remains one of our major interests. Attempts to work out the classification of spider groups invariably lead to thinking about their evolution, and field collecting brings behavior to one's attention. The relationship between behavior and evolution in spiders is as obvious as it is anywhere in the animal world, so it is not surprising that people who are primarily interested in evolution turn also to behavior. Coddington has become deeply involved in the evolutionary meaning of behavior, as he attempts to understand the taxonomy of the orb-weaving Theridiosomatidae.

Eberhard and Stowe are ethologists whose work exemplifies the comparative method. Eberhard has published numerous papers on remarkable types of webs—making discoveries and fitting them into the larger scheme of behavioral analysis. He found the strange web of *Synotaxus*, and wrote the best study yet of a bolas spider—a new species that he named *Mastophora dizzydeani*, in honor of a baseball hero. Stowe spends long nights in the field, watching and recording the behavior of spiders that use moth attractants, and discovered the "asterisk" web of *Wixia*. Lubin has concentrated on the behavior of orb weavers, and in her chapter surveys a little-known family of this group.

Coyle is well known for his detailed, painstaking systematic work on the difficult mygalomorph spiders, and he has gained from this study a good understanding of the systematic value of behavior and natural history. Silk use by the mygalomorph spiders has been little investigated, and some remarkable results surely will be forthcoming. Carico has only just begun the first intensive study of how spiders take their webs down, and reports on his results here. I hope that my own contribution provides an adequate summary of all the material in this volume, though it treats some topics only indirectly. My interests are broad, often extending to other groups of arachnids, to the myriapods, and more recently even to paleontology. I have tried to provide an evolutionary overview of spider webs.

We have many people to thank, and to attempt to list them all here would be inevitably to omit the name of some helpful person. As editor of this volume, I especially appreciate the excellent work of Karen Harris, of the

J. B. Fuqua Computing Center at Hampden-Sydney College, in preparing the manuscripts. Final alterations to the figures were made by Donna Breckenridge, under the editor's direction. Initial typing of my own contributions was done by Jean Hudson, with characteristic efficiency and good cheer. Preparation of the manuscript was aided by major grants from Hampden-Sydney College and the American Arachnological Society. Additional financial assistance was provided by Western Carolina University, the University of Cincinnati, and Harvard University.

The intense interest and constant cooperation of William Carver, John S. Feneron, and David Hathwell at Stanford University Press is something I appreciate more than I can say. My wife, Noelle, provided a great deal of inspiration and aid by her unfailing energy and support.

I shall close this preface by acknowledging a debt owed by all of us in spider studies, and that debt is to the taxonomists. In many cases it is they who, by their sweeping surveys of whole families of spiders, have drawn attention to problems in the interpretation of behavior, and who by their diligent labors (often in isolation) have given us the evolutionary framework that we seek to test, modify, and test again. Because orb-weaving spiders figure so prominently in this book, I must single out one influential figure still forging ahead on a decades-long taxonomic revision of that group. Professor Herbert W. Levi, of the Museum of Comparative Zoology, has contributed enormously to our understanding of spiders, through his patient work and great store of biological knowledge. He has served as mentor and graduate advisor to five of the authors of this book, and has been an important influence on most of the others. With the greatest affection and respect, this volume is dedicated to him.

W.A.S.

Contents

Contributors

JAMES EDWIN CARIÇO is Professor of Biology at Lynchburg College in Virginia. He received his Ph.D. from Virginia Polytechnic Institute and State University. His main research activity is in the systematics of the family Pisauridae in the Western Hemisphere. Other research activities include a worldwide revision of *Dolomedes* (Pisauridae) and various behavioral studies of orb-weaving spiders.

JONATHAN CODDINGTON received his Ph.D. from Harvard University in 1984 and is now Associate Curator of Arachnids and Myriapods at the Smithsonian Institution, National Museum of Natural History. His research includes morphological and behavioral studies of orb-weaving spiders, especially the family Theridiosomatidae. His fieldwork has been mainly in Central and South America.

FREDERICK A. COYLE received his Ph.D. in biology from Harvard University in 1969. His primary research interests are the systematics, behavior, and ecology of mygalomorph spiders. He has conducted field research on these spiders throughout the United States, Mexico, and Costa Rica. He is Professor of Biology at Western Carolina University and Research Associate in the Department of Entomology of the American Museum of Natural History.

WILLIAM G. EBERHARD is a member of the staff of the Smithsonian Tropical Research Institute and Profesor Asociado at the Universidad de Costa Rica, where he teaches biology. After receiving his Ph.D. from Harvard University in 1969 he moved to Colombia and later to Costa Rica, where he has worked on the ecology and behavior of spiders and insects, and the evolution of intracellular organelles and animal genitalia.

ROSEMARY G. GILLESPIE is a graduate student of zoology at the University of Tennessee. She received her B.Sc. in zoology from Edinburgh University, Scotland, in 1980 and moved to Tennessee in 1982 to study under Susan E. Riechert. The subject of her doctoral research is habitat selection in web-building spiders. She is also involved in a field study at Highlands Biological Station in North Carolina, where she is working on the long-jawed spider *Tetragnatha elongata*.

ROBERT R. JACKSON is a Senior Lecturer in Zoology at the University of Canterbury in New Zealand, where he teaches animal behavior and evolutionary biology. He received his Ph.D. from the University of California at Berkeley in 1976. Born in North Carolina, he has studied spiders in Africa, Asia, Australia, Europe, New Zealand, and North America.

ANTHONY C. JANETOS received his Ph.D. from Princeton University in 1980. Besides the behavioral ecology of spiders, he is interested in the population dynamics of assemblages of species. He is currently a Research Fellow at the Harvard School of Public Health.

YAEL D. LUBIN was educated in Israel and the United States. She received her Ph.D. in zoology from the University of Florida in 1972 and is now a Research Associate at the Smithsonian Tropical Research Institute in Panama and Visiting Assistant Professor at the University of Florida. Her studies of the behavioral ecology of tropical spiders and the biology of social insects have taken her to Central and South America, Papua New Guinea, and Australia. At present, she is investigating sociality in spiders.

HUBERT S. MARKL received his doctorate in zoology from the University of Munich in 1962. He has taught at the Universities of Munich, Frankfurt, and Darmstadt and is presently Professor of Biology at the University of Constance, West Germany. He has done fieldwork on social insect behavior in the United States, Trinidad, and Australia. His major research interests are communication in animals and the evolution of social behavior.

W. MITCH MASTERS received his Ph.D. from Cornell University in 1979. Along with his wife, Anne J. M. Moffat, he then spent three years at the University of Constance, West Germany, in the laboratory of Hubert Markl studying vibrations in orb webs and spiders' responses to web-borne stimuli. He and his wife are presently at the University of Oregon.

ANNE J. M. MOFFAT is interested in the neurophysiology of hearing, which she has studied in frogs, mice, and, currently, bats. While at the University of Constance she worked with her husband, W. Mitch Masters, on studies of vibration in spider webs.

SUSAN E. RIECHERT is a Professor of Zoology at the University of Tennessee. She received her Ph.D. in zoology with a botany minor from the University of Wisconsin in 1973. Riechert is a behavioral ecologist with interests in the evolution of social structure, in within- and between-species competition, and in the potential of spiders as agents of biological control. She has completed field studies in the desert southwest, Tennessee, Wisconsin, the Galápagos Islands, and equatorial west Africa.

WILLIAM A. SHEAR is Professor of Biology at Hampden-Sydney College. He received his doctorate in evolutionary biology from Harvard University in 1971. He has published more than 40 articles and book chapters on the taxonomy, evolution, and behavior of arachnids and millipedes. His current research is on fossils of the oldest terrestrial animals from America, 380-million-year-old arachnids from the Devonian period.

MARK K. STOWE is a graduate student at Harvard University working at the Museum of Comparative Zoology and the USDA Insect Attractants Laboratory in Gainesville, Florida. His thesis research on the moth-attracting spiders stems from long-time interests in spider behavior and chemical ecology.

WILLIAM JAMES TIETJEN is Assistant Professor of Biology at Lindenwood College. He received his Ph.D. in zoology in 1980 from Ohio University, where he worked on chemical communication in spiders. His current research in sociobiology was begun while he was a postdoctoral researcher at the North Carolina Foundation for Mental Health Research.

GEORGE W. UETZ received his Ph.D. in ecology at the University of Illinois in 1976. His research is centered on the behavioral ecology of spiders, particularly their courtship and social behavior. He has made numerous trips to Mexico to study the social behavior of communal spiders, funded by the National Geographic Society and the American Philosophical Society. He is currently Associate Professor of Biological Sciences at the University of Cincinnati.

SPIDERS

Webs, Behavior, and Evolution

Introduction

WILLIAM A. SHEAR

THIS IS A BOOK of questions. For some of them we have at least tentative answers, but for others there only seem to be a number of conflicting hypotheses. All of the questions concern a fascinating fact that, through familiarity, has become mundane: there exists a group of arthropods distinguished by their ability to make structures from protein fibers produced from their own bodies, and to use these structures in highly adapted ways. I refer, of course, to spiders.

Though other arthropod groups (such as lepidopterous insects) are capable of making silk constructs, in no other animals are these structures such a dominating fact of life for their makers. So much is this true that a recent book on spiders, published in German, bears the title *Leben am seidenen Faden*: life on a silken thread (Stern and Kullmann 1981). Biologists have long recognized that, for spiders, silk is a multiuse material. They trail it out continuously behind them as a dragline, and the path so marked serves not only the marker but also potential mates (Tietjen and Rovner 1982). It lines their hiding places. Many species construct tents of silk in which to spend their resting hours. Egg masses are covered with silk that frequently is of a special kind, tough, leathery, fibrous, or even woody (Kullmann et al. 1975).

But most characteristic of spiders, predators all, is the use of silk to make webs. The major function of a spider's web is the detection and capture of prey, though some webs, such as the nurseries of the pisaurids, have secondarily taken on other functions. The earliest and simplest webs were mere accidental extensions from a hiding place. As spiders darted out from their burrows or from silk-lined natural crevices, they trailed their draglines. At first serving as triplines to warn of approaching prey, these few threads, with the action of natural selection over many millennia, became the elaborate constructs that we see in spiders today. (See the chapters in this volume by Coyle, Coddington, and Shear.)

Certainly, this account is an oversimplification. To begin with, do we not too often assume adaptation without really testing for its presence? The naturalistic fallacy has proved very tempting to evolutionary biologists, and has led to what the sympathetic call "scenarios" and the critical "just-so stories." The new field of sociobiology seems particularly prone to this sort of speculation. In this book, there is indeed speculation, and several scenarios. But many of them are distinguished by having been subjected to clever tests, and so have validity added to plausibility (Stowe, this volume; see also the chapters by Eberhard and Carico). And, of course, though webs seem to contribute most to the predatory activities of spiders, they also serve other requirements, such as communication with conspecifics (Witt and Rovner 1982), and especially protection, because a web is frequently an obvious feature of the landscape, and tiny predators fall victims to larger ones, many of which hunt visually. The problem of predation on spiders that make webs is most acute in the communal spiders, the vast masses of silk of which sometimes enclose many cubic meters of space and contain thousands of potential prey for birds, wasps, lizards, and other spider enemies. In their chapters in this volume, Tietjen and Uetz examine the trade-off that must take place between the predatory activities of the spiders and potential predation on the spiders by other animals.

Webs are not always passive traps. The concept of the orb web, for example, as a mere sieve or filter for collecting aerial plankton and nekton is being seriously challenged (Chacón and Eberhard 1980; Eberhard, this volume). And it has been clear for some time that many spider species use their webs more like weapons. For example, *Theridiosoma gemmosum*, which can be observed under the overhanging banks of streams in the eastern United States, pulls up its little orb into a taut cone and snaps it loose on feeble flies. In the tropics, *Dinopis* species with eyes like headlamps flip squares of sticky silk at passing crawlers, pinning them to leaves and tree trunks. Even the common cobweb spider *Achaearanea tepidariorum* and its relatives use thrown sticky silk to subdue prey caught in their messy webs.

Nor do all spiders sample the local insect populations randomly. Many have become highly specialized to take only certain sorts of prey. Stowe (this volume) discusses the consequences for spiders of specializing on moths, which has led to the almost unbelievable adaptations (that word again!) of the bolas spiders. Some whole groups of spiders have even specialized in attacking other spiders. Members of the family Mimetidae have several strategies for invading webs and killing the owners, and virtually the whole theridiid genus *Argyrodes* consists of kleptoparasites, little spiders that live in the webs of bigger ones and steal prey from them. More than a few have "learned" to attack their hosts as well. There is even a parasitic spider. *Curimagua bayano* is a minute member of the Symphytognathidae from Panama, where it lives in the webs of sheet-weaving tarantulas. This strange animal crawls over its host's body and, as the giant feeds, literally drinks from its lips. *C. bayano*'s own chelicerae are so weak that it could never capture prey or feed on its own. We thus have an example of an obligate parasite among the spiders.

And what of those spiders that apparently do not use webs for prey capture? Members of the family Salticidae, for example, are the quintessential visual predators among all the invertebrates on land. Their tubular eyes have separately allocated functions, telescopic capacity, and even some color sensitivity (L. Forster 1982a). They seem almost like small robotic intelligences, as they look up at us with their two big, glossy, black main eyes. Yet, as Robert Jackson argues in Chapter 9, even these spiders have probably come down from web-building ancestors, perhaps by the route of invading the webs of other spiders to steal prey or to attack the owners. In the tropics, new discoveries are reported of web-building members of families of spiders previously thought not to use silk for predation.

This catalog of wonders could go on, but much more detail is put forth in the chapters that follow. Let me just say at this point that most biologists are committed to the paradigm of evolution by means of natural selection. Remarkable creatures, then, urge us to search for explanations. We put our questions to nature in the form of falsifiable hypotheses, discarding them only where they are shown to be untrue. In this book you will, as I say, find many questions, and a few answers.

How do spiders choose a spot to build a web? Do they simply disperse at random, and live or die wherever they happen to end up? Or can they test the important parameters of the environment and decide to move on? Imagine that a spiderling of a web-building species climbs to the top of a withered grass stem in early spring, and, letting out a long line of silk to the wind, is carried many miles—a process called ballooning. Do spiders balloon just once, or can they sample a habitat and balloon again if the habitat is unsuitable? And if they can, what is it they are sampling; what makes a good web site? In Chapter 1, Anthony Janetos attempts to define such questions more clearly. In Chapter 2, Susan Riechert and Rosemary Gillespie address the question of what happens after spiders choose their web sites and build their webs. They have studied a desert sheet-web spider, *Agelenopsis aperta*, in a highly variable environment where good web sites are scattered, and spiders defend them. How does this interaction affect the way the spiders use their habitat? How do they deal with other animals in their world and with physical factors such as heat and water balance? These are some of the questions that Riechert and Gillespie pursue.

We know that for many spiders the web stands between them and their environment, both as a protector and as a transducer of information. To be effective as a predatory device, the web must have characteristics that allow a spider to be accurately informed about prey. And not just about their presence—orb weavers in particular use different strategies to handle different kinds of captives. Studies of how spiders perceive vibrations through their webs are still scarce. In Chapter 3, Mitchell Masters, Hubert Markl, and Anne Moffat describe how they are approaching this problem of understanding the spider's *Umwelt*. Using sophisticated laser vibrometry, they have found that the web and the spider together are a responsive system that filters the vibratory stimuli in the environment. Certainly one of the main constraints on web architecture must be the spider's use of the web as an extension of its senses.

There are many other constraints. The nature of silk itself provides certain physical givens (Langer 1969)—among other things, all parts of the web must be under tension if any degree of rigidity is to be maintained. There are only so many ways such a structure can be made, and though the behavioral limitations are there, we know little enough about them. William Eberhard is publishing a series of papers (see the Bibliography) describing and analyzing the ways in which orb webs are built, and has found that the details of how a spider attaches one thread to another, and measures where the next thread goes, can be used to see relationships among the thousands of orb-weaving species. Some of his work is reviewed by Coddington in Chapter 12.

In his contribution to this volume (Chapter 4), Eberhard extends ideas put forward a few years ago (Chacón and Eberhard 1980). He challenges the common assumption that an orb web works like a simple sieve. (That idea may still be tenable, by the way; Riechert and Gillespie in Chapter 2 and Uetz in Chapter 8 find it useful.) Eberhard tests his ideas by constructing model webs and simulating the impact of prey items. Here is a good example of a testable hypothesis in evolutionary biology, and when we know how webs work, we shall be able to understand their adaptiveness (or lack of it) much better.

Another common assumption as yet relatively poorly tested is that spiders that use silk sparingly compared to their relatives do so because silk is expensive. In my contribution (Chapter 13) will be found the suggestion that reduced webs are attractive, where permissible, if they allow relatively inexpensive movement to be substituted for relatively expensive silk synthesis. Mark Stowe adds to this the factor of prey specialization. It has sometimes been observed that members of the spider family Araneidae catch biased samples of available insects. Are webs attractive to certain kinds of insects because of something the spiders do? In Chapter 5, Stowe reviews the evidence for the use of aggressive chemical mimicry by certain araneids, and concludes that this extreme form of prey specialization (catching, e.g., only male moths of one species by mimicking the sex attractant of the female moth) has evolved in such a way that the typical orb web is reduced. Phylogenetic series can be arranged, and put forward as hypotheses to be tested by their congruence with other lines of development. Evidently both prey specialization and the more efficient use of energy are involved in the phylogenetic reduction of orb webs. Again, by learning about the functions of such webs in detail, we gain understanding of their evolution.

The orb web seems to us such a perfect piece of work that the likelihood of its having evolved twice seems slight. But that is exactly one of the evolutionary hypotheses, if not the dominant one: that in two spider families, the Uloboridae and the Araneidae (plus annectent families), similar selection pressures led to the invention of the orb by unrelated spiders. The Uloboridae have a spinning plate, or cribellum; the Araneidae do not. The significance of this difference has been greatly overestimated in the past, and current thinking is that cribellate and ecribellate species may occur in the same genus; the cribellum was a feature of the primitive araneomorphs and has been lost many times independently (Lehtinen 1967, Platnick 1977). This new conception of the cribellum opened the way for the revival of Thorell's (1886) hypothesis that the orb web was invented just once, by a cribellate common ancestor of the modern Uloboridae and Araneidae. This hypothesis is taken up in detail and thoroughly analyzed by Coddington in Chapter 12. He supports it. As I indicate in Chapter 13, I am not so sure.

In Chapter 6, Yael Lubin characterizes the uloborids as "poor relatives" of the more intensively studied araneids. Uloborids are secretive, most kinds occur in deep tropical forests, and they are rather small and hard to study. Lubin surveys all that is known of their web-related behavior and predatory versatility, and demonstrates that they are no less interesting than their larger, more common (possible) sister group. Of special interest is the fact that among the uloborids are several groups with reduced webs, providing an opportunity for an independent test of hypotheses about phylogenetic web reduction.

The "poor relatives" of all the araneomorph spiders are the mygalomorphs and mesotheles. The former include what we in the United States call tarantulas, and their relatives, the trap-door spiders. But even here in North America these retiring, mostly burrow-dwelling spiders show unsuspected diversity. There are species making camouflaged silk tubes against the trunks of trees, and tiny species making filmy webs in moss mats on tops of the high Appalachians. Worldwide they have exploited their fossorial habits to diversify remarkably, yet few arachnologists study them. The mesotheles have fared better, only because they seem so exotic. With segmented abdomens and spinnerets near the middle of the venter, they look like living Coal Age fossils. And one must travel to Borneo, Malaysia, or Japan to study them in their native habitats. There has been a need for a survey of the behavior of these two spider groups with regard to silk use, and in Chapter 10 Frederick Coyle provides it. He also discusses factors related to predatory behavior that have been balanced against requirements for protection and concealment to produce some remarkable architecture.

Only one mygalomorph, *Macrothele darcheni*, of Africa, has adopted communal habits. This may be accounted for in part by the fact that few mygalomorphs make proper webs, the presence of a web being almost a prerequisite for social life in spiders (Shear 1970a). Though the young of the wolf spider cluster together on their mother's back, when they move apart they "lose touch" in the figurative sense as well, and without a medium of constant communication the formation of a society would seem unlikely. By contrast, the young of web-building spiders emerge in large numbers from egg sacs and quickly make a tangle of threads. The spiderlings may form a tight cluster within their tangle, but only touch one thread and all immediately react in panicky dispersal. If such an association is prolonged, a spider species is on the path to subsociality. Like other animals exhibiting forms of sociality, communal spiders are currently the subject of intensive study. Two chapters concerning their webs are to be found here. In Chapter 7, William Tietjen provides a useful survey of silk constructs in communal spiders, and reviews his own research on the life of *Mallos gregalis*, a small communal dictynid from Mexico. Such spiders again raise questions about the trade-offs necessary under the impact of natural selection. For example, communal spiders gain additional protection by being deeper in the mass of silk making up their nest. But insect prey are unlikely to penetrate far inside the nest, so the chances of eating well are sacrificed for a greater measure of security. How is this dilemma resolved? Tietjen's research provides a partial answer.

Another communal spider has evoked comment for years from naturalists traveling in Mexico. The "power-line spider" became legendary for its massed, adjacent webs extending from tall electrical transmission or telephone lines to the ground and going on for many miles along rights-of-way. In Chapter 8, George Uetz identifies the spider as *Metepeira spinipes*, and reports that it can live solitarily in small colonies, or in massive communal webs that last for years and may grow ever larger. The web-related adaptations of *M. spinipes* are quite different from those of *Mallos gregalis*, because it is hard to conceive of more than one spider building a single orb. So each *Metepeira spinipes* has its own orb in the common mass of lines, interacts territorially with its neighbors, and attacks and subdues prey on its own. As well as summarizing his work on this spider in the field, Uetz reviews the predatory strategies of other communal orb weavers.

The individual orbs of *M. spinipes* are renewed daily, as are those of many (but certainly not all) orb weavers. It is peculiar that no one has thought of web removal as a part of the overall web-building program, but until the idea was investigated by James Carico, little was known even about how removal was accomplished. Webs go up and down at odd hours—one must be willing to miss some sleep. When we have descriptions of the methods that spiders use to take webs down, we shall be able to construct hypotheses about why they do so. The key original research in this new area of study is presented in Chapter 11, by Carico.

The last two chapters—Jonathan Coddington's (Chapter 12) and my own (Chapter 13)—have been mentioned here and there in the preceding paragraphs. Coddington's contribution will, it seems to me, be regarded as one of the most valuable essays in spider biology for some time to come. He not only summarizes and gives order to a great deal of scattered information about orb webs, but also presents a model of a rigorous, classical evolutionary argument to support the hypothesis that the orb arose but once. This hypothesis is a challenge to spider taxonomists and anatomists to test it against their own data. Pay particular attention to the question that he ends with: what, if anything, is an orb web?

I shall introduce my own contribution just by suggesting that the reader unfamiliar with the variety of silk constructs made by spiders may find it useful to read Chapter 13 first. For the arachnologist, it may be a report on the state of the art; its title should make it clear that it is speculative and meant to stimulate. I hope also that it succeeds in summarizing much of the material in the preceding 12 chapters in a phylogenetic framework.

As to the question of background material: we have little space to present it here, and recommend one or two of the many excellent popular accounts, some regional in scope, of spider natural history. For Americans Willis Gertsch's *American Spiders* (1979) is good, and for Europeans William Bristowe's *The World of Spiders* (1958) is difficult to surpass for its graceful prose and wealth of natural-history observations. The recently published *Biology of Spiders* by Rainer Foelix (1982) is concise and up-to-date; its strengths are in its clear accounts of spider anatomy, physiology, and basic behavior. The Bibliography of the present volume can be searched for introductory books suited to many regions, including Australia, New Zealand, Africa, and Japan.

To reiterate, this is a book full of questions, some answered, some not—a book containing guesses, speculations, and "scenarios," as well as rigorously tested hypotheses. Our aim here is to summarize what current research suggests about function and form in spider webs, and how they have evolved. We hope that, in addition, we stimulate our readers—biologists and nonbiologists alike—to investigate further the questions that we raise.

I

Web-Site Selection: Are We Asking the Right Questions?

ANTHONY C. JANETOS

THE PROBLEMS INVOLVED in studying web-site selection in spiders are numerous, as any field researcher knows. Not least among them is the problem of choosing the right approach. So far, major contributions have been made from the points of view of natural history, physiology, and behavioral ecology. All of these approaches have strengths; none can claim preeminence.

Web-site selection is a complex phenomenon, calling for complex analysis. The first task is descriptive. Which habitats do particular species favor, and where in those habitats do they spin webs? The next level of inquiry is concerned with the question, why are those locations favored? These questions involve consideration of the physical structure of the habitat, the physiology of the spider, the presence of food, and protection from predators.

The behavioral ecologists' contributions are of a different nature. Given a set of suitable places to build webs, how does a spider choose to spin in a particular place? Alternatively, one might ask, why does a spider ever leave a web site, having found it? These questions involve subtleties of behavior and data on habitat variability not previously considered. The latter question, in particular, leads to some surprising insights into the ecology of spider communities, and makes it possible to compare foraging studies on spiders with studies on other groups.

This chapter has four goals. The first is to suggest the diversity of the studies that have been made on web-site selection. The second is to examine a few of the important partial answers to the question of why spiders spin webs in some places and not in others. The third is to describe in some detail the newer contributions from behavioral ecology. The last is to examine current ideas about web-site selection and to suggest ways in which these ideas might be amended.

It will be apparent that my review is selective, and that I emphasize the recent contributions of ecologists. This is due in part to my own bias as an ecologist, but in part also to the fact that these newer findings are only beginning to attract attention, whereas earlier studies are already familiar. I hope to create the impression, not that earlier studies are somehow incomplete or flawed, but that the field is advancing swiftly in all areas.

The Suitability of Web Sites

I shall begin by discussing the wide variety of factors that determine the suitability of a web site.

Habitat

The first step in the selection of a suitable web site is, of course, the selection of a habitat appropriate for web building. For some species "selecting" a suitable habitat is not a problem. For example, *Cyclosa conica* does not balloon in riparian woods in canyons in the West; rather, the juveniles disperse by walking and climbing away from the spot where the egg sac was deposited (pers. obs.). Presumably, this behavior evolved because of the extremely patchy nature of the habitat. Ballooning spiders would have a high probability of landing in an unsuitable environment, namely, desert. As a result, the young spiders are not confronted with crucial choices about the suitability of habitats for web building.

The problem faced by long-distance dispersers, especially ballooning species, is more complicated. Duffey (1956) cataloged the aerial dispersal of species in censused populations from grasslands and woods. Species typical of each habitat were found to balloon, yet there is no a priori reason to expect departure from woods to ensure arrival in woods, or departure from meadow to ensure arrival in meadow. What happens to spiders that land in an unsuitable habitat is largely unknown. They may have behavioral mechanisms for finding a suitable habitat, or they may simply die. The extent to which habitat choice is an active process or a passive, probabilistic endeavor is simply not known.

Physical Support

A potential web site within a suitable habitat must above all possess the appropriate physical supports for the architecture of the web. The number of available web sites is usually determined by the percentage of vegetation cover and the variety of its growth forms. Simple grassland and desert habitats certainly provide the least diverse range of web sites. As succession proceeds, the physical complexity of the habitat increases. However, a mature, closed-canopy forest may not have the diversity of available sites found in a young, open forest or an old field of late successional stage.

The number of available sites influences the appearance of web-spinning species. In an intriguing study of spiders in a warm desert, Chew (1961) collected a total of 20 species in an area with a total vegetation cover of 21 percent, most of which was creosote bush, and found that only six of the 20 species were web spinners. Comparing these numbers with those of other studies, Chew found that preforest successional stages around Lake Michigan had similar numbers of species (Lowrie 1948), but that forest climax communities had not only more species (122–68) but also a greater proportion of web spinners (29–48 percent). The proportion of web-spinning species thus increases with increasing vegetational complexity. Duffey (1962a, 1962b, 1966, 1968) provides strong evidence of similar trends in different habitats. Of particular interest is his observation that the greatest number of spider species occurred in the most floristically diverse habitat, which was not necessarily that with the greatest percentage of cover.

Recently, Robinson (1981) provided striking experimental confirmation of these correlative studies. In field sites in montane canyons in northern Utah, he investigated the effects of increasing the physical complexity of a habitat on the number of spider species present. Using chicken wire cages strung with jute to provide varying degrees of spatial heterogeneity, he found that increased diversity of attachment points resulted in the presence of larger numbers of web-building species.

In all of the preceding studies attachment sites and architectural support are defined in a broad way, since these studies consider broad patterns, such as species richness. In fact, the appropriateness of the physical structure varies with the type of web. For example, orb webs require only very simple supports; they can be spun in a nearly empty space that has vertical surfaces on either side for the attachment of bridge threads. On the other hand, the bowl-and-doily web of *Frontinella pyramitela* requires a particular structure of supports: a fork in which to spin the bowl and an overhanging projection from which to hang the tangled threads of the knockdown trap (Janetos 1980a, 1982b). *Neriene radiata* requires similar supports for its three-dimensional filmy-dome web.

This is not to say that all orb weavers or all sheet weavers require identical supports for webs. Variation in appropriate supports exists within, as well as between, web types. For example, among orb weavers, many *Araneus* species in old fields require broad-leaved plants in which to construct retreats, but *Argiope* species of similar size are able to construct webs in nearly any vegetation sturdy enough to bear their weight (Sholes and Rawlins 1979; Olive 1980).

Support requirements may change during the growing season as a result of the spider's growth (Sholes and Rawlins 1979). This is especially so for some of the large-bodied araneid species, which undergo explosive growth (increases in weight of three to four orders of magnitude).

Physiological Requirements

Physiological requirements are also important. The simple statement that a spider will not build at a site for which it is physiologically ill adapted hardly suggests the complexity of the spider's actual behavior. For one thing, the spider may have a wide range of physiological tolerances, and the width of that range may depend on the spider's ecological relationships. Colonizing species are often hypothesized to have wider tolerances for different environmental conditions than do species typical of climax communities (MacArthur and Wilson 1967). Duffey (1968) and Meijer (1977) found this to hold true for spiders typical of two stressful and newly opened habitats: seashore and the surrounding dunes (Duffey) and land newly reclaimed from the sea (Meijer).

Another important point is that a spider cannot predict what the physiological parameters of a new site will be before arriving there and attempting to spin a web. Thus, it is indeed possible for a spider to spin a web at a site for which it is physiologically ill adapted. A spider can only react to environmental conditions, not predict them.

This point is best exemplified in a series of studies by Riechert and coworkers, who documented in impressive detail the interplay between habitat structure, environmental stress, physiological requirements and tolerances, and food supply for the desert-dwelling species *Agelenopsis aperta* (Riechert 1974a, 1976, 1978a, 1981; Riechert, Reeder, and Allen 1973; Riechert and Tracy 1975). An agelenid, this species builds its sheet web in depressions in the ground. The walls of the depression are used as structural support for the sides of the sheet. Small crevices, when present, are used for the construction of the funnel. Other important structural and physical features are the presence of shrubs and of flowers, animal dung, and other insect attractants. The placement of the web on high ground is important during the rainy season, for webs on low ground are flooded.

The physiology of the spider interacts intimately with its web site. The habitat can be harsh; insolation and radiant energy are both extremely high during the hot months of the year. Riechert and Tracy (1975) present a detailed analysis of the heat load experienced by the spider at different times of the day and compare it to the thermal tolerance of the species. They find that the heat load does make some sites less advantageous than others, not because the heat load prevents the spider from building a web there, but because the extreme thermal stress during part of the day prevents the spider from being active on its sheet, the only place where it can catch prey. Because it must stay in the funnel, it loses opportunities for prey capture, and cannot feed as much or as quickly as conspecifics in more advantageous spots.

A. aperta lives in a very harsh habitat: the deserts of the southwestern United States. For obvious reasons, much of our understanding of the interplay of physiology and web-site selection comes from studies of such habitats. However, the same processes may occur in less stressful habitats. Of particular interest in this regard is the recent study by Biere and Uetz (1981) that shows the importance of behavioral thermoregulation in the orientation of the webs of *Micrathena gracilis*. This species is an orb weaver characteristic of the understory of deciduous forest in the eastern United States. It builds webs where sunflecks occur because of gaps in the canopy. Biere and Uetz found that all the webs in their study area had the same compass orientation, despite what would seem to be a relatively benign environment. The best explanation for this phenomenon is that the spiders are able to reduce heat stress by adjusting the orientation of their webs.

Tolbert (1979) and Pointing (1965) discuss the orientation of spiders in webs, rather than of the webs themselves, with respect to the sun as a means of thermoregulation, in *Argiope trifasciata* (Tolbert) and *Frontinella pyramitela* (Pointing).

The studies conducted in less harsh habitats are particularly interesting in the light of the presumed relationship between tolerance and colonizing ability, or tolerance and typical habitat, noted above. Perhaps species that typically inhabit older, more complex communities not only have reduced ranges of physiological tolerances but also are capable of more fine-tuned responses to environmental fluctuations. This could well prove to be a fruitful area for further study.

Supply of Prey and Protection from Predators

Supply of prey and protection from predators are also important in characterizing web sites. The importance of food supply has been emphasized in many recent studies, including those of Turnbull (1964), Janetos (1980a, 1982a, 1982b), Olive (1980, 1981a, 1981b, 1982a), Wise (1975, 1979), and Riechert and Tracy (1975). Protection from predators, though obviously important, has received less attention (Eberhard 1971a). I shall treat both of these subjects more fully later.

Web-Site Desertion

We have examined some of the characteristics of potential web sites that make them suitable for occupation. It is clear that the suitability of a web site depends on which species and which habitat one is considering. It is also clear that there will be some variability in these requirements within habitats; but the question of how a spider finds a suitable web site has long been a thorn in the side of arachnologists.

Enders (1973, 1975b, 1977) has given this question considerable attention for one species, *Argiope aurantia*, a common orb weaver. In simple choice experiments, he established that adults and juveniles of this species preferred the drier end of a humidity gradient. The spiders also chose the lighted end of boxes in a light-dark choice, but Enders's inability to alter the spiders' behavior (they preferred the light even after their eyes were destroyed) throws some doubt on this result. It is likely that the spiders were actually responding to a heat gradient. Juveniles responded to wind by building lower in the vegetation than during a calm period.

Enders's work shows that *A. aurantia* is not very selective about where it stops to build a web. None of the behavior that Enders investigated allows the spider to do any more than spin a web in an open space near the top of the vegetation. How the spider decides where to stop and start climbing is completely unknown.

I suggest that we are looking at the problem from the wrong perspective. A spider's initial relative indifference about where to start building is a real phenomenon, and it is futile to attempt to find very specific cues for the recognition of good sites. What is important, but as yet not fully studied, is a spider's decision to leave a web site. If we regard spiders as animals that must exploit a habitat whose quality they cannot predict, the importance of the decision to leave becomes clear.

In what follows, I shall discuss several topics. First, I shall examine the evidence that spiders actually do leave web sites. Then, I shall present a critique of current ecological theories of "giving-up" rules, and summarize a few studies that deal with variation among web sites. Finally, I shall describe a method for analyzing web-site desertion.

Evidence

That movement away from web sites does occur with some frequency has begun to be appreciated only with the advent of studies in which marked individuals are followed in a relatively limited area. Such movement has been known to occur in the laboratory since Turnbull (1964) demonstrated that juveniles of *Achaearanea tepidariorum* leave sites at which they receive few prey. *Agelenopsis aperta* females commonly leave web sites and either usurp the webs of other individuals or build new webs (Riechert 1974a, 1976, 1978a, 1981; Riechert, Reeder, and Allen 1973; Riechert and Tracy 1975). Wise (1979) reported that *Mecynogea lemniscata* and *Metepeira labyrinthea* abandon web sites on artificial supports and rebuild elsewhere. Wise (1975) also reported that adult females of *Neriene radiata* increase their residence times at natural web sites in response to augmented food supplies. Robinson and Robinson (1973a) reported that *Nephila maculata* occasionally leaves a web site and moves to another spot. In a study of the orb- and the sheet-weaving guilds, Janetos (1980a, 1982a, 1982b) documented that orb weavers frequently leave web sites and reestablish themselves elsewhere. Sheet weavers also move, but far less frequently. Variation in the frequency of movement exists within guilds (Janetos 1982a, 1982b), but the significant differences are between guilds.

Many, and probably most, disappearances from web sites are the result of motion to a new site and not predation. This can be documented in two ways. The first is to find marked individuals that have moved from marked sites and reestablished themselves elsewhere. However, most habitats are structurally too complex to make this a practical field method. The second is to census the same area every day, so that new arrivals can be marked as soon as they appear. As long as there is no reason to suspect that the incidence of predators in the study area is significantly higher than in the source areas, the comparison of appearances and disappearances will yield an estimate of the relative importance of predation.

Giving-Up Rules

Giving-up rules have received much attention in the ecological literature in the last few years. They have been used primarily to analyze problems in foraging ecology, although they certainly have other uses (Janetos 1980b). I shall introduce the most widely used of these rules, indicate its limitations, and suggest how it might be modified to apply to spiders.

Charnov (1976), following earlier work by MacArthur and Pianka (1966), Schoener (1971), and MacArthur (1972), formulated the marginal-value theorem: a forager leaves a patch when its rate of return in that patch falls to the overall average rate of return in the habitat. This theorem has generated a number of empirical studies on small passerine birds (Krebs, Ryan, and Charnov 1974; Cowie and Krebs 1979) and parasitic wasps (Cook and Hubbard 1977). The crux of the problem is the estimation of when the expected gain (in terms of the rate of net energy gain) elsewhere exceeds the expected gain from remaining in the same place. Because the forager cannot know the precise quality of a new patch in advance, the timing of the decision to leave the present patch is crucial.

Although the marginal-value theorem is a powerful insight, it cannot be applied to the behavior of spiders at web sites because it assumes that the predator causes the depletion of its prey resources in a patch. This assumption is reasonable for the animals already studied, for they obviously have the characteristic of predictably depressing the rate at which they find food in a patch (Charnov, Orians, and Hyatt 1976). However, the effect that spiders and other predators that filter food items from wind or water currents have on the abundance of their prey is not at all clear (Charnov, Orians, and Hyatt 1976; A. C. Janetos, in prep.). The supply of food at a web site is likely to be nearly independent of the presence of a spider. If, in addition, the site changes quality, the spider is faced with a problem that is fundamentally different from the one previously analyzed.

Other weaknesses of the marginal-value theorem are its implicit assumptions of continuous time and instantaneous estimation of averages. The first assumption probably holds for vertebrates and some insects, but spiders almost certainly make their decisions in discrete time units, namely, days. This assertion is based on observations of web building behavior, especially that of orb weavers. Once an orb web is spun in the morning, the occupant will remain there for a day, unless the web is entirely destroyed. Observations of Linyphiidae in the field indicate that such behavior is also typical of sheet-weaving families. All of the linyphiid webs observed were built in the early morning, and all spiders of both guilds moved to new web sites at night. Moving at night may be primarily an adaptation to avoid predators (Enders 1977), but it has the effect of making spiders behave in discrete blocks of time.

The problem with assuming that foragers can estimate averages instantaneously with little sampling is obvious and has been the focus of several recent studies. Notice, however, that the assumption of discrete rather than continuous time simplifies the problem for spiders. If the spider is constrained to stay at least a day at the web site once it arrives there, it only has to be capable of judging the site in one day's time at the shortest.

Thus, the giving-up model most widely used in the literature of foraging ecology is only weakly applicable to web-spinning spiders, if at all. However, the necessary adjustments in the theoretical background are straightforward, and in fact simplify the analytic treatment (Janetos 1982a, 1982b). I shall discuss this further below.

Variation among Web Sites

The keystone of this or any similar theory is the spatial and/or temporal variation in the quality of web sites. That spatial variation exists in the architectural features of a habitat is obvious; sites differ from each other in terms of the suitability of supports. Riechert and co-workers have shown that site-to-site variation exists in the physiological stresses associated with web sites

(Riechert, Reeder, and Allen 1973; Riechert 1976, 1981; Riechert and Tracy 1975). However, it is not always clearly recognized that spatial variability in food supply is also prevalent in many different habitats.

Although few studies have addressed this problem explicitly, data do exist in the literature. The numbers and proportions of web sites of orb and sheet weavers that had zero, one, or more than one prey per day at the times of sampling are shown in Table 1.1 (Janetos 1982a). Significant variation exists for both guilds, but the orb weavers show a stronger pattern of boom or bust than do the sheet weavers. Results culled from two field studies (Turnbull 1964; Robinson and Robinson 1973a) and one laboratory study (Wise 1979) in which data have been collected on the number of insects caught per day at web sites are shown in Table 1.2. Although the periods of time over which the measurements were made are vastly different, so that the studies are not strictly comparable, in each case significant variation among web sites occurs.

What is missing from these and other studies is measurement of the temporal component of variation. Insect attractants are known to influence prey abundance at web sites in some systems (Riechert 1976). The longevity of these attractants should therefore affect the longevity of the web site as a place to forage. The temporal component of web-site variability may be particularly important in old-field habitats, where much of the spiders' prey is made up of Hymenoptera, Diptera, and Lepidoptera, which are either flower visitors or parasites of flower visitors (Olive 1980; Janetos 1982b; Biere and Uetz 1981). In these systems, the pattern of flowering plants around the web sites will strongly influence the flight patterns of potential prey and thus the supply of prey available to the spiders. As flower patches die out and new ones arise, both temporal and spatial variability are added to the system. The sequence of flowering in old fields in the Northeast suggests that the spatial patterns change on a temporal scale of approximately 2 weeks. In a series of experiments with *Argiope trifasciata*, Olive (1982a) showed that this typical old-field orb weaver is able to respond to such changes. Janetos

TABLE 1.1
Numbers (and Proportions) of Webs with Zero, One, or More Than One Prey per Day

Number of prey per day	Orb weavers 1978	Orb weavers 1979	Sheet weavers 1978	Sheet weavers 1979
0	162 (.76)	309 (.68)	142 (.63)	118 (.54)
1	21 (.10)	51 (.11)	60 (.28)	77 (.35)
>1	31 (.14)	98 (.21)	22 (.09)	25 (.11)

SOURCE: Janetos 1982a.
NOTE: $p < 0.001$ both years, chi^2 test.

TABLE 1.2
Variation of Catches at Web Sites

Species	Mean	S.D.	Coeff. of variation
1. *Achaearanea tepidariorum* (6 web sites)			
Fruit flies per day	12.7	1.9	15.5
	13.3	1.6	12.2
	13.8	2.8	20.1
	13.5	3.2	23.8
2. *Mecynogea lemniscata* (6 web sites)			
Fruit-fly equivalents per day	8.1	5.3	65.4
3. *Metepeira labyrinthea* (6 web sites)			
Fruit-fly equivalents per day	14.5	13.6	94.0
4. *Nephila maculata* (10 web sites)			
Caught in a year:			
Odonata	1.2	1.1	94.6
Orthoptera	20.1	5.6	27.8
Hemiptera	28.1	8.4	29.8
Lepidoptera	143.5	28.8	20.0
Diptera	56.3	11.5	20.4
Hymenoptera	18.6	4.1	21.8
Coleoptera	158.2	23.5	14.8
Others	3.4	2.6	76.2
Unidentified	173.3	26.9	15.5
TOTAL	603.9	57.2	9.5

SOURCES: 1, Turnbull 1964; 2–3, Wise 1979; 4, Robinson and Robinson 1973a.

(1982b) showed that this ability is typical of the orb-weaver guild as a whole.

The spatial and temporal variability of web sites is the raw material for the spiders' exploitation of their habitats. In habitats with little variability, the question of when to leave a web site is moot. Measuring the variability is thus of crucial importance.

Amendments to the Giving-Up Rules

Earlier, in discussing general giving-up rules, I suggested that, to make the rules applicable to web-spinning spiders, discrete-time solutions replace continuous-time solutions. I now suggest, in addition, that new formulations be devised in which temporal variation of prey supply at a web site and spatial variation among web sites replace depletion of resources by the predator. Finally, I suggest that, in analyzing the decision to leave a site, we not seek optimal solutions. One can never prove that an animal is behaving optimally unless nearly all the constraints on its behavior are understood (Oster and Wilson 1978). We do not yet understand the energetic and metabolic requirements of even a small number of spider species, nor do we yet appreciate the complexity of behavior they are capable of. At this time, the more prudent course seems to be to design models that ask, not what behavior is optimal, but when one behavior is favored over another. In many cases, such models based on simple rules that spiders actually follow will have solutions close to those of optimality models (Janetos and Cole 1981).

A Model for Analysis

There are obviously many stimuli that could induce movement from a web site, not all of which have to do with foraging decisions. The diverse range of possible stimuli can be divided into two general classes, randomly and nonrandomly occurring, according to the statistical pattern of their occurrence.

Random stimuli are those that occur with a constant small probability in any short period of time. Disturbance to the supports of the web, damage from excessive wind or rain, and attacks by predators and parasites all should follow this pattern. There is no reason to expect that any of these will occur more or less frequently at any particular time during a spider's tenure at a web site.

Nonrandom stimuli are those that occur with varying probabilities during the tenure of the spider. Primary among such stimuli is the pattern generated by changes in the food supply at a web site. Because not all sites yield equal supplies of insect prey, a spider arriving at a new site will experience either good or poor rewards. Thus, if the stimulus for movement is some function of the food supply, it will generally occur at the very beginning of the spider's tenure, or much later (Janetos and Cole 1981).

Assuming that the spider moves in response to any stimulus, we can easily calculate the expected distribution of residence times of spiders at web sites (Janetos 1980a, 1982b). The distribution of times between events that occur randomly will be exponential (Ross 1970), and this will be the same as the distribution of residence times of spiders (Janetos 1982b). By itself, deviation from random expectation reveals little, since randomness is an extreme hypothesis in a biological system. However, if the spiders deviate from expectation in an interpretable fashion, the type of stimulus and the behavioral rules for moving can be determined.

An example of a study informed by these ideas is that of Janetos (1982a, 1982b). The theoretical basis for the work is contained in a very simple model that captures the parameters outlined above, all of which are measurable in the field. They are (1) the difference in energetic payoff between good and poor web sites, perhaps more easily measured as the variation in payoffs at typical web sites; (2) the efficiency of the behavioral decision that tells a spider when to leave a site; and (3) the cost of moving to a new site, which for web-spinning spiders is mostly the energetic and material cost of spinning a new web from stored body reserves.

These parameters can be combined in the following manner: Consider a habitat composed of good and poor sites, where G and B are the respective payoffs. Good and poor sites occur with frequencies p and q, and have small probabilities of changing quality at the end of each day, which are independent of the spider's behavior. An optimal exploitation pattern for this type of habitat is not sought; rather, two different movement tactics are compared: sitting at one site and averaging functions, and moving to a new site after experiencing one poor day. Moving incurs a cost C.

The following simple expression yields the criterion for moving: $(G - B)(p' - p) > q'C$. When this expression is true, the expected gain from moving exceeds the expected gain from remaining at the web site.

Censuses of marked individuals at web sites revealed that the average residence time of orb weavers was half that of sheet weavers. Several predictions made by the simple model were testable in the field:

1. Orb weavers should have a simple decision rule to give up at a web site.
2. Orb weavers' web sites should be more variable in payoffs than sheet weavers'.
3. Orb weavers should have a lower cost of moving.

Analysis of the full distribution of residence times reveals marked deviation from random expectation among orb weavers, and lesser deviation among sheet weavers. Moreover, combining field data with experimental feeding data reveals that orb weavers have a simple rule for leaving web sites: having arrived at a site, the spider departs the next day if the site has been of poor quality; otherwise the spider remains. Sheet weavers may also have rules for leaving web sites, but they are not as evident.

The web sites of orb weavers are more variable than those of sheet weavers. No study has yet provided a recipe for how spiders themselves judge sites, so it is possible that either the number or biomass of prey is important. By both measures there is greater web-to-web variation among orb weavers than among sheet weavers. The amount of temporal variability in web sites is still to some degree an open question, although there are a priori reasons for expecting orb weavers' sites to exceed sheet weavers' in this respect too.

The cost of moving for web-spinning spiders is certainly more strongly related to the energetic and material costs of building the web than to locomotion through the habitat. Because many orb weavers recycle the silk from old webs, their energetic costs are at least eightfold lower per gram body weight than sheet weavers'. Including the material cost of the silk undoubtedly increases this estimate, since sheet webs contain far more silk than do orb webs spun even by much larger spiders.

All three predictions stemming from the simple model are upheld by data from the field. Further studies will investigate the responses of single species more closely (A. C. Janetos, in prep.). This step is important, since Olive (1980) has shown that specialization within a guild can be considerable.

Problems for Further Study

Two major questions are left unanswered by this view of web-site selection.

First, how does a spider judge the quality of a web site, and how long does it take to do so? The importance of this question cannot be overestimated. A study of the foraging of two web-spinning guilds (Janetos 1982b) suggests that availability of food is an important factor. Riechert and Tracy (1975) have demonstrated that in a desert species, *Agelenopsis aperta*, the quality of a web site is measured by some combination of food abundance and the thermal environment. Spiders in less stressful habitats may judge web sites either by the total number of prey or, more probably, by the total biomass of prey. If total biomass is the criterion, the occasional occurrence of large prey at a web site will be the determining factor in judging the site (Olive 1980). Studies of the decision rules that different species employ in leaving web sites will be an invaluable addition to our knowledge of the movement patterns of web-spinning spiders. In particular, the influence of the variation of prey supply, and the time scale over which spiders respond to different stimuli, are almost completely unknown.

Second, when are foraging decisions inadequate as an explanation for movement? In analyzing web-site desertion I have stressed the role of food supply, but it is clear that other stimuli for leaving web sites may be important, especially when the payoffs at web sites do not vary greatly. Disturbance to the supports, damage to the web from debris or weather, or changes in microclimatic conditions are other possible stimuli; as already discussed, these will tend to result in distributions of residence times of spiders that are indistinguishable from those expected from a hypothesis of randomness. Another possible stimulus, the importance of which is only beginning to be realized, is disturbance from predators and parasitoids. Aggregational response in spiders has been clearly demonstrated (Olive 1982a; A. L. Rypstra, pers. comm.), and the aggregational responses of predatory and parasitic wasps are well known. Since both prey (spider) and predator (wasp) show such responses, there may be a conflict in the responses of spiders to prey density. We may well see that spiders respond to prey but are also extremely sensitive to disturbances from their own predators. However, at present this is purely speculative.

Conclusion

The answer to the question posed in the title of this chapter—Are we asking the right questions?—is, I believe, finally yes. We have progressed from purely descriptive catalogs of the characteristics of suitable web sites to a more predictive, analytical view of web-site selection and desertion as an adaptive response of a spider to changes in its resources. As a consequence, the problems involved in studying web-site selection are now being attacked from many points of view at once: physiological, ecological, and behavioral. The results are proving to be useful in addressing broader questions of habitat selection and community organization, and in adding to our knowledge of how animals adapt to environmental vagaries. This pluralistic approach will ultimately yield a more complete understanding of the foraging behavior of spiders and their mechanisms for avoiding predators as reflected in their most obvious behavioral characteristic: where they build their webs.

2

Habitat Choice and Utilization in Web-Building Spiders

SUSAN E. RIECHERT AND ROSEMARY G. GILLESPIE

CHOOSING A FORAGING SITE is a decision of vital importance to a spider, since, for a spider, as for other animals (Fretwell and Lucas 1969; Fretwell 1972; Stearns 1977), such decisions have a profound effect on growth, survival, and reproductive success (Riechert and Tracy 1975). In this chapter, we deal with various aspects of habitat selection in web-spinning spiders. First, we consider the extent to which active selection of web sites occurs, as opposed to random dispersal and the subsequent differential survival of individuals chancing upon adequate sites. We then discuss the factors that influence the decision to leave a web site and seek a new one. We find that differences in the frequency of moves and in habitat selection strategy can be correlated with the different patterns of foraging behavior that spiders exhibit.

In analyzing the actual habitat selection process, we identify two separate phases. The first involves random search for a suitable habitat patch; the second involves active search for a suitable microhabitat within the patch. A "suitable" microhabitat must satisfy a spider's particular physiological and web-construction requirements; it must also provide a supply of prey and protection from predators.

Actually, as a result of competition for web sites, spiders may not be free to use the habitats of their choice (Morse 1980). We consider whether habitat utilization more closely approximates and ideal "free" distribution, in which all individuals settle in preferred habitats, or a "despotic" distribution (Fretwell 1972), in which habitat utilization is influenced by competition for sites.

We conclude by describing habitat selection and utilization in the desert spider *Agelenopsis aperta* to demonstrate the complexity of this behavior in spiders.

Active Habitat Selection vs. Random Dispersal

Unlike plants, animals have an element of choice in their mobility, and hence one might expect that they would be constantly testing their environments and improving on conditions by moving (as do flying insects: Johnson 1969). There is some evidence, however, for random dispersal in spider populations. For instance, many of the araneids that colonize old fields and early-successional-stage vegetation produce large numbers of offspring (more than a thousand per female) that disperse as spiderlings by ballooning (Tolbert 1977). In these species, less emphasis is placed on the survival of offspring than on the colonization of new habitats. Dispersal by mass ballooning has also been linked with the predictability of the habitat occupied. In several *Pardosa* species, for example, ballooning frequency increases with decreasing habitat predictability (Richter 1970a; Greenstone 1981).

Data on the mortality that results from aerial migration are difficult to obtain and are as yet unavailable. It has been observed, however, that even in those species that produce large numbers of offspring and consequently enhance their ability to colonize new habitats (MacArthur and Wilson 1967) dispersal is not totally random: spiders are known to reinitiate ballooning, presumably after sampling the environment encountered following initial dispersal. Tolbert (1976), for example, observed multiple ballooning bouts in spiderlings of the orb weaver *Argiope trifasciata*, which colonizes road cuts and old fields in eastern Tennessee.

Thus, unlike plant propagules, which, when dispersed by wind, must attempt to survive wherever they chance to fall, ballooning spiders can quickly abandon unsuitable habitats by reinitiating dispersal. Certain spider species are known to exhibit aerial dispersal at later stages in the life cycle. For example, aerial dispersal in linyphiids has been linked with increasing litter temperature (Duffey 1956, 1963) and with changing biotic conditions (Van Wingerden and Vughts 1974). Aerial dispersal may also be initiated by conditions of low humidity (Braendegaard 1937) or melting frost (Nielsen 1932). Certainly these are all examples of behavioral responses to habitat quality. What are the factors that might influence a spider's "decision" to abandon a particular foraging site and seek a new one?

Incentives to Change Sites

The habitat selection process may be said to be initiated by the decision to change sites. The timing of the site change is crucial in terms of the spider's fitness. The various parameters that have been observed to affect the decision to move are enumerated in Table 2.1. In the case of spiderlings, dispersal from the vicinity of the egg sac has been shown to be associated with the development of agonistic behavior toward conspecifics (Riechert 1978a), with age (Barnes 1953; Enders 1975b; Toft 1976), with nutritional history (Valerio 1975), and with specific temperature and wind conditions (Tolbert 1977). These factors may also operate in combination, of course. For example, agonistic behavior developing with age may be the underlying cause of spiderling dispersal, which then occurs under appropriate weather conditions.

As spiders mature, physiological changes may render habitats that were once optimal now unsuitable. Kronk and Riechert (1979), for instance, show that, following maturation, *Lycosa santrita* abandons grass-sod areas that permit efficient detection and capture of prey, and moves to those areas along creek banks where leaf litter and rock substrates predominate and prey are most abundant. This change in habitat occurs at the same time that females must maximize prey intake for the production of egg sacs. The authors conclude that the males migrate to the new areas only because the females do. Riechert and co-workers consider how web-size needs change with age in the funnel-web spider *Agelenopsis aperta* (Riechert, Reeder, and Allen 1973; Riechert 1974a). In the desert grassland habitat studied, juveniles often build their web traps in hoofprints made by cattle after rains. As *A. aperta* matures, however, it must build larger webs, so that larger depressions are required. Similar findings are reported for orb weavers (Risch 1977; Robinson and Lubin 1979a).

Agonistic interactions with conspecifics or heterospecifics may also lead spiders to change sites. In recent studies of interspecific competition, A. B. Cady (pers. comm.) and D. K. Hoffmaster (pers. comm.) observed moves caused by agonistic interactions in various web-building species. The loss of contests over web sites is also a major cause of movement in *A. aperta* (Riechert and Luczak 1982).

Attacks by predators may also stimulate spiders to abandon sites. Moves motivated by the activity of predators have been observed in, e.g., *Argiope trifasciata* (Tolbert 1976, 1977) and *Agelenopsis aperta* (S. E. Riechert, unpub.).

Web damage caused by rain, wind, or mammal and bird movement has been found to stimulate the moves of spiders of various taxa (Eberhard 1971a; Enders 1975b; Lesar and Unzicker 1978; Marson 1974). The extent to which web damage influences a spider's decision to abandon its web site varies with the expected life span of the particular web. As a stimulus for moving, web destruction should be less important for those orb weavers that rebuild their webs each day than for the agelenids, which maintain the same structure for extended periods.

Changes in temperature or humidity may also stimulate spider migration (Almquist 1973a, 1973b; Cloudsley-Thompson 1957; Edgar and Loenen 1974; Cherrett 1964).

Since web-site choice is closely linked with foraging success, one must

TABLE 2.1
Stimuli for Site Abandonment

Family and species	Agonism	Age	Temperature	Web damage[a]	Humidity	Predation	Hunger	Space	Reference
Amaurobiidae									
Ciniflo sp.					X				Cloudsley-Thompson 1957
Uloboridae									
Uloborus diversus				X					Eberhard 1971a
Scytodidae									
Drymusa dinora		X							Valerio 1974b
Theridiidae									
Achaearanea tepidariorum					X		X		Turnbull 1964
Achaearanea wau								X	Lubin and Robinson 1982
Latrodectus hesperus							X		Mackay 1982
Steatoda fulva							X		Hoelldobler 1970
Enoplognatha ovata		X							Stevenson and Dindal 1982
Linyphiidae									
Erigone arctica						X	?[b]		Van Wingerden 1978
Erigone arctica				X					Van Wingerden and Vughts 1974; Vughts and Van Wingerden 1976
Erigone atra	X								Duffey 1956
Linyphia marginata							X		Wise 1975
Meioneta mollis	X								Duffey 1956
Savignia frontata	X								Duffey 1956
Stemonyphantes lineatus			X		X				Almquist 1970
Tapinocyba praecox	X								Duffey 1956

Araneidae								
Argiope aurantia		X	X	X				Tolbert 1977; Enders 1975
Argiope trifasciata		X	X	X		X		Tolbert 1976, 1977
Cyclosa insulana				X				Marson 1974
Micrathena gracilis				X				Biere and Uetz 1981
Nephila clavipes		X						Hill and Christenson 1981
General (two spp.)		X						R. L. Lindley, pers. comm.
Tetragnathidae								
Tetragnatha laboriosa				X				Lesar and Unzicker 1978
Agelenidae								
Agelenopsis aperta	X	X						Riechert 1978a
Lycosidae								
Lycosa santrita		X						Kronk and Riechert 1979
Pardosa sp.			X	X				Richter 1970a
Pardosa sp.			X		X			Nielsen 1932
Pardosa lugubris			X					Edgar and Loenen 1974
Clubionidae								
Phrurolithus festivus			X		X			Almquist 1970
Thomisidae								
Misumena vatia							X	Morse and Fritz 1982
Salticidae								
Euophrys frontalis			X		X			Almquist 1970
General communities		X						Barnes 1953
		X						Toft 1976

[a]For example, by wind.
[b]Factor with which site abandonment might be associated.

also consider the roles prey availability and hunger play in determining when spiders abandon old sites to search for new ones. The classic example of the effect of hunger on spider movement was presented by Turnbull (1964) for the comb-footed spider *Achaearanea tepidariorum*. Turnbull released spiders into an empty room and found that the spiders continued to change web sites until they had all built webs in the vicinity of a window where flies were attempting to escape the room. Turnbull then placed spiders in open boxes, some of which housed prey and some of which did not. He observed that only the spiders that had been placed in boxes lacking prey moved out of their boxes. In a field study of the ant specialist *Steatoda fulva*, Hoelldobler (1970) noted that *S. fulva* abandoned a web site at a particular ant mound entrance when ants ceased to use that entrance. Mackay (1982) found that *Latrodectus hesperus* migrated from an ant nest when the ants had ceased to forage. In all these studies it is difficult to determine what actually influenced spider movement—hunger level or the frequency of encounters with prey.

The point at which, in deciding to migrate, a spider's level of hunger reaches a critical level was recently investigated by Janetos (1982a). He examined the two extremes of the continuum of foraging behavior in web-building spiders. One extreme is represented by the sit-and-wait forager, which builds a relatively permanent web; the other extreme is represented by the active forager, which builds a transitory web. The foraging behavior exhibited by orb weavers differs according to species and spans the entire continuum (Enders 1974), whereas sheet and line weavers are for the most part true sit-and-wait foragers (Janetos 1982a). Janetos borrowed from optimal-foraging theory to predict the lengths of time sit-and-wait and active foragers should remain at given web sites. According to the marginal-value theorem (Charnov 1976), an animal should stay longer in an area that provides a rich supply of food, especially if the distances to other areas that provide suitable prey are great. Further, site tenacity should be favored if there has been an initial investment in the form of modification of the site and if the resources available at the site are not significantly more abundant elsewhere (Morse 1980). Thus, according to Janetos (1982a), the following factors delimit the move-stay decision: (1) the expected payoff in food at a good site, (2) the expected payoff at a poor site, (3) the probability of encountering a good site, (4) the probability of encountering a poor site, (5) the cost of moving, and (6) the period of time before a good site becomes a poor one.

It seems unlikely that spiders would be capable of completing the complex calculations required for optimal habitat choice (Oaten 1977; Janetos and Cole 1981). However, natural selection should favor those spiders that, by responding to appropriate cues, obtain more food and produce more offspring. Moreover, one would expect that the decision rules followed by various web-building species take the factors cited above into account. According to the marginal-value theorem, a spider should move only when it estimates that the expected gain from moving will exceed the expected gain from staying at the same site, after the cost of the move has been included (Schoener 1971). Sit-and-wait foragers (e.g., linyphiids and agelenids) should therefore spend more effort in searching for suitable patches, and settle at sites affording high concentration of prey—and this they indeed do (Riechert and Luczak 1982; Janetos 1980a, 1982a, 1982b). They also show persistence at these sites: for linyphiids, $\bar{x}$ = 5 days, with high variance (Janetos 1982b); for the agelenid *Agelenopsis aperta*, $\bar{x}$ = 42 days (S. E. Riechert, unpub.). More active foragers, on the other hand, should do more sampling of habitat patches by building temporary web traps, and decide daily whether to stay at a site or to move (Olive 1982a). The araneids, uloborids, and tetragnathids for which Janetos (1982a) investigated the timing of moves changed web sites with a mean frequency of every 3 days. Further, the araneids and tetragnathids ingested their webs before moving, which minimizes the energy expended in changing sites. The energy investment per web trap built is therefore considerably less for araneids and tetragnathids than for linyphiids and agelenids (Peakall 1971; Prestwich 1977). Initial silk-laying costs, as measured in milligrams of silk produced relative to milligrams of body weight, are lower for the orb weavers studied, by a factor of 3 (Janetos 1982a). Olive's studies on orb weavers (1981a, 1981b) yield similar results.

The Habitat Selection Process

Our knowledge of habitat association in spiders indicates that spider fitness is correlated with web-site location in most patchy habitats (e.g., Riechert and Tracy 1975; Kronk and Riechert 1979). We have just shown, in our discussion of web-site abandonment, that the timing of moves may be correlated with the efficient utilization of resources, and thus is subject to selection pressure. We assume that natural selection encourages development of those responses to habitat-quality cues that will place spiders in microhabitats promoting maximum fitness. This is achieved by the differential selection of spiders that use active habitat choice. We shall now consider the habitat selection process itself: what are the conditions of a habitat that enable a spider to maximize its foraging success and how are habitats affording these conditions located?

Habitat Requirements

For most animals, a suitable habitat must satisfy various physical constraints and provide both sufficient prey and protection from predators. As members of the cryptozoan arachnid lineage (Savory 1971), spiders are subject to problems of water loss and must therefore seek favorable thermal environments. Physical conditions also affect foraging success. The thermal environment, for instance, limits foraging activity to certain periods of the day (e.g., Edgar and Loenen 1974; Riechert and Tracy 1975; Biere and Uetz 1981), and wind speed and direction affect what kinds of webs can be built (P. R. Craig, pers. comm.). Wind speed and direction also influence prey contact with webs, which can serve as filters. (See Rubenstein and Koehl 1977 for a general discussion of filters, and Chacón and Eberhard 1980 for a discussion of webs as filters.) A suitable habitat must also provide appropriate web supports, and in a configuration that permits that orientation of the trap which maximizes prey contact. The selection of sites on the basis of foraging considerations may conflict with the need to avoid predators, since a trap providing maximum prey interception will generally be highly exposed. This possibility has yet to be investigated for spiders. These are the kinds of considerations that spiders, especially the more sedentary taxa with a high investment at a particular site, must weigh in choosing suitable habitats.

Random Search

The first phase of the habitat selection process involves movement away from the habitat patch occupied. In the case of aerial dispersal, no individually induced movement is involved. Rather, wind velocity and direction at the time of ballooning determine how far and in what direction the spider travels. However, one might hypothesize that ground dispersal is also initially random—that a spider leaves a site in the direction it is facing when the move is initiated and that the path it follows is determined by the obstacles it encounters as it moves. Movement continues until the spider encounters a favorable macrohabitat—i.e., a macrohabitat that satisfies some of the minimal requirements of the particular spider (e.g., the presence of leaf litter, shade, or prey).

A random pattern of migration is known to occur in caddis fly larvae, which also build web traps (Hildrew and Townsend 1980). Gillespie (1981) believes that the cribellate spider *Amaurobius similis* exhibits a random migration pattern. The pattern of movement observed in *Agelenopsis aperta*, however, is indicative of initial migration that is systematic, not random (see below).

Selection of a Site

The second phase of the habitat selection process involves active search for a suitable web site within a favorable macrohabitat. Cues resulting in cessation of the search have been identified for spiders of various taxa (Table 2.2). Among the active hunters, wolf spiders are known to associate with particular types and strata of vegetation that permit efficient perception and capture of prey (Hallander 1967, 1970; Kronk and Riechert 1979) and with those types and strata of vegetation that provide an abundance of prey (Greenquist and Rovner 1976; Kronk and Riechert 1979). Some crab spiders are also known to settle in vegetation that offers significantly higher numbers of prey (Morse and Fritz 1982). In the case of *Agelenopsis aperta*, both olfactory and vibratory cues stimulate movement to patches affording sufficient prey, whereas movement into shade, followed by movement along a temperature gradient, ensures that spiders settle in patches affording a suitable thermal environment.

As indicated earlier, different families of web-building spiders use different habitat selection strategies. Some species, it seems, can estimate the availability of insects in a particular habitat patch only by building a web trap in that patch; to find a suitable patch they must build "trial" webs in various patches. Gillespie (1981) found that the cribellate spider *Amaurobius similis* lays silk at apparently random locations for a period of, on the average, 12 to 14 days before settling at a site. Prey at this site are more abundant than at the previous sites. Turnbull (1964) reported similar web-building activity in the theridiid *Achaearanea tepidariorum*. The question arises whether this pattern is unique to these spiders or a more general phenomenon—whether spiders as a group obtain only inconclusive estimates of site quality from various cues that elicit foraging behavior in particular locations. It may be that for all spiders, foraging activity at a site alone provides a true estimate of site quality, and that the search for a site is either stopped or reinitiated depending on the numbers of prey and the physical conditions encountered during actual foraging. This possibility requires further investigation. For instance, not enough data are available at present to determine whether the species discussed above (*Amaurobius similis* and *Achaearanea tepidariorum*) are capable of responding to olfactory or vibratory cues without the assistance of silk threads. And for only a few species are data available on the frequency of web-site abandonment relative to the abundance of prey.

One of the major problems encountered in studying habitat selection in spiders is distinguishing between ultimate and proximate factors. The structural features of a habitat, for instance, may serve as proximate cues to the presence of ultimately limiting factors. *Schizocosa ocreata* shows a positive association with leaf litter of a particular depth and density, which serves as a

TABLE 2.2
Bases for Site Selection

Family and species	Vegeta-tion	Bark/rock	Grass	Leaf litter	Prey	Tempera-ture	Wind	Humid-ity	Orienta-tion	Light	Reference
Amaurobiidae											
Amaurobius similis					X+						Gillespie 1981
Ciniflo sp.								X+			Cloudsley-Thompson 1957
Ciniflo spp.		X−									Curtis and Morton 1974
Uloboridae											
Uloborus diversus							X+				Eberhard 1971a
Psechridae											
Psechrus spp.	X−										Robinson and Lubin 1979b
Hypochilidae											
Hypochilus thorelli		X+									Riechert and Cady, in press
Theridiidae											
Achaearanea tepidariorum					X+						Turnbull 1964
Achaearanea tepidariorum		X+									Riechert and Cady, in press
Enoplognatha lineata					X+						Dabrowska-Prot, Luczak, and Wojcik 1973
Enoplognatha ovata	X−										Stevenson and Dindal 1982
Euryopis funebris					X+						Carico 1978
Steatoda fulva					X+						Hoelldobler 1970
Linyphiidae											
Floronia bucculenta	X−										Schaefer 1978
Linyphia triangularis					X+						Dabrowska-Prot, Luczak, and Wojcik 1973
Oedothorax apicatus	X−							X−			Almquist 1973a
General	X−				?						Cherrett 1964
Araneidae											
Arachnura melanura	X−										Robinson and Lubin 1979a
Araneus cavaticus	X+										Riechert and Cady, in press
Araneus diadematus	X+							?			Colebourne 1974
Archemorus roosdorpi	X−									X−	Robinson and Lubin 1979a
Argiope aurantia	X−					X−					Enders 1973, 1975b
Argiope ocyaloides	X−										Robinson and Lubin 1979a
Argiope trifasciata	X−										Enders 1975b
Argiope trifasciata									X+		Coventry 1967; Tolbert 1976
Argiope sp.										X−	Cherrett 1964
Cyrtophora molluccensis	X−										Robinson and Lubin 1979a
Herennia ornatissima	X−										Robinson and Lubin 1979a

Mangora placida	X−			?				Uetz, Johnson, and Schemske 1978
Mecynogea lemniscata				?				Wise 1979
Metepeira labyrinthea				?				Wise 1979
Micrathena gracilis	X−			?				Uetz, Johnson, and Schemske 1978
Micrathena gracilis							X+	Biere and Uetz 1981
Nephila clavipes	X−			?	?			Carrel 1978
Nephila clavipes							X+	Krakauer 1972
Nephila clavipes							X+	Robinson and Robinson 1974
Tetragnathidae								
Tetragnatha laboriosa	X+							Lesar and Unzicker 1978
Tetragnatha montana				X+				Luczak 1970
Agelenidae								
Agelenopsis aperta				X+	X+			Riechert 1976
Lycosidae								
Lycosa nigriceps	X−							Duffey 1966
Lycosa monticola	X−					X−		Duffey 1966
Lycosa pullata	X−							Duffey 1966
Lycosa santrita		X+		?				Kronk and Riechert 1979
Pardosa sp.			X+	?				Greenquist and Rovner 1976
Pardosa chelata			X+	?				Hallander 1967, 1970
Pardosa lugubris			X+		X+			Edgar and Loenen 1974
Pardosa prativaga	X−							Hollander and Lof 1972
Pardosa pullata	X−							Hollander and Lof 1972
Pardosa ramulosa				X−				Greenstone 1978
Schizocosa ocreata			X+					Cady 1978
General	X−					X−		Dondale 1977
Oxyopidae								
Oxyopis scalaris	X−							Cutler, Jennings, and Moodie 1977
Clubionidae								
Clubiona trivialis	X−				X−			Almquist 1973
Thomisidae								
Misumena vatia	X+			X+				Morse and Fritz 1982
Misumenops coloradensis	X−							Jennings 1971
General communities	X−							Barnes 1953
	X−							Riechert and Reeder 1972
	X−							Waldorf 1976

NOTE: X+ indicates features selected by spiders, X− indicates features with which the habitat of the spider is associated, and ? indicates factors with which the choice of site might be associated.

cue to local humidity (Cady 1978). The wolf spider *Pardosa lugubris* also associates with leaf litter, but uses it to escape cold temperatures (Edgar and Loenen 1974). Various other spider species are known to associate with vegetation of particular densities and physiognomies (Barnes 1953; Duffey 1966; Riechert and Reeder 1972; Uetz and Burgess 1979; Stevenson and Dindal 1982). These features may act as proximate cues to wind strength (Enders 1973), humidity, temperature, or the presence of preferred prey (Riechert and Reeder 1972). Or it may be that the association is merely secondarily correlated, as wind speed at a particular height in vegetation is directly proportional to the density of vegetation. Eberhard (1971a) noted that *Uloborus diversus* cues directly on wind without the aid of vegetation. S. E. Riechert (in prep.) found that significant associations of *Agelenopsis aperta* with flowers and fecal material, and with shrubs and depressions, are cued by the presence of prey in the former case and by the presence of shade in the latter. In order to be able to distinguish between proximate and ultimate factors, we must first determine the extent to which habitat association in spiders is cued by habitat structure or by various parameters of the physical environment and the presence of prey.

Certainly habitat structure can be an important consideration in selecting a web site—especially for the many orb weavers that require the presence of strategically positioned supports (Colebourne 1974; Schaefer 1978). Habitat structure may be even more important for social species that form long-lived colonies and hence must find relatively permanent physical structures (Lubin 1974; Buskirk 1975a; Uetz and Burgess 1979). Certain species, however, are more flexible in their structural requirements, since they can modify their traps to fit the available space and orient the traps such that favorable macrohabitats are achieved and access to prey is enhanced. Riechert and Tracy (1975), for instance, showed that individuals of *A. aperta* that have webs in more exposed locations orient the web funnels so that the hot desert sun penetrates the funnels for the shortest time possible. Biere and Uetz (1981) found that the web orientations chosen by the orb weaver *Micrathena gracilis* allow it to escape high temperatures and hence to have more time for foraging. Orb weavers must also consider orientation with respect to the wind in placing their webs, because orb webs act as filters, and wind velocity and direction will influence how successfully the webs trap prey (Rubenstein and Koehl 1977). In studies of *Argiope trifasciata*, Coventry (1967) and Tolbert (1976) both found a significant correlation between changes in web orientation and changes in wind direction.

Few workers have considered the way in which avoidance of predators affects habitat selection in spiders. However, spiders have been observed to associate with tree bark, which affords camouflage (Curtis and Morton 1974; Robinson and Lubin 1979a; Robinson 1982b).

Competition for Habitat Use

In our discussion so far, we have assumed that spiders are free to occupy the habitats of their choice. However, as pointed out by Morse (1980), habitat utilization may not necessarily follow from habitat selection: if populations are large and resources limited, intra- or interspecific competition may exclude animals from preferred habitat patches and force them into less suitable patches. The population distribution so formed has been called a despotic distribution (Fretwell 1972; Fretwell and Lucas 1969; Whitham 1980). In this system, all individuals select the most favorable microhabitats available at a particular period in time. The first individuals to arrive at these microhabitats, or the better competitors, force other individuals to occupy less favorable microhabitats. They do so through either "exploitation" or "interference" competition (Jaeger 1974). Exploitation competition involves differential efficiency in the use of available resources. Among spiders, individuals that build larger web traps may deplete the prey available to nearest neighbors, and thereby force them to abandon their sites. In interference competition, site owners actively prevent other spiders from foraging in the vicinity of their sites. Both types of competition can lead to the deviation of habitat utilization from that predicted from habitat selection—specifically when population densities are high and resources are in limited supply.

There is considerable evidence for despotic distribution in spider populations, though most of it comes from observations of the local patterns of dispersion of various species, and is thus indirect. (See Riechert 1982 for a review of the literature.) King (1973) thinks that a regular pattern of dispersion is nearly always the result of intraspecific competition, and many spider species exhibit this kind of dispersion (Riechert 1982). In the detailed population studies completed for some of these species, intraspecific agonistic behavior has been found to underlie the observed regular spacing. Evidence to date suggests that interspecific competition plays a minor role in habitat utilization, though, in a study involving the orb weaver *Micrathena gracilis*, Uetz, Johnson, and Schemske (1978) were able to demonstrate that the presence of other species—particularly congeners—leads to a small but definite change in web placement. The more usual lack of significant interspecific competitive effects is the result, in part, of the presence of a high degree of intraspecific competition. A recent study by Riechert and Cady (in press) indicates that both intraspecific competition and intraspecific predation limit the effects of interspecific competition on habitat utilization and population size in a spider community. Horton and Wise (in press) found interspecific competition to play an insignificant role in habitat utilization for two species of *Argiope*, but attributed this to variation in vegetation and certain physical factors rather than to intraspecific competition. The low degree of intraspecific competition observed in this case may be explained by the fact that the study was confined to older juveniles and adults, and therefore the effects of intense competition among juveniles prior to dispersal were ignored.

Limiting Factors

The dynamics of spider populations depend ultimately on the availability of resources. For instance, spider population density is known to vary with prey density (Turnbull 1966; Riechert 1979, 1981, 1982). In particular, spider population densities tend to increase (within limits) in response to increases in local prey numbers (the aggregational response: Readshaw 1973); the increases in population densities may in fact reflect increases in reproductive output (the reproductive response: Holling 1966). Turnbull (1966), Riechert (1974a, 1976), Mackay (1982), and Uetz (this volume) provide examples of aggregational responses in spiders; Wise (1975, 1979) and Riechert and Tracy (1975) provide examples of reproductive responses. That these increases in spider population numbers may be limited by intraspecific competition has been shown by Riechert (1981) for *Agelenopsis aperta*. Despite the fact that *A. aperta* exhibits elements of both the aggregational and the reproductive response, increases in spider population numbers in general, and within favorable habitat patches in particular, are not realized because of the operation of a territorial system that assures the better competitors of reproductive success in even the worst years.

Spider population numbers have also been found to be limited by the availability of suitable habitat structures. For example, population growth in the linyphiid *Floronia bucculenta* is limited by the number of web sites available in grass tussocks in a dry sphagnum bog (Schaefer 1978). The availability of structural features necessary for the attachment of orb webs is also known to limit spider population sizes of, e.g., *Araneus diadematus* (Colebourne 1974), *Araneus cavaticus* (Riechert and Cady, in press), and *Micrathena gracilis* (Hartsock 1983). The population numbers of *Agelenopsis aperta* are limited by the availability of particular habitat features that provide a high degree of foraging success (Riechert 1981).

These studies suggest that when certain environmental constraints are released, spider populations show an increase within the limits set pretty much by the availability of suitable web sites. It is unlikely, however, that single factors operate alone to limit population densities. More probably, the resource that limits population growth is that which is in most limited supply at a particular time (i.e., Liebig's law of the minimum). Accordingly, if one limiting factor is removed, a population increases to the level set by the resource present in the next most limited supply. It is thus difficult to isolate all of the factors that might limit spider population densities. It is evident, however, that when a resource is present in limited supply only the best competitors occupy optimal sites. The other spiders are forced into habitats of marginal quality or fail to obtain sites altogether.

The Case of *Agelenopsis aperta*

It is painfully clear from the foregoing review that we have little understanding of the significance of habitat selection and utilization for spider foraging success. We have even less understanding of the mechanisms that underlie habitat choice. To understand habitat selection and utilization in spiders, we must first make detailed studies of the habitat selection and utilization systems of particular species. Only one such study has been completed: that of Riechert and co-workers on the desert agelenid *Agelenopsis aperta*. A brief summary of the results of this study follows.

Agelenopsis aperta (Gertsch) is a member of the funnel-web spider family Agelenidae. Its web consists of a flat sheet and an attached funnel that extends into some feature of the surrounding habitat (Fig. 2.1). Occasionally a scaffolding is present. The sheet has no adhesive properties, and serves merely as an extension of the spider's legs and hence of its sensory apparatus. *A. aperta* carries out much of its activity within the sheltered environment of the web funnel; it comes out onto the web sheet only to secure prey and make repairs. The web is a semipermanent structure, to which silk is added daily.

This species occupies grassland habitats in desert areas throughout the southwestern United States and northwestern Mexico. Its life cycle is closely linked to seasonal changes in the desert (Riechert 1974a). Young emerge from the egg sac in August and September, when insect numbers are highest following summer rains. Ground dispersal is accomplished within a few weeks after emergence from the egg sac; web building begins after dispersal. Spiders overwinter as immatures and complete development the following spring. Maturation and the production of offspring occur in May and June in exposed habitats and in July and August in riparian habitats. Multiple egg sacs may be constructed, but in most habitats few adults survive long enough to produce more than one clutch. For instance, in an area in south-central New Mexico composed of grassland and a recent lava bed, most of the adults are drowned in floods produced by the summer rains that begin in July (Riechert 1974a). Because of flooding and the distinct seasonality of the desert environment, the spider exhibits no age-class overlap in this area. The population thus consists of a single age class with an annual life cycle. We restrict our discussion to this population of *A. aperta*.

Fig. 2.1. Web of *Agelenopsis aperta: a*, web sheet and attached funnel; *b*, web funnel.

Habitat Association and Reproductive Success

Study of adult patterns of dispersion showed the distribution of *A. aperta* to be closely correlated with the distribution of depressions in the grassland and of shrubs on the lava bed (Riechert, Reeder, and Allen 1973). Closer analysis revealed significant correlations between spider distribution and the presence of the following microhabitat features: large depressions (with diameters exceeding 1 m), shrubs in bloom at the time of sampling, other flowering plants, litter, and deer, rabbit, and cattle feces (Riechert 1976). Juveniles demonstrate less specificity in their choice of web sites (Riechert 1974a). They frequently build their webs in small hoofprints made in the dirt by cattle after rains. However, the significant association with large depressions that was observed in adults was not observed in juveniles, nor was the association with flowering plants, litter, and fecal material. These findings do not necessarily indicate that spiderlings are incapable of exercising active habitat choice. Rather, since they build very small webs, they have no need for the large depressions used by the adults; and because juveniles are present in the fall, when temperatures are cooler and insect numbers high following the summer rains, they do not require the presence of the other features, all of which have been shown to be important to the fulfillment of the physiological and prey-availability requirements of the adults (Riechert and Tracy 1975).

The depressions and shrubs used as web sites, and the litter that may be present at both types of sites, have important effects on spider thermal balance. Depressions and shrubs provide *A. aperta* with shade for at least part of the day, and significantly decrease spider body temperatures: the body temperatures of spiders in the shade provided by these features are lower by as much as 15° C than the body temperatures of spiders exposed to full solar radiation. Depressions and shrubs also provide structural support for scaffoldings, and protect webs from destruction by wind. Litter decreases the thermal radiation that would be received if a web were located directly above a dark dirt or black lava substrate. Association with litter substrates decreases spider body temperatures by as much as 12° C during the hottest part of the day.

The presence of flowering plants or fecal material at web sites increases the numbers of available prey (Riechert and Tracy 1975). These features probably serve as insect attractants, the plants attracting pollinators and the fecal material attracting mainly flies and crickets.

Riechert and Tracy (1975) developed a model to explain the effects of thermal environment and prey availability on spider reproductive success. The productivity of females that occupy sites affording all of the features mentioned was estimated to be 13 times greater than the productivity of females that occupy sites lacking these features. Model results suggest that more food is obtained, and hence greater fitness achieved, at a site affording a favorable thermal environment (eightfold difference between high- and low-quality sites) than at a site offering greater numbers of prey (twofold difference). Insect attractants increase the probability of occasionally encountering high prey densities, whereas depressions, shrubs, and litter, by helping to maintain body temperatures within a favorable range (i.e., 21–35° C), give spiders considerably more time to forage. All of the model predictions were corroborated by subsequent empirical studies (e.g., Riechert 1978a, 1981, 1982).

Only 4 percent of the available habitat provides those features that permit a high degree of reproductive success (Riechert 1979). Further, only 12 percent of the available habitat provides suitable microhabitats—i.e., depressions or shrubs with or without litter and insect attractants (Riechert 1979, 1981). In most years, reproduction is limited to spiders that occupy these microhabitats. If these spiders were to share space with conspecifics, the prey available to them would be decreased by approximately 40 percent (Riechert 1978a), and as a result food consumption would drop below the optimal level for the species population. Hence there is active competition for web sites and associated territories, and the population distribution pattern is despotic: the availability of web sites limits population growth, and individual reproductive success varies with the type of habitat patch occupied.

Evidence for Active Habitat Selection

Using marked individuals, Riechert (1976) obtained estimates of spider movement. Spiders were observed to reposition their webs, or reorient the funnels, at given sites, or to move to new sites within the study area, as many as seven times in a 3-month study period. The maximum distance a spider was observed to move was 200 m, though *A. aperta* may be capable of moving farther; habitat beyond the confines of the census area was not checked for marked individuals. No difference was observed between the number of moves made by adults and the number of moves made by juveniles, although adults were observed to move greater distances than juveniles. (Adults moved a mean distance of 19.58 ± 4.63 m, juveniles a mean distance of 1.73 ± 0.92 m.) Clearly, we are not dealing in this case with differential survival. Rather, the behavior observed indicates that *A. aperta* practices active habitat selection, throughout its life cycle.

Incentives to Change Sites

A. aperta spiderlings are tolerant of conspecifics. Spiderlings may even share in the capture of prey and in feeding on them. Agonistic behavior develops with growth to the third instar, at which time individuals are observed to circle areas and defend them by engaging in pushing-and-shoving matches with intruding individuals (Riechert 1978a). It is at this time that initial dispersal from the vicinity of the egg sac is observed.

As noted, *A. aperta* may change web sites several times during its life cycle. Riechert and Luczak (1982) tallied the causes of migration from web sites in this spider. Web destruction by rain or wind and disturbances by animals ranked highest among the causes of migration, and accounted for 31.2 percent of the moves observed. Aside from trampling by mammals, spiders lost their webs to pompylid wasps, caterpillars, lycosid spiders, centipedes, and marauding ant columns. The wasps place wolf spiders they have parasitized in the web funnels of *A. aperta*. Caterpillars use the funnels as cocooning sites, and wolf spiders use entire webs for foraging. Centipedes use the funnels for shelter after chasing off the web owners. With the exception of the caterpillars, these animals will eat the web owners if they catch them. There is some question, however, as to whether the activity of the ants is predatory or merely disruptive to the spider's activity on the web, much as wind or rain can be disruptive.

Loss of territorial disputes and molting were also identified as important causes of migration in this species, and accounted, respectively, for 21.1 and 5.8 percent of the moves observed (Riechert and Luczak 1982). The factors underlying the remaining 41.8 percent of the moves remained undetermined, though unfavorable thermal conditions might have been a contributing factor. It is hard to imagine that *A. aperta* abandons a web site on the basis of hunger, since this spider is known to occupy a poor site for 22 days on the average—a long time for an annual species (S. E. Riechert, unpub.).

Habitat Selection

Because *A. aperta* occupies markedly different types of web sites—i.e., shrubs on the lava bed and depressions in the adjacent grassland—it is unlikely that spiders select sites on the basis of vegetation structure or substrate type. Likewise, it is difficult to imagine what structural features fecal material and flowering plants share that might cue spider association with them. It was thus assumed that *A. aperta* responds, not to the physical attributes of these features, but to the way in which these features change the local environment.

Knowing that the habitat features encountered by *A. aperta* significantly affect local thermal and prey environments, Riechert conducted a series of experiments to test spiders' responses to exposure to sun versus protection by shade, to temperature differences, and to differences in the availability of prey. The open-field experimental design used is shown in Fig. 2.2. The suitability of the habitat structure for web placement was controlled: poultry netting laid over the sand substrate permitted web attachment anywhere within the enclosed area, and 16 stations were provided with screen cones to encourage the construction of funnel webs.

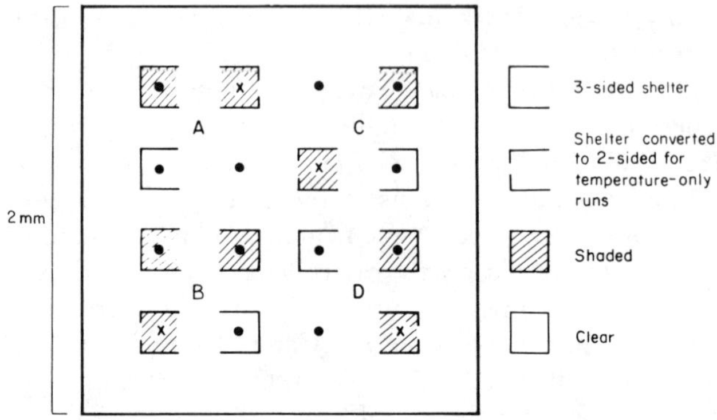

Fig. 2.2. Enclosure used in open-field habitat selection experiments on *Agelenopsis aperta.* The sand and poultry-netting substrate was bounded by a sheet-metal flashing barrier 30.5 cm high. Framed-in areas represent Plexiglas shelters 4 cm in height and 14 cm^2 in area. ● and x indicate screen-cone locations; x indicates which shelters housed prey during prey-detection trials. Temperature was controlled by heat lamps placed above release points *A–D.* Spiders were released one at a time into one of the four arenas designated by the release points. A trial was ended and scored either when silk laying was initiated or when a pause of 15 minutes occurred.

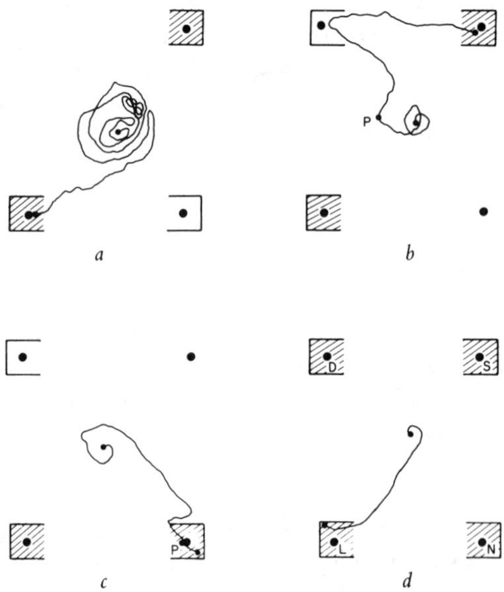

Fig. 2.3. Search patterns of *Agelenopsis aperta* in open-field habitat selection experiments: *a–c,* patterns exhibited during temperature trials; *d,* pattern exhibited during a prey-detection trial. *P* marks points where pauses in movement of greater than 15 seconds occurred; *D* identifies a screen cage containing dead flies, *L* a cage containing live flies, *S* an empty cage, *N* no cage. The framed-in areas are described in the legend to Fig. 2.2.

Examples of the search patterns exhibited by spiders in the enclosure are shown in Fig. 2.3. If temperatures vary but prey availability is controlled (i.e., if no prey are offered in the boxes at various stations), *A. aperta* initially moves in short, jerky steps around the point of release in a pattern of increasingly larger arcs; it then moves directly to one of the boxes, presumably in response to the detection of shade (Figs. 2.3a–c). Once in a box, the spider either initiates web building (Figs. 2.3a,c) or abandons the box and resumes the search process (Fig. 2.3b). The spider's movement, within or between boxes, is frequently interrupted by pauses in locomotor activity (Figs. 2.3b,c).

The temperature preferred by *A. aperta* is 30° C (Riechert and Tracy 1975). In a significant number of cases, movement from pause points was directional, from less preferred temperatures towards the preferred temperature. Further, spiders settled in significant numbers at stations that afforded the preferred temperature regardless of the structural features of the station (Fig. 2.4).

A spider's choice of thermal environment can be influenced by its thermal history. A group of spiders raised at 23° C exhibited a significantly broader range of temperature associations than did a group of spiders raised at approximately 30° C (Mann-Whitney test, $p < 0.005$). In particular, the spiders raised at the lower temperature tended to settle at lower temperatures (Fig. 2.5).

In another set of experiments, temperature was controlled (i.e., held within the favorable range of 21–35° C) and prey availability was varied: screen cages containing live or dead flies were placed at certain stations. In these experiments, the circular pattern of search was not observed; rather, after release, spiders in most cases moved directly to stations housing prey (Figs. 2.3d, 2.6). The significant number of "no moves" recorded in this set of experiments suggested that settling behavior could be triggered by olfactory or vibratory cues effective over the 50-cm distance between shelters and release points. Since spiders were observed to respond to both dead prey and live prey (Fig. 2.6), it was assumed that olfactory cues influence site choice in the vicinity of prey. When only dead prey were used, spiders still exhibited significant association with shelters affording prey (frequency = 0.40, $\bar{x} = 15.5$, $p < 0.005$). Likewise, shelters fitted with vibrators elicited settling behavior: spiders proceeded to these shelters in significant numbers (frequency = 0.48, $\bar{x} = 16.4$, $p < 0.001$). These experiments clearly indicate that for *A. aperta* both olfactory and vibratory cues are important in locating sites affording suitable prey.

In the next set of habitat selection experiments, both temperature and prey availability were varied—and thus the conditions of these experiments more closely resembled the conditions actually encountered by *A. aperta* in

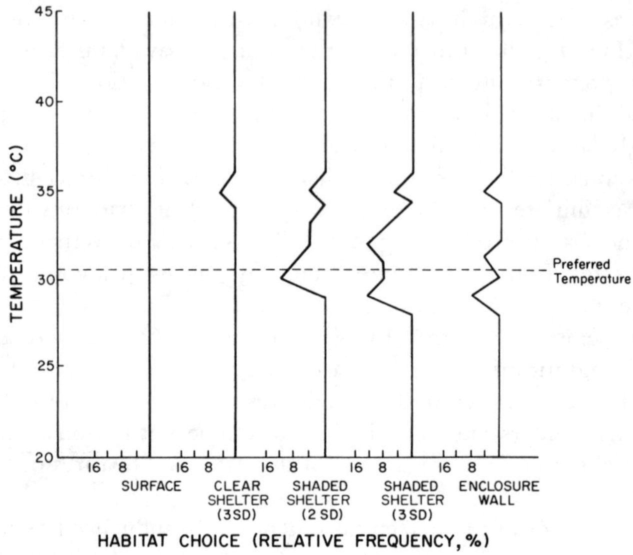

Fig. 2.4. Results of open-field habitat selection experiments on *Agelenopsis aperta*: the effects on habitat choice of variation in the thermal environment. The experimental relationship is shown in Fig. 2.2. Habitat choice is measured by relative frequency of association. The dashed line indicates the temperature preference shown in the field by this spider (Riechert and Tracy 1975).

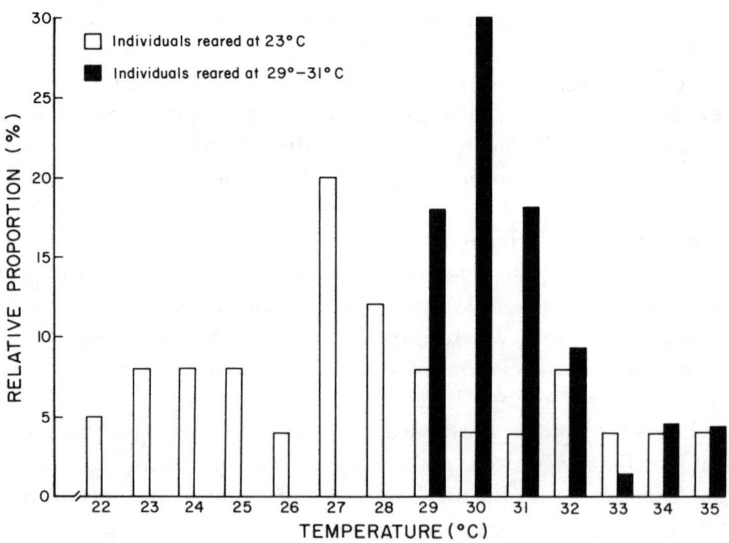

Fig. 2.5. Temperature associations exhibited in runs of the open-field habitat selection experiment on *Agelenopsis aperta* by spiders raised at 23° C (□) and at approximately 30° C (■).

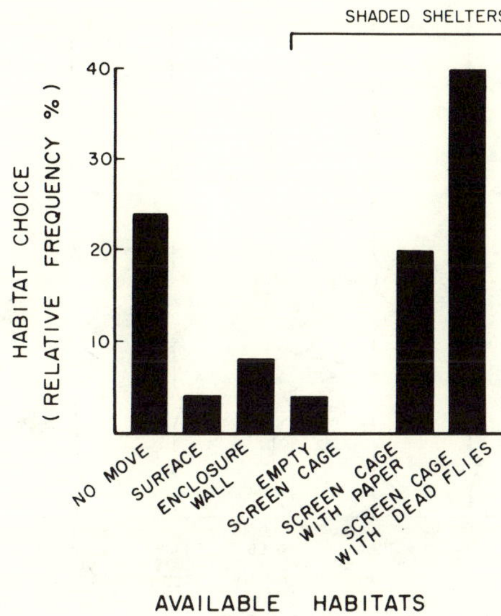

Fig. 2.6. Results of open-field habitat selection experiments on *Agelenopsis aperta*: the effects on habitat choice of variation in the prey environment. The experimental arrangement is shown in Fig. 2.2. Shaded shelters were present at all stations, and the temperature was maintained at 29–31° C. Screen cages were placed in three of the four shelters in each arena: one cage contained live flies (*L*), one cage contained dead flies (*D*), and one cage was empty (*S*). See Fig. 2.3d.

the field. In these experiments, optimal habitat choice was no longer exhibited; i.e., spiders did not move to those stations that provided both the preferred temperature and prey. Rather, they settled at sites that afforded the preferred temperature but no prey 43 percent of the time and at sites that afforded prey but less than optimal temperatures 46 percent of the time (Fig. 2.7). Further, in multiple runs involving the same spiders there was no evidence for consistent choice: a given spider sometimes settled at a site offering a favorable temperature and at other times settled at a site affording prey.

Why did the search not continue until a site was found that provided both a favorable temperature and prey? One might hypothesize that high prey densities and favorable thermal conditions are linked in desert environments—that in these environments insects, like spiders, seek favorable thermal conditions. Riechert and Tracy (1975) in fact found that in the desert area under investigation prey availability was significantly higher in depressions and under shrubs than in other microhabitats. Thus, given potential risks involved in searching, perhaps *A. aperta* does best to accept the first location that satisfies either its prey-availability or its temperature requirements. Further, once settled at a location, this spider can influence its ther-

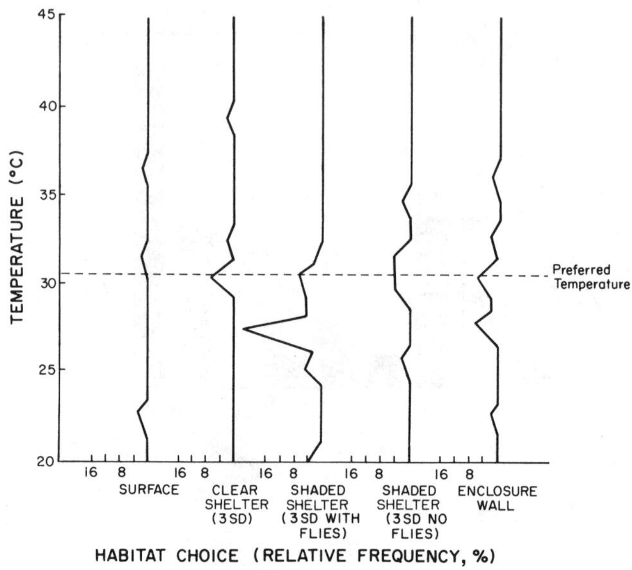

Fig. 2.7. Results of open-field habitat selection experiments on *Agelenopsis aperta*: the effects on habitat choice of variation in both the thermal and the prey environment. The experimental arrangement is shown in Fig. 2.2.

mal environment by choosing an appropriate funnel orientation. It can also adjust prey intake by varying the area of its web trap. Spiders at optimal sites—depressions with insect attractants and litter—exhibit random funnel orientation (Riechert and Tracy 1975). They also build smaller web traps than do spiders at average-quality sites (Riechert 1981, 1982). Spiders at poor-quality sites—sites on the grassland or lava-bed surface—exhibit a significant funnel orientation: their funnels face towards the northwest. As a result, the sun penetrates the funnels only in the late afternoon—when the temperature is dropping and spiders are in need of external heat. Spiders at average-quality sites have been observed to receive as little as 15 mg of dry weight per day, as compared to the 22 mg *A. aperta* is known to take in the laboratory (Riechert 1978a). Spiders at poorer-quality sites apparently do not have the energy reserves needed to build large web traps. Their webs are approximately the size of the webs of spiders at excellent-quality sites (Riechert 1981, 1982).

Competition for Web Sites

Because competition for available web sites is fierce in the habitat studied (Riechert 1978a, 1981, 1982) and reproductive success varies markedly with site quality, one would expect natural selection to favor the adjustment of agonistic behavior to the quality of the disputed sites. Riechert (1979) in fact observed a significant logarithmic increase in estimated contest cost with increasing site quality in this habitat. How, in a dispute over a web site, is site quality assessed?

Whereas both contestants in a dispute can, by making weight assessments, estimate their relative potential for winning (Riechert 1978b, in prep.), only the site owner possesses knowledge of site quality at the beginning of the dispute. When the site owner was removed, for instance, the significant relationship between contest cost and site quality ceased to exist (Riechert 1979). A positive correlation was observed between web area and the agonistic behavior of site owners, but no correlation was observed between web area and the agonistic behavior of intruders (S. E. Riechert, in prep.). Further, no correlation was observed between spider body temperature during an agonistic interaction, or the presence of prey on the web during an interaction, and the behavior of either contestant. The cues used by spiders in initial habitat selection thus do not seem to influence persistence decisions in disputes over web sites.

At first analysis, one might surmise that, in defending their sites, site owners are cueing in on web size, as indicated by the positive correlation between web size and owner behavior. However, in the habitat studied the webs of owners of excellent-quality sites are *smaller* than the webs of owners of average-quality sites, and it is difficult to envision the mechanism by which spiders might be able to distinguish between the two ends of the site-quality continuum on the basis of web size. At least a two-step decision process would be required, and such processes are out of keeping with the conservative nature of adaptation. By Occam's razor, the simplest decision rule available to an animal should be the one favored by natural selection.

Further study shows that in the area under investigation web size is actu-

TABLE 2.3

Contest Statistics for Disputes at Poor-Quality Sites of Agelenopsis aperta Involving Site Owners That Either Have or Have Not Received Food Supplementation

Category	Site owner: Supplemented	Site owner: Unsupplemented	Significance level	Test
Estimated contest cost rank score (mean ± SE)	62.8 ± 15.0	46.4 ± 18.1	$p < 0.05$	Mann-Whitney
Average number of bouts	2.8 ± 0.04	1.8 ± 0.08	$p < 0.25$	chi^2
Probability of loss given no size advantage (%)	76.9%	80.0%	$p < 0.50$	chi^2

SOURCE: S. E. Riechert, in prep.

ally correlated, not with site quality, but with another factor: the feeding history of a site. That this is the factor that affects agonistic behavior was demonstrated by a simple experiment involving owners of poor-quality sites, i.e., sites on the grassland surface (S. E. Riechert, in prep.). A group of site owners that had first been weighed and marked were provided with supplementary prey (moths) daily for a minimum period of 1 week and then entered into contests with intruding spiders. As indicated in Table 2.3, agonistic behavior changed significantly with food supplementation, mainly with respect to the length of time site owners persisted in contests. However, although the persistence of site owners increased with food supplementation, the probability of their winning contests did not—which perhaps indicates that in the habitat under study it is the poor competitor that is forced to occupy poor-quality sites.

Conclusion

The dynamics of habitat choice and utilization are complex and difficult to analyze. A spider's behavior at any point in time represents a compromise between the many needs and selection pressures felt by the spider at that time in its life cycle. Understanding the suite of adaptations involved in habitat choice and utilization requires knowledge of the history of local populations, an understanding of the relationship between spider population densities and the availability of important resources, and knowledge of the sensory capabilities of particular species. A detailed study has been made of the habitat choice and utilization system of *Agelenopsis aperta*. Whether the behavior exhibited by this spider is typical of web-building spiders generally cannot be determined until similar studies are completed on other spider taxa.

3

Transmission of Vibration in a Spider's Web

W. MITCH MASTERS, HUBERT S. MARKL, AND
ANNE J. M. MOFFAT

AS IS CLEAR to anyone who has watched a spider capture an insect in its web, vibration signals traveling through the web from the prey to the spider appear to be of great importance in directing the prey-catching behavior of the spider. There is also ample evidence that vibration signals play an important role in other types of spider behavior, such as courtship and avoidance of predators (e.g., Vollrath 1979b; Krafft 1982). It is probably safe to assume that vibration signals play a role in most of the activities of spiders, especially since, in many cases, their visual abilities are minimal.

As a consequence of the obvious importance of vibration signals for web-building spiders, numerous investigators have attempted to clarify how spiders obtain and use vibratory information. The first study of web vibration was made over a century ago by Boys (1880), who used a tuning fork to stimulate the web; later studies include that of Barrows (1915), who made the first noncontact measurements of vibration transmission by observing the blurring of stimulated web strands in photographs, and such recent studies as those of Suter (1978) and Vollrath (1979b). Although measurement techniques have been continually improved, nearly all of these studies have been handicapped to a greater or lesser extent by lack of a satisfactory way of measuring the vibration in the web—and without such a method there is little hope of determining the true input signal to the spider. However, recent advances in optoelectronics, especially the development of lasers and broad-bandwidth photodiodes, have opened new ways to attack this recalcitrant problem.

In this chapter we present the results achieved by using two optical, noncontact instruments to measure vibrations in the web of the orb weaver *Nuctenea sclopetaria.* Our goal is ultimately to describe the signal transmitted from the signal generator (e.g., an insect trapped in the web, or a courting male) through the transmitting medium (the web) to the receiver (the spider). However, to facilitate analysis we break up the complete system into its constituent parts. We begin with a description of the transmission properties of an empty web. (The description presented here is largely a summary of results already published: Masters and Markl 1981.) We also present preliminary descriptions of the effect of a spider's presence on signal transmission in a web, and of the signal produced by a fly trapped in a web. We then address the question of what our results suggest about the limitations of other methods of measuring web vibration, and we conclude by discussing some of the consequences of our results for future studies of vibration signals in orb webs.

Measuring Web Vibration

Vibrations along strands in orb webs are of three different types, each of which has different transmission characteristics (Masters and Markl 1981). A fourth type—torsional vibration—is possible, but no good method exists for detecting this type of vibration and we neglect it. (See Frohlich and Buskirk 1982 for a theoretical discussion of web vibration.)

If we consider the vibration of a single radius, then the vibratory motion can be (1) parallel to the axis of the radius (longitudinal vibration), (2) perpendicular to the axis of the radius and perpendicular to the plane of the web (transverse vibration), or (3) perpendicular to the axis of the radius but parallel to the plane of the web (lateral vibration) (Fig. 3.1). Transverse and lateral vibration are perpendicular to each other, and consequently transverse vibration of a radius produces transverse vibration of the sticky spiral cross strands connected to the radius, whereas lateral vibration produces longitudinal vibration of the sticky spirals. For web threads not joined by cross strands, such as the signal strand that often runs from the spider's retreat to the hub of the web, lateral and transverse vibration are equivalent. Because of the way in which an orb web is constructed, vibration of one type at one point in the web may be converted to vibration of other types at other points in the web (Liesenfeld 1956).

An "ideal" instrument for measuring vibration in an orb web would satisfy three principal requirements.

1. It would not load the web; i.e., use of the instrument would not involve adding mass to the web or otherwise interfering with its motion. An orb web is so light (on the order of milligrams) that almost any object

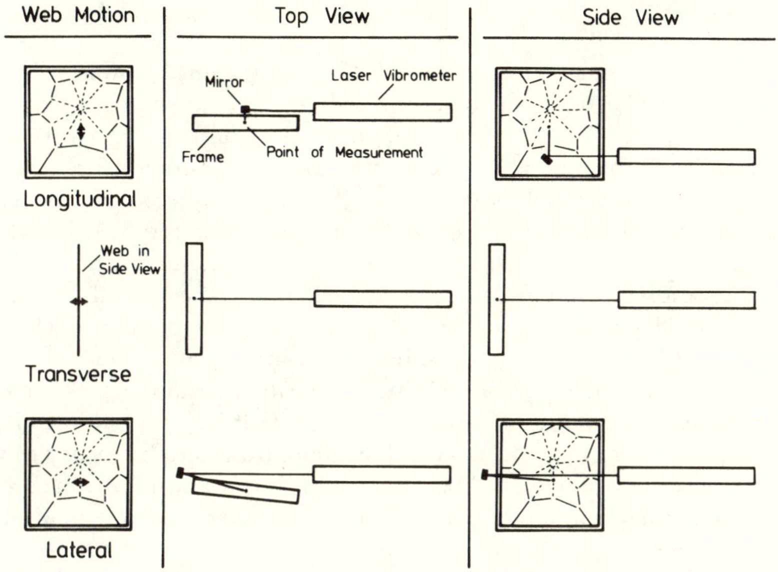

Fig. 3.1. *Left*, direction of movement (*arrows*) of an orb-web radius as it undergoes longitudinal, transverse, or lateral vibration. *Middle and right*, setups of the laser Doppler vibrometer required to measure the three types of web vibration.

attached to it will affect its vibration characteristics to some extent. This requirement practically necessitates use of an optical—i.e., noncontact—instrument.

2. It would operate over a wide frequency range. Spiders have been reported to be sensitive to vibrations from well below 1 Hz to over 10 kHz (Walcott and Van der Kloot 1959; Bohnenberger 1981). An ideal instrument would function over the frequency range, even though it is not yet known whether spiders actually use vibratory information throughout this range.

3. It would operate over a wide dynamic range; i.e., it would work as well for small-amplitude motions as for large-amplitude motions. A spider's vibration threshold has been reported to be on the order of nanometers at certain frequencies (Walcott and Van der Kloot 1959; Liesenfeld 1961), which sets the desired lower limit of measurement. The upper limit is on the order of millimeters, since some insects can produce web motions this large when they are trapped in webs.

Unfortunately, no web-vibration detector exists that completely fulfills all three requirements. However, two instruments, the laser Doppler vibrometer and what we call a spider-web vibrometer, satisfy at least the first requirement and have been used successfully for noncontact web-vibration measurement.

Laser Doppler Vibrometer

The laser Doppler vibrometer (LDV) operates on a simple principle. When light strikes a moving object the reflected light undergoes a frequency shift owing to the Doppler effect, and the amount of shift is proportional to the velocity of the object. The velocity of the object can therefore be calculated from the measured Doppler shift. (See Buchhave 1975 and Michelsen and Larsen 1978 for further discussion of laser vibrometry.) Although the principle of operation of the LDV is simple, practical use of the LDV for web-vibration measurement is not. Fig. 3.1 shows schematically the setups of the LDV required to measure the three types of web vibration, and Fig. 3.2 shows the LDV arranged to measure lateral vibration of a radius.

The LDV completely satisfies the first of the three requirements of an ideal web-vibration detector, since the vibrometer itself has absolutely no influence on web vibration. The LDV also operates from zero frequency to very high frequencies (above 100 kHz), so in principle it also fulfills the second requirement. However, since the LDV detects velocity rather than displace-

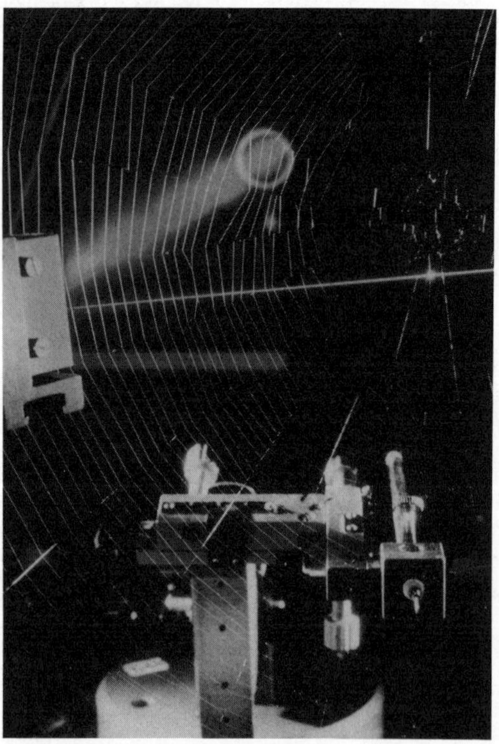

Fig. 3.2. Laser Doppler vibrometer arranged to measure lateral vibration of a radius at the edge of the hub. The laser beam emerges from the vibrometer lens (ring in the upper part of the photograph), situated approximately 1 m behind the web, and strikes a mirror (left), where it is reflected to the right nearly in the same plane as the web. The beam strikes the web at the intersection of the radius and the edge of the hub, at which point it passes through the plane of the web. The focal point of the beam is the bright spot where it hits the web, approximately 10 cm to the right of the mirror. Light reflected back from this point along the path of the beam to the lens is compared with the outgoing beam to detect shifts in the light frequency that are caused by the movement of the web. The micromanipulator in the lower part of the photograph holds the piezoelectric bimorph used to produce lateral vibration in the radius.

ment, its sensitivity to low-frequency vibrations, which may have small velocity amplitudes even though they have large displacement amplitudes, is relatively poor. We have found that for web-vibration detection the LDV works best at frequencies of a few hundred hertz or higher, although it can be used down to 1 Hz in certain situations. The LDV in principle also satisfies the third requirement of an ideal vibrometer—that of a wide dynamic range. In practice, however, because the laser beam must be focused precisely on a very small spot—a spot with good surface properties so that sufficient light is returned to the detector, and returned in good phase coherence—it is very difficult or impossible to use the LDV to measure web vibration when the motions are large. Large motions cause the web to move out of the focal region of the laser beam, or into a position where the surface characteristics of the web are unsuitable for correct light reflection. In either case, the LDV loses track of the vibration and "signal dropout" occurs. Since the instrument can detect signal dropout, false signals are not processed. Nevertheless, these difficulties restrict use of the LDV to situations where the web is moving with very small amplitude (on the order of tens of micrometers).

Despite these problems, the LDV is extremely useful, especially for measuring high-frequency signals, which usually have quite small amplitudes. It can also be used to measure vibration of objects other than spider webs, and we in fact used it to measure the vibration of a spider's tarsus. Signal dropout is less of a problem when the object is large, since the chances that sufficient light will be reflected back to the vibrometer are improved. Finally, the LDV has the advantage of being available commercially. (We used an instrument manufactured by DISA Elektronik A/S, DK-2740 Skovlunde, Denmark.)

Spider-Web Vibrometer

A second instrument, which overcomes most of the problems the LDV has in measuring large web motions, is our spider-web vibrometer (SWV), shown in Fig. 3.3. The SWV is an improvement of a device first used by Graeser (1973). This instrument requires a strand of the web to be illuminated off-axis from behind, which produces a brightly lighted strand against a dark background. An image of the strand is focused by a lens onto a position-sensitive photodiode (United Detector Technologies, type PIN-LSC/5D). The diode produces an output electrical signal directly proportional to the position of the lighted strand on its face, i.e., proportional to web displacement. By measuring the magnification of the focusing lens and the proportionality factor of the photodiode, one can determine the motion of the portion of the web strand that is imaged on the photodiode.

A potentially serious problem exists in that the proportionality factor of the photodiode, that is, the change in output voltage for a given movement

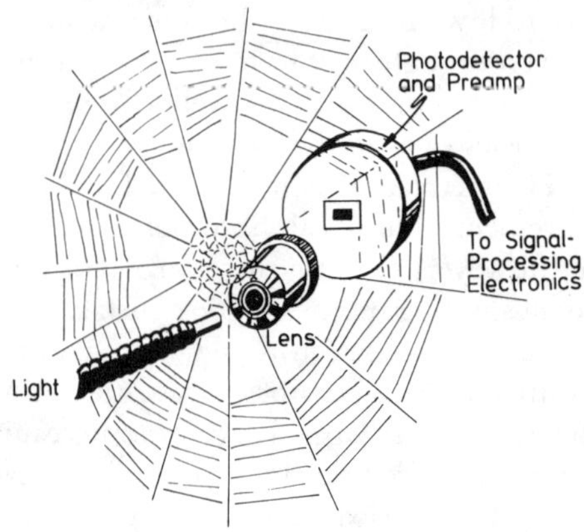

Fig. 3.3. Spider-web vibrometer arranged to measure lateral vibration of a radius. The lens (a microscope objective) is focused on the radius at the point of measurement and projects an image of the radius onto the photodetector, the black rectangle mounted in the round housing at the rear. Since light from the optic-fiber bundle (left) strikes the radius off the lens-detector axis, the image projected by the lens appears as a bright line against a dark background. Lateral movement of the radius (sideways in the diagram) causes the image to move on the photodetector, and this movement is transduced to an electrical signal by the detector. Not shown are the manipulators used to position the light guide, lens, detector, and web.

of the image, depends on the intensity of illumination, i.e., on the brightness of the image of the web strand, and this can change even from one point to another on a strand. Fortunately, the photodiode can also measure the intensity of the light that reaches it, and it is therefore possible to remove the intensity dependence of the position output of the photodiode, by dividing the position output by the intensity output. We perform this division electronically, with a divider circuit. (Before performing this division one must first null out the background illumination of the photodiode.) We have calibrated the SWV so that a 1-mm movement of the web strand imaged on the face of the photodiode produces a 1-V output that is independent of the brightness of the image.

The SWV completely fulfills the first of the three requirements of an ideal web vibrometer in that it in no way affects web vibration. The SWV partly fulfills the second requirement: its frequency response is adequate, extending from 0 to 7 kHz (with a rolloff of 3 dB at 7 kHz and 5 dB at 10 kHz). The SWV also partly fulfills the third requirement. It measures relatively large web motions easily. The linear section of the position-sensitive photodetector is 4 mm long, and since we ordinarily use a magnification of 10, web motions of up to 0.4 mm can be measured. A detector circuit senses when the image of the web strand moves out of the linear range of the photodiode; the output of the SWV is then turned off by an analogue switch with a rise-fall time of 10 msec, to prevent false readings from appearing in the output signal. The lower limit of detection by the SWV is set by the inherent noise level of the electronic circuitry and of the photodiode and by the brightness of the image of the strand. Typically, without any filtering or averaging, web movements of about 1 μm are measurable with an oscilloscope. The measurement of smaller movements requires the use of filtering, averaging, lock-in detection, or power-spectrum analysis (i.e., fast-Fourier-transform analysis).

The SWV is less sensitive than the LDV to high-frequency motions, but more sensitive to low-frequency and large-amplitude motions. By using the SWV we were able to record the vibration produced by a fly trapped in a web. The SWV has the further advantage of being inexpensive. One disadvantage of this device, however, is that, compared to the LDV, the SWV requires use of a more complicated setup and more position-manipulating equipment. The web must be positioned between the lens and the light source (we use a fiber-optics light guide), and to calibrate the SWV (i.e., to null out the background illumination) the web must be moved into and out of the field of view of the system. It is also somewhat difficult to use the SWV to measure transverse vibration of a strand, since this requires viewing the strand from the side, i.e., in the plane of the web. However, this viewing angle can be achieved to within about 20° to 30° by using a lower power objective (which has a greater working distance) and a mirror mounted at about 45° on the front of the objective. With this arrangement, however, the sensitivity of the SWV is reduced, because the illumination of the strand is reduced.

Experimental Method

We used the LDV and SWV to measure vibrations in strands of orb webs of *Nuctenea sclopetaria* (Fig. 3.4). Our procedure was simply to stimulate a web with a known vibration at one point and measure the signal transmitted to another point. We stimulated the radius of a web, at a point about in the middle of the prey-catching zone. By appropriate arrangement of the stimulator, vibration of principally one type (longitudinal, transverse, or lateral) can be produced in the radius, and the same type of vibration measured at the point where the radius joins the hub of the web. Comparison of the amplitude of vibration at the point of measurement with that at the point of stimulation gives the transfer function, or transmission, between the two points. We have determined transfer functions at frequencies from 1 to 10,000 Hz in empty webs. A value of 0 dB on a transmission curve indicates that the amount of vibration at the point of measurement is equal to that at

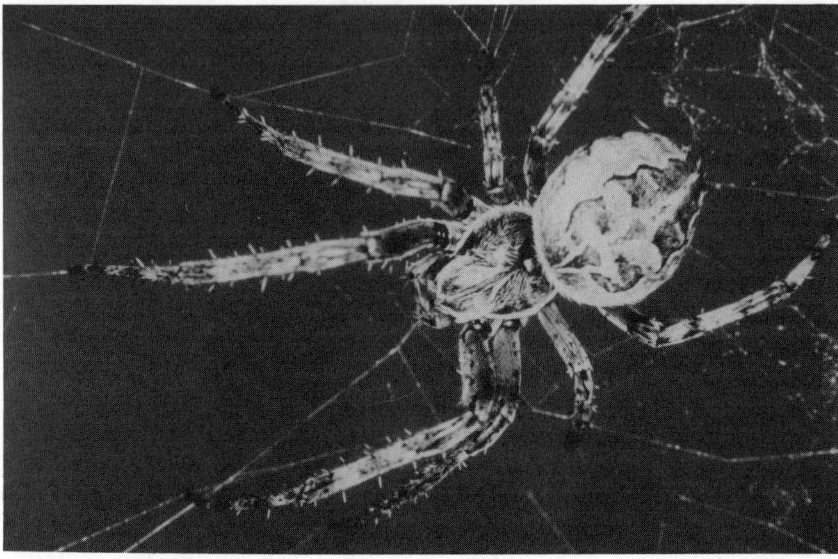

Fig. 3.4. Adult female *Nuctenea sclopetaria* in her web. The spider, in response to a stimulus produced by touching a vibrator to the web-frame thread at about the 8 o'clock position, has just turned and grasped with leg pair I, the radii leading towards the point of stimulation. Legs II hold the neighboring pair of radii; legs IV hold threads in the hub of the web.

the point of stimulation (no attenuation); a transmission of, e.g., −20 dB indicates a loss of 20 dB in the signal (20 dB attenuation), which corresponds to a factor-of-10 reduction in signal amplitude. Normally the measured vibration is less than the input vibration, so the decibel value of the transmission is usually negative.

In an empty web, transfer functions were determined by the single-frequency method. A sinusoidal voltage was applied to the web stimulator and the amplitude of web motion was compared with that of the vibrator. This procedure was repeated until the frequency range from 1 to 10,000 Hz had been sampled. When a spider was in the web, however, this approach could not be used, because the spider attacked the stimulator rather than waiting patiently for a series of readings to be made. In order to measure vibration transmission to the spider's tarsus, therefore, we needed a method for determining the transfer function that could be carried out quickly and that would not provoke the spider to attack.

We solved this methodological problem by using impulses rather than sine waves as input to the web. A brief voltage pulse applied to the web stimulator produced a sudden deflection in the web. The power spectrum of the motion of the vibrator was computed by fast-Fourier-transform (FFT) analysis and compared with the power spectrum of the motion pulse measured in the web. The difference in amplitude at each frequency component in the FFT spectrum is the transmission from the vibrator to the measured point on the web at that frequency.

In contrast to the sine-wave technique, the impulse technique gives transmission information at a number of different frequencies simultaneously, but it has the disadvantage that only a small amount of energy can be put into the web at each frequency, and consequently the signal-to-noise ratio for the impulse technique is worse than for the sine-wave technique. The signal-to-noise ratio can be improved by averaging several impulses together before performing the FFT analysis, but the effectiveness of this operation decreases rapidly as the number of waveforms averaged increases (we usually averaged four), and, in addition, all measurements must be performed quickly before the spider changes position. Furthermore, because with the impulse technique only a small amount of energy can be put into the web at each frequency, a particular web stimulator is effective over only a limited bandwidth. Gathering information over a broad frequency range therefore requires using two or more stimulators successively. For empty webs, however, it is usually possible to measure over the entire frequency range using only one stimulator, since there is time for filtering or averaging. When, for comparison, the impulse and sine-wave techniques were used to measure signal transmission in an empty web, differences in measured transmission did not exceed 3 dB.

Input vibration to the web was produced by piezoelectric ceramic bimorphs taken from phonograph cartridges and driven by voltages of about 80 V (peak to peak). The bimorphs were of two lengths, 4 and 13 mm. The longer bimorphs produced a larger displacement (about 7 μm at low frequencies) but were effective up to only about 2 kHz; the shorter bimorphs produced a smaller displacement (about 1 μm) but were effective up to 10 kHz. These bimorphs were glued inside glass tubes, which were positioned by micromanipulators. The web radius was attached to the tip of a bimorph with beeswax; this procedure did not affect the equilibrium position of the radius.

The spiders were collected as adults near Constance. They were kept in the laboratory in wooden frames with inside dimensions of 45 x 45 x 6 cm in which they built webs.

Results

We began our web-vibration studies by measuring longitudinal, transverse, and lateral vibration in empty webs. (For a full account of the results see Masters and Markl 1981.) We also measured vibration transmission to a light object attached to the web, and to the tarsus of a spider on the web. Finally, we measured the vibration produced by a fly trapped in the web.

Transmission in an Empty Web

Figure 3.5 shows typical examples of longitudinal, transverse, and lateral vibration transmission along a radius from the middle of the prey-catching zone to the edge of the hub in an empty web. Longitudinal vibration shows little attenuation (typically only about 1–3 dB) over the entire measured frequency range. At high frequencies (above 2 kHz), longitudinal vibration shows positive transmission owing to resonances in the radius or web, but no regular decrease in transmission occurs. In contrast, the transmission of both transverse and lateral vibration decreases with increasing frequency. Transverse vibration transmission often falls off more or less constantly with increasing frequency (Fig. 3.5), but may sometimes show a clear corner frequency (Fig. 3.6, solid curve), or even drop in a series of steps. At low frequencies, attenuation of transverse vibration is usually about 10–20 dB; at very high frequencies, attenuation increases to about 20 to 50 dB. Resonances occur at frequencies above about 500 Hz, and though they may sometimes be pronounced, they are not always so.

At frequencies up to about 300 Hz, attenuation of lateral vibration is typically about 15–30 dB and the transmission curve is flat (Figs. 3.5, 3.6). Above 300 Hz, transmission drops rapidly until about 1 kHz. Above this frequency, resonances may halt the decline in transmission. Nevertheless,

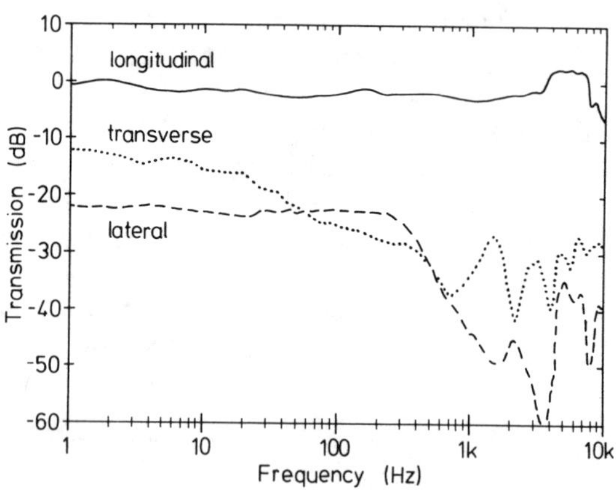

Fig. 3.5. Transmission of longitudinal, transverse, and lateral vibration along a radius from the middle of the prey-catching zone to the edge of the hub in an empty web. The amplification of longitudinal vibration between about 4 and 7 kHz is due to resonances. The distance from the point of stimulation to the point of measurement was approximately 6, 7.5, and 5.5 cm (top to bottom).

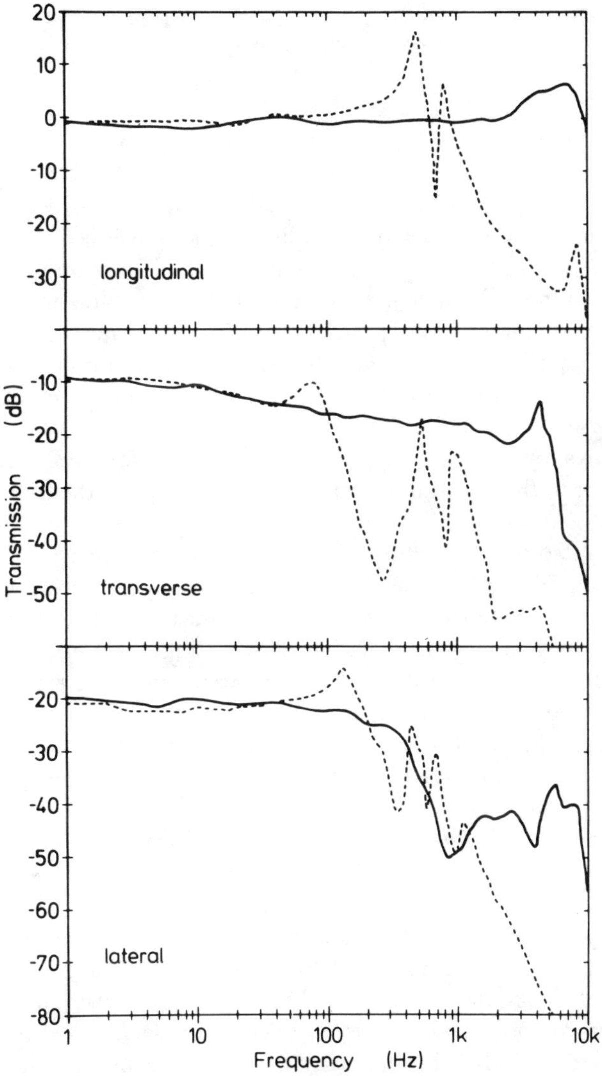

Fig. 3.6. Transmission before (*solid curves*) and after (*dashed curves*) loading the point of measurement at the edge of the hub with a 0.3 ± 0.03-mg object. Note that the intercept on the *y* axis differs for the three types of vibration, and that at low frequencies attenuation increases by approximately 10 dB in going from longitudinal to transverse and from transverse to lateral vibration. The distance from the point of stimulation to the point of measurement was approximately 5.5, 7.5, and 4.5 cm (top to bottom).

above about 1 kHz, lateral vibration is relatively poorly transmitted; it usually shows an attenuation of 25–50 dB or more.

Transmission to a Light Object

By measuring vibration transmission in an empty web, we have, in effect, determined the vibration transmission characteristics of the medium. In nature, of course, the web is loaded by the spider and sometimes by prey, and so we also need to know how loading of the web affects vibration transmission. We might begin by asking, what are the effects on vibration transmission of adding a light object to the web? Are they drastic or can they be ignored? These questions are particularly relevant because some of the techniques that have been used to measure web vibration involve adding a tiny object to the web and measuring the motion of this object. Is this procedure acceptable?

Figure 3.6 shows the change in vibration transmission produced by adding a light object to the point of measurement at the edge of the hub. The object (a 1 × 1-mm piece of index card glued to the web with fingernail polish) weighed 0.3 ± 0.03 mg—approximately 20 percent of the web's weight. The solid curves in Fig. 3.6 show vibration transmission measured before adding the object, and the dotted curves show that measured after adding the object. At low frequencies the added object has a negligible effect on transmission, but above a certain frequency, which varies with the type of vibration being measured, the object begins to exert an influence on the motion of the web. For each of the three types of vibration, the object and web form a spring-mass resonance system with a characteristic frequency determined by the mass of the object and the restoring force (the spring constant) of the web for that type of vibration. The first peak in transmission occurs at this resonance frequency. The higher the spring constant, the higher the resonance frequency, and the wider the range over which the vibration of the web remains uninfluenced by the added object. Above the resonance frequency, however, the vibration of the web is increasingly affected by the added object. The results shown in Fig. 3.6 suggest that, in general, web vibration up to about 100 Hz may be adequately determined by measuring the motion of a light object (about 0.3 mg) attached to the stimulated strand, but that above this frequency the effects on web vibration of the attached object become more pronounced and such a method becomes less acceptable for measuring web vibration. Above 1 kHz the effects of the added object are drastic for all three types of vibration.

Another problem that arises is that the amplitude of the higher resonance peaks (e.g., the 800-Hz peak for longitudinal vibration or the 500-Hz peak for transverse vibration) changes depending on the point of measurement on the object. This probably occurs because the object has several rotational modes of motion open to it (despite its being attached at the intersection of two strands) that additively combine with the purely translational motions we desire to measure.

Transmission to a Spider's Tarsus

As we have seen, the presence of even a small object can markedly affect web vibration. Since a spider may weigh 200 mg or more, we might expect its presence to have a much greater effect on web vibration than that of a 0.3-mg object. However, the two cases—that of a 200-mg spider connected to the web by its legs and that of a 0.3-mg object glued to the web—are not really comparable. The principal mass of the spider is not directly affixed to the web, but rather is coupled to the web via eight thin, light and springy legs. The motion of the body of the spider will be rather small, or, put another way, vibration transmission to the body of the spider will be poor, except below the resonance frequency (less than a few hertz: Graeser 1973) determined by the mass of the spider and the restoring force of the web. However, though the spider's body may not move appreciably, motion of the tarsus that actually holds the vibrating strand is likely to be considerable, since the tarsus is a much smaller mass for the web to drive. Furthermore, most of the vibration receptors of a spider are located on its legs (Barth 1981, 1982); among these receptors is a very sensitive lyriform organ that lies at the tarsal-metatarsal joint and is thought to be the principal receptor of small web-borne vibrations (Walcott and Van der Kloot 1959; Liesenfeld 1961; Barth 1972). Thus, we should very much like to measure vibration transmission to a tarsus holding a stimulated strand.

In our attempts to measure vibration transmission to a spider's tarsus we used the impulse technique described earlier. Typical results are presented in Fig. 3.7. The solid curve shows longitudinal vibration transmission along a radius from the prey-catching zone to the edge of the hub in an empty web; the dotted curve shows transmission to a tarsus (leg I) grasping the radius at the point of measurement at the edge of the hub. The dashed curve shows longitudinal transmission (above 200 Hz only) to the tarsus of a different spider in a different web. The break in the dotted curve at 300 Hz indicates the point at which we changed stimulators, as required by the impulse technique; for the dashed curve we used only one stimulator. The transmission curve for the empty web is not complete, but since other measurements of longitudinal vibration transmission in an empty web have consistently shown little attenuation between 1 and 10,000 Hz, the solid curve can probably be extended in both directions.

We see from the dotted and solid curves in Fig. 3.7 that at lower frequencies longitudinal vibration is transmitted to the spider's tarsus with an overall attenuation of only about 8 dB, and that at these frequencies the spider's

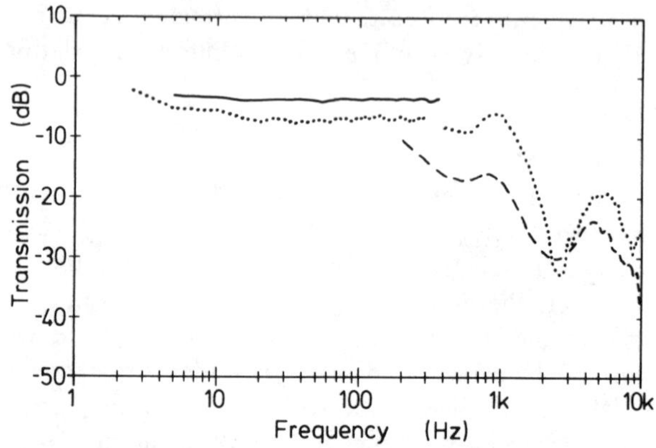

Fig. 3.7. Transmission of longitudinal vibration to a spider's tarsus. *Solid curve*, transmission along a radius from the middle of the prey-catching zone to the edge of the hub in an empty web; *dotted curve*, transmission to a tarsus grasping the radius at the point of measurement at the edge of the hub; *dashed curve*, transmission to the tarsus of a different spider in a different web. The distance from the point of stimulation to the point of measurement was approximately 7.5 cm for the solid and dotted curves and approximately 6 cm for the dashed curve.

presence increases the attenuation by about 4 dB. At higher frequencies, however, the attenuation produced by the leg touching the vibrating strand increases markedly, and transmission falls 20–30 dB below that in the empty web. This is still reasonably good transmission, however, especially when we consider that at high frequencies transverse and lateral vibration are usually attenuated by this much in an empty web.

We have not yet succeeded in measuring the transmission to a spider's tarsus of transverse or lateral vibration, partly because web transmission of these two types of vibration is poorer than that of longitudinal vibration.

Prey-Produced Vibration

In order to gain a complete understanding of signal transmission in a spider's web, we need to have some idea of the vibration produced by trapped prey. Once we know the characteristics of the stimulus provided by a trapped insect, we can estimate the input stimulus to the spider, i.e., the vibration of the spider's tarsus, by taking account of the effect of the spider's presence on vibration transmission. Although many examples of prey-produced web vibrations will be required before we can make generalizations, we have begun by examining a stimulus known to be highly effective in inducing spiders to attack: the vibration produced by a buzzing fly trapped in a web (Boys 1880; Barrows 1915; Meyer 1928; Peters 1931; Burgess 1979b).

Figure 3.8 shows power spectra for the longitudinal, transverse, and lateral vibration produced by a buzzing fly trapped in a web. The vibration was

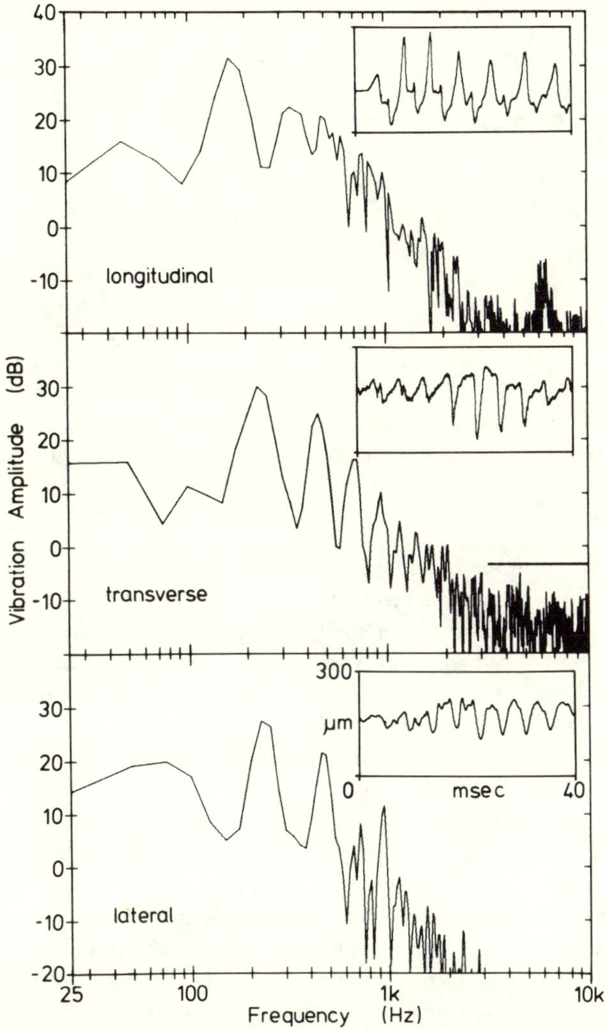

Fig. 3.8. Power spectra for the longitudinal, transverse, and lateral vibration produced by a buzzing fly trapped in a web. The vibration at the intersection of the radius closest to the fly and the edge of the hub was measured with the spider-web vibrometer. The insets show the time waveforms for the power spectra (0 dB = 1 μm rms). From the digitized, Hanning-weighted time waveforms, the power spectra were computed at 25-Hz intervals by fast-Fourier-transform analysis. The solid line in the spectrum for transverse vibration indicates the region where the vibration signal was obscured by background electrical noise. For longitudinal and lateral vibration the signal-to-noise ratio was better and the entire spectra are above the background-noise level. All spectra have been corrected for the rolloff in frequency response of the spider-web vibrometer. The distance from the fly to the point of measurement was approximately 8, 10, and 7 cm (top to bottom); the fly weighed 30, 29, and 25 mg (top to bottom).

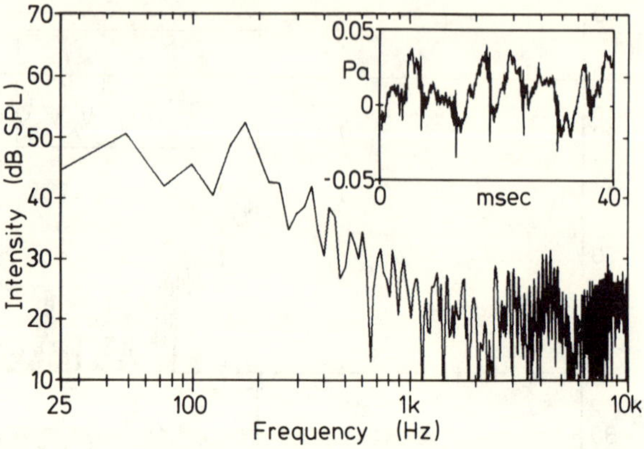

Fig. 3.9. Power spectrum for the airborne sound produced by a buzzing fly trapped in a web. The inset shows the pressure waveform (recorded with a Bruel and Kjaer 12-mm microphone, type 4131) from which the power spectrum was computed by fast-Fourier-transform analysis. The reference intensity level (0 dB) is 20 μPa. The fly weighed 26 mg.

measured at the intersection of the radius nearest the trapped fly and the edge of the hub. In the spectrum for each type of vibration, the maximum energy occurs at about 200 Hz. This corresponds to the wing-beat frequency of the fly, which is also responsible for the major peak in the airborne sound spectrum of a trapped fly (Fig. 3.9). Peaks at multiples of the frequency in Fig. 3.8 and 3.9 are harmonics of the wing-beat frequency. Above 200 Hz, the energy in the vibration power spectra decreases. Our measurements of vibration transmission in an empty web suggest that this decline in energy is partly attributable to the effects of the web, at least for transverse and lateral vibration. However, since we see a similar decline in the airborne sound intensity, evidently the vibration energy of the source (i.e., the fly) is also less at high frequencies than at low frequencies, as we would expect. The airborne sound produced by a trapped fly was recorded with a pressure microphone held only about 3 cm away from the fly to obtain a good signal-to-noise ratio; holding the microphone close to the source results in underestimation of the velocity of vibration of the source at low frequencies compared to high frequencies. In addition, since displacement is the integral of velocity, the displacement amplitude for a given velocity amplitude drops by 20 dB per decade. Both of these considerations indicate that the displacement amplitude of the vibration produced by a trapped fly declines even more steeply than the pressure amplitude shown in Fig. 3.9. This fact, and the attenuation due to web transmission, account for the steeper falloffs of the vibration power spectra shown in Fig. 3.8.

Since measurements of transmission in an empty web have shown that longitudinal vibration is transmitted along a radius with much less attenuation than transverse or lateral vibration, we would expect to find more high-frequency energy in the longitudinal vibration produced by a trapped fly than in the other two types. Our measurements do in fact indicate that the longitudinal vibration produced by a trapped fly contains more energy in the upper frequency range than the other two types—although for definitive comparison we would, of course, have to be certain that the three types of vibration are produced in equal amounts by the fly, and that they are all equally well coupled into the web. Energy in the longitudinal vibration produced by a trapped fly is often measurable at frequencies as high as 10 kHz (Fig. 3.8, top). Energy in lateral vibration, in contrast, is hardly ever detectable above 3 kHz (Fig. 3.8, bottom) and usually not above about 2 kHz. This limit is quite possibly a consequence of the poor web transmission of lateral vibration above about 1 kHz. Transverse vibration is difficult to compare with the other two types since it is not possible to obtain quite as good a signal-to-noise ratio when recording this type of vibration. Energy in transverse vibration can nevertheless be detected up to at least 2 kHz (Fig. 3.8, middle).

Limitations of Existing Techniques

The new techniques that we have used to measure web vibration are not without drawbacks and limitations. However, the results obtained with these techniques give us new insight into the relative usefulness of other methods.

With few exceptions, previously devised methods of measuring web vibration have involved some loading of the web. Barrows (1915) ingeniously avoided this problem by photographing webs being stimulated and noting the blurring of vibrating strands. Liesenfeld (1956) used microscopic examination with stroboscopic illumination to measure sinusoidal web vibrations. Graeser (1973), using a device much like the SWV, measured transverse vibrations of the signal thread in the webs of *Zygiella x-notata*, but the low-frequency response of his detector seems to have limited the range of his measurements.

Most of the methods previously employed to measure web vibration have used some kind of relatively massive (as compared to the web) vibration pickup, usually a phonograph cartridge or ceramic bimorph element (Tretzel 1961; Walcott 1963; Parry 1965; Krafft 1978; Suter 1978; Burgess 1979b). Although these methods of detecting web vibration have the advantage of being simple and inexpensive, they have the great disadvantage of interfering substantially with web vibration, and therefore they do not give accurate measurements of web-vibration amplitude. It is legitimate to use such methods to detect the presence of energy at certain frequencies, since the pickup itself ought not to contain spurious frequencies in its output signal; i.e., if there is energy in the output of the pickup at a given frequency, then there is energy in the web at that frequency. Unfortunately, it is not also true that if there is no energy in the output of the pickup at a given frequency, there is no energy in the web at that frequency; and furthermore, such a pickup cannot be used to compare signal amplitudes at different frequencies.

These difficulties arise for two main reasons. First, it is unlikely that the pickup will be equally sensitive to energy at all frequencies; i.e., its frequency response will not be flat since it will almost certainly have mechanical resonance characteristics that improve detection at some frequencies and suppress it at others. Moreover, some investigators have merely touched the pickup to the web rather than fixing it securely, and it is doubtful whether the web-pickup connection transmits equally well at all frequencies in this situation. Second, placing the pickup in contact with the web essentially creates a displacement node at the point of measurement. This will cause certain web resonances to appear as artifacts and other natural resonances to be destroyed. For this reason such pickups cannot be used to compare web vibration at different points (in order to determine vibration transmission between two points, for instance), since the vibration pattern of the whole web is likely to change when the pickup is moved from one point to another.

To reduce the problems that result from loading of the web, several investigators have devised measurement techniques that involve attaching to the web only very light objects, such as tiny mirrors (Liesenfeld 1956; Finck, Stewart, and Reed 1975; Suter 1978; Vollrath 1979b). However, the results shown in Fig. 3.6 indicate that even a 0.3-mg object can markedly affect web vibration at frequencies above about 100 Hz. If the upper frequency of interest is under this limit, then there seem to be no serious drawbacks to placing a light object in the web; and if the object is made light enough, the upper cutoff frequency can be raised. As noted earlier, however, another problem with such techniques is that measuring different points on the object attached to the web may produce different values because the object itself has certain resonant modes of motion (rotational, for instance) that may be excited by the web vibration. Since this problem occurs principally at high frequencies, where these measurement techniques are clearly unsatisfactory anyway, it cannot normally be deemed serious. However, if a mirror is used (e.g., Finck, Stewart, and Reed 1975), exaggeration of such effects by an optical lever could cause this problem to occur even at low frequencies.

Questions in Need of Answers

Although obviously incomplete, the results presented here at least give us an idea of what information is available to a spider to help it decide how to respond to web disturbances. Our results also raise some questions that affect the findings of previous studies. For instance, there seem to be clear differences in the transmissibility of different types of vibration along a web radius from the prey-catching zone to the hub, longitudinal vibration being transmitted much better than either transverse or lateral vibration. Only a few previous investigators appear to have recognized the possibility that different kinds of web vibration might exist (Frings and Frings 1966; Walcott 1969; Graeser 1973; Burgess 1979b; Vollrath 1979b), and fewer have compared different types (Liesenfeld 1956). As a result, very little experimental information is available for determining whether a spider differentiates among the various vibration types or whether, possibly, to the spider they are all equivalent. Furthermore, although there have been numerous neurophysiological investigations of vibration reception in spiders (for a review of the literature see Barth 1981, 1982), there has been no comparison of the effectiveness of different vibration types in exciting receptors (but see Barth 1972). The two investigations of the sensitivity of the tarsal-metatarsal lyriform organ (Walcott and Van der Kloot 1959; Liesenfeld 1961) used only one type of stimulus, probably transverse vibration. We have as yet, then, no good physiological evidence whether a spider is capable of distinguishing among vibration types.

It is, of course, reasonable to ask whether the ability to differentiate among vibration types would in fact be of any advantage to a spider. We have found (Masters and Markl 1981) that longitudinal vibration is more directional than transverse or lateral vibration, i.e., that a spider at the hub of a web would, in comparing the vibration of a stimulated radius and that of a neighboring radius, detect a greater difference in amplitude for longitudinal vibration than for transverse or lateral vibration. This suggests that, in locating vibration sources, a spider might do well to pay particular attention to longitudinal vibration. There is, therefore, some reason to think that it might be advantageous to a spider to be able to distinguish among vibration types.

The results presented here also raise questions about the importance to spiders of high frequency energy in prey-generated signals. When we measured the vibration signal transmitted by a buzzing fly trapped in a web, we found energy to at least 2–3 kHz for transverse vibration and to at least 10 kHz for longitudinal vibration. However, previous investigators studying vibration in sheet webs and cobwebs found little energy above 1 kHz in prey-produced signals (Walcott 1963; Burgess 1979b), and those studying vibration in orb webs found essentially no energy at such frequencies (Liesenfeld 1956; Graeser 1973; Suter 1978). Although there may indeed have been little or no high-frequency energy in the prey-produced signals in these studies, it is also quite possible that energy was merely not detected because of methodological limitations. One of these investigators (Graeser 1973) used a photocell with a relatively low upper cutoff frequency as a detector; the others used techniques that involved some degree of web loading, and, as we have shown, even small amounts of loading can severely damp high-frequency vibrations.

Although the results of the present study suggest that high-frequency energy can be present in prey-generated signals, there is as yet no direct evidence to show whether a spider uses this information in detecting or evaluating prey-produced vibration. Many investigators have observed that a buzzing fly trapped in a web provides a more attractive stimulus than a fly that is merely swinging in the web or struggling without buzzing (Boys 1880; Peters 1931; Burgess 1979b; pers. obs.). Since a buzzing fly is a rich source of high-frequency energy, we might hypothesize that high frequencies are particularly effective in alerting a spider or in releasing prey-capturing behavior. Furthermore, electrophysiological measurements of the whole leg nerve of various spiders have shown that the vibration threshold appears to be lowest at frequencies above 1 kHz (Walcott and Van der Kloot 1959; Walcott 1963; Liesenfeld 1961). It is tempting to speculate that a spider is able to use the presence of high-frequency energy in a signal as a cue to the presence of a potential food item, since a signal containing high-frequency energy would probably have a biological origin. In contrast, since a signal containing low-frequency energy might often have a nonbiological origin, (it might be generated by wind, for example), it would be more difficult for a spider to distinguish between biologically and nonbiologically produced signals at low frequencies.

Another question that remains to be answered is whether the high-frequency energy that reaches a spider is above its detection threshold. In the power spectrum for longitudinal vibration shown in Fig. 3.8, the small peak in energy at 5 kHz corresponds to a vibration amplitude of roughly 0.3 μm rms. From the results shown in Fig. 3.7 we can estimate that this amplitude will be further attenuated by 20 or 30 dB in being transmitted to a spider's tarsus, so that the amplitude of tarsal motion will be about 10 nm rms. If it is assumed that the data given by Liesenfeld (1961) apply to longitudinal vibration, the detection threshold of the orb weaver *Zygiella x-notata* may be 1 nm or even less at 5 kHz—20 dB less than the energy we have measured in the signal of a buzzing fly at this frequency. Thus the energy is almost certainly detectable.

Acknowledgments

This work was supported by Deutsche Forschungsgemeinschaft Grants Ma 374/13 and Ma 374/15-2 to Hubert Markl and a NATO postdoctoral fellowship to W. Mitch Masters.

4

Effects of Orb-Web Geometry on Prey Interception and Retention

WILLIAM G. EBERHARD

MANY SPIDERS construct silk webs to snare prey, and the geometric designs of these webs vary widely. The most highly organized design is that of an orb web, which consists of a sticky line laid in a more or less spiral pattern on a nonsticky scaffolding of radial, frame, and hub lines (Fig. 4.1). There have been a number of largely speculative attempts to determine the advantages and disadvantages of this type of web compared to others (Witt 1965; Eberhard 1972b; Lubin 1973; Denny 1976). Chacón and Eberhard (1980) showed that, contrary to the usual assumption that an orb web works like a sieve (e.g., Witt, Reed, and Peakall 1968), sticky threads intercept prey most efficiently when the spaces between them are larger than the diameter of the largest prey. Aside from the work of these authors, however, no study has addressed the functional implications of the fact that orb webs vary widely in design, some having relatively small spaces between both radii and sticky threads, others having large spaces, and still others being intermediate (Fig. 4.2). Since the primary function of an orb web is prey capture, and since the ability of a web made of a given type of line to capture prey must be a function of its geometrical design, it is possible to study the selective advantages of different geometrical designs by simulating prey impact on orb webs of different design.

This study attempts to determine the effects of different orb-web designs by simulating the interception of prey. To do this it first was necessary to calculate the amount of both silk and energy that a spider expends in building an orb web, so that the numbers of prey captured by webs that differ in design but that represent equal investments of material and energy could be

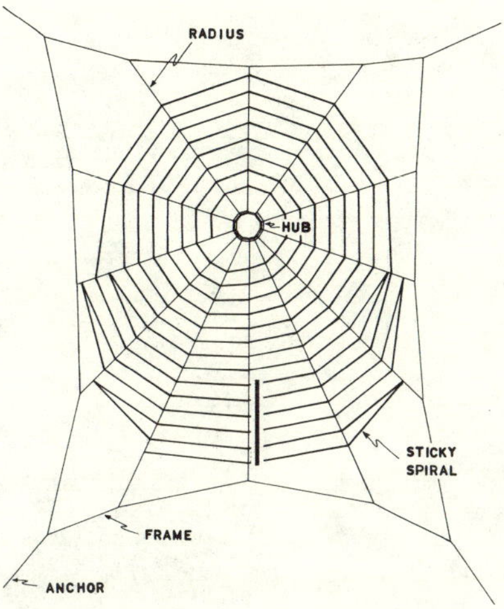

Fig. 4.1. Stylized orb web. The thicker lines indicate sticky web lines. The line beside the 6 o'clock radius indicates the area in which weights were dropped to determine the stopping abilities of lines.

compared. Calculation of these costs turned out to be complicated, since a single change in web design involves a change in several different design parameters. The first part of this chapter describes how these cost factors relate to each other; then expected payoffs from different designs are examined in terms of expected numbers of prey intercepted and retained in webs of equal cost. Finally the possible effects on the expected payoffs resulting from several other factors such as prey nutrient value, web angle with horizontal, angle of prey impact, prey shape, etc., are discussed.

Materials and Methods

Simulation of prey impact on orb webs of different designs was used to determine capture rates. Although simulations could have been performed with a computer, a simpler but no less precise technique was used. Patterns of evenly spaced radii and concentric regular polygons (representing the sticky spiral) were drawn on large pieces of paper (Fig. 4.3). Each piece of paper was also ruled into a grid, and a small transparent disk (representing the prey) was then placed on the paper at a series of 450 points; a random-number table was used in combination with the grid to ensure that the points were randomly distributed over the entire surface. At each point, the numbers of radii and sticky spiral lines passing under the disk were noted. Different-sized disks (with diameters of 0.5, 1.15, 2.0, and 3.0 cm) were used to simulate different-sized prey. In all, nine simulation webs were used, three each with spaces of 0.5, 1, and 2 cm between sticky spiral lines where they crossed radii. Each web received the same array of prey, for a total of 16,200 presentations of prey.

a

c

Fig. 4.2. Orb-web designs. In each photograph the scale line indicates 4 cm. Numbers of specimens deposited in the Museum of Comparative Zoology at Harvard University are given for spiders that cannot be identified to species, and the angle of the plane of the web with the horizontal is given in parentheses. *a*, *Ogulnius* sp. 802 (90°, not strictly planar); *b*, *Wagneriana* sp. 815 (88°); *c*, *Chrysometa* sp. 1,232 (82°); *d*, *Eriophora edax* (87°); *e*, *Hypophthalma* sp. 1,714 (82°); *f*, *Alpaida tuonabo* (81°); *g*, *Tetragnatha* sp. 2,045 (70°); *h*, *Alpaida niveosagillata* (85°).

e
f
g
h

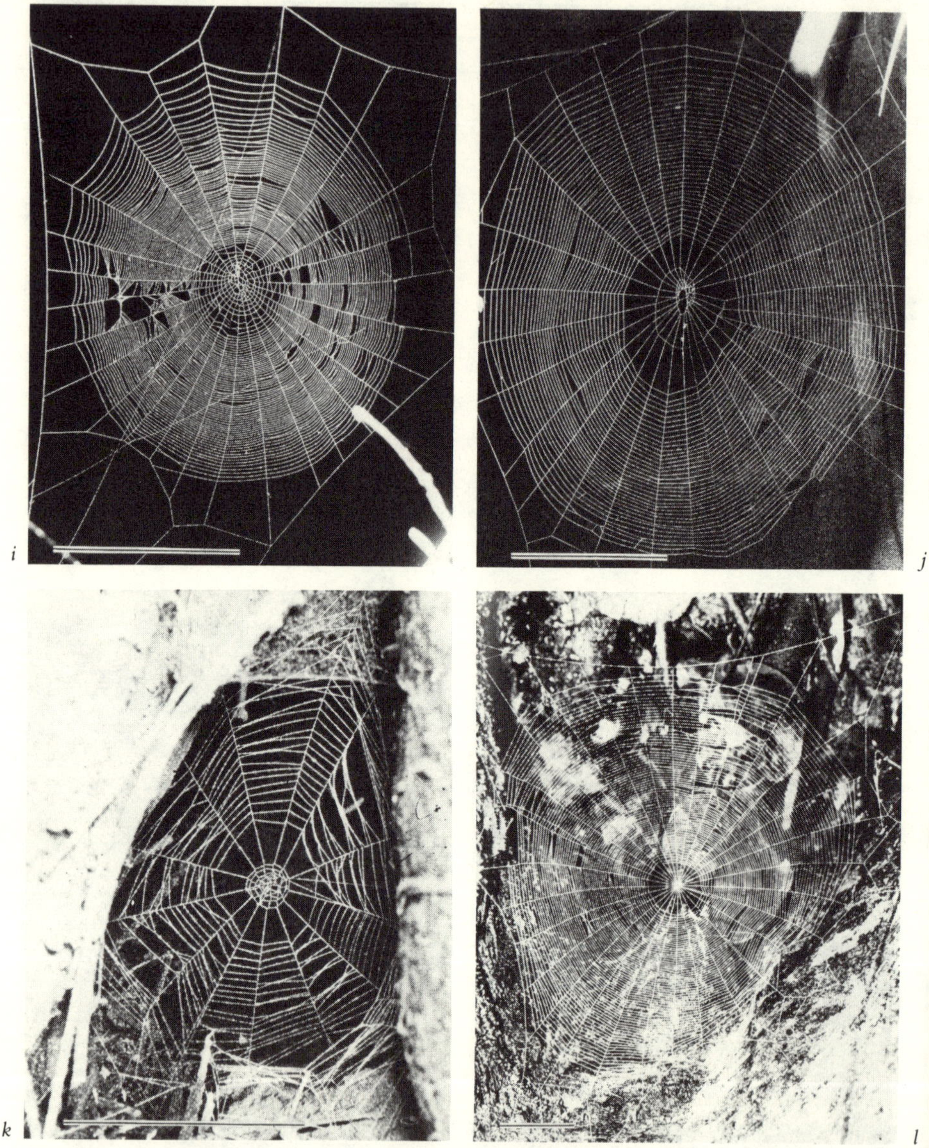

Fig. 4.2 (*continued*). Orb-web designs. *i*, *Wixia* sp. 573 (90°); *j*, *Gasteracantha cancriformis* (71°); *k*, *Cyrotognatha* sp. 537 (approximately 5°); *l*, *Dolichognatha* sp. 1,007 (17°); *m*, *Cyclosa* sp. 2,068 (89°); *n*, *Micrathena triangularis* (56°); *o*, *Spilasma artifer* (25°); *p*, *Uloborus* sp. 242 (0°).

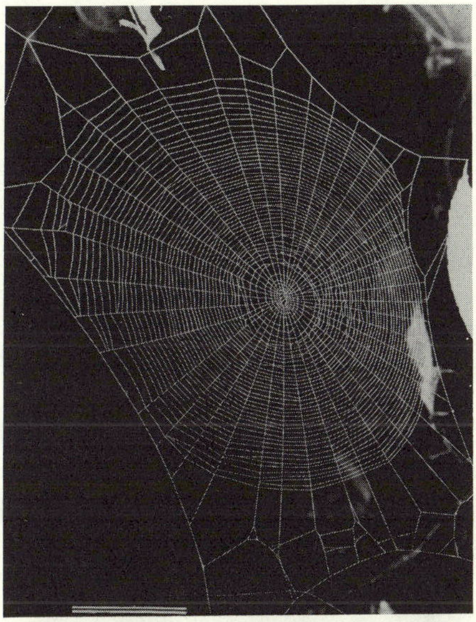

m

n

o

p

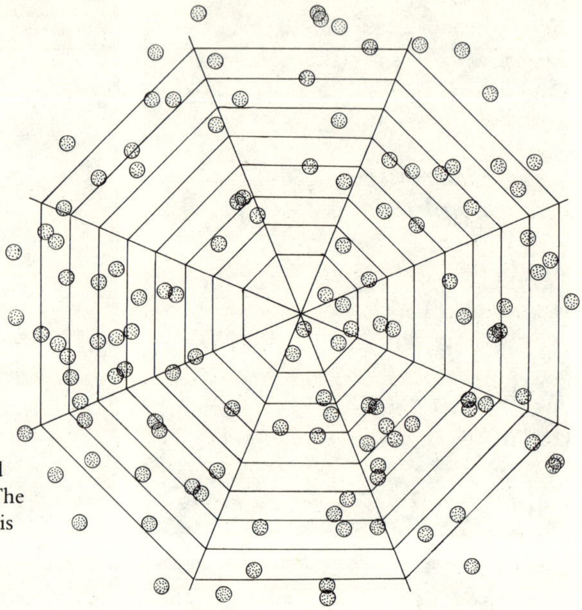

Fig. 4.3. Simulation web and impacts of simulation prey. The distance between sticky lines is 1.0 cm; the prey diameter is 0.5 cm.

The simulation webs were not perfect replicas of any known spider webs, but their geometric properties are probably similar to those of many species' webs. Although sticky lines in real webs are not arranged in concentric regular polygons, they often closely approximate this pattern (compare Figs. 4.2 and 4.3). The patterns used for the simulation webs had the advantages of being easy to draw, describe, and measure precisely, and of representing a general rather than a special case.

Relative strengths of sticky spiral and radial lines in real webs were determined experimentally by dropping small weights (U-shaped pieces of fine copper wire) from a distance of 1 cm above the web onto lines in intact, newly spun, and nearly horizontal orb webs (±20°) of adult female *Leucauge* sp. near *venusta* and *Cyclosa bifurca* (specimens numbered 2,124 and 2,117 deposited in the Museum of Comparative Zoology in Cambridge, Massachusetts). The webs of *Leucauge* were tested in the field. The *C. bifurca* females were induced to build webs in the laboratory in oval wire frames, and the frames were held horizontal for the tests. Each sticky thread was tested only once, but some radii were tested repeatedly. All tests were performed in the outermost fourth of the area covered by sticky spiral (Fig. 4.1), and weights were dropped near the center of each sticky segment tested.

Several of the conclusions derived in this study depend on interpolation of intermediate values from graphic representations of calculated or experimental results. This seems justified in view of the fact that all of the properties are direct consequences of web geometry, and thus must change continuously as geometric parameters change. Fortunately most of the relationships involving prey interception are close to linear, which makes for relatively confident determination of intermediate values.

Unless otherwise noted, all statistical tests were two-tailed chi^2.

The Material Costs of Webs

A spider expends both material and energy in building a web. Since the data concerning the material cost of webs are more complete, I shall discuss this type of cost first.

There is direct evidence from *Araneus diadematus* (Peters 1937; Witt, Reed, and Peakall 1968) and indirect evidence from *Uloborus diversus* (Eberhard 1972b), *Leucauge* sp. near *venusta*, and *Uloborus trilineatus* that an orb weaver uses more or less all of its reserve of sticky silk in building each web. It appears, however, that an orb weaver does not use its entire supply of nonsticky silk: by cutting out radii as they were laid, *A. diadematus* and *Leucauge* sp. near *venusta* were induced to lay about three times the normal number of radii (Koenig 1951; W. G. Eberhard, in prep.), and *Gasteracantha cancriformis* was induced to lay about twice the normal number (W. G. Eberhard, in prep.). The size of a spider's material investment in a web may thus be limited by the amount of sticky rather than nonsticky silk it has available.

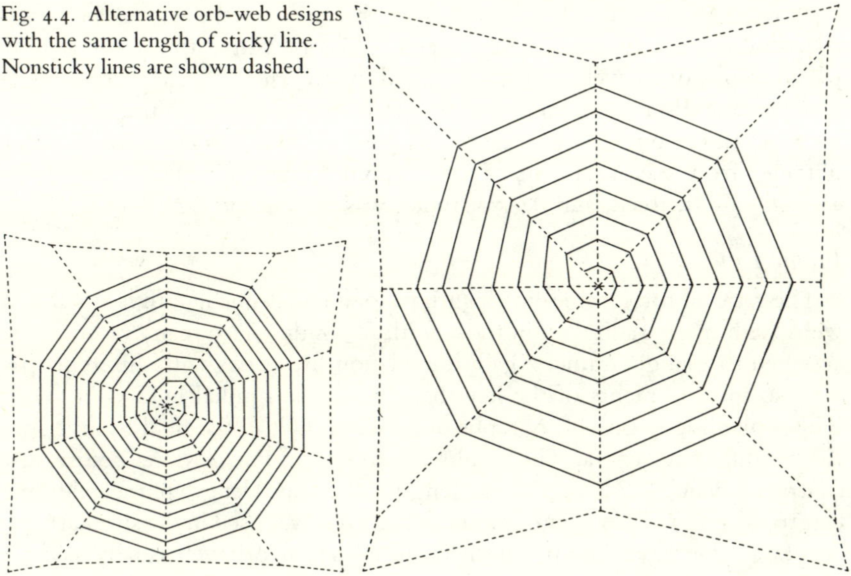

Fig. 4.4. Alternative orb-web designs with the same length of sticky line. Nonsticky lines are shown dashed.

Consider a spider with enough material in its aggregate and flagelliform (coronate) glands to produce a given length of sticky line. It could choose to make a small web, with the loops of sticky line spaced close together, or a larger web, with fewer loops of sticky line spaced farther apart (Fig. 4.4). (Early, mechanistic beliefs that mesh size is determined by the spider's own size and thus that an individual spider has no such option have proved unfounded: Peters 1937, 1954; Witt, Reed, and Peakall 1968.) A spider with a given amount of sticky silk that opts for, say, a larger web will simultaneously opt for particular lengths and arrangements of radial and frame lines (nonsticky lines shown dashed in Fig. 4.4). The reasons for this are the following.

Radii

There is a positive correlation between the number of radii and the number of sticky spiral loops in the webs of most orb weavers (Eberhard 1972b, in prep.). The number of radii is somewhat greater than the number of sticky spiral loops in some genera, such as *Cyclosa*, *Micrathena*, and *Uloborus*, and somewhat less than the number of sticky spiral loops in other genera, such as *Dolichognatha*, *Mangora*, and *Acacesia*, but on the average there are about equal numbers of both (W. G. Eberhard, in prep.; see also Fig. 4.2). Equal numbers of radii and sticky spiral loops were therefore used in the simulations.

Since radial lines extend beyond the outermost loop of sticky spiral to an attachment on the frame line, the lengths of radii in orb webs are always somewhat greater than the distance from the center of the hub to the outermost loop of sticky spiral. This difference was measured for all the radii in photographs of 25 webs built in the field by one species each of 17 genera. The average distance to the frame was 116 percent of that to the outermost loop of sticky spiral. The value of 115 percent, which is close to the value obtained from measurements of radii in photographs of webs of *Araneus diadematus* (Witt, Reed, and Peakall 1968), was used in the simulations.

Frame Lines

The ratio of total frame-line length to total radial-line length is smaller in webs with many radii than in those with few radii, since many radii can be attached to a single frame. Again the relationship varies with different species, so measurements of the lengths of frame and radial lines were made with a planimeter on photographs of 51 webs built in the field by five species of different genera. The results are shown in Fig. 4.5. As before, the central tendency—log (total frame length/total radius length plus total frame length) = 1.127 − 1.099 log (number of radii)—was used in the simulations. Several studies have shown that the frame lines in an orb web usually have up to four times more fibers than the radii (DeWilde 1943; Denny 1976; Work

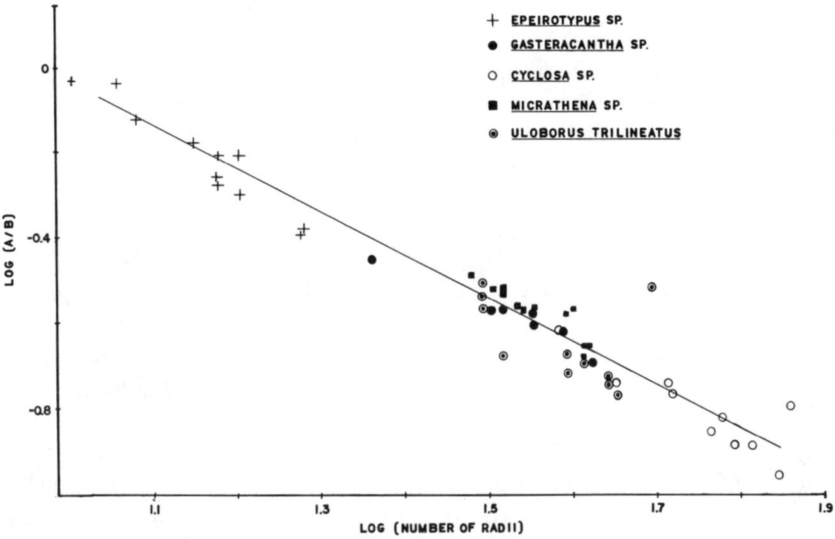

Fig. 4.5. Relationship between number of radii and the ratio of total frame line length (*A*) to total length of radii and frame lines (*B*).

1981). In the simulations it was assumed that frames had twice as many silk fibers as radii; i.e., the length of the frame lines was multiplied by 2.

Anchors

The extensions of the frame threads to attachments on the substrate represent another expenditure, although the magnitude of this expenditure varies widely among species. Some species, such as *Acacesia hamata*, *Gasteracantha* spp., *Hypophthalma* spp., and some *Micrathena* and *Neoscona* species, build their webs in large open spaces and lay relatively long anchor lines; others, such as *Uloborus* spp., *Azilia* spp., and *Argiope* spp., build in much smaller spaces and lay relatively short anchor lines. Since the objective of the simulation was to examine typical orb-web patterns, the species *Araneus diadematus*, which lays anchor lines of intermediate length, was chosen as a model, and measurements were made of the relative lengths of anchor and frame lines in *A. diadematus* webs from a photograph (Witt, Reed, and Peakall 1968; this photograph of a juvenile's web is apparently the only published photograph of an orb web that includes all the anchors in their entirety) and drawings (Mayer 1952; Koenig 1951). The length of the anchors was, on the average, 20 percent of the length of the frames, and this value was used in the simulations. It was assumed, moreover, that the anchors had as many fibers as the frames—a conservative assumption, since e.g., Denny (1976) found more fibers in anchors than in frames in the webs of *Nuctenea sclopetaria*.

Temporary Spiral

Most orb weavers build a spiral of nonsticky silk just before building the sticky spiral. They then destroy this temporary spiral as they lay the sticky spiral. Careful observation of the web-building behavior of some large species (*Eriophora* spp., *Argiope argentata*, *Micrathena* spp., and *Leucauge* sp. near *venusta*) shows that spiders break the temporary spiral but do not ingest it, and thus the temporary spiral represents an additional outlay of silk. In the simulations, the length of line needed for the temporary spiral was estimated using the fact that in real webs the space between loops of temporary spiral is about five times that between loops of sticky spiral. Values were measured from figures showing webs of *Araneus diadematus* (Witt, Reed, and Peakall 1968, Fig. 44; Peters 1970), *Scoloderus tuberculifer* (Eberhard 1975), and *Nephila clavipes*, *Gasteracantha cancriformis*, *Argiope argentata*, *Micrathena gregalis*, *Micrathena duodecimspinosa*, and *Cyclosa trifida* (Peters 1954). The average value was 5.6; the values ranged from 3.4 to 8.0. It was assumed in the simulations that the innermost loop of temporary spiral is laid at the same distance from the hub as the innermost loop of sticky spiral, as has been found to be the case in a number of species (W. G. Eberhard, in prep.).

Hub

There is no obvious correlation between the number of hub loops in an orb web and other variables, such as the number of radii. The lengths of hub lines in the simulation webs were measured from drawings with three loops, as in Fig. 4.1. This number of loops is more or less typical (see Fig. 4.2).

Sticky vs. Nonsticky Lines

Another consideration in determining the material cost of an orb web is that the sticky and nonsticky lines in a web are made of different materials, and the sticky line has balls of adhesive material on it. It appears, from studies by Lubin (1973) and Prestwich (1977), that spider silk has a relatively constant caloric value of around 4,600 cal/g. Lubin's data include both webs with entirely nonsticky lines (*Cyrtophora molluccensis*) and webs with large amounts of sticky line (*Nephila clavipes*). The relative weights of the two types of line are not known, however. In a few cases the diameters of sticky and nonsticky lines have been measured; e.g., Denny (1976) found that in the webs of *Nuctenea sclopetaria* the sticky fibers have an average diameter of 1.5 μm and the radial fibers an average diameter of 1.9 μm. However, the balls of sticky material are not included in these measurements, and thus the relative costs of the two types of line are still unknown. For the simulations, this uncertainty was taken into account by calculating two different costs for each web on the basis of two extreme assumptions: (A) a length of sticky line is 0.67 times as expensive as an equal length of nonsticky line, and (B) a length of sticky line is 1.5 times as expensive as an equal length of nonsticky line.

Effects of Reingestion of the Web

Some orb weavers carefully eat their previous webs before building new ones (Hingston 1922; Breed et al. 1964; Carico, this volume), and in *Araneus diadematus* reconversion of ingested webs into silk for later webs reaches the extraordinary level of 95–97 percent efficiency (Peakall 1971). Other species, however, do not reingest old webs. *Metazygia gregalis* (Araneidae), for example, carefully rolls up silk into balls and flicks them away with its front legs when it destroys webs built just a few hours previously (M. Barreto and W. Pfizenmaier, pers. comm.; W. G. Eberhard, pers. obs.), and *Uloborus diversus* discards silk that accumulates as it renews web lines. Moreover, even species such as *A. diadematus* probably do not reingest 100 percent of previous webs, since losses result from holes torn in the webs and rain washing away adhesive material.

These differences and uncertainties do not directly affect the simulations, since it was assumed that tendencies to reingest webs did not change with changes in web design. They do, however, influence the relative importance of material and behavioral costs. Without reingestion the behavioral cost of a web is probably only about 20 percent of the material cost, whereas with complete reingestion the behavioral cost may reach about 500 percent of the material cost. Since the estimates of material cost are probably more accurate than those of behavioral cost, the results are less certain for species that ingest a large fraction of previous webs.

The Behavioral Costs of Webs

There is only one study of the energy expended by a spider as it constructs an orb web, that of Peakall and Witt (1976) on *A. diadematus*. These authors used two independent methods: measurement of distances moved and steps taken, and measurement of oxygen consumption with a respirometer. The "step" analysis could easily be applied to other species, but its original application involved so many relatively arbitrary (and at the moment unjustifiable) assumptions that the agreement with measured values in this one case does not constitute convincing proof that the technique is valid. Problems include the criteria used to decide whether a spider "walks" or "runs" as it works; application of the term "step" to an activity in which different legs move in different patterns; the assumption that one step during radius construction is equivalent to one step during temporary spiral or sticky spi-

TABLE 4.1

Costs of Pairs of Orb Webs with Equal Lengths of Sticky Line but Different Sticky Spiral Spacing

(centimeters of silk line)

Category	Length of sticky line (cm); sticky spiral spacing (cm)											
	300		400		600		700		800		900	
	2.0	1.0	2.0	1.0	1.0	0.5	1.0	0.5	1.0	0.5	1.0	0.5
Number of sticky loops	6.6	7.6	7.6	10.0	12.4	16.4	13.6	18.0	14.6	19.3	15.6	20.8
Material costs												
Hub line	14	14	14	15	15	15	15	15	15	15	15	15
Temporary spiral	36	35	50	56	92	101	110	120	125	136	131	156
Frame	318	222	375	339	325	232	343	243	367	256	391	270
Radii	94	77	130	126	193	188	225	217	261	247	299	283
Anchors	64	44	75	68	65	46	69	49	73	51	78	54
Total nonsticky silk	526	379	630	590	674	567	747	629	826	690	899	763
Total silk												
Web not reingested, A	726	579	898	858	1,074	967	1,216	1,098	1,362	1,226	1,499	1,363
Web not reingested, B	976	829	1,230	1,190	1,574	1,467	1,797	1,679	2,026	1,890	2,249	2,113
Web reingested, A	29	23	36	34	43	39	49	44	54	49	60	55
Web reingested, B	39	33	49	48	63	59	72	67	81	76	90	85
Behavioral cost, A	145	116	180	172	215	193	243	220	272	245	300	273
Behavioral cost, B	195	166	246	238	315	293	359	336	405	378	450	423
Total cost												
Web not reingested, A	871	695	1,078	1,030	1,289	1,160	1,459	1,318	1,634	1,471	1,799	1,636
Web not reingested, B	1,171	995	1,476	1,428	1,889	1,760	2,156	2,015	2,431	2,268	2,699	2,536
Web reingested, A	174	139	216	206	258	232	292	264	326	294	360	328
Web reingested, B	234	199	295	286	378	352	431	403	486	454	540	508
Average cost of smaller web / average cost of larger web	82.5%		96.0%		91.5%		92.0%		91.5%		92.5%	

NOTE: Condition A assumes that sticky silk is 0.67 times as costly as dry silk; condition B assumes that sticky silk is 1.5 times as costly. Behavioral costs were calculated on the assumption that behavioral energy expenditures equaled 20 percent of the energy expended in web material.

ral construction, when it is clear that more legs are active and performing movements other than simple locomotion during the latter activities; lack of consideration of the force exerted by the spider to pull out silk as it walks (Work 1977); and the assumption that the energy expended in "one trip around the cage" is equivalent to that used in the preliminary exploratory activities preceding web building proper and in the placement of frame and anchor lines. Different, equally plausible assumptions give results that do not agree with the results obtained with the respirometer.

Measured oxygen consumption, based on 30 minutes of web building, was 504 ± 32 liters of oxygen per gram per hour ($N = 9$), or 18 percent of the calculated total caloric value of the silk in the web. The authors assumed that the webs of *A. diadematus* weigh 0.1 mg, but gave no data to support this assumption. They also assumed that the caloric value of the silk in these webs is equal to that of the silk in webs of *Nephila clavipes*—probably a legitimate assumption, since caloric values of silk from different species are similar (Lubin 1973; Prestwich 1977). The value of 20 percent was used in calculating the behavioral costs of webs for the simulations.

It is striking that the metabolic rate of *A. diadematus* while it built a web was found to be only about double the resting rate. It cost the spider the equivalent of only an extra half hour of resting to build an entire web—a remarkably low cost. It should be noted, however, that owing to its relatively large size this species may expend an atypically small amount of energy in building a web, since larger araneids have proportionally lower metabolic rates (Anderson 1970; Greenstone and Bennett 1980; see also Eberhard 1979b).

Total Costs of Alternative Web Designs

Table 4.1 gives calculated costs (in lengths of silk line) for the alternative web designs used in the simulations. Costs are given for pairs of orb webs with equal lengths of sticky line but different geometrical designs. The values were determined as follows. The number of loops of sticky line was interpolated from plots of the number of loops versus the length of sticky line for given spacings between loops. The length of hub line was measured from a drawn figure. The length of temporary spiral was calculated from the number of sticky loops on the assumption of one loop of temporary spiral per five loops of sticky spiral. The length of frame line was calculated from the regression equation relating total frame length to total radius length (Fig. 4.5) and multiplied by 2. The length of radii was interpolated from Fig. 4.6 and multiplied by 1.15. The length of anchors was taken as 20 percent of the length of frame lines and multiplied by 2. Silk cost if the spider

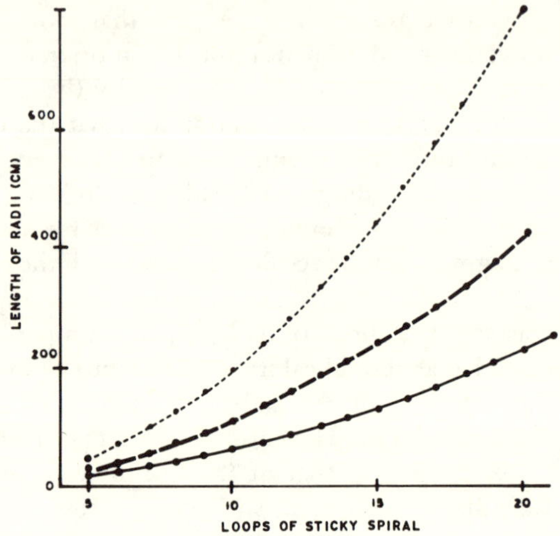

Fig. 4.6. Total length of radii vs. number of loops of sticky spiral when the number of radii equals the number of loops and each radius extends to the outer loop of sticky spiral. The sticky spiral spacings are 2.0, 1.0, and 0.5 cm (top to bottom).

reingests the web was calculated assuming 4 percent loss from recycling (Peakall and Witt 1976). Behavioral cost was taken as 20 percent of the total silk cost. Condition A assumes that a given length of sticky line is 0.67 times as costly as an equal length of dry line; condition B assumes that a given length of sticky line is 1.5 times as costly as an equal length of dry line. The numbers in parentheses are the average cost of the web with smaller spacing between sticky lines expressed as a fraction of the average cost of the web with larger spacing.

When the length of sticky spiral is held constant, webs with smaller spaces between loops are consistently slightly less costly than ones with larger spaces; the cost of webs with smaller spaces is, on the average, 91 percent (with a standard deviation of ±5 percent) of the cost of webs with larger spaces. This is because webs with smaller sticky spiral spacing have relatively less nonsticky silk. Thus the advantage of smaller sticky spiral spacing is greater when sticky silk is assumed to be 0.67 times as costly as dry silk (condition A in Table 4.1) than when it is assumed to be 1.5 times as costly (condition B); the average costs are 89.4 ± 4.9 and 92.6 ± 3.8 percent, respectively. In webs with longer lengths of sticky line, the relative amount of nonsticky silk decreases slightly; e.g., the total length of nonsticky silk is 1.26 greater than the total length of sticky line in the web with 300 cm of sticky line with loops spaced 1 cm apart, but is only 1.00 as long as the web with 900 cm with the same spacing.

Of the total of nonsticky silk, a larger percentage is dedicated to frame and anchor lines in smaller webs. Thus, for instance, 70.2 percent of the nonsticky silk is in frames and anchors in the web with 300 cm of sticky line and 1.0-cm spaces between loops, but only 52.2 percent is in frames and anchors in the web with 900 cm of sticky line at the same spacing; webs of intermediate size have intermediate percentages. It is well to remember that the number of fibers in frame and anchor lines may have been underestimated, and therefore both the total investment of nonsticky silk and the fraction of nonsticky silk dedicated to frames and anchors may be low.

Prey Interception

The number of prey of different sizes that were intercepted by different numbers and types of web lines were tabulated for each of the simulation webs. Some of these data are given in Fig. 4.7. The numbers of prey intercepted by webs that differ in sticky spiral spacing but that represent equal investments of material and energy were then determined by interpolating from the graphs in Fig. 4.7. The results are presented in Tables 4.2 and 4.4. To correct for the relatively greater cost of webs with wider spacing between sticky lines, the numbers of prey intercepted by the more costly webs were multiplied by the numbers in the last row of Table 4.1. (Actually, Table 4.1 gives ranges of values; the averages are used to simplify the presentation of the data. None of the calculations of significant differences in Tables 4.2 and 4.4 is altered by substituting other values from Table 4.1.)

In order to serve as effective traps, orb webs must perform several different functions: they must (1) intercept the prey with a line, (2) absorb the momentum of the prey (stop it), and (3) retain the prey until the spider can arrive to attack (Chacón and Eberhard 1980). Since different functions are performed by different threads, the results of the simulations are presented according to the type of thread that intercepts the prey; the patterns that emerge are then summarized and combined.

Interception by Sticky Lines

When interceptions by one or more sticky lines are examined (Table 4.2, column 4), two sets of patterns are evident. The first pertains to pairs of webs that are equally costly but differ in sticky spiral spacing. In all cases in which the diameter of the prey was larger than the sticky spiral spacing in both webs, the web with the wider sticky spiral spacing intercepted more prey. This difference was due to the larger areas covered by these webs.

When interception data that had not been corrected for different relative investments in nonsticky thread and construction behavior were compared in cases in which the diameter of the prey was larger than the smaller sticky spiral spacing but smaller than the larger spacing, the web with larger spacing made significantly more interceptions ($p < 0.05$). This result supports the argument of Chacón and Eberhard (1980) that sticky threads trap prey

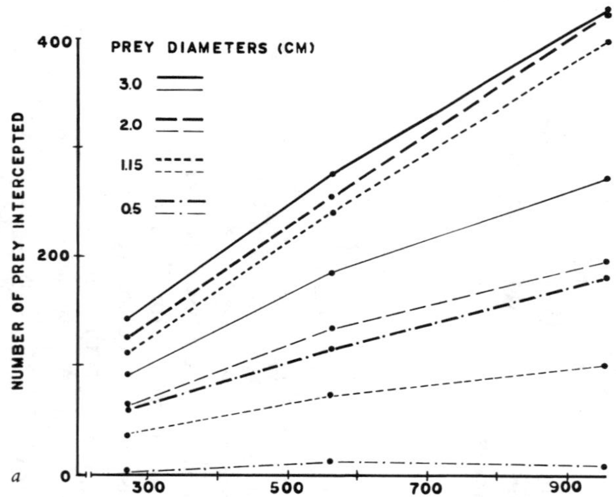

a

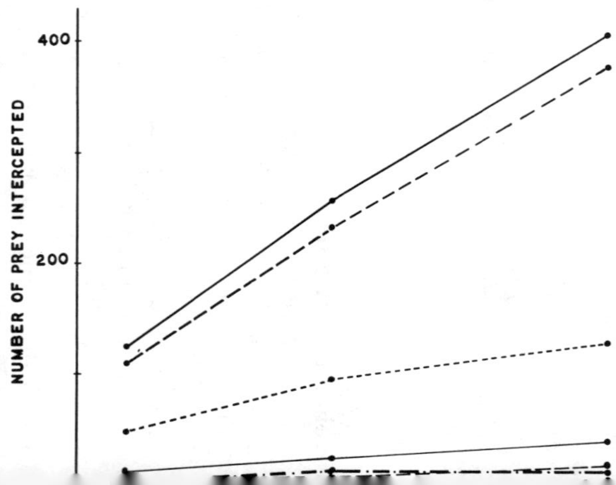

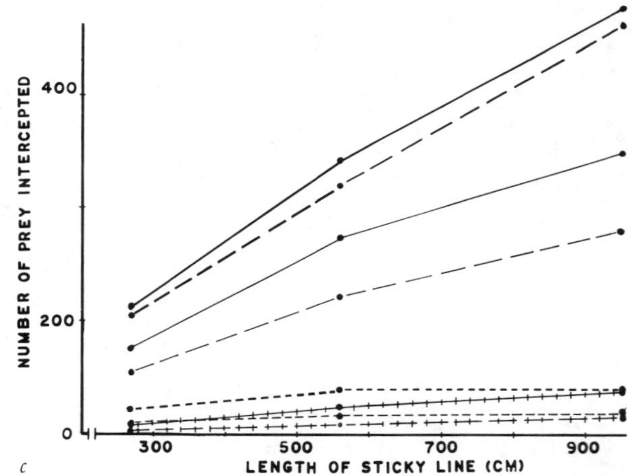

c

Fig. 4.7. Number of prey intercepted by different numbers and types of lines in webs with different lengths of sticky line. *a*: thick lines for prey that hit ≥1 sticky line; thin lines for prey that hit ≥1 radius as well as ≥1 sticky line. *b*: thick lines for prey that hit ≥2 sticky lines, or ≥1 radius as well as ≥1 sticky line; thin lines for prey that hit ≥2 radii as well as ≥1 sticky line, or ≥1 radius as well as ≥5 sticky lines. *c*: thick lines for prey that hit ≥2 sticky lines; thin lines for prey that hit ≥1 radius as well as ≥2 sticky lines; hatched lines for prey that hit ≥2 radii as well as ≥1 sticky line.

TABLE 4.2
Number of Prey Intercepted by Sticky Lines in Pairs of Webs That Differ in Sticky Spiral Spacing but Are Equally Costly

Length of sticky line in smaller web (cm)	Spacings compared (cm)	Prey diameter (cm)	Number of prey intercepted[a]	Number of prey/total cost	
				Larger web	Smaller web
300	2.0/1.0	3.0	206/158^{+} (1.30)	.687	.527
		2.0	191/140^{+} (1.36)	.637	.467
		1.15	120/125 (0.96)	.400	.417
		0.5	54/65 (0.83)	.180	.217
400	2.0/1.0	3.0	315/204^{+} (1.54)	.788	.510
		2.0	300/184^{+} (1.63)	.750	.460
		1.15	204/169$^{(+)}$ (1.21)	.510	.422
		0.5	100/83 (1.20)	.250	.208
600	1.0/0.5	3.0	267/170^{+} (1.57)	.445	.283
		2.0	251/155^{+} (1.62)	.418	.258
		1.15	236/142^{+} (1.66)	.393	.237
		0.5	112/129 (0.87)	.187	.215
700	1.0/0.5	3.0	305/195^{+} (1.56)	.436	.279
		2.0	290/179^{+} (1.62)	.414	.256
		1.15	272/165^{+} (1.65)	.389	.236
		0.5	128/152 (0.84)	.183	.217
800	1.0/0.5	3.0	338/216^{+} (1.56)	.422	.270
		2.0	328/202^{+} (1.62)	.410	.252
		1.15	309/188^{+} (1.64)	.386	.235
		0.5	143/175$^{(+)}$ (0.82)	.179	.219
900	1.0/0.5	3.0	377/240^{+} (1.57)	.419	.267
		2.0	372/225^{+} (1.65)	.413	.250
		1.15	351/210^{+} (1.67)	.390	.233
		0.5	159/197^{+} (0.81)	.177	.219

[a] A superscript + indicates that the difference between the numbers compared is significant at $p < 0.05$; (+) indicates $p < 0.1$. The numbers in parentheses are the preceding ratio expressed as a decimal fraction.

more efficiently when the spacing between them is larger than the prey diameter, and contradicts the common idea that an orb web works like a sieve. However, when the interception data were adjusted to take into account the relatively greater cost of webs with larger sticky spiral spacing, this difference was either eliminated (see, e.g., the data for a sticky-line length of 300 cm and a prey diameter of 1.15 cm in Table 4.2) or reduced to marginal significance (see, e.g., the data for a sticky-line length of 400 cm and a prey diameter of 1.15 cm).

Another tendency in Table 4.2 seems to lend support to the same argument of Chacón and Eberhard. When the prey diameter was greater than the sticky spiral spacing in both of the webs being compared, the difference in

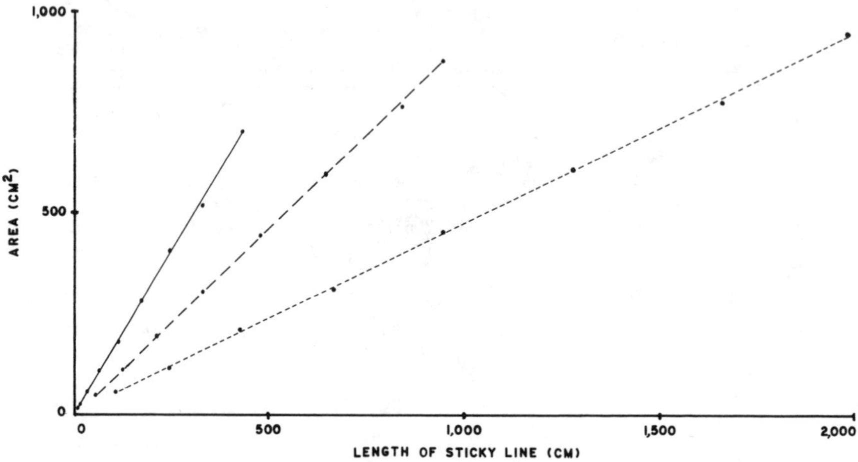

Fig. 4.8. Total area covered by sticky spiral vs. length of sticky spiral in webs of different design. The sticky spiral spacings are 2.0, 1.0, and 0.5 cm (top to bottom).

the number of interceptions was in all cases greater when the larger of the two spacings was more nearly equal to the prey diameter. (Compare, e.g., the data for prey diameters of 3.0, 2.0, and 1.15 cm in a web with 600 cm of sticky line.) Finally, when the prey diameter was smaller than the sticky spiral spacing in both webs, there was no consistent significant difference in the number of interceptions.

The second set of patterns pertains to webs that differ in cost but have the same sticky spiral spacing. There was a consistent tendency for smaller webs to be more efficient than larger webs of the same design ("efficient" as measured by the ratio of prey interceptions to total cost given in columns 5 and 6 of Table 4.2). For example, the results for 3.0-cm-diameter prey intercepted by webs with sticky spiral spacings of 1.0 cm indicate a steady decrease in efficiency, from 0.527 (for the web with 300 cm of sticky line) to 0.419 (for the web with 900 cm of sticky line)—a decrease of 20 percent ($p < 0.05$ when the costs of the two webs are compared). Stated another way, a threefold increase in investment resulted in the interception of only 2.39 times as many prey. Some of the results in Table 4.2 are exceptions to this pattern (compare, e.g., the results for 0.5-cm-diameter prey intercepted by webs with sticky spiral spacings of 2.0 cm and sticky spiral lengths of 300 and 400 cm), but in general the pattern is present (though not always statistically significant) throughout the table. The reason for this pattern is not clear. The total area covered by the sticky spiral of an orb web increases linearly with the length of sticky spiral (Fig. 4.8), so differences in area covered are not the explanation. Nor are differences in the relative costs of nonsticky lines, since, as noted above, larger webs of a given design require relatively shorter lengths of nonsticky line.

The greater efficiency of smaller webs may have been a factor in the evolution of body size in orb weavers, and may also have played a role in the independent evolution of reduced webs in genera such as *Hyptiotes* and *Miagrammopes* (Uloboridae) (Gertsch 1949), *Ogulnius* and *Wendilgarda* (Theridiosomatidae) (Coddington and Valerio 1980; Coddington, this volume), and *Pachygnatha* (Araneidae) (Bristowe 1958). This does not mean that larger webs do not intercept more prey—they do. Rather, the ratio of payoff to investment decreases. Thus spiders can expect to trap more prey by making larger webs of a given design, but the total investment increases more rapidly than the number of prey intercepted.

Interception by Other Lines

There are data in the literature (e.g., Denny 1976) comparing the strengths of single lines of sticky and radial silk, but these are not sufficient for understanding how orb webs function, since they do not take into account that some lines in orb webs are longer than others and that the physical properties of connected lines in intact webs vary. It was therefore necessary to test the strengths of lines in intact webs. Known weights were dropped on lines in webs of mature females of *Leucauge* sp. near *venusta* and *Cyclosa bifurca*, and it was found that radial lines in the webs of both species could stop objects with about ten times more momentum than could single sticky spiral lines (Table 4.3). Interception of prey with a radial line in addition to a sticky line is thus much more likely to result in the prey's being stopped and not breaking through the web. It is worth noting that in none of the tests did a hub or frame line break, so the strengths of these lines are probably not limiting factors in determining a web's ability to absorb prey impacts.

Since of two webs with the same length of sticky line but different sticky spiral spacing the web with the smaller sticky spiral spacing will usually have more radii, it would seem logical that more of the prey that hit such webs would hit a radius. It is not intuitively clear whether this advantage would make up for the disadvantage of webs with smaller sticky spiral spacing that was documented above. The simulation results (Table 4.4) show that the advantages and disadvantages nearly balance when one counts the number of prey intercepted by both one or more sticky spiral lines, and one or more radii, and that webs with smaller sticky spiral spacing intercept on the average 1.88 more prey with both one or more sticky spiral lines and two or more radii than do webs with larger sticky spiral spacing. Thus, for intercepting prey that are difficult to stop (i.e., that are heavy or fast-moving), webs with smaller sticky spiral spacing are slightly superior to webs with

TABLE 4.3
Number of Lines That Broke per Number of Lines That Were Tested by Dropping Weights on Them from About 1 cm Above

Spider	Test weight (mg)	Radial lines	Sticky spiral lines
Leucauge sp.			
(near *venusta*)	11.4	0/10	5/10
	114	3/13	10/10
Cyclosa bifurca	15.1	0/20	7/20
	18.9	0/20	14/20
	114	1/20	20/20
	146	5/20	20/20

larger spacing, and the superiority is more pronounced as the prey become more difficult to stop.

The retention, or holding, abilities of orb webs are much more difficult to assess than the interception or stopping abilities, since they depend not only on thread strengths and numbers but also on factors such as the stickiness and extensibility of threads, the geometric relations among threads, the slant of the web with respect to gravity (Nentwig 1982), and the idiosyncratic struggling movements and "slipperiness" of different prey species (Nentwig 1982). Measurements of the lengths of time that small sepsid flies were held by horizontal webs of *Leucauge* sp. near *venusta* (W. G. Eberhard, in prep.) showed that even for a single type of prey there is great variation in holding times ($\bar{x}$ = 22 seconds, coefficient of variation = 119 percent, N = 125). Similarly, Nentwig (1982) found great variation in the escape times of *Drosophila melanogaster* from orb webs of *Araneus diadematus* and *Zygiella x-notata* (coefficients of variation as high as 370 percent). Nentwig (1982) found that the strength of adhesion of sticky tape to the wings of a wide variety of insects was relatively uniform, with the exception of lepidopterans, trichopterans, and culicid and psychodid flies, for which the strength of adhesion was significantly lower.

Nevertheless, one qualitative statement concerning the holding abilities of orb webs can be made. The greater the number of sticky threads that contact the prey once it has been stopped in a web, the greater the likelihood that the prey will not escape before the spider arrives to attack. Rypstra (1981) studied retention of *Drosophila* in orb webs, and although her results are difficult to interpret because she combined data from different spider species, the predicted tendency for retention times to increase in webs with closer sticky spiral spacing did occur. Even though most of the orb-weaving spiders studied are substantially larger than *Drosophila* and thus probably produce sticky lines with relatively strong holding abilities compared to the escape abilities of *Drosophila*, nevertheless from 5 to 30 percent of the flies that hit webs were

TABLE 4.4

Number of Prey Intercepted by Multiple Lines in Pairs of Webs That Differ in Sticky Spiral Spacing but Are Equally Costly

Length of sticky line in smaller web (cm)	Spacing compared (cm)	Prey diameter (cm)	Number of prey intercepted[a] by: ≥1 sticky line and ≥1 radius	≥1 sticky line and ≥2 radii	≥2 sticky lines and ≥1 radius	≥2 sticky lines
300	2.0/1.0	3.0	93/100 (0.93)		56/93^{+} (0.60)	117/130 (0.90)
		2.0	68/69 (0.99)			16/120^{+} (0.13)
		1.15	26/40 (0.65)			
		0.5	5/3 (1.67)			
400	2.0/1.0	3.0	132/134 (0.99)		74/122^{+} (0.58)	157/173 (0.91)
		2.0	96/94 (1.02)			18/155^{+} (0.12)
		1.15	38/53 (0.72)			
		0.5	3/5 (0.60)			
600	1.0/0.5	3.0	178/159 (1.12)	23/37$^{(+)}$ (0.62)	163/138 (1.18)	231/154^{+} (1.50)
		2.0	128/109 (1.17)	8/16 (0.50)	112/97 (1.15)	199/136^{+} (1.46)
		1.15	69/64 (1.08)			37/120^{+} (0.31)
		0.5	11/22^{+} (0.50)			
700	1.0/0.5	3.0	199/179 (1.11)	25/46^{+} (0.54)	181/160 (1.13)	263/173^{+} (1.52)
		2.0	144/131 (1.10)	11/19 (0.58)	126/116 (1.09)	245/158^{+} (1.55)
		1.15	77/78 (0.99)			37/143^{+} (0.26)
		0.5	22/28 (0.79)			
800	1.0/0.5	3.0	219/199 (1.10)	28/55^{+} (0.51)	197/185 (1.06)	296/195^{+} (1.52)
		2.0	158/146 (1.08)	12/23$^{(+)}$ (0.52)	140/137 (1.02)	278/182^{+} (1.53)
		1.15	83/88 (0.94)			37/168^{+} (0.22)
		0.5	21/31 (0.68)			
900	1.0/0.5	3.0	240/234 (1.03)	32/65^{+} (0.49)	219/171^{+} (1.28)	333/215^{+} (1.55)
		2.0	175/170 (1.03)	14/27^{+} (0.44)	191/160 (1.19)	314/208^{+} (1.51)
		1.15	90/104 (0.87)			37/192^{+} (0.19)
		0.5	9/39^{+} (0.23)			

[a] A superscript + indicates that the difference between the numbers compared is significant at $p < 0.05$; (+) indicates $p < 0.1$. The numbers in parentheses are the preceding ratio expressed as a decimal fraction.

not retained for more than 3 seconds. Clearly, the holding function of an orb web is of vital importance in prey capture.

When one compares interceptions by two or more sticky lines in pairs of webs with different sticky spiral spacing (Table 4.4, last column), the same patterns emerge as those seen for interceptions by a single sticky line: webs with larger spaces between sticky lines capture more prey (an average of 52 percent more). This is true only for large prey, however. When the diameter of the prey is less than two times the larger of the two spacings, webs with smaller spacings capture 400 to 800 percent more prey.

Summary of Interception Tendencies

Smaller webs are consistently more efficient (number of prey/cost) at intercepting prey with sticky lines than larger webs with the same spacing between loops of sticky spiral. When investment in webs is held constant, webs with larger spaces between sticky lines are more efficient at intercepting prey with one or more sticky lines only as long as the prey diameter is greater than the space between sticky lines.

When one takes into account retention of prey as well as interception, relationships become complicated. Webs with smaller spaces between sticky lines are as good as alternative designs for intercepting prey with a radius as well as a sticky line, and superior for intercepting prey with more than one radius in combination with at least one sticky line. Interception with multiple sticky lines is more efficient in webs with larger spaces between sticky lines, but only for relatively large prey.

Discussion

Energetic Payoffs

The energetic return a spider obtains from its web depends on the sizes of the prey captured as well as on their numbers. In general, larger prey should give larger returns, although this will not always be strictly true, since spiders have limited capacities and well-fed spiders do not completely consume even relatively small prey (Turnbull 1962). To a first approximation, the food value of a prey item will be a function of its diameter cubed, and thus the largest prey item in the simulations would have about 72 times more food value than the smallest. (See Beaver and Baldwin 1975 for a more precise relationship derived from measurements of brachyceran and schizoceran Diptera.)

It seems clear, from data on the size of the prey in various studies of orb weavers (Robinson and Robinson 1970a, 1973a; Nyffeler and Benz 1978, 1979; Uetz and Biere 1980; Olive 1980; Castillo and Eberhard, in press) and a knowledge of the typical sticky spiral spacings in the webs of species in the genera involved in these studies, that most of the prey of orb weavers are larger than the sticky spiral spacings in their webs. (This may not be true for, e.g., some Theridiosomatidae and in the outer portions of some *Tetragnatha* webs; see, e.g., Fig. 4.2g.) The simulation of prey interception by single sticky lines may therefore have less biological importance than the simulation of interception by multiple sticky lines. As shown above, the same trends hold for single- and multiple-line interceptions.

Since heavier prey are probably in general both harder to stop (Nentwig 1982) and, at least in horizontal orb webs, also harder to retain (Nentwig 1982), it would seem to be important that such prey be intercepted by radial as well as sticky spiral lines, and by more than one sticky spiral line. Unfortunately, the holding function of webs is not yet well understood, and it is not possible to quantify the advantages of interception by multiple lines. As Olive (1980) points out, retention properties probably vary in vertical webs according to whether the prey usually escapes by falling through the web parallel to its plane, or by actively pulling free by moving perpendicular to the plane. There are data showing differences in the number of prey captured by species with different web designs in similar habitats (Nyffeler and Benz 1978, 1979; Olive 1980), but since, e.g., spider response time, attack behavior, thread numbers, diameters, and strengths, thread stickiness, and microhabitat all may have varied, the data cannot be used to test the effect of web geometry on retention. Not until the effects of such factors on retention are known will it be possible to weigh the changes in the interception, stopping, and retention abilities of orb webs resulting from larger or smaller sticky spiral spacing.

Effects of Angle of Interception and Prey Shape

In the simulations it was assumed that prey moved perpendicularly to the web plane. Although some species may succeed in placing their webs across insect flyways (e.g., *Nephila maculata*: Robinson and Robinson 1973a), most species appear to orient their webs in response to a variety of considerations, such as the insolation at the web hub (Krakauer 1972; Robinson and Robinson 1978; Carrel 1978; Cherrett 1964; Hieber 1979; Biere and Uetz 1981) or wind (Eberhard 1971a), and in general it seems likely that most prey do not strike orb webs perpendicularly. When a prey hits a web at an angle of less than 90° it is likely to encounter more threads, and the spacing between lines is effectively reduced. In addition, the amount of reduction in spacing (i.e., the effective spacing relative to the spacing measured in the web plane) will depend on where the prey strikes. Assume, for example, that a prey hits the web shown in Fig. 4.1 while flying toward the right in a direction that makes an angle of 45° with the web plane. If it hits the web directly to the left or right of the hub, the effective sticky spiral spacing will be reduced by up to 50 percent; if it hits directly above or below the hub, the effective sticky spiral spacing will be unchanged (but the effective *radial* spacing will be decreased); and if it hits at other points the effective sticky spiral spacing will be reduced by intermediate amounts. The average effective spacing thus will always be smaller in nature than the spacing measured in the web plane, but the amount of reduction will vary with the placement of the hub.

The effects of interception angle do not modify the generality of the results of the simulations, since these were based on comparisons of prey interceptions by alternative webs that were struck at equal angles by the prey. (The results would be affected only if prey tended to hit one of the alternative webs, or different parts of webs, at consistently different angles; there is no reason to suppose that this would occur.) They do, however, influence predictions about the sizes of prey that will be most effectively trapped by webs in nature. The measured mesh sizes of webs (i.e., measured from photographs taken perpendicular to the web plane) will consistently overestimate the average effective mesh sizes. If it is assumed that both the angles of incidence and the points of impact are random, the average angle of incidence is 45° and the average reduction of sticky spiral spacing calculated by summing the extremes mentioned above and dividing by 2 is (0.5 + 1.0)/2, or 0.75. Thus, if these assumptions are at least nearly correct, and the sticky spiral spacing is uniform throughout the web, the average effective spacing will be about 75 percent of the average measured spacing.

Prey impacts that are not perpendicular to the web plane will put stresses on web lines that differ from the stresses applied experimentally to determine the relative strengths of sticky and radial lines, since part of the prey's momentum will be parallel rather than perpendicular to the line. In general, nonperpendicular prey impacts will result in more severe stresses on web lines. This consideration does not bear directly on the present study, in which only the *relative* abilities of the two types of line to absorb momentum, and of web geometries to intercept prey, were determined.

Another consideration not taken into account in this study is that most flying insects have oval or figure-8 rather than circular outlines (see, e.g., the photographs in Nachtigall 1974). As shown in Fig. 4.9, such outlines are intercepted by different numbers of sticky and radial lines when they hit different parts of the web. If, however, one assumes a random distribution of prey impacts on a symmetrical web, the effects of prey shape balance each other out and give the same results as those achieved with circular outlines. The comparative results of this study are thus not affected, but it will be difficult to extrapolate to real situations.

A further consideration to be kept in mind is that interception in which a line strikes the prey far from its center of gravity or on only one side of it (i.e., a glancing blow) will be less likely to result in the prey's being detained,

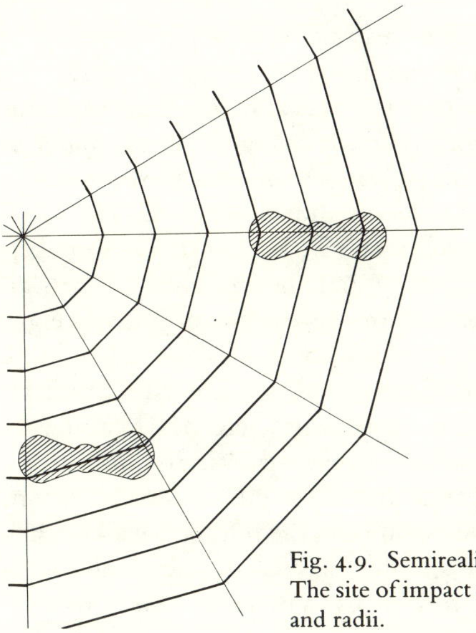

Fig. 4.9. Semirealistic outline of a prey hitting a web in two places. The site of impact affects the likelihood of hitting sticky lines and radii.

at least when impact is perpendicular to the web plane. In the simulations all interceptions were assumed to be equally effective. This assumption is obviously unrealistic, but probably does not affect the general patterns found, since the effect of taking this consideration into account should be equivalent to reducing the size of the prey. Thus captures of real prey 3.0 cm in diameter would be more nearly given by the captures of prey 2.0 or 1.15 cm in diameter in the simulations. Again, though the comparative conclusions of this study are unaffected, extrapolation to real situations will be difficult.

Limitations Due to Simplifying Assumptions

The classic problem with simulation studies in biology is that the simplifying assumptions that are made in order to eliminate or ameliorate complexities that make simulation difficult ultimately restrict the biological value of the study. It will therefore be useful to discuss the limitations that result from the simplifying assumptions made in this study, and at the same time to point out promising directions of future research.

Some limitations have already been mentioned. These include the limitations resulting from the use of concentric polygons rather than spirals, from the assumption that the total material investment is limited by the amount of silk available in glands, and from the assumptions just discussed concerning the angle of interception, prey shape, and the manner of collision. Such assumptions have been shown to be either reasonable approximations or at least unlikely to influence the patterns found in the simulation results.

Others are not so easily taken into account. It was assumed that the diameters of both sticky and nonsticky lines were not changed by the spider in alternative web designs. Preliminary measurements of lines in the webs of the large orb weaver *Eriophora ravilla* have shown, however, that fiber diameters are not uniform even within a given web (W. G. Eberhard, unpub.). Moreover, Christiansen, Baum, and Witt (1962) showed that *Araneus diadematus* systematically changed thread diameters when its body weight changed. (Work (1976, 1977) failed to find a correlation between body weight and fiber diameter, however.) Edmunds (1978) showed that sticky spirals in two vertical webs of *Paraneus cyrtoscapus* were nearly twice as strong as those in the horizontal web of a similar-sized individual of the same species, presumably because thread diameters differed. In addition, Work (1981) found that there is some variability in the number of fibers in a given class of line such as temporary spirals. Careful studies of the variability of thread diameters, and of the relative costs of sticky and nonsticky lines, would be very useful.

All simulation webs were symmetrical around the central point, but few spider webs are strictly symmetrical (Fig. 4.2). This asymmetry probably has little effect on the likelihood of interception by sticky lines (Fig. 4.10), but since in asymmetrical webs spiders modify the web design by putting more radii in the larger sectors (see Eberhard 1972b for a summary of evidence on this point, and also Fig. 4.2), the likelihood of interception by radii is altered. This subject will be treated in a separate study.

As noted above, the use of behavioral costs as measured for the relative large species *A. diadematus* may give a serious underestimate of behavioral costs in other, smaller species. It should also be noted that the length of the web-building process in *A. diadematus* may vary depending on the time of day at which the web is built (Ramousse 1980). In addition, no adjustment was made for the possibility that the behavioral cost of laying non-sticky lines differs from the cost of laying sticky lines. Moreover, behavioral costs are probably reduced somewhat in spiders that build horizontal rather than vertical webs as does *A. diadematus*; the calculations of Peakall and Witt (1976) suggest a 19 percent savings in behavioral costs, but this may be an overestimate since at least some species that make horizontal or nearly horizontal orb webs, such as *Uloborus diversus* and *Leucauge* sp. near *venusta*, descend and ascend several times in the area where they will build, apparently to check for objects in the vicinity (Eberhard 1972b, unpub.). Studies of the metabolic costs of different building activities would be very useful.

It was assumed that all webs were equally difficult for prey to avoid, regardless of differences in design, but there is plentiful field evidence that insects see and avoid webs (Turnbull 1960; Lubin 1973; Olive 1980; Rypstra

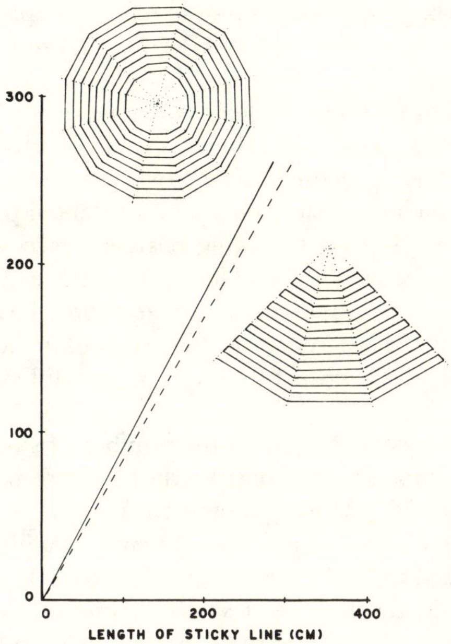

Fig. 4.10. Effect of web asymmetry on the area covered by given lengths of sticky line. The solid line is for symmetrical webs, the dashed line for asymmetrical webs, as pictured in the insets.

1981); those with more closely spaced lines are generally easier to spot visually, at least with the human eye. Web visibility probably does not affect prey captures for spiders that spin at night, but differences in web designs between species that spin in the morning and during the day and those that spin in the evening might be due to this factor. Some species, such as *A. diadematus* (Ramousse and LeGuelte 1979), *Leucauge* spp., and *Epeirotypus* spp. spin both in the morning and in the evening, and it might be rewarding to check for differences in web designs. It was also assumed that the spacing between sticky lines was constant throughout a web, when it is well established that sticky spiral spacing is seldom if ever constant and that in many—perhaps most—species it varies in a consistent way from the edge of the web to the hub. (See, e.g., LeGuelte 1966 on *Zygiella x-notata* and Eberhard 1972b on *Uloborus diversus*; see also Fig. 4.2.)

No adjustments were made to take into account probable differences in the mechanical stability of different web designs when subjected to stresses other than those imposed by prey. In general, more compact webs are probably more resistant to damage by wind, because they sag less and thus are less subject to the especially high stresses resulting from the billowing and then snapping tight that occurs when the wind changes direction (Langer 1969). They are probably also less susceptible to damage from falling or wind-blown detritus, since they intercept fewer objects, and are also more likely to absorb the momentum of the objects that hit them. Effects of resonant frequencies of vibration in webs were also ignored, but these are probably not important in prey capture (Finck, Stewart, and Reed 1975; Frohlich and Buskirk 1982).

Frame and anchor lines are sometimes reused rather than being discarded or reingested (Biere and Uetz 1981; Carico, this volume; W. G. Eberhard, unpublished observations of *Metazygia gregalis*, *Cyclosa caroli*, and *Leucauge* sp. near *venusta*). Since frame and anchor threads constitute substantial portions of the nonsticky lines in webs (Table 4.1), taking reuse into account could appreciably alter the results. In general it would lower material costs as compared to behavioral costs, and make total cost correlate more closely with sticky-line cost, thus increasing the differences in efficiency of interception by sticky lines in webs with equal lengths of sticky line but different sticky spiral spacing.

In addition, the number of radii was kept equal to the number of sticky spiral loops, when it is clear that some species consistently lay more radii than sticky loops (e.g., *Cyclosa* spp., *Micrathena* spp., and *Uloborus* spp.; see Fig. 4.2) and others consistently lay fewer (e.g., *Mangora* spp., *Dolichognatha* spp., and *Acacesia hamata*; see Fig. 4.2). It is not possible to estimate how inappropriate the simulations are for these species without knowing the relative costs of sticky and nonsticky silk, and of the behaviors used to lay them. On the other hand, the functional significance of these deviations seems clear in the light of the results presented here. The radius-rich designs probably serve to increase the webs' stopping abilities, and the sticky-spiral-rich designs to increase the webs' holding abilities. Thus I predict that species that build radius-rich webs capture relatively high percentages of large and/or fast-moving prey, and those that build sticky-spiral-rich webs capture relatively high percentages of prey that are difficult to retain. Stowe (1978, this volume) has already shown that one species that makes an extraordinarily sticky-spiral-rich web (*Scoloderus cordatus*) does indeed capture especially large numbers of moths, a type of prey known to be difficult to retain because of their covering of loose scales (Eisner, Alsop, and Ettershank 1964). Unfortunately, there are few other data to test these predictions, and, given the fact that orb weavers capture a wide variety of prey (Robinson and Robinson 1970, 1973a; Nyffeler and Benz 1978, 1979; Horton and Wise 1981; Castillo and Eberhard, in press) and the difficulty of knowing how similar the mix of prey species at any given site is to the mix(es) that were present as the spider evolved, such data will be difficult to obtain.

No adjustment was made for the fact that some species respond more rapidly than others when prey hit their webs (Suter 1978), or for the fact that some species attack more rapidly (Robinson 1975; Robinson, Robinson, and Graney 1972; Olive 1980). Furthermore, different species have different attack behaviors that are more or less effective against different types of prey

(Robinson, Mirick, and Turner 1969; Robinson 1969b, 1975; Olive 1980). Thus the expected payoff for one species of spider would not necessarily be the same as that for another species, even though their webs intercepted and retained the same prey; and different prey that were intercepted in equal numbers and retained for equal times would not necessarily be equally available to all spiders. The first limitation is not important when comparing different web designs of a single spider, as was attempted here, but it is important when comparing web designs of different species of spiders.

It has been assumed, finally, that, confronted with alternative web designs, spiders will adopt whatever design turns out to give the best yield on investment. This apparently reasonable assumption runs afoul of the well-established fact that webs of spiders of a given species vary substantially in design details both in the field (I. Maroto, in prep.) and in captivity when identical supports are provided (Witt, Reed, and Peakall 1968; LeGuelte 1966; Eberhard 1972b; Ramousse and LeGuelte 1979). Even individual spiders often spin substantially different webs from day to day in the laboratory. The causes of some of the variation in webs that spiders actually spin remain mysterious.

Summary

The relative material and behavioral costs of orb webs of different design have been estimated (Table 4.1) from data in the literature and measurements of webs in photographs. The relationships are complex, since spiders change a number of design features synchronously. In general, when the amount of sticky silk is held constant, webs with smaller spaces between sticky spiral lines are somewhat less costly overall. When the sticky spiral spacing is held constant, webs with shorter sticky lines contain a larger percentage of nonsticky silk, and a larger portion of the nonsticky silk is dedicated to frame and anchor lines.

Expected prey captures by webs of different design have been estimated by simulation (Tables 4.2, 4.4). The results confirm the "non-sieve" model of prey interception of Chacón and Eberhard (1980), but the predicted advantage of webs with wider spaces between sticky lines was counterbalanced by the comparatively greater costs of such webs. Calculations of web efficiency (as measured by the ratio of the number of prey intercepted to the total cost) indicate that larger webs are less efficient than smaller webs with the same sticky spiral spacing; the reason for this is not clear. Since radial lines in intact orb webs can absorb about ten times more kinetic energy than sticky lines, webs with especially large numbers of radii are probably more efficient at stopping large and/or fast-moving prey. Webs with smaller spaces between sticky lines are probably more efficient at retaining hard-to-hold prey. These findings may explain some of the differences in the orb-web designs of different species and genera (Fig. 4.2).

The simplifying assumptions made in this study are sufficiently numerous and profound to preclude quantitative predictions of design efficiencies in nature. Nevertheless some of the qualitative tendencies documented here are probably of general significance.

Acknowledgments

Dr. H. W. Levi identified the spiders I observed, and J. Coddington, C. Craig, Dr. Y. D. Lubin, and Dr. M. J. West Eberhard read preliminary drafts of this chapter. I was supported by grants from the Comité de Investigaciones of the Universidad del Valle in Cali, Colombia, and the Vicerrectoría de Investigación of the Universidad de Costa Rica. I thank all for their help.

5

Prey Specialization in the Araneidae

MARK K. STOWE

THE ARANEIDAE is one of the largest spider families, comprising roughly 10 percent of the approximately 30,000 known spider species. Most naturalists are familiar with what I shall call conventional araneids. These spiders spin orb webs and feed on a wide variety of airborne prey. A large number of araneids have departed to a greater or lesser extent from this mode of existence, which is presumably ancestral for the family. Many spin modified webs or no webs at all, and are specialized—some extremely specialized—in the prey they consume.

In this chapter I explore the different patterns of prey specialization in the Araneidae. I begin by considering briefly the factors that bias the rates of capture of different types of prey by conventional araneids. I then present a detailed discussion of prey capture in those araneids that I (somewhat arbitrarily) have deemed to be specialized enough to warrant consideration separate from conventional araneids. I also consider interesting parallels in prey specialization in other groups of spiders (Appendix A), what is known of the behavior of younger individuals of specialized araneids (Appendix B), details of web construction for certain specialized araneids (Appendix C), and, finally, the taxonomic relationships of some specialized araneids (Appendix D).

Conventional Araneids

Conventional araneids capture an enormous variety of prey. The prey of large spiders belonging to tropical genera such as the New World genus

Eriophora may occasionally even include small birds and bats. However, what an araneid captures is far from being a random sample of what is available. The pattern of capture rates is a complex function of the behavior and morphology of the spider and the behavior and morphology of its prey. In my discussion below of the relevant features of spiders and prey, I refrain from discussing the complex effects of web geometry on intercepting and retaining prey, since this is taken up by Eberhard (this volume).

For a more complete overview of the predatory behavior of conventional araneids, readers should start by consulting Robinson 1975 and Eberhard 1982; for the types and number of prey captured, Kajak 1965, Olive 1980, and Robinson 1982b; and for spinning behavior, Eberhard 1982.

Nonrandom Placement of Webs in Space and Time

Placement of the web in the environment (distance from vegetation, height above the ground, etc.) affects the types of insects that may intersect it (e.g., Chacón and Eberhard 1980). Araneids are able to create lines that span distant points by letting out silk and waiting until the wind wafts the thread to another location. As a first approximation, most points in the environment are potentially available for web attachment, and most spaces between attachment points are potentially available as web sites. The species-characteristic pattern of web placement in environments (e.g., Uetz, Johnson, and Schemske 1978) probably reflects, to some extent, selective pressure on members of different species to increase the probability of intercepting insects that they can capture effectively. (See Olive 1980 for a discussion of this and factors other than prey availability affecting web placement.)

The time of day that a spider hunts affects both the composition of available prey and the ability of prey to see webs and avoid them (see below). Some araneids (e.g., many *Nephila* species) are active for most of the day and night. Others (e.g., many *Argiope* species) are strictly diurnal. Many are strictly nocturnal, spinning their webs after nightfall, and taking down—and sometimes ingesting—their webs before dawn (Carico, this volume). Still others (e.g., *Verrucosa arenata*) are active at night and may or may not take down their webs at dawn.

Janetos (1982b) and Olive (1982) have shown that, in some araneid species, individuals change web sites more frequently when they are feeding less. (For a detailed discussion of web-site selection, see Janetos, this volume.) The evenness with which a particular prey species is distributed in space and time might be expected, therefore, to affect the rate at which it is captured by araneids. Janetos (1982b) suggests that araneids at web sites near flowers generally experience higher feeding rates, that they therefore remain longer at such sites, and that, as a result, for many araneids flower-foraging insects form overall a larger part of the diet than would be the case if these insects were more evenly distributed.

Different araneids exhibit different patterns of patch use. In some araneids, an important determinant of web-site change is the buildup of kleptoparasites in the web (Robinson and Robinson 1976b; Vollrath 1979a). For nocturnal araneids, web-site change is linked with web-removal behavior (Carico, this volume).

Nonrandom Movement of Prey with Respect to Webs

Some insects are attracted to the vicinity of spider webs, particularly kleptoparasites (see the relevant references in Sivinski and Stowe 1980), egg-case "parasites," and web commensals. Many of these insects appear to be adept at avoiding entanglement in webs, but others (e.g., male scorpion flies: Thornhill 1975) are often captured.

There is much anecdotal evidence that some insects (e.g., bees and syrphid flies) are skillful at avoiding webs, while others (e.g., jumping Orthoptera) are less so (see the relevant references in Olive 1980). Unfortunately, the mechanisms by which prey sense webs and maneuver around them have only begun to be studied in a systematic way (Robinson and Robinson 1973a; Lubin 1974; Nentwig 1982; C. Craig, in prep.; see also Eberhard, this volume). Although most web avoidance during the day is presumably visually mediated, some Nematocera may be able to sense webs by touch and back away (Chacón and Eberhard 1980).

A number of araneids build webs close to substrates. Robinson and Lubin (1979a) discuss two "arboricolous" New Guinea araneids, *Herennia ornatissima* and *Argiope ocyaloides*, that build webs exclusively against the sides of trees and catch insects that attempt to land on the bark. The remarkable webs of *H. ornatissima* match the curvature of the tree trunks against which they are built. W. G. Eberhard (pers. comm.) has observed a number of arboricolous araneids in the New World genus *Eustala*. Other araneids, such as *Araniella displicata* (McCook 1890; pers. obs.), seem to build webs preferentially in the cups of leaves. Webs in such situations are probably less visible to prey. It is also possible that insects attempting to alight on the substrates involved are less able to sense webs or maneuver to avoid them than they would be during normal flight.

A number of araneids build webs in aggregations. (For a review of the literature, see Buskirk 1981; see also Uetz, this volume.) Web aggregations are probably more visible and easier to avoid than isolated webs for many prey (Lubin 1974). On the other hand, multiple layers of webs are presumably more successful than single webs at catching insects with high momentum (Lubin 1974).

Differences in Retention Efficiency for Different Types of Prey

Insects differ considerably in the degree to which they adhere to viscid silk. Those with detachable scales, such as moths, adhere only slightly (Eisner, Alsop, and Ettershank 1964). In addition to the insects that are known to hang from nonviscid threads in spider webs (see references in Sivinski and Stowe 1980), there are flies that are able to hang from viscid spiral threads in araneid webs without being caught (Sivinski and Stowe 1980; J. Coddington, pers. comm.). Araneids differ in the amount of adhesive they apply to the sticky spiral, and it is possible that webs with large amounts of adhesive (e.g., those of *Eriophora* species: pers. obs.) retain large prey or moths longer. In artificial webs, increased quantities of adhesive result in increased capture of most insect taxa but especially of Lepidoptera and Nematocera (Chacón and Eberhard 1980).

The behavior of an entangled insect has a strong effect on the length of time it is held in the web. Suter (1978) observed the struggling movements of five different insects in an araneid web, and found that whereas the bee *Halictus* sp. vibrates its wings only after it has freed them with leg movements, the moth *Sitotroga cerealella* beats its wings from the moment of impact. These qualitative differences were reflected in significant differences in the temporal patterns of the web vibrations generated by the struggling movements. The moth's struggling movements are stereotyped; the legs are held close to the body and the wings are beaten without being fully extended. Suter suggests that this may be a specific web-escape behavior. More studies employing high-speed photography would be helpful in understanding how different insects attempt to escape from webs and how different webs hold struggling insects. It is likely that escape behaviors that are effective for certain types of webs are not effective for others—e.g., attempting to drop straight down might extricate an insect from an inclined-plane orb web but not from the "knockdown" maze above a *Cyrtophora* orb web.

Behavior of the Spider Once Prey Has Entered the Web

The speed with which the spider arrives at the struggling prey is critical to capture success, particularly for quickly escaping prey. Casual observation suggests that there is considerable variation among araneids in the maximum speed spiders can achieve in their webs. For any particular species the speed of approach is mediated, in part, by the vibrations produced by the struggling insect. The relationship between approach behavior, struggling behavior, and capture success has been studied in only a few species (Barrows 1915; Suter 1978; Olive 1980; W. G. Eberhard, in prep.).

Probably the feature of araneid behavior that has the largest effect on the pattern of prey capture is the differential treatment of prey by the spider after it has arrived in the vicinity of the prey. Different araneids show different patterns of attack behavior. (For a review of the literature, see Robinson 1975.) Some araneids attack all entangled prey by biting. Other araneids attack some prey by biting and other prey by throwing silk. Wrap attacking makes it less likely that the spider will be injured by the insect's struggling movements or any chemical defenses the insect may possess (Robinson 1975). Araneids that wrap attack are much more successful at capturing aggressive prey such as stinging ants than those that only bite attack (Eisner and Dean 1976; Robinson 1980, in prep.). The wrap attacking is not usually employed against Lepidoptera; instead, the spider clasps and bites the prey. This behavior secures the prey quickly, and presumably increases capture success for Lepidoptera, which are held in webs for only a short time. The ability of araneids to distinguish Lepidoptera from other prey is tactilely mediated and has been shown to be innate (Robinson and Robinson 1976a). Olive (1980) argues that there is a tendency for araneids that prey principally on quickly escaping prey (e.g., *Araneus trifolium*) to have wider and longer chelicerae, which aid in the bite attack, and for araneids that prey principally on large, dangerous prey (e.g., *Argiope trifasciata*) to have longer legs IV, which aid in the wrap attack (see also Robinson 1980, in prep.).

Araneids do not attack all of the insects that become entangled in their webs. For example, spiders do not attack insects that are particularly large or aggressive. Some chemically defended prey may be cut out of the web immediately after the spider's first contact; others may be rejected after being wrapped (Robinson and Robinson 1973a; Eisner 1980). There is some evidence that araneids can learn to distinguish the vibrations produced by insects that they cannot attack successfully and refrain from approaching such insects—although it is not clear that an araneid's "memory" lasts long enough to be ecologically important (Bays 1962; Robinson 1980, in prep.).

Araneids can effectively handle only one prey at a time, and an insect that flies into a web where the spider is already engaged in attack behavior is more likely to escape (Robinson 1975). It is possible that an insect that flies in a swarm is less likely to fall prey to an araneid than an insect that flies alone.

A final consideration is how many of the captured prey are actually eaten by the spider and how much is taken by kleptoparasites. The kleptoparasitic spiders in the theridiid genus *Argyrodes* steal many of the wrapped prey that the host spider sets aside to eat later (Vollrath 1979a).

Minimal-Web Specialists

Relatively few araneids have been studied extensively, and the behavior of many is completely unknown. Many species are restricted to tropical areas not frequented by araneologists. Many—possibly the majority of the species in the tropics and subtropics—are strictly nocturnal, and cryptic during the day. Some species may be restricted to tree canopies (Levi 1973). We owe much of what we do know of araneid behavior to the observations of non-academic naturalists. Although many Asian genera have been observed by Japanese workers, language barriers have unfortunately kept Eastern and Western investigators unaware of one another's work.

A number of araneids have been discovered to be quite specialized in their prey, and it is these that I discuss here. For some spiders I discuss genera without specifying species; in such cases I assume that the predatory behavior of the species is representative of the genus, as has generally proved to be true for araneids. Where only one reference is cited for a particular genus or species, the cited reference is the source of all the information presented about that genus or species.

In selecting spiders for consideration in this chapter, I have used a traditional view of the Araneidae. Although recent work on the phylogeny of the family and its relatives will probably change the family placement of some genera, it is likely that future phylogenetic schemes will, like the more traditional schemes, support the view that all the spiders considered here are descended from orb-weaving ancestors (Coddington, this volume, in prep.). It is interesting to note that, despite many modifications of their predatory behavior, all of the specialized araneids engage in typical araneid postimmobilization prey wrapping.

Webs for Pedestrian Arthropods

The nocturnal spider *Wixia ectypa* (of the Eastern United States) has the anomalous behavior for an araneid of preying, not on flying insects, but on pedestrian arthropods (Stowe 1978). Its "asterisk" web (Fig. 5.1a) has no sticky elements, and serves only to signal the presence of an insect on one of the narrow branches between which the web is built. The insect is captured by what must be regarded as a modified wrap attack: the spider runs around the branch and the insect while trailing swathing silk. The insect is thus bound to the branch and the spider is able to administer a bite. It remains to be determined whether an insect must actually contact the end of a radius to alert the spider or whether the spider can detect the insect's vibrations through the branches to which radii are attached. If the spider does detect vibrations through branches, one wonders whether it can distinguish vibrations of potentially dangerous arthropods from those of prey. The spider seems to attach radii only to narrow branches (not to wide branches or leaves), and it would be interesting to know whether its attack behavior is effective only on branches of less than a certain size.

Hickman (1975), working in Tasmania, studied *Paraplectanoides crassipes*, another web-spinning araneid that catches pedestrian arthropods. This spider makes its web near the ground. It spins a small number of radii and a

Fig. 5.1. Webs of three nocturnal arboreal spiders (in different families) that prey on pedestrian arthropods. *a*: "asterisk" web of *Wixia ectypa* (Araneidae), a reduced orb web consisting only of radii and a hub. The web shown is approximately 10 cm across. The spider (not shown—usually at hub) attacks insects that come near the radius attachment points. *b*: web of a *Dinopis spinosa* (Dinopidae) female (body length 2 cm), a small, cribellate silk square held by the spider with legs I and II. The spider expands the web and throws it over insects that appear on the surface (out of the picture) in front of the spider. *c*: inverted Y web of a *Spintharus flavidus* (Theridiidae) female (body length 1 cm). The spider rests at the fork of the Y. It wrap-attacks insects that contact a line.

hub, and then surrounds these with an ovoid mesh sack of silk, into which leaves and detritus are incorporated (Fig. 5.2). The spider spins no spiral or viscid silk. There is a single, small opening to the exterior. The spider rests on the underside of the hub (Fig. 5.2), and attacks arthropods that enter the nest and run on the inside of the mesh sack. (Why they enter is unclear.) In the field older spiders catch mostly cockroaches; in captivity younger spiders seem to feed only on Collembola. Prey are bitten and then dealt with "in the usual araneid manner." These peculiar spiders are the longest-lived araneids studied—they take 5 years to mature in captivity.

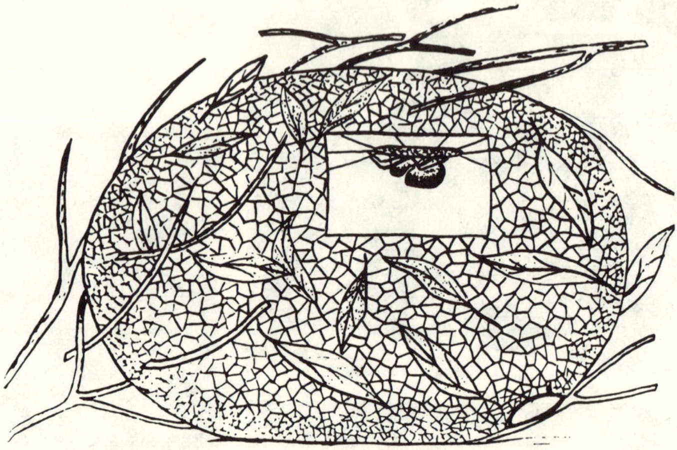

Fig. 5.2. Web of *Paraplectanoides crassipes*, consisting of a hub and radii surrounded by a mesh enclosure. Webs of adults measure up to 8 × 5 × 5 cm. Reprinted, by permission, from Hickman 1975.

Preying Without Webs

The genera *Neoarchemorus*, *Archemorus*, and *Arcys* (all from tropical Australia) are hunting spiders that have abandoned web spinning completely. According to Clyne (1979: 153–54), *Neoarchemorus* rest on a silk pad on the upper side of a leaf, the flattened posterior end of the abdomen against the leaf surface, the legs held close to the side of the cephalothorax, and the body axis perpendicular to the leaf surface. (The same posture is assumed after prey capture; see Fig. 5.3a.) The spiders, apparently using visual cues, leap at, and seize with legs I, flies and other insects that land on the leaf. In the unnamed species observed by Clyne, most of the diet seems to consist of the stratiomyid *Neoexaireta spiniger* (Fig. 5.3a), but other flies are caught as well. The behavior of *Archemorus* is apparently identical (Robinson 1980, 1982b; Mascord 1980; M. H. Robinson and B. Robinson, in prep.). Because the vision of araneids is thought to be quite poor, the use of vision by these spiders is surprising, but has been confirmed experimentally (M. H. Robinson, pers. comm.). After subduing the prey with venom, an *Archemorus* individual may suspend itself under the leaf and wrap the prey in typical araneid fashion before returning to the silk pad to feed. Spiders in the genus *Arcys* hold their spiny legs I and II out to the side in a characteristic fashion (Fig. 5.3b). The spiders presumably grab prey which comes within reach, and have been observed feeding both during the day, when they spend their time on leaves and are not associated with webs, and at night, when

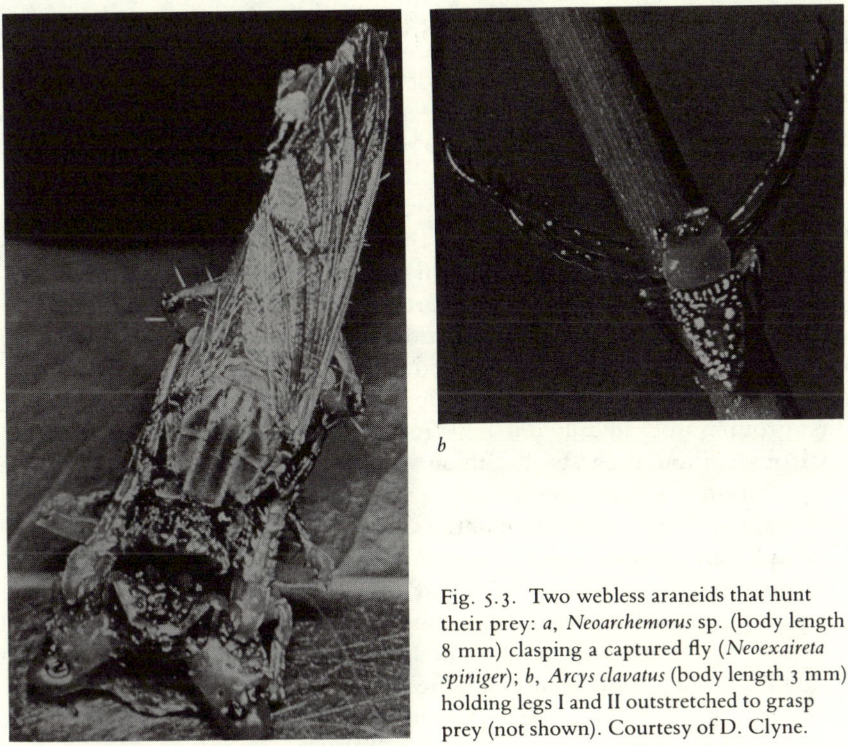

Fig. 5.3. Two webless araneids that hunt their prey: *a*, *Neoarchemorus* sp. (body length 8 mm) clasping a captured fly (*Neoexaireta spiniger*); *b*, *Arcys clavatus* (body length 3 mm) holding legs I and II outstretched to grasp prey (not shown). Courtesy of D. Clyne.

they are usually found hanging on "suspension webs"—webs consisting of one or two non-viscid lines which may or may not play a role in prey capture (Main 1982). All spiders included in this section wrap at least some of their prey. For spiders such as those in the genera *Archemorus* and *Arcys* this is the only readily recognizable araneid behavior in which the spiders engage.

Spiders in the genus *Chorizopes* (of tropical Asia) are specialized spider predators (Matsumoto, Shinkai, and Ono 1976; Eberhard, in press). They do not make webs of their own, but invade the webs of other araneids and attack the owners (which may be much larger spiders). Araneid individuals frequently attempt to usurp the webs of other araneid individuals, and Eberhard (in press) suggests that the web owners mistake attacking *Chorizopes* for small, would-be usurpers that they try to chase away. In the ensuing encounters, *Chorizopes* are able to administer bites to their victims; presumably the massive chelicerae of these spiders give them an advantage. After the venom has taken effect, the victims are wrapped and transported out of their webs.

Moth Specialists

Lepidoptera escape readily from most araneid webs. Robinson and Robinson (1970a), in a year's census of the prey of *Argiope argentata*, found that Lepidoptera constituted only 2.9 percent by number or 5.4 percent by weight of the total prey (N = 4,672), and estimated that 54 percent of the Lepidoptera that struck *A. argentata* webs escaped. In a study of the prey of six araneids, Nyffeler and Benz (1981) found that the highest monthly percentage of Lepidoptera captured by any of the species was 1.3 percent. Lepidoptera probably form a small proportion of the diet of many araneids. On the other hand, there are a few araneids for which moths are the only prey (see below). Many araneids occupy the continuum between these two extremes. I divide these into two groups. The behavior of the spiders in the first group is not radically different from that of conventional araneids. Casual observation suggests that these spiders capture more moths because they attack quickly and/or their orb webs hold moths longer. (The radii and viscid spiral form a dense mesh or the spiral is heavily coated with adhesive.) Examples are spiders in the American genus *Acacesia* (J. E. Carico, pers. comm.; pers. obs.) and in the Australasian genus *Poltys* (Mascord 1970). These spiders are nocturnal, spin webs with a dense mesh (Fig. 5.4a), and capture large numbers of moths. The spiders in the second group (discussed below) spin webs with features unique to the genera involved, and these features are apparently responsible for the large number of moths the spiders capture. Moths form a high percentage of the prey of these spiders, but in fact many of the spiders in the first group probably capture comparable numbers of moths.

Ladder Webs

Spiders in two distantly related araneid genera make extremely long, vertical "ladder" webs. An as yet unidentified New Guinea ladder-web spider spins an extended orb web up to 1 m long (Fig. 5.5a), with the hub near the top (Robinson and Robinson 1982). (H. W. Levi (pers. comm.) places this spider in a genus close to *Meta* and *Tylorida*. In previous reports, e.g., Robinson and Lubin 1979a, it has been inadvertently misidentified as a *Tylorida* species.) Spiders in the American genus *Scoloderus* also spin an extended orb web up to 1 m long (Fig. 5.5b), but with the hub near the bottom (Eberhard 1972a; Stowe 1978). Although many araneids occasionally make long, narrow webs, these are the only spiders I am aware of that make such webs consistently (the average ratio of length to width is 7 for *Scoloderus cordatus*, N = 166) and make them consistently in spaces that permit wider webs. Moreover, I know of few other spiders, other than the arboricolous species mentioned above, that consistently make strictly vertical webs. (No deviation from verticality is perceptible to the eye in *S. cordatus*

a

b

Fig. 5.4. Webs of two araneids that are comparatively successful at catching moths. *a*: orb web of *Acacesia hamata*. The viscid spiral is closely spaced and holds moths somewhat longer than the more widely spaced spiral in other araneids' orb webs—long enough for this quick spider to arrive at the prey. The web shown is 20 cm in diameter. *b*: web of *Cyrtarachne inaequalis*. The viscid "spanning" threads are coated with a unique glue capable of adhering to moths. Spanning threads are often separated by as much as 10 cm. Courtesy of E. Shinkai.

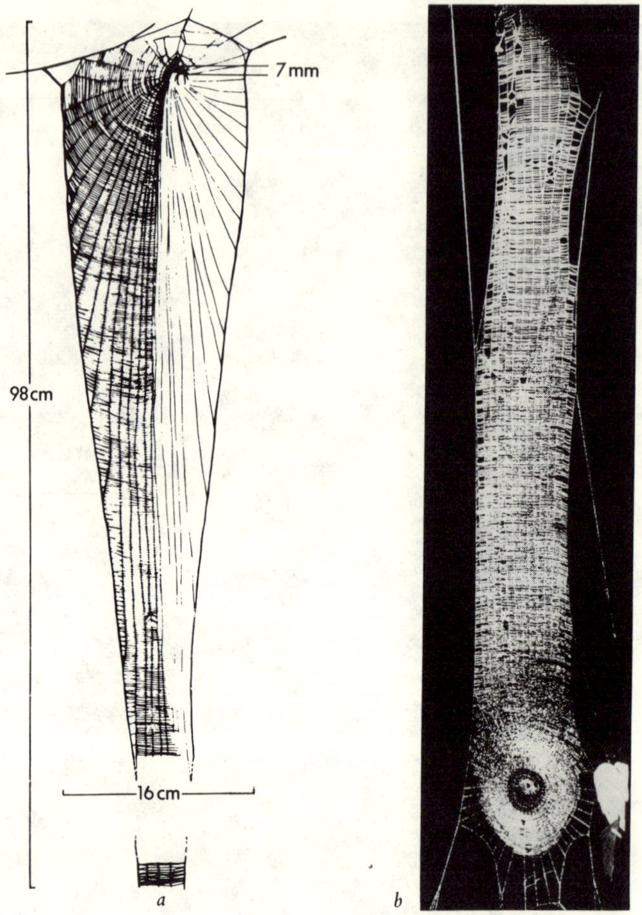

Fig. 5.5. Long, vertical "ladder" webs (up to 1 m) made by two araneids in distantly related genera. *a*: ladder web of a New Guinea spider (genus near *Meta*). Reprinted, by permission, from Robinson and Robinson 1972. *b*: "inverted" ladder web of *Scoloderus cordatus* (spider not shown). The web shown is 70 cm long.

webs when compared to a plumb line.) The incline of most orb webs, even if it is slight, seems to facilitate the spider's movement around the web by allowing the spider to hang below, and therefore out of the plane of, the web. *Scoloderus cordatus* moves in its web with difficulty. The spider is able to avoid contact with the closely spaced viscid spiral lines only by distorting the web. As the spider moves, this distortion causes the spiral lines that the spider passes over to adhere to one another. (In doing so they presumably become less effective for prey capture.) This difficulty of movement in the web is unusual but not unique to these spiders (W. G. Eberhard, pers. comm.).

Ladder-web spiders catch a large number of moths. For example, in field studies of the prey of *S. cordatus*, moths constituted 68 percent of the prey at one field site (166 spider nights, 988 web hours, 212 prey items; Stowe 1978) and 40 percent of the prey at another site (78 prey items; C. S. Hieber, in prep.). (At both sites prey were recorded by checking webs and spiders hourly through the night.) Field observation indicates that the unique features of ladder webs—their length and verticality—are responsible for the spiders' success in capturing moths (Eberhard 1972a; Stowe 1978; Robinson 1982b). When a moth flies into a web, the fine mesh prevents it from using its wings effectively. It cannot flutter free of the web, and as its scales cover the viscid spiral in the area where it entered the web, it slides downwards into the viscid spiral below the point of contact. The moth continues sliding down the web, leaving a vertical trail of scales. The speed of the moth's progress decreases, and if the moth's initial point of contact was a sufficient distance from the bottom of the web, the moth comes to rest apparently having lost so many scales that it is held in place by the viscid threads. Unlike some of the spiders discussed below, *S. cordatus* catches moths of both sexes and of many different species. Moths less than 6 mm in body length predominate (C. S. Hieber, in prep.); perhaps larger moths can flutter free of the web.

It would be valuable to measure the percentage of scales lost and the adherence of the moth cuticle at different points in the capture process. Experiments should be conducted in which moth and nonmoth prey are introduced into ladder webs at different heights, and into some webs that have been inclined and some that have not. I expect that such experiments will show that capture success for moths, but not for other prey, increases with increasing web length and verticality, which would indicate that these two features do in fact represent adaptations for increased capture of moths specifically.

Spanning-Thread Webs

Spiders in the Old World tropics tribe Cyrtarachneae (Simon 1892) spin large, slightly inclined to horizontal webs with small numbers of radii and viscid elements (Figs. 5.4b, 5.6). The descriptions of these spiders presented here are based on observations of four genera: *Cyrtarachne* (Kayashima 1952; Okuma 1957; Ono and Yaginuma 1963; Arita 1963; Suginaga 1963), *Poecilopachys* (Clyne 1973; Robinson 1980, 1982b), *Pasilobus* (Robinson and Robinson 1975), and *Paraplectana* (Chigira 1978); where I do not refer to particular genera, the information given applies to all four genera. The unique web-construction behavior of these araneids is described in Appendix C. The viscid elements, or "spanning" threads (Clyne 1973), in these webs are quite different from the sticky spiral segments in the webs of conventional araneids. Also, the viscid coating of these threads seems to differ from that em-

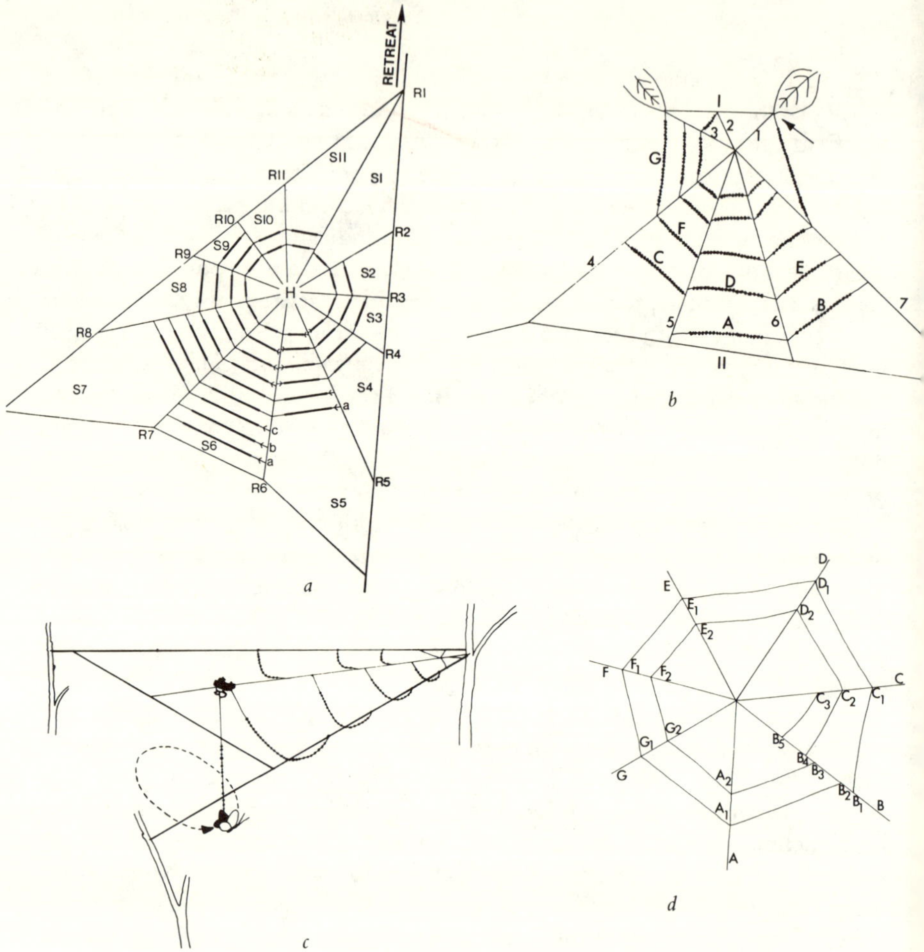

Fig. 5.6. Structure, method of construction, and/or function of webs of four genera in the Cyrtarachneae. *a*: web of *Poecilopachys australasia*. Viscid lines were spun from the outside in, usually starting on *R*6 but occasionally on *R*5. Each was spun in the direction indicated by the arrows in at most a half circle, i.e., never extending around *R*1. As indicated by the arrows, the innermost viscid lines were spun in alternate directions. Webs are approximately 80 cm in diameter. Reprinted, by permission, from Clyne 1973. *b*: web of *Paraplectana tsushimensis*. The lettered spanning threads were spun in the order *A* to *G*. Webs are approximately 70 cm in diameter. After Chigira 1978. *c*: two-sectored web of *Pasilobus* sp. (oblique view). The web is approximately 50 cm in length. A moth has contacted a spanning thread, which has broken at the low-shear joint. The spider has traveled down the central line and is pulling up the moth. Reprinted, by permission, from Robinson and Robinson 1975. *d*: center of the web of *Cyrtarachne bufo* (frame not shown). Viscid lines form full circles. The spanning threads were spun as follows: *B*1–*C*1, *C*1–*D*1, around to *A*1–*B*2; then *B*3–*A*2, *A*2–*G*2, around to *C*2–*B*4; then *B*5–*C*3. Webs are approximately 1 m in diameter. After Suginaga 1963.

ployed by other araneids: it is distributed in unusually large, visible droplets and it sticks effectively to moths. A moth contacting a single line is invariably held, and held indefinitely. The spanning threads are also unique in that one end—the "low-shear joint" (Robinson and Robinson 1975)—breaks when prey contacts the thread (*Cyrtarachne*: Arita 1963; Suginaga 1963; *Poecilopachys*: Robinson 1980; *Pasilobus*: Robinson and Robinson 1975). An insect that contacts a spanning thread goes into tethered flight until it is hauled up by the spider (Fig. 5.6c). Each spanning thread is good for capturing only one prey item.

The spiders feed largely, but not exclusively, on moths. For example, Robinson and Robinson (1975) found that moths formed 92 percent of the 25 prey items captured by *Pasilobus* sp. at one field site. Experimentation would probably confirm that the difference between the capture efficiency of these webs and that of conventional araneid webs is greatest for lepidopteran prey, and that this difference is primarily due to the special adhesive. It seems likely that, as Robinson (1980) suggests, the low-shear joint reduces the momentum of the prey and increases the likelihood of capture, but whether this is especially important for capturing moths is unclear.

It is interesting to compare the webs made by these spiders with those made by ladder-web spiders. In ladder webs, where individual viscid elements are ineffective in capturing the principal prey (moths), the number of viscid elements and their proximity to one another is near the upper limit for araneids (W. G. Eberhard, pers. comm.). In Cyrtarachneae webs, where one viscid element is sufficient to capture a moth, the number of viscid elements is at the lower limit, and the viscid elements are much more widely spaced than in other araneid webs. Eberhard (this volume) predicts that in an "optimal" orb web viscid spirals are spaced only as far apart as the flight diameter of the largest prey. Spanning threads are spaced much farther apart than this; the wide spacing presumably reflects the fact that each spanning thread is good for capturing only one prey item, and that an insect in tethered flight might contact and damage other spanning threads besides the one it originally encountered if the spanning threads were more closely spaced.

Moth-Attracting Araneids

Spiders in a number of araneid genera capture only one type of prey: male moths. Several lines of evidence indicate that these spiders are able to produce compounds that have, for the males of the prey species, the same effect as the sex-attractant odors (sex pheromones) produced by the females of those species. The male moths that are lured to the spiders are captured in a manner that depends on the genus to which the spider belongs. I shall discuss these spiders in three groups: Celaenieae, Mastophorinae, and *Kaira*. It is reasonably certain that each group is monophyletic, but the relationships between the groups are unclear, and it is possible that the ability to mimic moth sex pheromones evolved more than once (see Appendix D).

These spiders are quite inconspicuous even at night—the only time they are active. Although they are present in many biomes in tropical and temperate areas all over the world, they seem to be patchy in their distribution and are among the most rarely encountered spiders. The most common species are represented in major collections by a few score of specimens, most species by only one or two.

As is the case for many araneids, the field conditions that elicit normal hunting behavior in these spiders seem to be impossible to reproduce in the laboratory. A major barrier to observing the hunting behavior of these spiders in the field is the fact that the sex-attractant-mediated approach behavior of many of the moth prey species is disturbed by visible (even red) light. (This is not uncommon in moths: e.g., Conner et al. 1980.)

Eberhard (1977a, 1980a) has worked with *Mastophora* extensively in Colombia. With the aid of night-vision equipment I have been studying *Mastophora* and *Kaira* in Florida and California. For most of the other genera, our knowledge is limited to scattered observations of single individuals.

The Celaenieae (Simon 1892) consist of two genera: *Celaenia*, a genus of Australia and New Zealand, and *Taczanowskia*, a genus of South America. The behavior of the studied species is similar (*Celaenia*: McKeown 1952; *Taczanowskia*: Eberhard 1981b). The spiders spin no web whatsoever and hang upside down from a line, or the underside of a leaf, with legs I and II spread (Fig. 5.7a). Moths are captured when they fly into the spiders' outstretched legs. After immobilizing and wrapping a moth, a spider may set the prey aside and resume hunting. The spider later eats the prey that were set aside, one by one.

The Mastophorinae (Gertsch 1955) consist of six genera: the New World *Mastophora* and *Agatostichus* and the Old World *Dicrostichus*, *Ordgarius*, *Cladomelea*, and *Acantharachne*. In all four genera in which hunting behavior has been observed, it is similar to that described first by Hutchinson (1903) and most recently by Eberhard (1980a) for *Mastophora*. The spiders spin a simple horizontal line, and then the "bolas": a line that is attached to the horizontal line and has a free end bearing one to several balls of viscid glue. The bolas line is grasped towards the middle by one of the legs (Fig. 5.8). The bolas is replaced periodically, presumably because it dries and loses its adhesiveness (Eberhard 1980a).

Interestingly, the genera differ in the manner in which they handle the bolas. In *Mastophora* and *Agatostichus*, the bolas is held with one leg I and is flicked at approaching moths. In *Dicrostichus* (Longman 1922; McKeown 1952) and *Ordgarius* (Mascord 1980; Shinkai 1981; Robinson 1982b) it is held

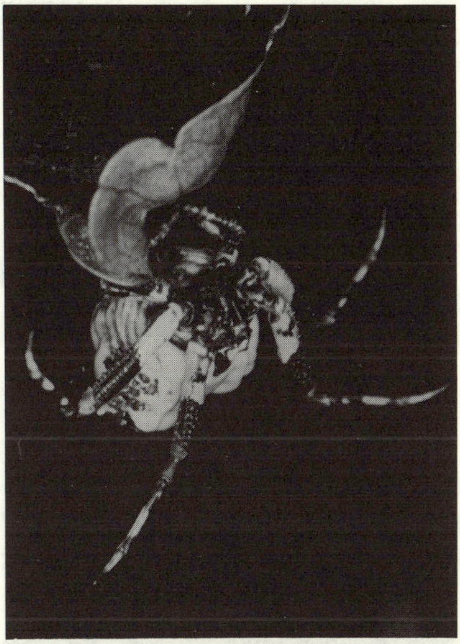

a

Fig. 5.7. Two araneids that capture moths in their outstretched legs: *a*, adult female *Celaenia kinbergi* (body length 1.7 cm) in its hunting position; *b*, adult female *Kaira alba* (body length 1 cm) hunting while hanging upside down from its "zigzag" web.

b

with one leg II and whirled as the moth approaches. In *Cladomelea* it is held with one leg III and whirled continuously (Akerman 1923). In *Mastophora*, spiders may occasionally spin a series of one to nine bolases on a single line and adopt the hunting posture, without holding a bolas, at one end of the line (*Mastophora* sp.: J. A. Beatty, pers. comm.; *Mastophora bisaccata*, *Mastophora cornigera*: pers. obs.; *Mastophora dizzydeani*: Eberhard 1980a).

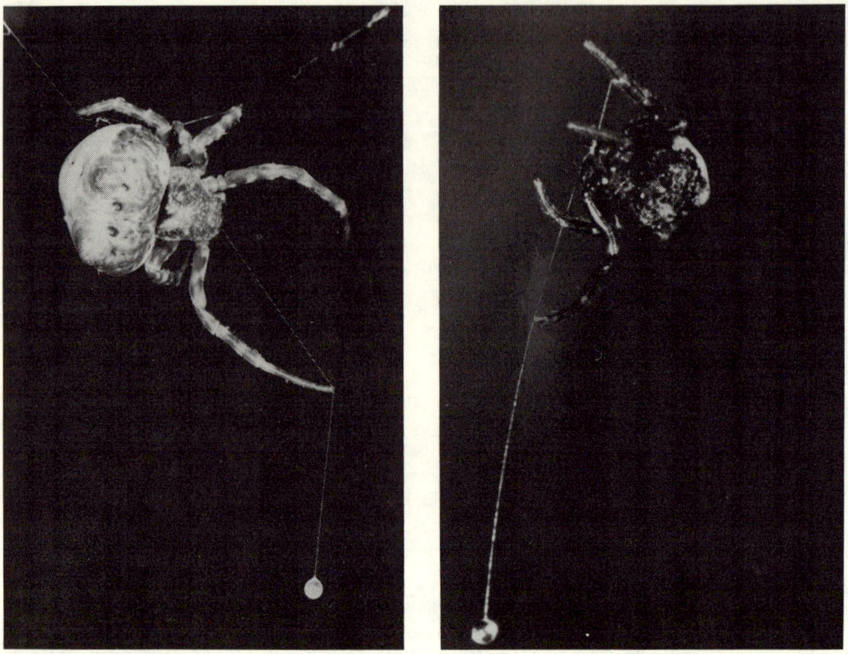

a b

Fig. 5.8. Hunting position of *Mastophora*: *a*, adult female *Mastophora cornigera* (body length 1.6 cm); *b*, newly emerged female (body length 0.2 cm) of the same species (photograph taken in captivity). Spiders grasp the bolas line with one leg I and flick the bolas at approaching moths.

The bolas thread possesses the same moth-holding properties as the Cyrtarachneae spanning threads. If a moth contacts the adhesive, it is invariably held. The moth goes into tethered flight and is hauled up by the spider in the same way that moths that have contacted spanning threads are hauled up by spiders in the Cyrtarachneae. (Reasons for believing that the Mastophorinae and Cyrtarachneae are closely related are given in Appendix D.) Like the spiders in the Celaenieae, bolas spiders often hang up wrapped prey and resume hunting (*Dicrostichus*: Longman 1922; *Mastophora*: Eberhard 1980a). I have seen *Mastophora bisaccata* catch nine moths in one bout.

The hunting behavior of the spiders of the New World genus *Kaira* has not been reported previously. Before hunting, and roughly every 20 minutes during hunting, the spiders spin a small web; the spiders hang upside down from this web with their legs outstretched (Fig. 5.7b; for details see Appendix C). The web plays no obvious role in prey capture; an attracted moth flies into the basket formed by the spider's outstretched legs, whereupon the spider drops on a short single line while clasping and biting the moth. After the moth stops struggling, it is wrapped in araneid fashion. Usually, the wrapped moth is placed on the trapeze line between the spider's legs IV, and the hunting posture is resumed. Unlike the spiders discussed previously, *Kaira* wraps together all the moths that it catches in one bout and later feeds on a single prey package. I have seen *Kaira* catch as many as eight moths and wrap them together.

The three groups of moth-attracting spiders are certainly among the most specialized spiders known. Whereas most araneids (and almost all other spiders) will accept a wide taxonomic range of prey, spiders in these three groups, in tests in captivity and in the field, appear to accept only moths and certain Nematocera (*Celaenia*: Hickman 1971; *Mastophora*: pers. obs.). The spiders in these three groups, and a few social species, are the only spiders I know of that will not feed on other spiders. (This statement is based on extensive experience with captive individuals of all three groups.) Spiders in these three groups feed exclusively on moths in nature. (Spiderlings may be an exception; see Appendix B.) Whereas almost any moth that is offered to the spiders is accepted, their natural prey consists entirely of male moths of a few species (Table 5.1).

The evidence for sex-pheromone mimicry by these spiders is of three kinds.

1. Evidence that male moths are attracted to the spiders. Whereas the 165 prey of the *Mastophora dizzydeani* individuals observed by Eberhard (1977a) were all male moths of two species, the 145 insects caught in artificial webs in the same area included only one individual of the moth prey species. In one study, I allowed *Kaira alba* to hunt inside large tubes. Fans pushed air past the hunting spiders, and out into the field. Moths flying into a tube were caught in a chamber downwind of the spider. In eight nights, tubes with spiders caught 45 male moths whereas tubes without spiders caught no moths at all. On many occasions I have found several moths at a time hovering downwind of one of these spiders (I have seen as many as four)—they have always turned out to be males of the same species.

2. Evidence that the moths are attracted olfactorily. Moths always approach the spiders from downwind, alternately flying upwind, and zigzagging back and forth perpendicular to the direction of the wind, as they approach; the amplitude of the zigzagging decreases as they draw closer to the spiders. (*Mastophora*: Eberhard 1977a; pers. obs.; *K. alba*: pers. obs.; the other authors listed in Table 5.1 and Hutchinson (1903), without describing the moths' flight in detail, indicated that they thought the moths were attracted to the spiders). This behavior is indistinguishable from typical anemotactic flight to an odor source.

Moths will fly to spiders around visual barriers. Eberhard (1977a) placed pieces of cardboard downwind from individuals of *M. dizzydeani*, and found that moths flew around the downwind edges to the spiders. On seven occa-

TABLE 5.1

Prey of Moth-Attracting Spiders

Spiders[a]			Season	Nights	Prey[b]	
Species/locality[c]	No.	Stage[d]			Species (in order of abundance)[e]	No.
C. kinbergi						
1. ? Australia	1	1	Nov.–Feb.	90	A *Heliothis zea* Boddie	34[f]
					B *Agrotis infusa*	
					C *Peripyra sanguinipuncta*	
2. Glen Innes, Aus.	1	1	Feb.	—	C *Peripyra sanguinipuncta*	—[f]
3. Sydney, Aus.	—	—	—	—	D *Pseudaletia unipuncta* Haworth	—[f]
4. ? Australia	300	0	—	—	E male psychodid flies	—[f]
Taczanowskia sp.						
5. Meta, Colombia	1	1	July	2	F unidentified Pyralidae	—[f]
D. magnificus						
6. Brisbane, Aus.	1	1	Apr.	—	G *Remigia frugalis* Fabr.[g]	—
M. dizzydeani						
7. Cali, Lago, Calima, Colombia	8	$\frac{2}{3}$–1	Oct.–Mar.	312	H *Spodoptera frugiperda* Abbott and Smith	309
					I *Leucania* sp.	18
					J unidentified Olethreutidae	18
					K unidentified Noctuidae	2
	—	0	Oct.	—	L unidentified nematocerous flies	2
M. phrynosoma						
8. Alachua Co., Fla.	1	1	July–Aug.	4	M *Eulithis diversilineata* Hubner	5
M. cornigera						
9. San Diego Co., Calif.	7	$\frac{1}{2}$–1	Jan.–Apr.[h]	42	N *Peridroma saucia* Hubner	23
					O Noctuidae (Agrotinae) sp. A	8
					D *Pseudaletia unipuncta* Haworth	3

M. bisaccata							
10. Alachua Co., Fla.	42	$\frac{1}{3}$–$\frac{2}{3}$	June–July[i]	120	P	*Acrolophus plumifrontinellus* Clemens	104
					Q	*Metalectra quadrisignata* Walker	1
	5	$\frac{1}{3}$–$\frac{2}{3}$	Aug.[h]	10	R	*Acrolophus piger* Dyar	7
					S	*Scolecocampa liburna* Geyer	7
					T	*Mocis texana* Morrison	2
K. alba							
11. Alachua Co., Fla.	3	$\frac{1}{2}$–1	June–July	10	Q	*Metalectra quadrisignata* Walker	8
					U	*Acrolophus* sp. A	2
					V	unidentified moth sp. B	2
	22	$\frac{1}{2}$–1	Aug.–Sept.[i]	160	R	*Acrolophus piger* Dyar	79
					W	unidentified moth sp. C	29
					X	*Acrolophus* sp. C	7
					Y	unidentified moth sp. D	3
					Z	*Acrolophus* sp. B	1
					a	unidentified moth sp. A	1

SOURCES: (1) Rayment 1942; (2) Lowry in McKeown 1952; (3) LeSouef in McKeown 1952; (4) Lauder in Eberhard 1981a; (5) Eberhard 1981b; (6) Longman 1922; (7) Eberhard 1980b; (8–11) Stowe, unpub.

[a] All the spiders were females. (The habits of the much smaller males are unknown.)

[b] All moth prey were males (but see note *f*). Since the prey of *Kaira alba* are wrapped together in prey packages and the moths are small, some of the genitalia in the packages may have been missed.

[c] For full names see Table 5.2.

[d] The numbers given are the approximate fraction of maturity, 0 representing spiderlings and 1 adults.

[e] In Eberhard's and my work moths were identified from rejectamenta caught in traps below the spiders and almost certainly represent the sole prey of the spiders.

[f] The author did not report the sex of the prey.

[g] In Longman 1922 the genus name is spelled *Remigra*, which is not a valid genus name. The moth is almost certainly *Remigia frugalis* (Guenée 1852).

[h] Spiders were studied for two field seasons.

[i] Spiders were studied for three field seasons.

sions I observed moths weaving their way through leaves to reach spiders (*M. bisaccata* and *K. alba*) that were hunting in the middle of vegetation and were hidden from the view of the moths.

3. Evidence that the spiders attract moths in a manner identical to a source of sex pheromone. All the known prey of these spiders are male moths (Table 5.1), and these moths represent a small proportion of available moth species. Eberhard (1977a), for example, found 71 species of moths in the same area where *M. dizzydeani* was capturing only two. A number of prey species are not capable of feeding as adults (e.g., moths in the genus *Acrolophus*: D. R. Davis, pers. comm.), and an attractive odor other than a sex pheromone is hard to imagine. Most moths are sexually active for a period of only a few hours, at a time of night that is species-specific (Roelofs and Carde 1974; Shorey 1974). The *Mastophora* and *Kaira* species I have studied catch their prey at characteristic times of the night—which presumably correspond to the periods of sexual activity of the moths. It is tempting to think that the tendency of these spiders to resume hunting after prey capture, rather than immediately eating prey and hunting later, is an adaptation selected for by the fact that prey are available only during relatively narrow time slots.

In my ongoing research with these spiders (too preliminary to report here), I am trying to determine the identity of the compounds they use and to answer related questions. For example, what is the glandular origin of the compounds? (Current evidence indicates that in the case of *Mastophora* the attractants emanate from the spiders themselves rather than from their "webs": Eberhard 1977a; M. K. Stowe, pers. obs.) When are the compounds released during hunting behavior? Do the spiders learn to concentrate their hunting behavior during the particular time of night at which their prey are sexually most active? Are bolas spiders actually releasing compounds and/or catching moths when they are in a hunting posture without a bolas (Eberhard 1977a)? And, finally, what is the succession of moth prey species over the course of the spider's life?

Conclusion

The specialized habits described here almost certainly represent a small proportion of the "alternative life-styles" in the Araneidae. If, as seems to be the case, bizarre morphology in araneids is associated with unusual behavior, then a quick look through Simon's *Histoire naturelle des araignées* (1892) will suffice to convince the reader that there are many exciting discoveries in nonstandard araneid behavior to look forward to.

Appendix A: Parallels in Prey Specialization in Other Families

At a number of sites in Florida where I have worked, the araneid *Wixia ectypa* is one species in a guild of three nocturnal, web-spinning spiders that feed on arboreal pedestrian arthropods. Another species is the dinopid *Dinopis spinosa*, which spins what is probably a reduced cribellate orb web (Fig. 5.1b; see Coddington, this volume) and catches pedestrian arthropods on branches, tree trunks, and leaves (Robinson and Robinson 1971; pers. obs.; see also Austin and Blest 1979). The third species is the theridiid *Spintharus flavidus*, which spins a simple Y-shaped web (Fig. 5.1c) on narrow branches and throws swathing silk with legs IV on arthropods that contact the sticky ends of the forks of the Y. (Its diurnal resting "web" has been mistaken for a capture web: Comstock 1948.) Similar behavior has been described for the theridiid genera *Chrosiothes* (J. Coddington, pers. comm.) and *Episinus* (Bristowe 1958; Clyne 1969). Since most theridiids spin extensive webs, it seems likely that the *Spintharus* web also represents a reduction of that spun by the spider's ancestors. How do such reduced webs catch a sufficient number of insects? Presumably, the branches to which the webs are attached provide a relatively small set of paths for pedestrian-arthropod movement and on any one branch that movement is restricted to essentially one dimension, which is much easier to sample than three dimensions (the task of conventional webs). A theridiosomatid that has abandoned the orb-spinning, airborne-prey-capturing life-style of its ancestors is *Wendilgarda* (see Fig. 12.10). This remarkable spider spins a web above running water in streams, with sticky elements that are actually attached to the water surface, and catches floating insects (Coddington and Valerio 1980).

It is likely that many groups of spiders (representing large and small taxonomic units) that hunt without webs are descended from web-spinning ancestors (e.g., Oxyopidae: Rovner 1980; see also Jackson, this volume). The existence of araneids such as *Archemorus* and *Arcys* that hunt without webs is therefore not surprising. Two other examples of isolated genera of hunting spiders in families consisting mostly of web-spinning spiders are the tetragnathid *Pachygnatha* (the spiders hunt as adults—preadults spin orb webs: Martin 1978) and the theridiid *Euryopis* (the spiders prey exclusively on ants: Carico 1978).

Spiders are preadapted to prey on other spiders. Groups of spiders that specialize in preying on other spiders have evolved at least six times. Like the araneid *Chorizopes*, spiders in the family Mimetidae (Bristowe 1958; Czajka 1963) and the theridiid *Theridion sterninotatum* (Matsumoto, Shinkai, and Ono 1976) invade other spiders' webs and attack the owners. Spiders in the family Archaeidae are reported to "spear" other spiders with their elongate chelicerae (Legendre 1961). Spiders in the remarkable salticid genus *Portia* attack web makers in their webs and stalk and leap on members of their own family (Jackson, this volume; Jackson and Blest 1982a). Members of the theridiid genus *Argyrodes* are kleptoparasites and/or spider predators. In some species individuals are purely predatory and attack spiders that crawl onto their single-line "web" (Hashimoto 1963; Clyne 1979; Eberhard 1979a).

I know of no moth specialists outside of the Araneidae. Eberhard (1981c) and Horton (1979) report tentative indications of attraction of certain other insects to the webs of some spiders.

Appendix B: Behavior of Younger Individuals of Specialized Araneids

In a number of araneids, the web form of younger spiders differs from that of the adults. In many cases the younger spider's web is closer in form to the simplest orb webs than that of the adult (e.g., *Zygiella*). In others the reverse is true (e.g., certain *Leucauge* species: pers. obs.). It might be expected that some of the younger individuals of specialized araneids show different, perhaps less specialized behavior than the adults. Although much less is known about the behavior of the younger individuals of these spiders (young spiders are hard to identify), this does not appear to be the case. I have seen *Wixia ectypa* individuals one-eighth the body length of adults in typical asterisk webs, but these were probably not small enough to be spiders in their first free-living instar. Craig Hieber and I have both seen young *Scoloderus cordatus* individuals in ladder webs—the spiders were the same size as spiderlings that emerged from egg cases laid by females in captivity. On the other hand, Eberhard (1972a) found spiders in conventional orb webs that he believed were *Scoloderus tuberculifer* spiderlings.

All the moth-attracting spiders appear to hunt in an identical manner throughout their lives. Newly emerged *Mastophora cornigera* spiderlings (personal observation of many spiderlings in captivity) make miniature bolases (Fig. 5.8b), although Eberhard (1980a) suggests that *Mastophora* spiderlings may hunt without a bolas. *Celaenia kinbergi* spiderlings (personal observation while raising spiderlings) and *Kaira alba* spiderlings (personal observation of two spiderlings in the field) assume the characteristic hunting posture—they hang from a horizontal line. (I am uncertain whether this line includes collapsed zigzags—see below.) The natural prey of the spiderlings is unknown except for two reports of *C. kinbergi* and *Mastophora dizzydeani* spiderlings feeding on Nematocera (Table 5.1). Preliminary evidence indicates that as the spiders grow they catch different moths, but that which moths are caught is not strictly determined by instar: individuals have been found that catch the same moths before and after an instar, and that switch moths in the middle of an instar (*M. dizzydeani*: Eberhard 1980a; *Mastophora* sp.: W. G. Eberhard, pers. comm.; *Mastophora bisaccata*, *K. alba*: pers. obs.).

Appendix C: Web-Spinning Behavior

I present here a detailed description of the web-spinning behavior of the genera in the Cyrtarachneae, and of *Kaira*. The description of the web-spinning behavior of the Cyrtarachneae is the first synthesis of all the available information. The web-spinning behavior of *Kaira* has not previously been described. In this section the applicable references for statements about a genus where no reference is given are all the references given for that genus at the start of the discussion of Cyrtarachneae above.

The accounts of web spinning by the members of the Cyrtarachneae are somewhat fragmentary; the spiders are rarely encountered, they may spin as late at night as 2 hours after darkness (*Cyrtarachne*: Okuma 1957; *Poecilopachys*; *Paraplectana*), and they spin only under favorable conditions, which seem to include high humidity (*Poecilopachys*, *Pasilobus*). The web-spinning behaviors of members of the different genera exhibit a number of shared features. The hub is always located away from the center of the web, closer to the diurnal retreat. The web is weakly bilaterally symmetrical around the axis formed by the retreat and the hub. Unlike most other araneids (Eberhard 1982), these spiders do not make a temporary spiral. In laying spanning thread, the spider makes the initial attachment on one radius, and travels along that radius to the hub and then out along an adjacent radius to make the second attachment point. As the spider travels between the two attachment points, legs IV alternately comb viscid silk from the spinnerets. The second attachment is always a low-shear joint. Spanning threads are spun from the outside in, in a series of "viscid lines" (Clyne 1973): runs of spanning threads made across contiguous sectors. The viscid lines are spun in a sequence that is nonspiral and distinctly different from that followed by other araneids (see below).

The webs of the genera differ in their geometry. In *Poecilopachys australasia* (Clyne 1973; 35 webs observed) webs are built within a triangular frame (Fig. 5.6a). Two lines run from the apex, near the retreat, to distant points of attachment, and a third stretches between these lines. There are 7 to 12 radii and sectors may have 0 to 13 spanning threads. In *Cyrtarachne*, the web may be in the form of a triangle or a higher-order polygon (Arita 1963; Eberhard

1980a), and may include radii with Y attachments to the frame (Arita 1963; Suginaga 1963). In *Paraplectana tsushimensis* (Chigira 1978; eight webs observed) there are four to seven radii, which may be spun between two frame lines; i.e., the frame may be incomplete. See Fig. 5.6b. To spin the web shown in this figure (web construction by *P. tsushimensis* was observed only once), the spider wafted silk both from the retreat, to establish the line that became radii 1 and 4, and from the point that became the hub, once to start radius 3 and again to start radius 7. In most araneids, a radius is spun with each frame thread (Coddington, this volume), but here the two frame threads I and II were made, and then the remaining radii. In *Pasilobus* the frame is the same as that described for *Poecilopachys*, but the hub appears to be extremely eccentric and all radii other than the radius that runs down the center of the web are rudimentary (Fig. 5.6c). The web essentially consists of the two sectors on either side of this radius; only these sectors have spanning threads.

In laying spanning threads, all of the spiders follow a similar scheme, although there is some discrepancy among the descriptions of this behavior for *Cyrtarachne* species. In *Cyrtarachne nagasakiensis* (Arita 1963) and *Paraplectana tsushimensis*, viscid lines are always begun on the radius opposite the radius closest to the retreat. Viscid lines span a variable number of sectors, but never extend around past the retreat radius; they form, at most, half circles. The spider alternates between viscid lines made clockwise and those made counterclockwise. The behavior of *Pasilobus* is the exact equivalent for a two-sector web: spanning threads are started on, and made on alternate sides of, the triangle-bisecting "radius." The behavior of *Poecilopachys* is essentially equivalent, except that the starting radius is occasionally one of the radii adjacent to the radius opposite the retreat radius. Suginaga (1963) describes similar behavior for *Cyrtarachne bufo*; the major difference is that viscid lines can form full circles instead of half circles (Fig. 5.6d). Like the other spiders, *C. bufo* spins viscid lines in alternate directions.

As the first step in constructing its web (Fig. 5.7b) *Kaira alba* spins a simple, slack, horizontal "trapeze" line. This line is not viscid. The spider then spins a zigzag series of lines in the V of the trapeze line, laying down the elements of the zigzag sequentially from the uppermost segment to the lowermost segment. The second attachment point of one zig is the first attachment point of the next zag. The spider spins the uppermost segments by walking back and forth between attachment points on the trapeze line. Towards the bottom of the web, as the distance between attachment points decreases, the spider merely moves the spinnerets between attachment points. The silk is paid out with one leg IV. Just before an attachment, the spider faces down and with the other leg IV taps the trapeze line near the eventual attachment point. The spider then grasps the line near the attachment point with the leg IV that did the tapping, and makes the attachment. The manner of their construction strongly suggests that the zigzags are homologous with the viscid spiral of typical orb weavers. However, the zigzags are minimally viscid. The zigzag segments often adhere to one another, and a freshly spun web may appear to consist of the trapeze line alone, all the zigzags having coalesced with one another and the trapeze line. The function of this web—which is rudimentary, variable in form, and plays no part in prey capture—eludes me.

Appendix D: Preliminary Ideas Concerning Taxonomic Relationships

Eberhard (1980a) and Robinson (1982b) in effect argue that the Cyrtarachneae, Mastophorinae, and Celaenieae form a monophyletic group (hereafter referred to as the larger group), and that behavioral data indicate that one scenario for the evolution of the different members of the group is particularly likely. I shall start by presenting (as fact) this scenario and its implied scheme of relationships. Brackets enclose my own embellishments.

The ancestral spider of the larger group spun a web similar to that of present-day *Cyrtarachne* and *Poecilopachys*. This spider gave rise to groups of spiders that were the ancestors of present-day *Cyrtarachne* and *Poecilopachys* [and *Paraplectana*] and to a group of spiders that spun a web similar to that of present-day *Pasilobus*. It was argued in Appendix C that this web is essentially a *Cyrtarachne* web in which all but two sectors are reduced. This group of spiders gave rise to present-day *Pasilobus* and to a group of moth-attracting spiders. The costs of web building (e.g., time and energy [and the water necessary for the thick viscid coating of the spanning threads]) favored web reduction in the group of moth-attracting spiders, and a group of spiders evolved that spun a string of bolases, as present-day *Mastophora* sometimes do. [Two simple behavioral changes, which presumably evolved in tandem, brought about this transition: (1) the spiders stopped making the time-consuming trip out from the initial point of attachment on the central line to the frame threads to make the low-shear attachment, and instead started detaching the spanning thread from the spinnerets to create freely hanging bolases, as present-day Mastophorinae do, and (2) the spiders built fewer and fewer threads other than the central line.] This group of spiders gave rise to the present-day Mastophorinae. Gertsch (1955) suggests that early Mastophorinae whirled the bolas in the manner of *Dichrostichus* or of *Cladomelea* and that *Mastophora* subsequently evolved their modern behavior, making directed flicks of the bolas. On the other hand, it is possible that the three methods of manipulating the bolas evolved independently, concurrently with the reduction of the web to a single bolas, and one group of bolas spiders gave rise to a group of spiders that showed increased ability to arrest and hold onto moths using their legs alone, and that eventually abandoned bolas spinning. These spiders may have been the ancestors of the Celaeniae.

The principal value of this scenario is that it provides reasonable intermediate steps in the evolution of present-day behaviors. Below I discuss the strengths and weaknesses of the scenario and of the evidence for it.

It is likely that the Cyrtarachneae spanning thread and the Mastophorinae bolas are homologous; both are covered with an adhesive with unique moth-holding properties, and both are spun with extensive combing action of legs IV. Future work should focus on specifying the biochemical and biophysical features common to the production and operation of these threads.

Eberhard (1980a) points out another feature that is probably a synapomorphy for the larger group: production of a distinctive, pungent odor by disturbed spiders. This odor (the presumed defensive function of which has not been tested) is associated with regurgitation and is produced by all *Mastophora* listed in Table 5.1 (Hutchinson 1903; Eberhard 1980a, pers. comm.; pers. obs.) and by unspecified *Cyrtarachne* species (Eberhard 1980a). Its homology would be proved or disproved by studies of its biochemistry and glandular origin. This character is probably not present in the Celaenieae; it is not mentioned in any of the behavioral descriptions of these spiders, and the *Celaenia kinbergi* individuals I have worked with did not produce a disturbance odor.

Eberhard (1980a) also believes that the production of egg cases with a tough outer layer and the bending of legs I and II in the diurnal resting posture are other synapomorphies. Another possible synapomorphy is the side-by-side accumulative production of egg cases by females at the diurnal retreat. All of these characters are present in all three members of the larger group. However, none of these characters at its present level of description is unique to the larger group, and I do not consider these useful as synapomorphies.

The principal reason for believing that the Mastophorinae and Celaenieae are descended from a common ancestor is that spiders in both groups attract moths. The validity of this character as a synapomorphy depends on a common glandular and biochemical basis for attraction—something for which we have no evidence at present. Males in both groups may emerge from the egg case mature. This may not be a useful character, since males mature in the egg case in at least one other araneid genus (*Arachnura*: Forster and Forster 1973), and males mature early in a number of others.

From the evidence available, I would tentatively agree that the Cyrtarachneae and Mastophorinae are descended from a common ancestor. However, I am not at all sure that the Celaenieae are necessarily even closely related. It would be of interest to know the hunting technique employed by the spiders in the Exechocentreae (*Exechocentrus* of Madagascar, *Coelossia* of Madagascar and Africa, and *Euglyptila* of Asia), which show morphological similarities to both the Celaenieae and the Mastophorinae (Simon 1892; Emerit 1978).

How does *Kaira* fit into this scheme? It seems unlikely that *Kaira* spinning behavior could be derived from the completely different spinning behavior of the Cyrtarachneae and Mastophorinae. In addition, *Kaira* individuals lack all but two of the characters discussed above. They do not produce a disturbance odor. The egg cases lack a hard outer covering and are made one on top of the other (Fig. 5.9). Legs I and II are not bent in the diurnal resting posture. Levi (1983) places all araneids with a terminal apophysis (a male palpal sclerite), including *Kaira*, in the subfamily Araneinae. The Mastophorinae lack this character.

I shall mention another conflicting character having to do with egg-case construction, even though its usefulness is questionable because it has not been studied in other araneids. Whereas *Mastophora* make the loops of the silk layer surrounding the eggs by holding down silk with both legs IV while moving the abdomen to the next attachment point, *Kaira* alternate legs IV. The only immediately apparent character that *Kaira* and the Mastophorinae share is the ability to attract moths, and the homology of the mechanisms of moth attraction has not been determined. I tentatively conclude that *Kaira* cannot be descended from the moth-attracting group of spiders that gave rise to the Mastophorinae, that its web-spinning ancestors are

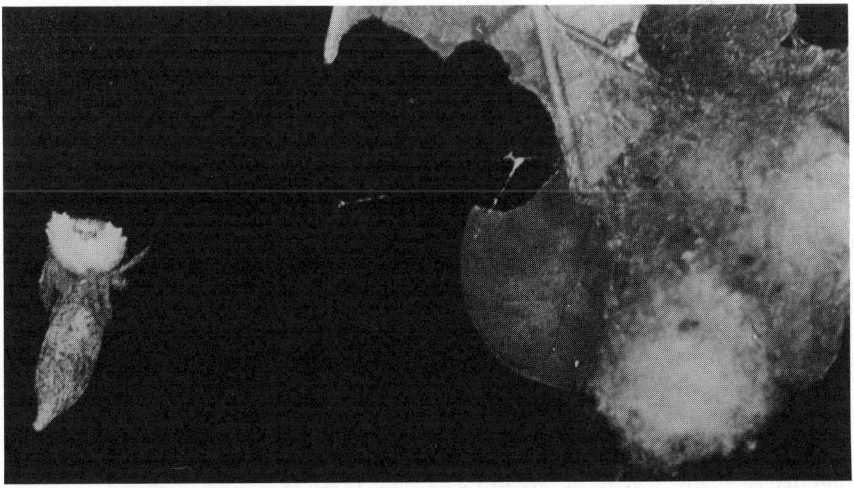

Fig. 5.9. Adult female *Kaira alba* (body length 1 cm) feeding on a moth. A typical conglomeration of several egg cases is to the right of the spider.

TABLE 5.2
Diversity of Prey Caught by Three Groups of Moth-Attracting Spiders

	Spider group and species							
	Celaeneae sensu Simon 1864		Mastophorinae sensu Gertsch 1955					Kaira
Prey	*Celaenia kinbergi* Koch	*Taczanowskia* sp.	*Dicrostichus magnificus* Rainbow	*Mastophora dizzydeani* Eberhard	*Mastophora phrynosoma* Gertsch	*Mastophora cornigera* Hentz	*Mastophora bisaccata* Emerton	*Kaira alba* Hentz
Lepidoptera								
Geometridae								
Larentiinae					M			
Noctuidae								
Agrotinae	B					NO		
Amphipyrinae	C			H				
Catocalinae			G				T	
Hadeninae	D			I		D		
Heliothinae	A							
Hypeninae							QS	Q
Unidentified Noctuidae				K				
Tineidae								
Acrolophinae							PR	RUXZ
Unidentified Pyralidae		F						
Unidentified Olethreutidae				J				
Unidentified Lepidoptera								VWYa
Diptera: Nematocera	E			L				

NOTE: This table summarizes the information in Table 5.1 and organizes the moth species into families and subfamilies. Letters refer to moth species in Table 5.1.

only distantly related to the web-spinning ancestors of the Mastophorinae, and that it therefore evolved the ability to attract moths independently from the Mastophorinae.

Research into the mechanism of moth attraction will undoubtedly provide additional clues to the relationships of the moth-attracting spiders. Although there are some trends in moth sex-pheromone chemistry that reflect moth taxonomy, many compounds are widespread (Mayer and McLaughlin 1975), and it is probably risky to infer spider relationships by chemistry, or to predict the diversity of the compounds used by the spiders from the diversity of the moths they catch (Table 5.2). However, it seems likely that the attractants of *Mastophora bisaccata* and *Kaira alba* are chemically similar since they both attract moths in the genus *Acrolophus* and in the noctuid subfamily Hypeninae, and there are two moth species that are caught by both spiders. The final picture of the relationships and the chemistry of the attractants of the spiders promises to be an interesting one.

Acknowledgments

Many people have helped me in my research. I thank, in particular, my parents, H. Levi and others at Harvard University, J. Tumlinson, J. McLaughlin, and many others at the USDA Insect Attractants Laboratory, Lt. J. Beckham and others at the Florida Department of Natural Resources, and S. and C. Johnson, W. Icenogle, and M. Greenfield. For their criticism of an earlier draft of this chapter, I thank H. Levi, W. Shear, D. Bowers, C. Craig, J. Coddington, W. Eberhard, M. Robinson, J. Carico, and Y. Lubin; for kindly letting me use their photographs, E. Shinkai and D. Clyne; and for identifying moths, E. Todd, R. Poole, D. Davis, and D. Ferguson. My research was supported by graduate fellowships from NSF and Harvard University, National Geographic Society Grant 2506-82, and personal funds.

6

Web Building and Prey Capture in the Uloboridae

YAEL D. LUBIN

MARPLES (1962) wrote that "nothing is known of the habits of the majority of species of *Uloborus*, but there is little variation amongst those which are known." Twenty years later, we are just beginning to appreciate the diversity of web types and behavior in the Uloboridae (Table 6.1). This has come about largely as a result of renewed interest in field observations of the webs, behavior, and natural history of spiders, and in the systematics of the Uloboridae.

Uloborids are often considered the "poor" sister group of a much larger family of orb weavers, the Araneidae. Most species of both families spin orb webs, but there are distinct differences in the structure and function of these webs. The basic morphological differences that influence web structure and predatory behavior in these two families are well known: (1) Uloboridae have a cribellum and a calamistrum and produce cribellate sticky silk, as opposed to the viscid droplets produced by the aggregate glands of the Araneidae, and (2) Uloboridae lack poison glands (Millot 1949; Opell 1979). It has generally been assumed that the orb webs of uloborids and araneids evolved independently and that the similarities in structure and function are therefore convergent (Kaston 1964; Kullmann 1972a), although Kovoor (1977c) presents convincing evidence based on the structure of the silk glands and spinnerets in support of earlier arguments that the Araneidae and Uloboridae share a common orb-weaving ancestor (Wiehle 1931), and Coddington (this volume) also argues persuasively for this view. I attempt to show here that the structure of a web may be strongly influenced by the spider's ecological relationships, and that it would therefore seem logical that similar ecological pressures could have selected for an orb-type structure in different groups of spiders. Orb like webs have evolved independently in the

Theridiidae (e.g., in *Synotaxus*: Eberhard 1977b), the Psechridae (e.g., *Fecenia*: Robinson and Lubin 1979b), and the Amaurobiidae (e.g., in *Titanoeca*: Szlep 1966).

For this review I draw on recent treatments of the taxonomy and functional morphology of Uloboridae (Opell 1979, 1981, 1982a, 1984a; Lubin et al. 1982; Yoshida 1981), on both early and recent accounts of uloborid webs, web building, and predatory behaviors, and on previously unpublished data on uloborid spiders from Papua New Guinea and Central America. My purpose is to bring together what is known about webs and predatory behavior in the Uloboridae, to examine certain basic questions concerning the function and evolution of webs and web-related behaviors, and to highlight areas in which further studies would be valuable.

Methods

Webs and predatory behavior of a number of uloborids in the genera *Zosis*, *Uloborus*, *Philoponella*, and *Lubinella* were studied in Papua New Guinea during 1979–81. Specimens have been deposited in the Museum of Comparative Zoology at Harvard University. Measurements of webs of *Zosis geniculatus* were made on Santa Cruz Island, Galápagos, in 1982. Colonies of *Philoponella republicana* (Simon) were observed on Barro Colorado Island, Panama, during 1972–73 (Lubin 1980a), and webs of other uloborids were observed in Australia, Java, and Galápagos. The following methods were used in studying web structure and prey capture. Statistical tests follow Siegel (1956).

Web Structure

Webs were powdered with cornstarch (Carico 1977), measured, and photographed (using Kodachrome-65 for color and Plus-X for black and white). The total length of silk in webs was calculated using the formula of Witt, Reed, and Peakall (1968): total length of silk = Σ lengths of radii + Σ lengths of frame threads + Σ lengths of spiral arcs. The following measurements were made of the silk in natural webs of adult females of *Z. geniculatus*: lengths of all radii, lengths of frame and guy threads, and distances from the center of the hub to each sticky spiral along four radii in each of the four cardinal directions. Since neither the hub silk nor the temporary (nonsticky) spirals were included in the calculations, the estimates of the total silk in *Z. geniculatus* webs are low. Mesh size of a web was calculated as: spiral area $\div$ (number of radii $\times$ mean number of spirals) (Risch 1977). The spiral area is the area between the outermost and innermost loops of sticky spiral.

TABLE 6.1

Distribution, Web Types, and Stabilimenta of Uloborid Genera

Genus	Distribution	Webs			Stabilimentum
		Adult	Juvenile	Egg-sac	
Tangaroa	South Pacific islands (1); Australia (2)	horizontal orb (3); orb + cone (2)	?	?	?
Waitkera	New Zealand (1)	horizontal or diagonal orb (4)	?	?	?
Ariston	Central America (1)	?	?	?	?
Polenecia	Mediterranean basin (1)	vertical, radial, with sticky silk on radii, frames (5)	?	egg sac on branch at hub (+adult web?) (5)	none
Hyptiotes	North America, Europe, Japan (1); India, Sri Lanka (6)	triangle (1, 5)	none	egg sac on branch, adult web (1, 5)	none
Siratoba	southwestern North America, Mexico (1)	?	?	?	?
Miagrammopes	cosmotropical (1)	one to several sticky threads (7, 8)	none	long, tubular egg sac in rudimentary orb (7)	none
Sybota	Argentina, Chile (1)	reduced web (triangle?) (9)	?	?	?
Orinomana	Ecuador, Chile (1)	?	?	?	?
Uloborus	cosmopolitan (1)	horizontal or diagonal orb ± barrier web (1, 5, 10); orb + cone (11)	instar 2, males (1, 12)	variable egg-sac shape, in rudimentary or adult orb (1, 5, 13, 14)	line, disk (15); spiral, debris, none (13, 16)

Purumitra	Philippines, Carolines (1)	?	?	?	?
Zosis	cosmotropical (1)	vertical or diagonal orb (1, 15)	instar 2 (12)	stellate egg sac in adult orb (1, 17)	line, disk (13)
Octonoba	North America, China, Okinawa (1)	horizontal orb (1, 18)	instar 2 (18)	?	spiral, line (18, 19)
Conifaber	Colombia (11)	juvenile orb + cone (11)	adult (11)	mace-shaped egg sac on radius of adult web (11)	line (11)
Daramuliana	South Pacific islands, Australia (1)	?	?	?	?
Ponella	Paraguay, Brazil (1)	?	?	stellate egg sacs in chain (1)	?
Philoponella	cosmopolitan (1)	vertical to horizontal orb ± barrier web, solitary, commensal, or colonial (1, 13, 20)	instar 2, juveniles (10, 13, 20)	stellate or tubular egg sac in rudimentary orb (1, 13)	line, disk, O, spots, spiral (1, 13)
Lubinella	Papua New Guinea (13, 21)	horizontal or diagonal adult or juvenile orb ± empty sector (13)	?	stellate egg sac, no web (13)	line, spiral (13)

SOURCES: (1) Opell 1979, (2) V. Davies, pers. comm., (3) Opell 1983, (4) Forster 1967, (5) Wiehle 1931, (6) Muma and Gertsch 1964, (7) Lubin, Eberhard, and Montgomery 1978, (8) Akerman 1932, (9) Darwin 1876, (10) Peters 1955a, (11) Lubin et al. 1982, (12) Wiehle 1927, (13) this study, (14) Comstock 1912, (15) Eberhard 1973, (16) Hingston 1927, (17) Marples 1962, (18) Yoshida 1981, (19) Peaslee and Peck 1983, (20) Smith 1982b, (21) Opell 1984a.

Prey Capture

Predatory behavior was observed using insect prey of various species thrown or placed in webs, and behavior sequences were timed with a stopwatch. The length of silk used to wrap insects was calculated as follows. From video recordings of prey capture, measurements were made of the average duration (in video frames) of a single wrap, *f*, i.e., the time it took for both legs IV to make one trip each over distance *d* between spinnerets and prey. Averages of *d* were based on three to five measurements and averages of *f* on 10 to 15 counts made at different times in the wrapping sequence. The total wrap duration at the capture site and at the hub, w_t, was measured with a stopwatch. The total length of silk, *S*, was calculated as $S = 2d(w_t \div w)$, where $w = f \div 30$ (one video "frame" = $1 \div 30$ seconds).

Orb Webs

Most known species of uloborids spin orb webs consisting of frame threads, radii, a hub, a nonsticky "temporary" spiral, and a sticky spiral made of cribellate silk (Marples 1962; Opell 1979). A "typical" uloborid orb web, such as that of *Uloborus diversus*, has the following characteristics (Figs. 6.1, 6.2, 6.4c; see also Eberhard 1972b, Pls. I, IV). Radii arise from the hub spiral and attach distally to primary frame threads, which in turn attach to the substrates. Secondary frame threads (frame threads that span corners formed by primary frame threads), which are characteristic of some araneid orb webs, rarely occur in uloborid orb webs (Eberhard 1972b; pers. obs.) The temporary spiral is laid from the hub outwards, is continuous with the hub, and is later removed as the sticky spiral is constructed centripetally. While constructing the temporary spiral, the spider sometimes inserts additional radii that, in the completed web, appear to branch off from the primary radii. In *U. diversus*, the sticky spiral begins at or on a frame thread, and the first spiral loop follows an irregular, jagged path, segments of sticky spiral being laid directly on radii and frame threads, as well as between radii. Switchbacks of the sticky spiral occur frequently; as a result the larger sectors of an asymmetrical orb are filled with sticky spiral loops.

Attachments of the sticky spiral to radii are of the "zero" junction type (Jackson 1971): the spiral is cemented to the radius and will not slide through its junction when stressed, as the spirals in araneid webs sometimes do (Eberhard 1976). Uloborid sticky spirals are also less elastic than araneid sticky spirals (W. G. Eberhard, pers. comm.; pers. obs.). As a result, uloborid orbs may be less able to withstand sudden, large loads (e.g., the im-

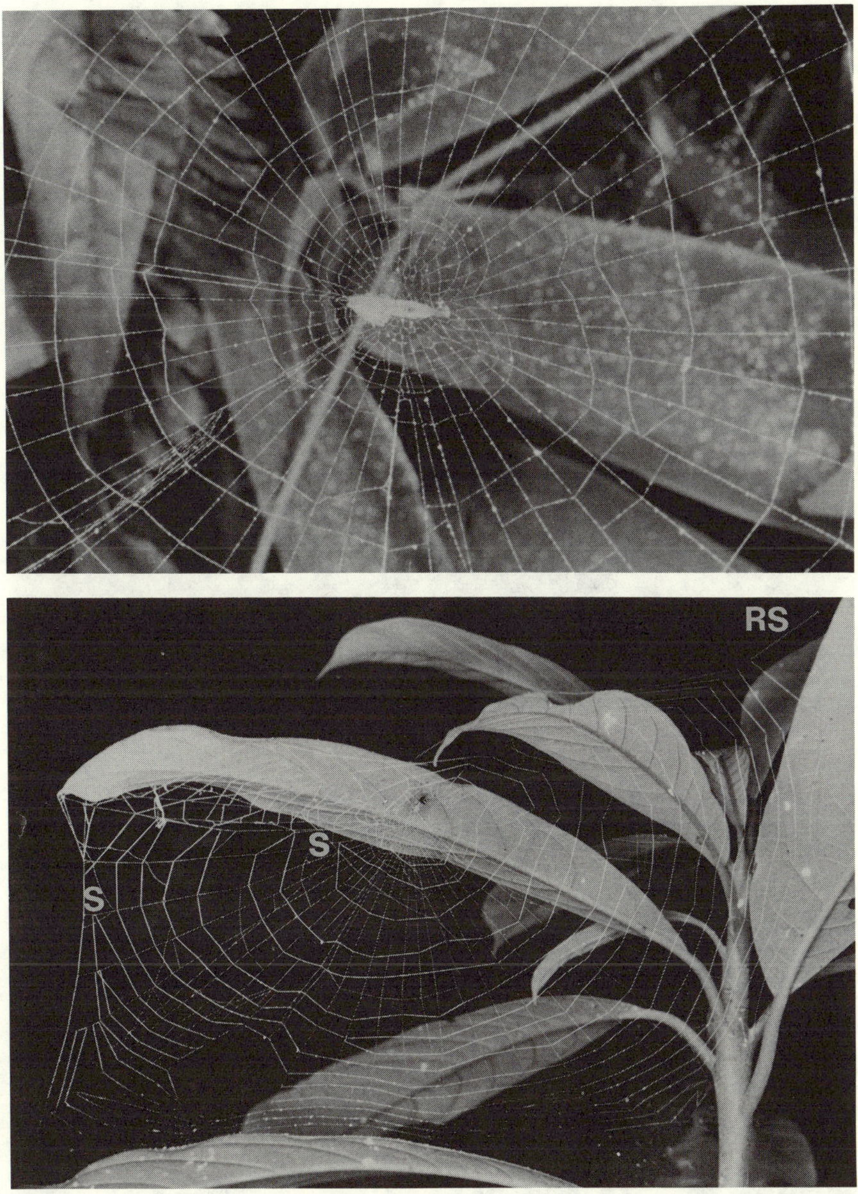

a

b

Fig. 6.1. Horizontal orb webs. *a*: web of *Uloborus barbipes* from Australia. The hub region is shown. *b*: web of *Uloborus* sp. J from Papua New Guinea. The hub is concealed beneath a leaf and is attached to it by a vertical tangle of barrier-web threads. Note the sticky spiral switchbacks (*S*) and lengths of sticky spiral placed on radii (*RS*) and frame threads.

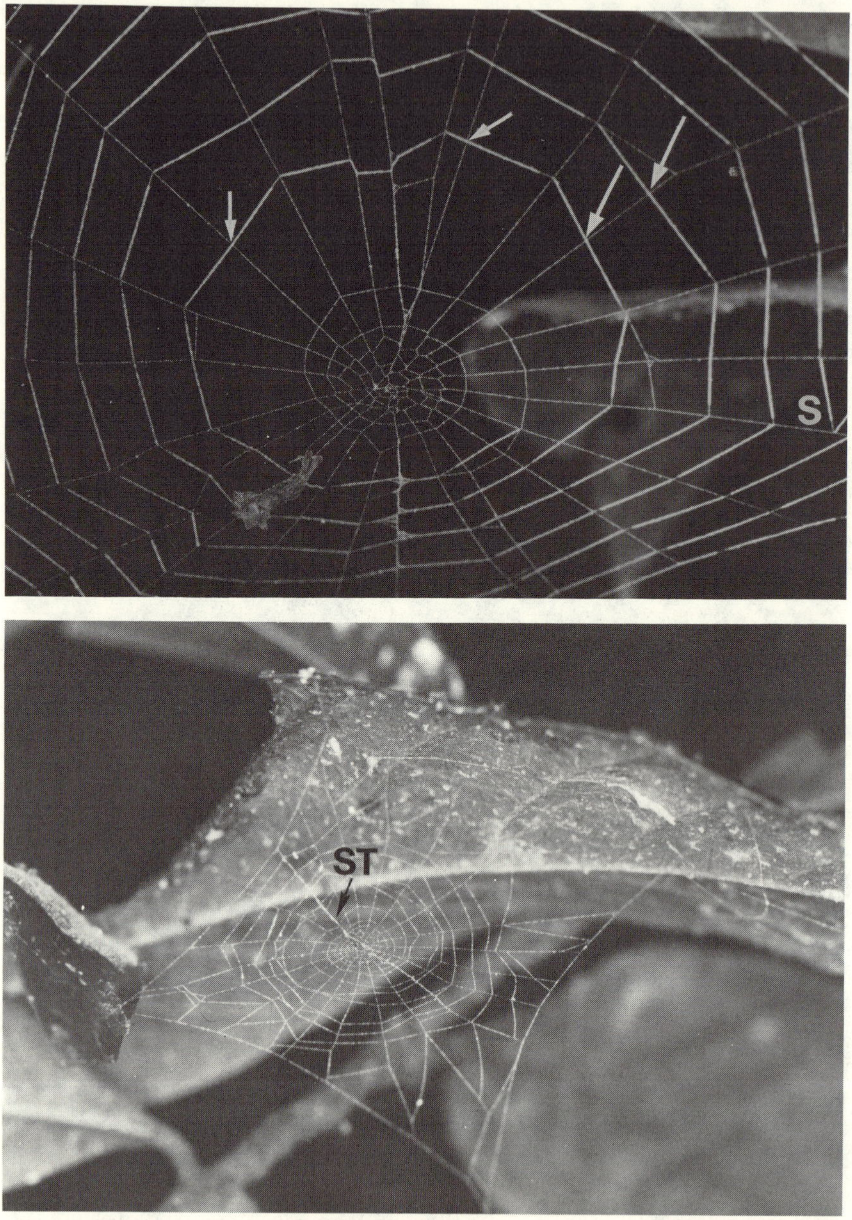

Fig. 6.2. Orb webs of *Uloborus* sp. MB from Papua New Guinea. *a*: web of an adult female. Note the sticky spiral switchbacks (*S*) and places where the sticky spiral crosses radii without attaching to them (*arrows*). The female hangs in a cryptic posture beneath the hub. *b*: horizontal web of a second(?)-instar spiderling. The web is beneath a leaf. Note the strongly zigzagged outer loop of sticky spiral and a linear stabilimentum (*ST*) extending through the hub.

pact of a large insect) than araneid orbs (Langer 1969; for further studies of the abilities of web threads to bear stress, see Eberhard, this volume). Two features of uloborid webs may improve their restraining power: (1) when stressed, the mat of cribellate threads pulls away from the less elastic, paired axial baseline thread (Eberhard 1976), and (2) some spiral loops cross radii without attaching to them (Figs. 6.2, 6.7a) (Eberhard 1972b; Lubin et al. 1982). Both features may allow the sticky spiral threads greater extensibility under sudden stress.

Detailed observations of web construction and measurements of various parameters of webs of *U. diversus* and other species provide insight into the decisions that spiders make during web building (Eberhard 1969, 1972c, 1981a; LeGuelte 1966; Peters 1939). Clearly, during the building process the orb weaver senses angles, distances, and tensions of threads as they are being placed, or of threads already in place, and makes decisions about placement of new threads based on these cues. Certain elements of web-building behavior and web structure appear to be highly consistent within species and even within genera and subfamilies, a fact that may prove useful in establishing phylogenetic relationships among higher taxa (Eberhard 1982). Such consistency notwithstanding, webs are of necessity adapted to the particular array of environmental conditions encountered by spiders, and a certain degree of within-species variation in web structure is to be expected (a factor generally ignored in comparative studies of orb webs: e.g., Risch 1977). Some of the ways in which environmental and biotic factors may influence intraspecific variation in webs (proximate effects) and interspecific variation in webs (ultimate effects) are examined next.

Uloborid orb webs vary in the degree to which they are asymmetrical and in the use of space-filling modifications, in the use of stabilimenta, in the use of certain modifications in webs of juveniles and males, and in the use of extra-orb structures. There is also intraspecific variation in orientation, size, and placement of webs, and in tendencies to build webs in aggregations or in association with other spiders.

Asymmetry and Sticky Spiral Modifications

In uloborid orbs, the entire area within the frame threads tends to be filled with sticky spiral loops (Eberhard 1972b). In highly asymmetrical webs this is accomplished by means of sticky spiral switchbacks (Figs. 6.1b, 6.2a, 6.4c), zigzag loops (Figs. 6.2b, 6.7, 6.9b), or both. Zigzag outer loops of sticky spiral occur in webs of *U. diversus* (Eberhard 1972b) and *Uloborus conus*, *Uloborus albolineatus*, and *Uloborus bispiralis* (Lubin et al. 1982), and can be seen in Comstock's (1912) photographs of *Uloborus glomosus* (= *Uloborus americanus*). Some *Philoponella* construct zigzag outer loops, e.g., *Philoponella* sp. B from Papua New Guinea and *Philoponella variabila* (= *Ulobor-*

a

Fig. 6.3. Orb webs of *Philoponella undulata* from Papua New Guinea. *a*: new orb web with part of an old orb web attached below. The outer loops of the nonsticky hub spiral of the old orb are still visible (*OH*). The spider sits off the orb, on a thread attached to the hub of the new web. Note the irregularity of the hub (*H*). The outer loops of sticky spiral are zigzagged, and short sticky spiral segments are laid along the radii throughout the spiral area (*arrows*). *b* (facing page): web of a *P. undulata* female. The spider sits on a thread attached to the hub. *c*: partially renewed web. Wadded-up silk divides the new inner section of the web from the old section (*arrows*). The spider sits at the hub. Three arms of a thin linear stabilimentum are visible, two above and one below the hub.

us variabilis) from Australia, and a species of *Tangaroa* is reported to do so as well (V. Davies, pers. comm.).

Though both switchbacks and zigzag loops increase the trapping area of asymmetrical orbs, zigzag loops undoubtedly do so less effectively; they produce a coarser mesh than that of normal spiral turns and switchbacks. Possibly zigzag loops are a compromise between the demand for a greater trapping surface and the cost of the greater quantity of cribellate silk needed for additional spirals and the greater amount of time needed to lay them. In webs of *U. diversus* (Eberhard 1972b) and *U. conus* (Lubin et al. 1982), which contain zigzag loops, the outer loop describes a complete turn around the orb; switchbacks, if they occur, appear inside the zigzag outer loop (Fig. 6.7). In highly asymmetrical webs of *Z. geniculatus*, *Philoponella undulata* and *Uloborus* sp. L, which do not contain zigzag outer loops, numerous switchbacks occur in the outer part of the orb (Fig. 6.3).

The tendency to fill in the "corners" of asymmetrical orbs makes many uloborid webs well suited to fit in confined spaces. *Lubinella morobensis*

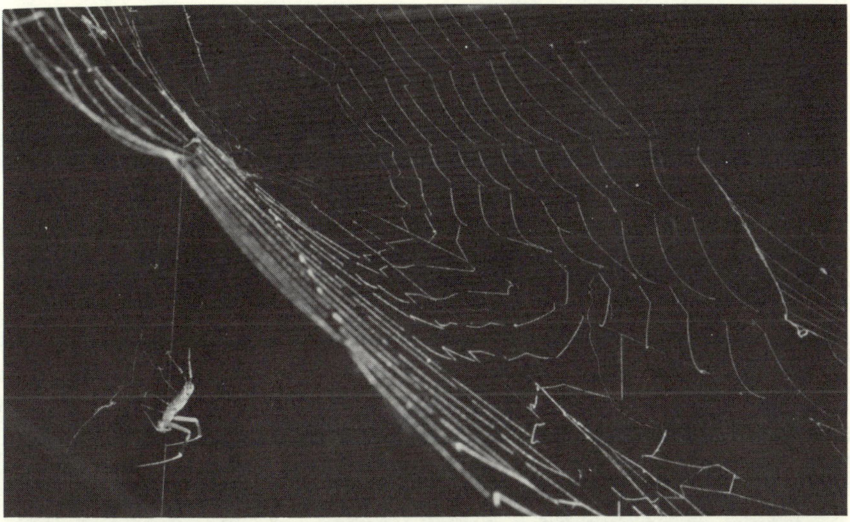

b

c

(Opell, 1984a) builds in tree buttresses and beneath rocky overhangs; many *Uloborus* species (e.g., spp. J, K, and L and *U. glomosus*) and *Philoponella tingena* in Central America (Opell 1979) construct asymmetrical orbs with the smaller section concealed beneath a leaf or other projecting surface and the larger section protruding from beneath the surface. One horizontal orb of *Uloborus* sp. J had only three complete turns of the spiral (the three innermost loops); all the rest were switchbacks, producing a highly asymmetrical orb with all corners of the frame threads filled with sticky spiral (Fig. 6.1b).

Stabilimenta

Hingston (1927) described some 22 different types of "protective devices" in orb webs, eight of which occurred in webs of uloborids. Although it no longer seems useful to produce a catalog of stabilimentum forms, the fact remains that stabilimenta are highly variable in form both within and between species and often even in different webs of the same individual. Stabilimenta observed in uloborid webs consist of: one or more linear arms of white silk radiating from the hub, sometimes crossing the hub, but usually with a spider-sized gap at the hub; a circle at the hub; an opaque disk or spiral disk of silk at the hub; or a combination of any of these (Figs. 6.2b, 6.3c, 6.4).

The forms of stabilimenta and their frequencies of occurrence may change with age (as was shown for an araneid orb weaver: Ewer 1972). Webs of *U. conus* immatures often had thin linear stabilimenta, but webs of adults rarely had stabilimenta (Lubin et al. 1982). Whereas webs of adults of *Z. geniculatus* in Papua New Guinea had disks at the hub more often than linear stabilimenta, the opposite was true for webs of immatures. Webs of *Philoponella* sp. G had several stabilimentum forms—(1) a circle ("O") at the hub, which could be opaque or translucent, (2) a line with or without a gap

a

b

c

Fig. 6.4. Stabilimenta in uloborid webs. *a*: spiral-disk stabilimentum in the web of a subadult female *Lubinella morobensis* from Papua New Guinea. The spider is nearly invisible beneath the disk. *b*: spiral-disk stabilimentum in the web of an adult female *Philoponella* sp. G from Papua New Guinea. The outline of the spider is dimly visible beneath the disk. *c*: partially filled O stabilimentum in the web of a juvenile female *Philoponella* sp. G. Note the switchbacks of the sticky spiral.

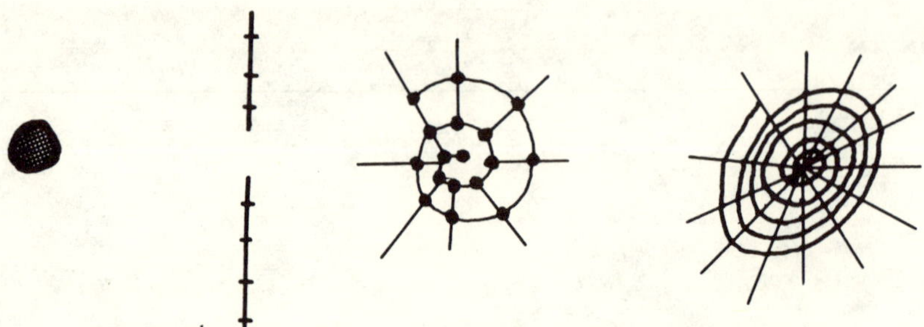

a *b* *c* *d*

Fig. 6.5. Stabilimenta in webs of *Philoponella* sp. G: *a*, circle or O at the hub; *b*, line with a gap at the hub; *c*, spots in a spiral at the hub; *d*, spiral disk at the hub. Flocculent bits of silk often occur at junctions between stabilimentum lines and radii (in spiral-disk stabilimenta) and between stabilimentum lines and hub threads (in linear stabilimenta).

at the hub, (3) a series of spots in an open spiral arrangement at the hub, or (4) a tight spiral disk covering the hub (Figs. 6.5, 6.4)—or no stabilimentum. Transition forms between O and spiral-disk stabilimenta were seen occasionally. O and spot stabilimenta occurred more frequently than expected (on the basis of the distribution of stabilimenta in the total population) in webs of males and of immature spiders of both sexes; line and spiral-disk stabilimenta occurred more frequently than expected in webs of larger juvenile (more than 2 mm long) and adult females (Table 6.2; $\text{chi}^2 = 16.3565$, $p < 0.01$, 2×5 contingency table).

Philoponella sp. G is apparently capable of building all forms of stabilimenta at any stage of development. Although "*Philoponella* sp. G" is used throughout the text, the generic placement of this species remains in doubt. Certain morphological characters distinguish the species from other *Philoponella* and place it close to *Lubinella* (B. D. Opell, pers. comm.). (It is possible, however, that adult females do not build spots.) Nonetheless, spiders of a given age class tend to use one or two forms preferentially. Because of the change in behavior that must occur when constructing a new type of stabilimentum, the ontogeny of stabilimentum building would be a rewarding area for further experimental study.

Various functions have been attributed to stabilimenta in spiders' webs, including camouflage or predator distraction (Ewer 1972; Eberhard 1973; Lubin 1975; Tolbert 1975), advertisement of the web and warning to large insects or birds that might blunder into it (Horton 1980; Eisner and Nowicki 1983), strengthening of the web (Robinson and Robinson 1970b, 1973b), and tuning or adjustment of tensions of web elements (Robinson and Robinson 1970b). Eberhard (1973) showed that *U. diversus* in the laboratory built more stabilimenta when kept under a high level of nighttime illumination and suggested that webs in well-lit places would be more visible to preda-

TABLE 6.2
Number and Percentage of Webs of Philoponella sp. G with Different Stabilimentum Types

Stabilimentum	Males and immatures		Juvenile and adult females		Total
	N	Pct.	*N*	Pct.	
O	*18*	30.0%	3	13.0%	21
Spots	*15*	25.0	2	8.7	17
Line	14	23.3	*11*	47.8	25
Spiral disk	7	11.7	7	30.4	14
None	*6*	10.0	0	0	6
TOTAL	60	100.0%	23	99.9%	83

NOTE: Spiders are grouped into two size classes based on sex and size: (1) immature and adult males, immature females (≤2 mm long), and second-instar spiderlings of unknown sex, and (2) juvenile (>2 mm long) and adult females. Italics indicate that a stabilimentum type occurred more frequently than expected (chi^2 test; see text).

tors. These data and observations of the location of stabilimentum lines in orbs of *U. diversus* and of the postures of uloborids at their hubs (Eberhard 1973; Opell and Eberhard, in press) lend support to the antipredator hypotheses for the function of stabilimenta in uloborid webs.

One can only speculate on the adaptive advantages of the different stabilimentum forms. To the human eye, linear stabilimenta that cross the hub and spiral or opaque disks tend to hide the spider (Figs. 6.4b,c), whereas O and spot stabilimenta seem to serve as distracting devices. Small juveniles of *Philoponella* sp. G are pale grey and (again to the human eye) difficult to distinguish from the matrices of their filmy webs (see below); they appear as greyish spots, somewhat fainter than those of spot stabilimenta.

Changes in stabilimentum form were recorded in webs built on five successive days by 15 *Philoponella* sp. G that had been transferred from the field—the understory of closed-canopy forest—to an insectary about half of the area of which was exposed to the sky. Of 43 recorded web changes, seven (16.3 percent) involved changes in stabilimentum form. Two of these were changes from O stabilimenta, two from linear stabilimenta, and three from spiral stabilimenta; there were equal numbers of changes (three) to and from spiral stabilimenta. Although more data are clearly necessary, the changes recorded for the captive spiders indicate that stabilimentum form is a labile feature of webs of *Philoponella* sp. G. The distribution of stabilimentum forms in the captive spiders on day 2 after their release was different from that in the original population from which the spiders were collected. In the original population, O and spot stabilimenta occurred in 59.3 percent of 32 webs, whereas spiral stabilimenta occurred in only 15.6 percent. In the day-2 captive spiders, only 13.3 percent had O stabilimenta, none had spot stabilimenta, and 53.3 percent had spiral stabilimenta. The change, generally, from less concealing stabilimenta (O and spot forms) to more concealing stabilimenta (spiral forms) lends support to the idea that light level and web visibility influence the choice of stabilimentum form.

An unusual "stabilimentum" was found in webs of *Uloborus* sp. L. It consisted of spider-sized bits of debris (leaves, twigs, and grass) suspended by short threads from radii and from barrier-web threads above the orb. During the day the spider rested at the hub of its horizontal, asymmetrical web in a cryptic posture, all legs folded and pressed against the body. When disturbed, the spider moved into the barrier web and hung among the bits of debris, from which it was virtually indistinguishable. The web, stabilimentum, and cryptic posture of *Uloborus* sp. L are strikingly similar to those of the araneid orb weavers of the genus *Dolichognatha*, which likewise construct horizontal, asymmetrical orbs with a tight mesh (closely spaced spirals and numerous radii) under leaves in forest understory, and suspend spider-sized bits of debris into which they move when disturbed (pers. obs.). *Uloborus*(?) *scutifaciens* incorporates debris in a stabilimentum consisting of a disk at the hub with or without linear extensions, which may be similar in form to that of some species of *Cyclosa* (Araneidae) (Hingston 1927).

Juvenile Webs

Wiehle (1927) noted that webs of newly emerged *Uloborus walckenaerius* and *Z. geniculatus* differed from adult webs in having, instead of a sticky spiral, a thin sheet of fine threads arranged in a more or less radial pattern. Wiehle attributed the construction of juvenile webs to the lack of fully developed calamistra and cribella in second-instar spiderlings. Adult males also lack functional calamistra and cribella and may construct "juvenile" webs (e.g., *Uloborus* sp.: Eberhard 1977c). Juvenile webs have now been observed in many species of *Uloborus* (Wiehle 1927; Szlep 1961; Eberhard 1971b, 1972b; Lubin et al. 1982), *Philoponella* (Peters 1953; Opell 1979; this study), *Octonoba* (Peaslee and Peck 1983), *Conifaber* (Lubin et al. 1982), and *Zosis* (Wiehle 1927). Neither *Miagrammopes* nor *Hyptiotes* build juvenile webs; young of these genera do not build webs until after the second molt, when they have functional calamistra and cribella (Opell 1982a, 1982b; Lubin, Eberhard, and Montgomery 1978).

Juvenile-web elements sometimes occur in webs of post-second-instar juveniles and adult spiders, e.g., in webs of adult females of *Conifaber parvus* (Lubin et al. 1982), post-second-instar juvenile females and males of *Philoponella* sp. G and *Uloborus trilineatus* (Fig. 6.6, Table 6.3), and old ("senile") virgin females of *U. diversus* (Eberhard 1971b). Adult females of *C. parvus* and senile virgins of *U. diversus* appear to have normal cribella and calamistra. Although *Philoponella* sp. G individuals of all sizes were found in juvenile webs, juvenile males and immatures ($\leq$2 mm long) built signifi-

a

b

Fig. 6.6. Juvenile webs. *a*: web of an immature *Philoponella* sp. G. The fine silk threads have a primarily radial orientation, although some lines with a spiral orientation are visible in the lower-left part of the web. The entire area within the frame threads is filled with juvenile-web silk. *b*: web of a juvenile female *Uloborus trilineatus* from Isabela Island, Galápagos. Radii, non-sticky hub spiral, and juvenile-web silk covering part of the web surface are visible.

TABLE 6.3

Number and Percentage of Webs of Philoponella sp. G of Juvenile and/or Adult Type

Web type	Males and immatures		Juvenile and adult females		Total
	N	Pct.	N	Pct.	
Juvenile	*52*	86.7%	10	43.5%	62
Adult	5	8.3	*11*	47.8	16
Juvenile + adult	3	5.0	2	8.7	5
TOTAL	60	100.0%	23	100.0%	83

NOTE: Spiders are grouped into two classes based on sex and size: (1) immature and adult males and immature females (≤2 mm long), (2) juvenile (>2 mm long) and adult females. Italics indicate that a web type occurred more frequently than expected (chi² test; see text).

cantly more juvenile webs than did young (>2 mm long) and adult females ($chi^2 = 18.3032$, $p < 0.001$, 2×3 contingency table).

Adult orb webs of *Philoponella* sp. G sometimes included areas of juvenile web; e.g., the orb's center or part of the outer web was of juvenile type. This might have come about in three ways.

1. The original adult orb was repaired with juvenile-web sheet.
2. The original juvenile web was replaced by an adult web with sticky spiral, but certain sections containing juvenile-web sheet were left intact. Aberrant webs of third-instar *Uloborus plumipes* (Szlep 1961) may have been constructed in this manner.
3. Juvenile-web sheet was superimposed on part of an adult web (especially the hub area), as in webs of senile virgins of *U. diversus* (Eberhard 1971b). Adult males of *Philoponella* sp. G may have deposited juvenile-web silk on adult webs constructed in the penultimate instar.

The adaptive advantages of constructing thin, easily damaged juvenile webs in later instars remain a mystery. Economy of silk or web-construction time can be ruled out, since juvenile webs take as long to construct as similar-sized adult webs or longer (pers. obs.), and juvenile webs are probably less effective prey traps than adult webs. One might predict that the silk of juvenile webs ruptures easily upon prey impact and is less effective in restraining large, trapped insects than are the sticky spirals of similar-sized adult webs. Perhaps the advantage lies in better concealment of the spider. Juvenile webs may be advantageous where the risk of predation is high and where small and weakly flying insects constitute the major prey. The horizontal orbs of *Philoponella* sp. G were found in lowland swamp forest where small, nematocerous flies abound. To the human eye at least, juvenile webs were unquestionably more difficult to see in the forest understory than were adult webs, and even when a juvenile web was visible the spider was hidden from view from above by virtue of the web's opacity.

Extra-Orb Structures

A frequent variation in orb-web structure involves the presence of more or less regular arrays of threads on one or both sides of the orb-web plane. Such threads occur in many genera of orb-weaving araneids (*Cyrtophora*, *Mecynogea*, *Nephila*, *Argiope*, *Metepeira*, *Arachnura*, and *Leucauge*, to name a few) as well as in uloborids. Among uloborids, supra- and sub-orb tangles are almost always present in *Philoponella*, and occur commonly in many *Uloborus* and sporadically in other genera.

Cone-shaped arrays of threads beneath horizontal or near-horizontal orbs are found in *U. conus*, *U. bispiralis*, and *U. albolineatus* in Papua New Guinea, and in *C. parvus*, in *Uloborus* sp. 2,072, and occasionally in *U. trilineatus* in Central and South America (Lubin et al. 1982). Two other species that apparently construct cones beneath their orbs are *Uloborus quadrituberculatus* in Malaysia (Workman 1896) and a species of *Tangaroa* in Australia (V. Davies, pers. comm.). Construction of the orb-plus-cone web, observed in *U. conus*, involves a number of behaviors not previously seen in uloborids: to form the cone, the spider cuts radii at their attachments to the hub, lengthens them by adding dragline, and attaches them below to a central guy thread; later, the spider constructs a new orb within the cone (Fig. 6.7a). Orb-plus-cone webs of all three species of *Uloborus* from Papua New Guinea and *Uloborus* sp. 2,072 have sticky spiral threads both on the horizontal orb and on the cone, whereas those of *C. parvus* and *U. trilineatus* have no sticky spiral at all. The orb-plus-cone web of *U. trilineatus* may be a specialized molting web.

Barrier webs (irregular tangles of threads) above or to one side of near-horizontal orbs have been described in *Philoponella vicina* (Peters 1953), *P. tingena* (Opell 1979; pers. obs.), and *Philoponella semiplumosa* (Lahman and Eberhard 1979). To this list we may add *Uloborus* spp. J, K, and L (Fig. 6.1b), *Philoponella* sp. B, and *P. undulata* (Fig. 6.3b) in Papua New Guinea, *Philoponella* sp. near *emarginata* in Java (Fig. 6.9a) and *Uloborus barbipes* and *P. variabila* (Fig. 6.9b) in Australia. No doubt numerous other uloborids build such webs. The barrier-web threads attach primarily to the hub of the orb, but also to radii and frame threads (Peters 1953; pers. obs.), thus pulling the orb out of a single plane (e.g., Fig. 6.9). *Philoponella tingena* and *P. undulata* generally have an empty sector in the upper part of the orb and one or more signal lines (resting threads) leading from the hub to barrier-web threads. Some barrier-web threads may have cribellate-silk threads laid on them. When presented with prey on one such sticky thread, a *P. variabila* female attacked it in the usual uloborid fashion. Cribellate silk is also laid on threads of rudimentary egg-sac webs such as those of many species of *Uloborus*, *Miagrammopes*, and *Philoponella* (Fig. 6.8b) (Lubin, Eberhard, and Montgomery 1978; Lubin et al. 1982).

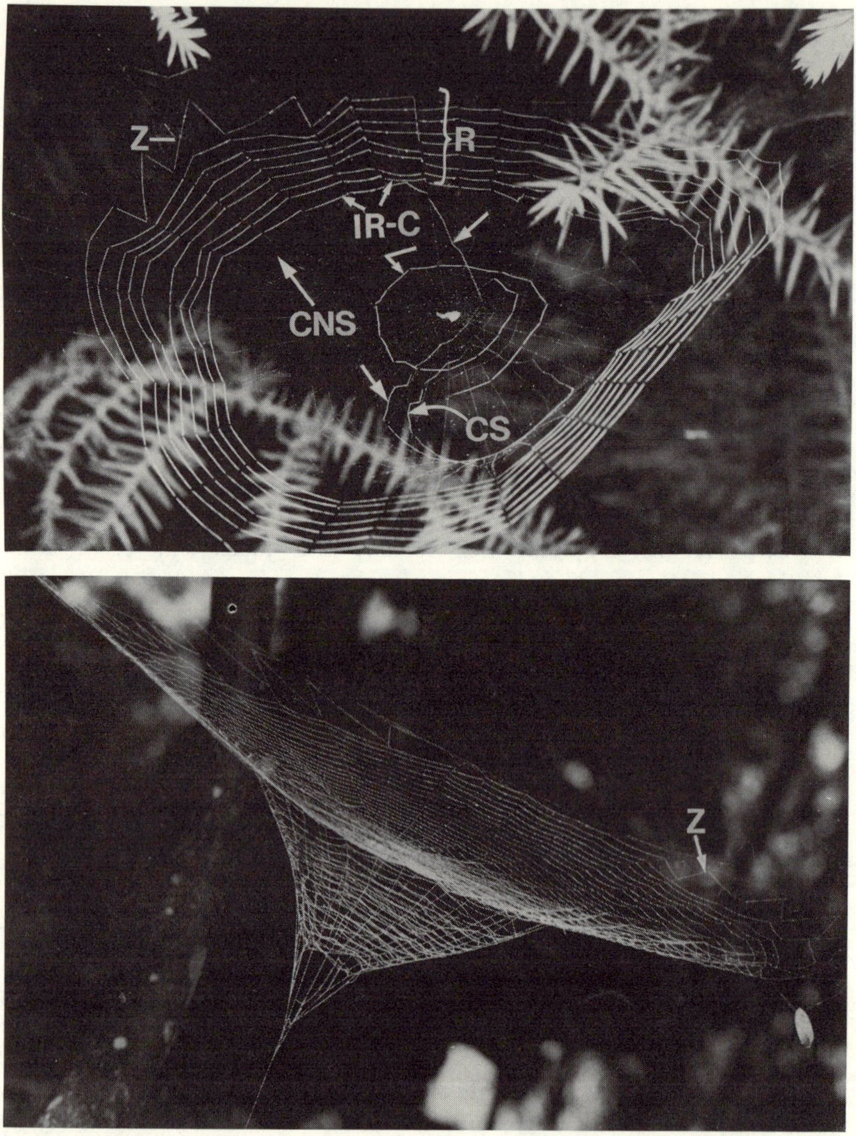

a

b

Fig. 6.7. Orb-plus-cone webs from Papua New Guinea. *a*: web of *Uloborus conus*, seen from above. Note the zigzag outer loop of sticky spiral *Z* and the jagged cone sticky spiral (*CS*), which is continuous with the rim sticky spiral (*R*). Note also the nonsticky spiral of the cone (*CNS*), points of attachment of inner-orb radii to the cone radii (*IR-C*), and the short inner-orb sticky spiral (*IS*). The sticky spirals of both the inner orb and the cone cross some radii without attaching to them (*arrows*). *b*: web of *Uloborus bispiralis*. The sticky spirals of the cone, rim, and inner orb are more regular and denser in this web than in the *U. conus* web. Note the zigzag outer loop of sticky spiral (*Z*) and the cone radii converging on a single guy thread.

Cones, barrier webs, and egg-sac webs are surely primarily defensive structures. In orb-plus-cone webs, the spider is essentially enclosed in a cage of silk, protected on all sides (Lubin et al. 1982). *Uloborus* and *Philoponella* species that build barrier webs tend to run up into the tangles when their webs are disturbed. (See the description of the stabilimentum of *Uloborus* sp. L given earlier). Sticky silk on cones and barrier-web threads may also originally have had a protective function, and secondarily evolved a prey-trapping function. The tendency to lay sticky threads on radii, frame threads, and barrier-web or egg-sac-web threads is widespread in Uloboridae. (See also the discussion of reduced webs, below.) In this respect, uloborids differ markedly from araneid orb weavers, which apparently never lay sticky silk on nonsticky threads.

Commensal and Colonial Webs

Philoponella species that live as commensals in webs of other spiders, or form colonies of their own, use the barrier-web threads of the host spider or each other's barrier-web threads both as supports for their orb webs and as retreats away from their hubs. In Papua New Guinea, commensal species are found in nonsticky barrier-web threads of theridiids (e.g., *Achaearanea*) and of *Cyrtophora* and *Nephila* (Araneidae), and in nonsticky tangles above agelenid and psechrid sheet webs. Although commensal uloborids do not appear to be strictly host-specific, they may prefer either a particular habitat or a particular type of host web. For example, in Papua New Guinea, *P. undulata* (Fig. 6.3) was found in shady, forest-understory habitats in webs of *Cyrtophora cylindroides*, in agelenid and psechrid webs, and occasionally solitarily (noncommensally). *Philoponella* sp. B occurred in communal webs of the social theridiid *Achaearanea wau*, and more rarely in webs of other species of *Achaearanea*. *Philoponella* sp. J was found only in agelenid webs. *Philoponella* spp. B and J may be obligate commensals.

Other commensal and colonial uloborids include *Z. geniculatus*, whose webs were often found in loose colonies in Papua New Guinea and Galápagos, *Philoponella congregabila* (= *Uloborus congregabilis*) and *P. variabila* (Fig. 6.9b) in Australia (Main 1976), *Philoponella* sp. near *emarginata* in Java (Fig. 6.9a), *Philoponella ferokus* (= *Uloborus ferokus*) in India (Bradoo 1979), and the New World species *P. tingena* (Opell 1979; Struhsaker 1969), *Philoponella oweni* (Smith 1982b), *P. semiplumosa* (Lahman and Eberhard 1979), *Philoponella vittata* (= *Uloborus servulus*) (Simon 1892), *Philoponella fasciata* (Smith 1982b), *P. vicina* (Fincke 1981), *Philoponella para* (W. G. Eberhard, pers. comm.), and *Philoponella republicana* (Simon 1891; Lubin 1980a). Of these species, *P. republicana* is apparently obligately colonial (Lubin 1980a) and *P. ferokus* is found only as a commensal in the communal webs of *Stegodyphus sarasinorum* (Bradoo 1979). The remaining species may be facultatively colonial or commensal.

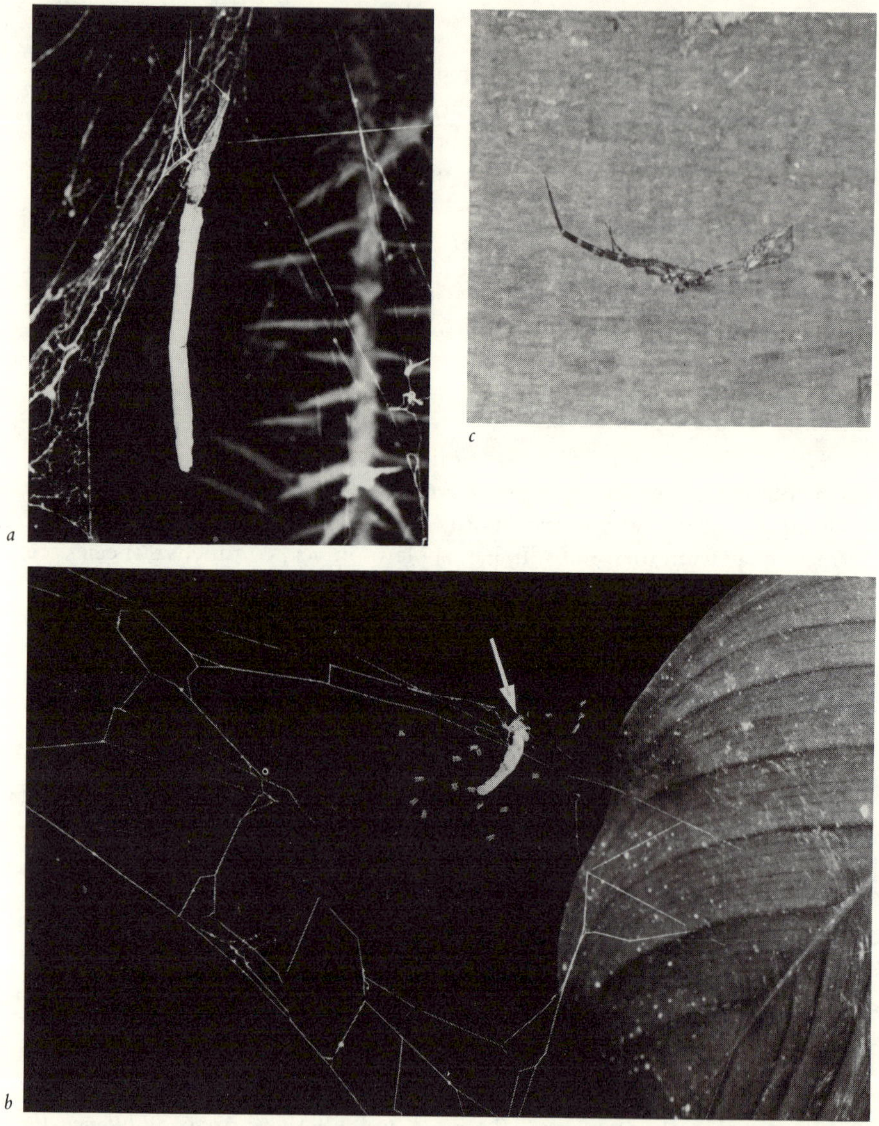

Fig. 6.8. Egg sacs and egg-sac webs. *a*: tubular egg sac of *Philoponella undulata*. The spider holds the egg sac with legs IV while hanging from the remains of a previous orb web. *b*: elongated egg sac of *Uloborus conus* in a rudimentary orb web. Sticky threads appear bright in the photograph. The spider (*arrow*) holds the egg sac with legs IV as the young begin to emerge. *c*: stellate egg sac of *Lubinella morobensis*. The spider hangs in a cryptic posture from a single thread while holding an egg sac with legs IV. *d*: two stellate egg sacs of *Philoponella* sp. near *emarginata* from Java. The egg sacs are suspended in a tangle of barrier-web threads (of a web of *Cyrtophora unicolor*) and rudimentary orb-web threads, some of which have sticky silk on them. The spider (*arrow*) sits in a cryptic posture with legs I and II folded against the cephalothorax.

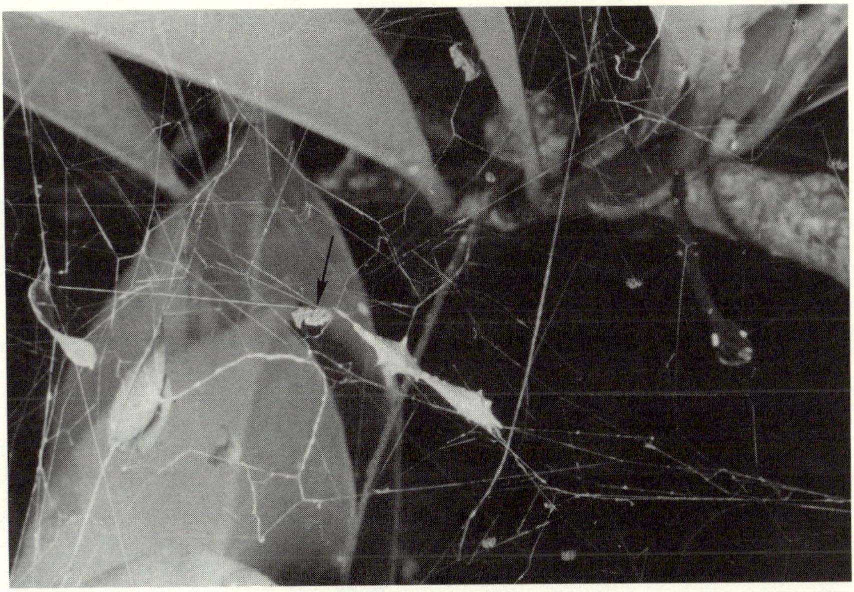

d

A great deal has been written about adaptations to group living in spiders. (For recent reviews of the literature see, e.g., Shear 1970a, Burgess 1978, and Buskirk 1981.) Certain points relevant to adaptations of uloborids to commensal or colonial living and the advantages of these adaptations are discussed next.

In commensal and colonial spiders, web construction is facilitated by the use of other webs as supports (Buskirk 1975b), but a likely disadvantage associated with this practice is an increase in the frequency of aggressive encounters with the hosts, or with neighbors, during web building (Lubin 1974; Buskirk 1975b; personal observation of *Philoponella* sp. B in webs of *A. wau*). To construct its orb, a commensal uloborid must remove a section of the host's barrier web, and runs the risk of encountering the host while doing so. Whereas most uloborids build their webs at night (Eberhard 1971a; pers. obs.), *Philoponella* sp. B commensals in webs of the communal theridiid *A. wau* avoid interactions with their hosts by shifting their activity to the day. They tear down their orbs or move onto frame threads of the communal web in the evening, when the *A. wau* colony becomes active in repairing its communal barrier web, and build new orbs in the morning, when the hosts are inactive.

Marples (1962) and Eberhard (1972b) noted that most uloborid orbs are horizontal or near-horizontal in orientation. For example, *U. diversus* webs on pack-rat nests in Arizona were most often less than 30° to horizontal (Eberhard 1971a). Exceptions are webs of commensal and communal spe-

a

b

Fig. 6.9. Commensal and colonial webs. *a*: commensal web of an adult female *Philoponella* sp. near *emarginata* in the upper barrier-web threads of a web of *Cyrtophora unicolor* (Araneidae). Another commensal web of the same species is visible in the upper right. An egg sac (*arrow*) is attached to the hub. *b*: colonial web of *Philoponella variabila* from Australia. The orb is more or less horizontal, but not entirely in a single plane. Barrier-web threads connect the orb to nearby orbs of other individuals. Note the zigzag outer loops of sticky spiral.

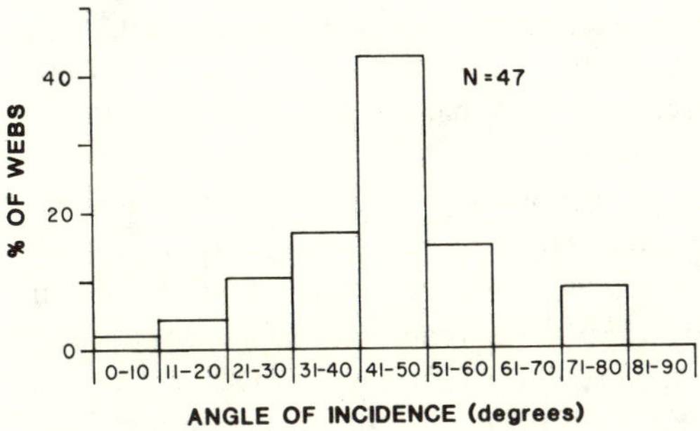

Fig. 6.10. Frequency distribution of inclinations (angles of incidence to horizontal) in webs of adult females of *Philoponella republicana* in a single colony on Barro Colorado Island, Panama.

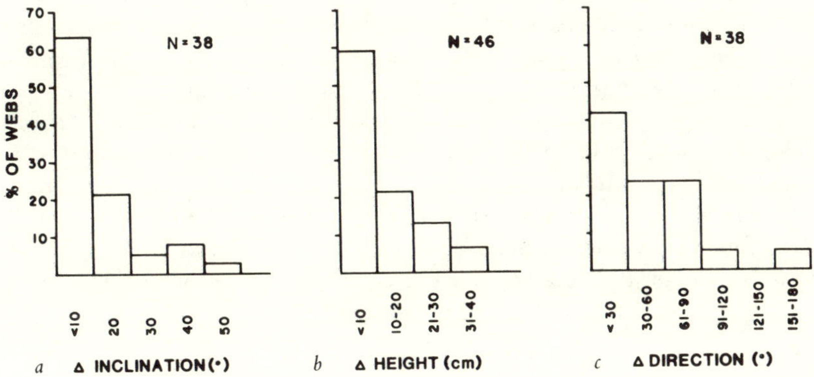

Fig. 6.11. Frequency distributions of daily changes in (*a*) inclination (angle of incidence to horizontal), (*b*) height above ground, and (*c*) compass direction of webs of adult females of *Philoponella republicana* in a single colony, measured on 18 dates in October 1973.

cies, which may be constructed at various angles to horizontal. *Zosis geniculatus* webs in colonies were rarely horizontal; they were usually inclined or nearly vertical (Marples 1962; pers. obs.). This was also true of webs of *Philoponella* sp. B and *P. undulata* that were in commensal associations with other web builders. *Philoponella republicana* orbs in colonies were most often inclined at angles of 40° to 60° (Fig. 6.10). Within colonies, spiders often changed the inclination, height above ground, and compass orientation of their orbs from day to day (Fig. 6.11). This suggests that intraspecific interactions during web construction and jockeying for more favorable positions in the colony may result in repeated adjustments of web position.

Increased prey-capture rates and improved defenses against predators and parasites are often cited as advantages in theoretical discussions of the pros and cons of coloniality (e.g., Buskirk 1975b, 1981; Lanham and Eberhard 1979; Lubin 1974). Increased prey-capture rates may be due to the larger volume and surface area of the colonial web as compared to that of a solitary web and to the ability of spiders in colonies to build webs spanning large gaps that often serve as insect flight paths. Improved defenses are due to the colony web matrix, which acts as a protective barrier and early warning system against predators and parasites. In addition, an intruder may become confused by the defensive or escape responses of so many spiders and may be unable to follow a single individual. Disadvantages concomitant with the greater visibility of colonial webs include the attraction of predators and parasites to a predictable resource and the warning off of potential insect prey (Buskirk 1975b, 1981; Lubin 1974). The balance of pros and cons may vary with habitat, season, or other factors. Although solitary and colonial individuals of *P. semiplumosa* caught the same amount of prey, solitary individuals built larger webs than did colonial ones (Lahman and Eberhard 1979). Solitary individuals also built stabilimenta more frequently than did colonial ones, which suggests that colonial individuals derive extra protection from the colony (it being assumed that stabilimenta function in defense against predators; see the discussion of stabilimenta). *Philoponella republicana* individuals in large colonies had a higher probability of producing egg sacs than did spiders in small colonies (Lubin 1980a). All of these findings indicate that living in colonies is distinctly advantageous. Nonetheless, Smith (1982b) found that colonial individuals of *P. oweni* suffered higher rates of egg-sac parasitism than did solitary individuals; in the colonial spiders this was compensated for by greater egg production.

Reduced Webs

Webs of *Polenecia* (formerly included in *Sybota*) are vertical and highly asymmetrical. The webs consist of a hub, which is on a branch, temporary spiral, and radii; sticky cribellate silk is laid on the radii and frame threads and there is no true sticky spiral (Wiehle 1931; H. Peters, pers. comm.). *Hyptiotes* construct vertical triangle webs, consisting of four radii attached to a single resting thread and cribellate-silk threads connecting the radii and, to some extent, laid along the radii as well (McCook 1889; Comstock 1912; Wiehle 1927, 1931; Marples and Marples 1937; Peters 1938). *Miagrammopes* (Opell 1984b) build webs consisting of one or more threads with sticky silk on them attached to a nonsticky resting thread (Akerman 1932; Lubin, Eberhard, and Montgomery 1978). Both *Polenecia* and *Hyptiotes* attach their egg sacs to a twig (Opell 1979), whereas *Miagrammopes* build tubular egg sacs similar to those of *Philoponella undulata* and *Uloborus bispiralis*, which they hang in a rudimentary orb web with or without a jagged spiral of sticky silk (Lubin, Eberhard, and Montgomery 1978). Morphologically, *Polenecia* and *Hyptiotes* are closely related (Opell 1979), and it is possible that web reduction occurred separately in this line and in *Miagrammopes*. Little is known,

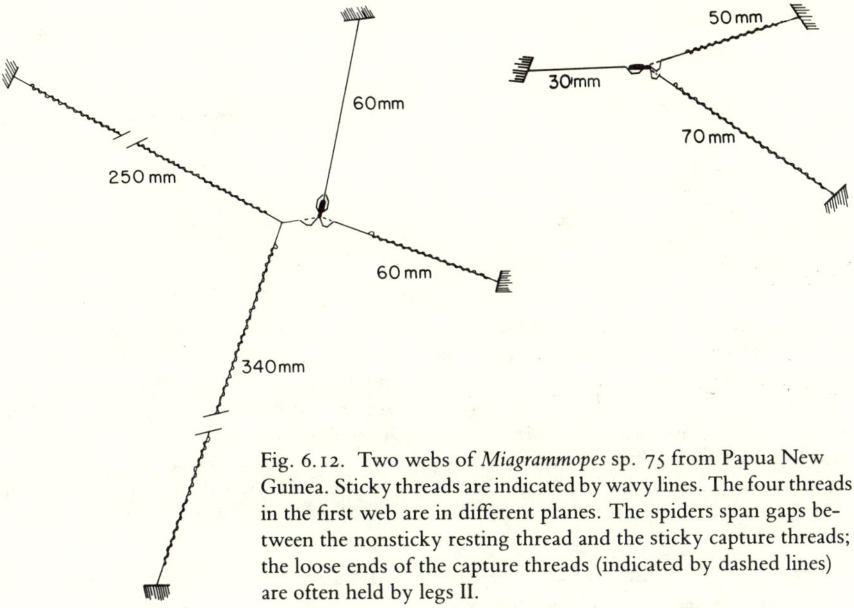

Fig. 6.12. Two webs of *Miagrammopes* sp. 75 from Papua New Guinea. Sticky threads are indicated by wavy lines. The four threads in the first web are in different planes. The spiders span gaps between the nonsticky resting thread and the sticky capture threads; the loose ends of the capture threads (indicated by dashed lines) are often held by legs II.

however, about webs and web-building behavior in *Polenecia*, and the webs of many other uloborid genera remain undescribed (Opell 1979; see Table 6.1), so that statements about the evolutionary pathways leading to web reduction in uloboridae are premature.

Characteristic of webs of both *Hyptiotes* and *Miagrammopes* is the use of a single resting thread that is held under tension and manipulated by the spider during prey capture (see below). Webs of six species of *Hyptiotes* have been described, and all appear to have the same structure (Opell 1982b). This contrasts markedly with *Miagrammopes* webs, which vary in both the number and the arrangement of sticky threads, although all apparently function on the same principle. Six species of *Miagrammopes* from Central and South America (Lubin, Eberhard, and Montgomery 1978), one from South Africa (Akerman 1932), and at least three species from Papua New Guinea (pers. obs.) had webs ranging in structure from a single resting and sticky thread combined (*Miagrammopes* sp. from South Africa and *Miagrammopes* sp. 75 from Papua New Guinea) to one to six sticky capture threads of variable length and orientation, often in different planes, attached to different points on a single resting thread (Fig. 6.12). The latter webs are, in effect, three-dimensional arrays of sticky threads, not unlike the webs of certain Theridiidae (e.g., some *Theridion* and *Achaearanea* species: pers. obs.) and Theridiosomatidae (Coddington, this volume).

Web building in *Hyptiotes* includes some behaviors similar to ones used in normal uloborid orb-web construction and others that are apparently unique

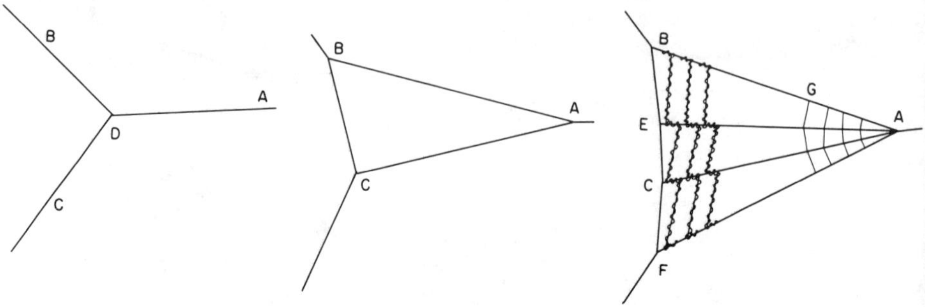

Fig. 6.13. Web construction in *Hyptiotes paradoxus* from England. Beginning with three threads intersecting at *D*, two radii are formed (*AB* and *AC*) that define the triangle apex (*A*). Two further radii are laid: *AE*, formed by going out on *AB*, and *AF*, formed by going out on *AC*. The nonsticky temporary spirals (*A–G*) are constructed centrifugally, starting at *A*; the sticky spiral is constructed centripetally, starting at the frame thread (at *B*). After each spiral turn (nonsticky or sticky), the spider returns to the apex. Note that sticky thread is laid both on and between radii, which gives the spiral a zigzag appearance. This was also observed in webs of *Hyptiotes cavatus* in North America (pers. obs.). After Marples and Marples 1937.

(Fig. 6.13) (Marples and Marples 1937). The single "frame thread" and four radii are constructed in much the same way as those of a normal orb web. Temporary spiral threads are laid from the apex of the triangle outward (there is no hub), but after laying each spiral loop the spider returns to the apex. Sticky threads, laid centripetally, start on a radius or on the frame thread and follow a zigzag path composed of alternating radial and interradial segments. Web construction has not been observed in detail in *Miagrammopes*, but it appears that sticky silk is laid directly on already existing nonsticky threads (Lubin, Eberhard, and Montgomery 1978), which suggests that these nonsticky baselines of *Miagrammopes* webs are homologous with radii or frame threads.

Prey Capture

Uloborids follow a relatively invariable and stereotyped sequence of predatory behaviors (Marples 1962; Opell 1979; Eberhard 1967; Lubin, Eberhard, and Montgomery 1978; Lubin et al. 1982). All uloborids attack and immobilize insects caught in their webs by wrapping them in silk. Before attacking, spiders may pull on threads in order to locate prey in the webs—possibly by its weight or by the web vibrations that it produces—and manipulate the tension in webs by jerking threads in such a way that the prey is further entangled in sticky web elements. In some uloborids, such as *Miagrammopes* and *Hyptiotes*, these movements are conspicuous and involve repeated jerking and sagging of the web (Akerman 1932; Marples and Marples 1937;

a

b

Fig. 6.14. Prey wrapping in *Uloborus conus*. *a*: initial immobilization wrap. The spider throws sheets of silk from below onto a small ant. The radius has already been severed on the proximal side (towards the hub) and the spider holds it with right leg I. *b*: postimmobilization wrap. The sticky spiral attachments to the prey have been cut and a new line connects the prey to the proximal end of the broken radius (*arrow*). The radius is still attached distally. The spider rotate-wraps the insect, using legs I–III to turn it and legs IV to apply silk.

TABLE 6.4

Predatory Behaviors of Philoponella undulata and Zosis geniculatus with Four Kinds of Prey

	Prey			
	Ants[a] (2.1 mg)	Fruit flies (5.8 ± 0.9 mg)	Blowflies (34.7 ± 0.9 mg)	Grasshopper nymphs (32.1 ± 5.8 mg)
P. undulata (22.2 ± 6.3 mg)				
WACS (min)	3.9 (49.4)	3.8 (75.3)	7.0 (34.3)	41.0 (76.4)
WAFS (min)	3.4 (108.2, $N = 5$)	4.5 (52.6, $N = 9$)	47.7 (54.8, $N = 5$)	47.7 (54.8, $N = 5$)
Total capture time (min)	7.5 (44.6)	8.7 (25.6)	40.4 (34.1)	65.2 (53.7)
Total silk used (m)	not calculated	18.0	87.1	192.3
Z. geniculatus (41.4 ± 5.1 mg)				
WACS (min)	2.9 (49.1)	4.0 (41.7)	8.0 (31.5)	19.4 (82.2)
WAFS (min)	not calculated[b]	8.1 (53.6)	39.3 (33.7)	74.2 (54.4, $N = 9$)
Total capture time (min)	5.3 (82.2)	12.3 (33.6)	47.6 (29.8)	86.5 (47.9)
Total silk use (m)	not calculated	10.2	39.6	82.1

NOTE: The average durations of the wrap at the capture site (WACS) and of the wrap at the feeding site (WAFS), the total capture time, and the total length of silk used in wrapping (WACS and WAFS combined) were calculated according to the methods described in the text. $N = 10$ trials, except where noted otherwise. Coefficients of variation (%) are given in parentheses. Average body weights of spiders and prey (± one standard deviation) are shown ($N = 5$). The average weight of *Anopolepis* ants was obtained by weighing ten ants together and dividing by 10.

[a] Most ants were left at the capture site; wrap durations are thus incomplete.

[b] Only four spiders wrapped at the feeding site. Wrap durations were 1.2–8.7 minutes.

TABLE 6.5

Average Duration of the Wrap at the Capture Site Given to Each of Five Fruit Flies Offered in Succession to Zosis geniculatus

Prey offered	Duration (min)	Coefficient of variation (%)	Range (min)
First			
WACS	2.9	116.1	0.7–12.1
WAFS	5.2 ($N = 9$)	47.4	3.2–8.5
Second	5.6	155.1	0.9–29.1
Third	2.7	71.8	1.1–7.0
Fourth	2.9	68.3	0.7–6.4
Fifth	2.8	81.1	1.1–8.5

NOTE: The average durations of the wrap at the capture site (WACS) and of the wrap at the feeding site (WAFS) were calculated according to the methods described in the text. Second to fifth prey were wrapped and left at the capture site. In three trials, however, second prey were wrapped together with the first prey and carried to the hub; in these instances only the immobilization wrap of the second prey is given (before the combined wrap with the first prey). $N = 10$ trials, except where noted otherwise.

Lubin, Eberhard, and Montgomery 1978; Opell 1982b). Before wrapping, the spider may collapse part of the web onto the prey (*Hyptiotes*), apply wadded-up hackled silk to the prey (*Miagrammopes*), or cut radii or sticky spiral loops above the prey so that it falls into sticky threads below (*Uloborus*, *Zosis*, and *Philoponella* species with vertical or slanted orbs). Large insects, or insects that struggle vigorously, are wrapped at first from a distance. The spider faces away from the prey and throws sheets of silk in a posterior direction onto the insect, using specialized combs of setae on the tarsi of legs IV to draw out the silk and apply it to the prey. Then the spider moves closer, cuts radial or spiral elements attached to the prey, and wraps with legs II and III (Fig. 6.14a). Usually the prey is cut entirely free from the web and, while being wrapped, is held by the spider in legs III and rotated with legs II. While wrapping, the spider holds the ends of the broken radius or radii with legs I and spans the gap with its body (Fig. 6.14b). Wrapped insects are transported to the feeding site (the hub or a resting thread) held over the cephalothorax in the palps in a manner characteristic of all uloborids and not observed in other spiders. Often a second wrap takes place at the feeding site. While feeding, the spider holds the prey package in the chelicerae and covers it with digestive enzymes that appear to digest the prey contents, leaving the prey exoskeleton and the wrapping silk intact (Peters 1982). The prey is rewrapped repeatedly during the feeding stage.

Although the basic sequence of behaviors is similar in all uloborids examined so far, there are differences that may be related to differences in web structure, web location, or prey type. Prey-capture behavior has been observed in detail in uloborids of several genera. The details presented here are derived from observations of *Miagrammopes* spp. (Lubin, Eberhard, and Montgomery 1978), *Hyptiotes* spp. (Opell 1982b; Marples and Marples 1937; Marples 1962; Comstock 1912; pers. obs.), *Uloborus conus* (Lubin et al. 1982), *Uloborus diversus* (Eberhard 1967), and *Lubinella morobensis*, *Philoponella undulata*, and *Zosis geniculatus* (this study). Prey-capture behavior of *L. morobensis* was observed with five fruit flies (Drosophiloidea), three houseflies (Muscoidea), and two ants (*Anopolepis* and unidentified Dolichoderinae) as prey. *Philoponella undulata* and *Z. geniculatus* were fed houseflies, blowflies (Calliphoridae), grasshopper nymphs (Acrididae), fruit flies, and ants (*Anopolepis*). Data on durations of prey-capture behaviors of *P. undulata* and *Z. geniculatus* are summarized in Tables 6.4 and 6.5.

Manipulation of the Web

When an insect strikes an orb web of *P. undulata* or *Z. geniculatus*, the spider turns at the hub to face the prey and jerks the web sharply by rapidly flexing legs I and IV (and perhaps II). The jerk is often followed by one or more gentle tugs at the radii with legs I and II. It is tempting to speculate that the initial jerk helps to entangle the prey in the web, whereas the tugs inform the spider of the exact location and perhaps of other characteristics of the prey. Sometimes the spider pauses to tug a radius as it approaches the prey, particularly if the prey remains immobile in the web.

Miagrammopes and *Hyptiotes* jerk the web with legs I (and perhaps II) and also sag the web by releasing a loose bundle of silk held anterior to legs IV. The latter behavior apparently has no counterpart in the behavioral repertoires of orb-spinning uloborids. Both *Miagrammopes* and *Hyptiotes* tense their webs by backing up and reeling in the resting thread with legs IV, and it is this loose thread that is released to sag the web. It is difficult to imagine whence this behavior is derived, since no other uloborids are known to manipulate web tension with legs IV. However, when females of *P. undulata*, *L. morobensis*, some *Uloborus*, and *Miagrammopes* rest in a cryptic posture in line with their egg sacs, they use legs IV to hold the egg sac or a radius attached to the egg sac (Fig. 6.8). Both *Miagrammopes* and *Hyptiotes* use legs IV to pull in the resting thread so as to position themselves in a cryptic posture at the tip of a leaf or twig, whether or not they have a capture web. The same leg movements may later have become important in adjusting the tension of the web. See Opell and Eberhard 1983 for a survey of resting postures in uloborids.

Another type of web manipulation involves rapid movement of thread attachments and replacement of the threads as the spider approaches prey in the web or returns to the hub or feeding site (Marples 1962). *Miagrammopes* reel in the capture thread above the prey as they approach to attack, and again as they return to the resting thread, each time replacing it with a new dragline. They also change attachment points of capture threads to the resting thread when switching positions on the resting thread (Lubin, Eberhard, and Montgomery 1978). *Hyptiotes* perform a complex series of maneuvers on the web (Marples and Marples 1937). Orb-weaving uloborids sometimes manipulate the dragline the same way as *Miagrammopes*, when returning on it to the hub of the orb. In most cases the spider doubles the dragline on the return trip, but *Z. geniculatus* and *L. morobensis* were both seen to roll up and replace the dragline. In dragline replacement, the wadded-up silk is either attached to the web (as at the apex of the triangle in *Hyptiotes*) or transferred to the mouthparts and presumably eaten (as in *Z. geniculatus* and *L. morobensis*).

Wrapping Behavior

After immobilizing the prey by wrapping it in silk and cutting it free from the web, the spider may continue to wrap the prey at the capture site, holding it free from the web in legs II and III, or it may carry the prey to the resting or feeding site and perform the same postimmobilization wrap there, or it may wrap the prey in both places. When postimmobilization wrapping occurs at the capture site, it follows the immobilization wrap as part of a single sequence. The cues that determine where the postimmobilization wrap occurs are not known. *Miagrammopes* omit postimmobilization wrapping at the capture site when the sticky line on which the insect is caught is to be discarded, i.e., when the prey is a large insect, or the spider already has one or more prey items in its palps (Lubin, Eberhard, and Montgomery 1978). On the one hand, wrapping at the capture site may be more awkward and more dangerous, because of exposure to predators, than at the hub or resting site; on the other hand, a large or struggling insect may escape if not given a long wrap at the capture site.

The duration of postimmobilization wrapping at the capture site in *Nephila clavipes* (Araneidae) is influenced by both the mass of the prey and its bulk (or by the presence of projecting body parts such as wings and legs), and probably functions to facilitate prey removal by trussing projecting appendages and to reduce the size of the prey package to be transported back to the hub (Robinson and Mirick 1971). Both *P. undulata* and *Z. geniculatus* gave blowflies (average weight 34.7 mg) a significantly longer wrap at the capture site (WACS) than fruit flies (5.8 mg) or ants (2.1 mg) (*P. undulata*: $p < 0.01$; *Z. geniculatus*: $p < 0.001$; *t*-tests), which suggests that prey mass may influence wrap duration in uloborids as well. There is evidence, however, that cues other than prey mass may also influence wrap duration. Fruit flies and ants were given the same-duration WACS, even though the flies weighed about twice as much as the ants. Furthermore, grasshopper nymphs (32.1 mg) were wrapped on the average two to six times longer than blowflies of approximately the same weight (*P. undulata*: $p < 0.01$; *Z. geniculatus*: $p < 0.02$; *t*-tests). These behaviors might be explained as follows.

1. *Anopolepis* ants secrete defensive chemicals that may repel spiders. When ants were presented to the spiders tested, in four out of ten trials with *P. undulata*, and in five out of ten trials with *Z. geniculatus*, spiders interrupted the prey-capture sequence to return to the hub, leaving the ants wrapped in the web. Possibly the spiders used more silk (i.e., a longer wrap duration) to reduce contact with the offensive secretion.

2. Grasshopper nymphs struggled more and were both more dangerous and less compact in body shape than blowflies. They were also less familiar as prey to both uloborid species tested.

There was no apparent correlation between the duration of WACS and the duration of the wrap at the feeding site (WAFS); i.e., a long WACS may or may not have been followed by a long WAFS. The WAFS tended to be longer than the WACS (*Z. geniculatus* and *P. undulata*: $p < 0.001$ for blow flies, $p < 0.02$ for fruit flies; *Z. geniculatus*: $p < 0.01$ for grasshoppers; *P. undulata*: $p < 0.01$ for ants; paired *t*-tests). WAFS durations were significantly longer for blowflies than for fruit flies (*Z. geniculatus* and *P. undulata*: $p < 0.001$; *t*-tests), which suggests that prey size or mass may determine the duration of the WAFS as well as the duration of the WACS.

All uloborids observed so far, including *Hyptiotes* and *Miagrammopes*, hang with legs I from a broken radius or dragline thread during postimmobilization wrapping at the capture site, and sometimes at the hub or feeding site as well. *Z. geniculatus*, *L. morobensis*, and *U. conus* tended to wrap prey on a short dragline thread beneath the hub, usually stretching from the hub to a radius at the edge of the hub. *Zosis geniculatus* sometimes put in a thread below the plane of a slanted orb, stretching from the hub to a radius above the hub, and wrapped hanging from this dragline. *Philoponella undulata* wrapped either at the hub or hanging from a signal thread.

Handling Multiple Prey

Unlike most araneids, uloborids do not leave the first prey at the hub while attacking another insect on the web. On one occasion only, an adult female *U. conus* was seen to attach the first prey item to the hub by means of a dragline thread before going out to attack the second prey (Lubin et al. 1982). When a spider feeding at the hub or on a resting thread is presented with another insect in its web, it transfers the first prey to the palps and carries it held over the cephalothorax while attacking the second prey. By carrying the first prey in the palps, the spider frees its mouthparts for cutting threads. Successive prey items are attacked by immobilization wrapping, in the usual manner, cut free from sticky spiral elements, and wrapped while held in legs II and III. Sometimes radius attachments on one or both sides of the prey are broken, and the prey may be given a short postimmobilization wrap at the capture site. The spider then reattaches the broken radius to the prey, first distally (on the side away from the hub) and then proximally, and returns to the hub or the resting thread to resume feeding on the first prey, leaving the second prey in the web.

With the exception of *Miagrammopes*, all uloborids in which prey capture has been observed can store successive prey items in the web. *Miagrammopes* wrap successive prey items with the previous ones, and carry the combined prey package back to the resting thread, where it is rewrapped. This places an upper limit on the number of insects that can be attacked in a given time span (Lubin, Eberhard, and Montgomery 1978). Other uloborids (e.g., *Z. geniculatus*; see Table 6.5) also occasionally wrap the second prey with the first, transporting them both back to the feeding site.

Storage of prey in the web presents several problems. First, since insects are immobilized solely by wrapping, some insects might be able to struggle free and escape. There was no significant difference in the durations of the wraps at the capture site given by *Z. geniculatus* to the first fruit fly and the other four fruit flies of five fruit flies offered in succession (Table 6.5; F_s = 0.76, analysis of variance). The spider, therefore, does not seem to use any extra restraint on insects that are to be left wrapped in the web. On several occasions, however, *Z. geniculatus* ran down to rewrap insects that struggled after having been left at the capture site.

Second, insects left in the web might be stolen by kleptoparasites. *Portia fimbriata*, a salticid that is sometimes kleptoparasitic (Wanless 1978; Jackson and Blest 1982a; Jackson, this volume), was observed to remove wrapped prey from a web of *Z. geniculatus* in Papua New Guinea (pers. obs.), and kleptoparasitic emisine bugs (Hemiptera: Reduvioidea) took prey from colonial webs of *Philoponella republicana* in Panama (Lubin 1980a). *Argyrodes* spp. (Theridiidae) do not build webs of their own and are often kleptoparasitic in webs of other species (e.g., Vollrath 1976b, 1979a). Trail (1980) found two species of *Argyrodes* feeding on eggs, young, and adults of *Philoponella oweni* in their webs; they were found more often in colonial than in solitary webs of *P. oweni*. These observations indicate that *Argyrodes* may behave as predators of spiders as well as kleptoparasites. *Argyrodes* spp. were observed in Papua New Guinea in or near the barrier-web threads of webs of *Philoponella undulata* and *Philoponella* sp. B, particularly when these spiders built their webs in association with webs of *Achaearanea* (Theridiidae), *Psechrus* (Psechridae), *Cyrtophora* (Araneidae), or Agelenidae. Prey theft from *Philoponella* webs was not seen.

Finally, one must ask how the spider at the hub keeps track of prey left in the web. Web-building uloborids and araneids do not generally respond to nonmoving objects in their webs (Robinson and Olazarri 1971), and one wonders how the spider distinguishes immobile, wrapped prey in the web from other objects such as debris.

Economy of Silk and Energetic Considerations

The energetic cost of building a web, C, includes the caloric value of the silk, S, the metabolic cost of web construction, M, and the metabolic cost of silk production, M_s (Prestwich 1977):

$$C = S + M + M_s$$

It has been argued that the metabolic cost of synthesizing silk is small relative to the energetic content of the silk and the cost of construction (Prestwich 1977), so that the cost of web building reduces to

$$C = S + M$$

In considering the total energetic cost of web building, five other factors must be taken into account: (1) the fraction of silk of the old web that is eaten and recycled, (2) the frequency of web renewal, (3) the proportion of the web that is repaired, (4) the frequency of repairs, and (5) the amount of time spent in web renewal and repair. Thus, the total energetic cost of web building will be

$$C = [M + (S - S_r)]p + [M + (S' - S_r')]q$$

where M is the metabolic cost of construction; S and S_r are, respectively, the energetic content of the total silk and the energetic content of the portion of silk recycled during web renewal; S' and S_r' are, respectively, the energetic content of the total silk and the energetic content of the portion recycled during web repair; and p and q are, respectively, the frequency of web renewal and the frequency of web repair. The metabolic cost of web building is proportional to the time spent in web construction, so that $M = M_a t$, where M_a is the metabolic cost of activity and t the time of activity (web building or repair).

The cost of web construction has been estimated in only a few species (Lubin 1973; Peakall and Witt 1976; Ford 1977; Prestwich 1977). None of these was an uloborid species. Neither metabolic rates nor caloric values of silk are known for any uloborid species. There are, however, a number of intuitive reasons to expect that web building in orb-weaving uloborids is energetically more expensive than in araneids of comparable size.

First, though the weights of equal-sized uloborid and araneid webs are not available for comparison, cribellate orb webs probably contain more silk than similar-sized araneid orb webs. There are two lines of evidence that suggest that this might be the case.

1. The orb webs of *Z. geniculatus* contained almost twice the length of silk per milligram body weight found in those of the araneid orb weaver *Neoscona domiciliorum*, although *N. domiciliorum* was more than twice the weight of *Z. geniculatus* (Table 6.6). Furthermore, the total length of silk used by *Z. geniculatus* in web construction was considerably underestimated because (*a*) the hub silk was not included in the calculations (see Methods) and (*b*) the cribellate-silk portion of the sticky spiral was underestimated, since it is at least 1.5 times the length of the baseline strands (Opell 1982a).

2. Several thousand cribellate-silk glands are involved in the production of fine threads, each only 200 to 300 Å in diameter (Friedrich and Langer 1969), to yield the sticky component of the uloborid hackled spiral (Kovoor 1977a), whereas in araneid orb weavers two pairs of aggregate glands produce a nonfibrous mucin that is the sticky spiral "glue" component (Peters 1955a; Kovoor 1972). This glue is probably produced in smaller quantities, and less costly to produce metabolically, than the highly organized, multiple-strand cribellate fibers.

TABLE 6.6
Measurements of Silk in Webs of Two Similar-Sized Orb Weavers: Zosis geniculatus, an Uloborid, and Neoscona domiciliorum, an Araneid

		N. domiciliorum	
Silk measurement	*Z. geniculatus* (31.5 ± 10.0 mg)	Sample A (66.7 ± 23.9 mg)	Sample B (94.7 ± 23.4 mg)
Number of radii	36.4 ± 7.9	27.0 ± 3.2	27.3 ± 3.3
Number of spirals	27.4 ± 5.8	no data	no data
Spiral area (cm^2)	870.4 ± 297.0	440.9 ± 123.6	491.5 ± 157.7
Mesh size (cm^2)	0.87 ± 0.23	0.55 ± 0.16	0.52 ± 0.13
Total thread length (m)	17.1 ± 5.5	18.6 ± 5.9	21.3 ± 6.1
Thread length/ body weight (m/mg)	0.54	0.28	0.22

SOURCES: *Z. geniculatus*: this study; *N. domiciliorum*: Risch 1977.

NOTE: Measurements and calculations were performed according to the methods described in the text. $N = 5$ webs for *Z. geniculatus*; $N = 30$ webs each for samples A and B of *N. domiciliorum* (samples A and B were from different locations). Body weights of spiders are given in parentheses.

Second, araneid orb weavers tend to eat their old webs before building new ones, recycling 92 to 96 percent of the silk protein (e.g., *Araneus diadematus*: Peakall 1971). Araneid orb webs that are not ingested include long-lived, nonsticky webs, e.g., those of *Cyrtophora* and *Mecynogea* (Lubin 1973) and of *Metazygia gregalis* (Eberhard, this volume), and webs of several species of coastal araneids (on beaches or in mangroves) that are exposed to salt spray (pers. obs.). The situation in uloborids is more complicated. *Uloborus diversus* eats relatively small accumulations of silk in the process of building a new web (e.g., the "protohub" silk), but throws out the old silk from portions of orbs being repaired (Eberhard 1972b). *Philoponella undulata* and *Z. geniculatus* behave similarly. Furthermore, the large amount of silk used by uloborids to wrap prey (Table 6.4) may be lost when the empty prey package is discarded (Peters 1982; W. G. Eberhard, pers. comm.). Thus, at least some uloborids lose a large proportion of both the web and the wrapping silk. *Uloborus conus*, however, appears to ingest the entire old web before building a new one.

Third, web construction in uloborids is a long and laborious process. One adult *U. conus* spun 11 to 16 sticky spiral loops (switchbacks account for the variable number) in 3.1 hours. The outermost loop, which had a radius of about 15 cm and was approximately 94 cm long, took 13 minutes to complete, yielding a sticky spiral spinning rate of 7.2 cm per minute. The total time required to build a web is longer for a uloborid than for an araneid: approximately 3.5 hours for *U. conus* compared with about 30 minutes for *Zygiella x-notata* (Witt, Reed, and Peakall 1968), both of which renew their orbs daily. I predict, furthermore, that the metabolic cost of activity

TABLE 6.7
Silk Costs and Web Energetics of Some Uloborid Spiders

Category	*Philoponella undulata, Zoris geniculatus, Uloborus diversus*	*Philoponella republicana*	*Uloborus conus*
Web type	orb ± barrier web	orb + barrier web, colonial	orb + cone
Silk cost[a]			
S	high	moderate	high
S_r/S	<1	<1	≃1
Web construction			
Duration	long (>1 hr)	moderate (≤1 hr)	long (>3 hr)
Renewal frequency p	<1/day	≤1/day	1/day
Repair frequency q	≃1/day	?	0
Stabilimenta	frequent	frequent	rare

Category	*Hyptiotes* spp.	*Miagrammopes* spp.
Web type	triangle	one to several threads
Silk cost[a]		
S	low	low
S_r/S	≃1	≤1
Web construction		
Duration	short	very short (<1 hr)
Renewal frequency p	1/day (?)	>1/day
Repair frequency q	0	0
Stabilimenta	none	none

[a] S = energetic content of silk of entire web; S_r = energetic content of recycled silk.

(M_a) is higher in uloborids, because cribellate silk has to be combed out with repeated movements of legs IV. Each combing movement involves a trip of both legs IV to and from the spinnerets, and there are several such movements for each step the spider takes on the web. By contrast, in araneids the sticky spiral silk is apparently drawn out of the spinnerets simply by the forward movement of the spider, legs IV being used only to hold the thread or to extend it before making an attachment (Eberhard 1981).

Like araneids, uloborids have evolved different strategies of web construction and use (see above), and these in turn have profound effects on the spiders' energy budgets. Some of the different web strategies used by uloborids are summarized and compared in Table 6.7. The orb weavers *P. undulata*, *Zosis geniculatus*, and *U. diversus* have a high investment in their webs, in terms of both silk content and construction time. Such webs are renewed infrequently and tend to be repaired or renewed in sections. A similar web strategy is found in araneids such as *Nephila*, which renew only parts of their large orbs daily, and *Cyrtophora*, which repair holes in their orbs resulting from prey capture but renew entire webs only infrequently (Lubin 1973). To reduce mechanical damage, orb weavers that build high-investment webs may build in sheltered places and with the orb plane parallel to prevailing winds (Langer 1969). The colonial orb webs of *P. republicana* are also high-investment webs, but the cost of web building may be reduced by the sharing of support threads. Commensal and colonial habits may provide increased protection both from mechanical damage and from predators, and thus enable spiders to build in more exposed locations.

Uloborus conus and *Miagrammopes* occupy opposite ends of the web-energetics spectrum. High-investment, frequently renewed webs such as those of *U. conus* may be efficient only in locations with abundant and predictable food resources, e.g., tropical rain-forest understory. Note that the major energetic cost of these webs is the metabolic cost of web building (M), since the web silk is probably entirely recycled ($S_r/S \simeq 1$). Webs of *Miagrammopes*, and to a lesser extent of *Hyptiotes*, are low-investment, renewable traps. *Miagrammopes* may spin several new capture threads a day (or a night), and capture threads are readily discarded when the spider has as many prey as it can handle, or when a distasteful insect is caught in the thread. Arachnologists have puzzled over the question of the effectiveness of one- or few-thread webs as insect traps. It may be that, per unit length of capture thread, webs with one or few lines trap more insects than multi-thread orb webs, and that few-strand webs have the advantage of being less detectable both by prey and by predators and less costly to construct (Lubin and Dorugl 1982; Chacón and Eberhard 1980).

Finally, one must consider the protective function of the web, an important factor in web energetics, but one that is difficult to measure quantitatively. Most uloborids are well camouflaged (indeed, unlike araneids, uloborids are never brightly colored or aposematic), and many adopt cryptic resting postures during the day, on or off the web (Opell and Eberhard 1983; Lubin et al. 1982). Furthermore, species with high-investment, long-lived webs tend to build stabilimenta, and, as discussed earlier, there is some evidence for correlation between stabilimentum building and exposure to visually orienting predators (Eberhard 1973; Lahman and Eberhard 1981). Barrier webs and cones are probably primarily antipredatory in function. Thus, species with high-investment webs have special defensive structures, whereas species with reduced, short-lived webs such as *Miagrammopes* and *Hyptiotes* have evolved cryptic body forms and resting postures (Akerman 1932; Lubin, Eberhard, and Montgomery 1978; Marples and Marples 1937).

Summary

Although uloborids characteristically spin orb webs, many modifications may be found involving either elaboration or reduction of the basic orb-web structure. The diversity of web adaptations of uloborid spiders parallels that of the better-known araneid orb weavers. Unlike araneids, uloborids often place sticky-silk lines over nonsticky radii, frame threads, and barrier-web threads, as well as in sticky spirals. This trait provides uloborids with a certain flexibility not found in araneids, allowing spiders to build zigzag loops of sticky spiral, sticky supra- and sub-orb structures (orb-plus-cone webs and barrier webs), and sticky-thread egg-sac webs, as well as the switchback loops of sticky spiral that are also found in araneid webs.

Many of these orb-web modifications may be seen as adaptations to reduced building space, such as is encountered in tree buttresses, under rocky overhangs, or beneath leaves in forest understory. In other instances, predation pressure may have selected for three-dimensional webs such as orb-plus-cone webs, egg-sac webs, and colonial or commensal webs. Defense against predators has also selected for an astonishing array of stabilimentum types, the form of stabilimentum being highly variable between species, within species, and even within different webs of the same individual. In species with a high degree of variability in stabilimentum type, the factors that determine which stabilimentum form, and which web form, will be constructed by a given individual at any given time, remain largely unknown. Protection from predators, the type of prey available, and wind force and direction may all play important roles.

Predation pressure may have also resulted in highly cryptic defensive postures and reduced webs—single- or few-thread webs and triangle webs—such as are found in *Miagrammopes, Hyptiotes*, and *Polenecia*.

Prey-capture behavior is, on the whole, more conservative (less variable) in uloborids than in araneids. All insects are attacked in the same manner—they are wrapped in large volumes of silk—and are transported to the hub or a resting thread in the pedipalps. The basic similarity of predatory behaviors in species with diverse web types may be a result of the lack of poison glands in the Uloboridae, and thus reliance on wrap attacking. The lack of poison glands may also restrict uloborids to handling only relatively small prey. The variability that does occur in uloborid prey-capture behavior is associated with the manipulation of the web upon prey impact and the handling of more than one prey at a time, differences that are clearly related to web structure.

The energetic cost of web building may be determined by measuring (1) the cost of the silk itself, (2) the metabolic cost of web construction, and (3) the metabolic cost of silk synthesis. (The second factor includes considerations of time as well as of energy output.) These costs have not been measured in any uloborid spider (and have only been estimated in certain other web-building species). Nonetheless, from our knowledge of uloborid web structures and of the characteristics of uloborid silk, we may predict that certain combinations of web-related characters will represent specific predatory strategies adapted to specific conditions. Three such strategies are: (1) large, high-investment webs with low frequencies of renewal, located in relatively exposed habitats, (2) small, high-investment webs with high frequencies of renewal, located in protected, prey-rich habitats, and (3) small, low-investment webs with high frequencies of renewal, located in protected, but not necessarily prey-rich, habitats. The first strategy is exemplified by the webs of *Z. geniculatus* and of many species of *Philoponella*, especially colonial and commensal species, the second strategy by orb-plus-cone webs and the webs of *Philoponella* sp. G and the third strategy by the reduced webs of *Miagrammopes* and *Hyptiotes*.

Acknowledgments

The Smithsonian Institution and Smithsonian Tropical Research Institute generously supported most of the research reported here, through an STRI postdoctoral fellowship and through grants to Dr. M. H. Robinson. The work in Papua New Guinea was done at the Wau Ecology Institute, where Dr. A. Allison, H. Sakulas, and the late Dr. J. L. Gressitt provided logistic support, pleasant accommodation, and good company throughout the 2-year study. The Charles Darwin Research Station supported my work on the Galápagos islands. Dr. B. D. Opell identified numerous specimens, reviewed drafts of the manuscript, and, with Dr. W. G. Eberhard, provided stimulating discussions and the impetus to assemble this review. The editor improved the manuscript considerably. I most gratefully acknowledge the help of all of these people and institutions.

7

Social Spider Webs, with Special Reference to the Web of Mallos gregalis

WILLIAM JAMES TIETJEN

ALTHOUGH MOST SPIDERS live a solitary existence, forming aggregations only early in life or during mating, a few species can be described as social. A number of reviews of the literature on social phenomena among spiders have been written (Burgess 1978; Kaston 1965; Kullmann 1972b; Shear 1970a; Breskirk 1981), but none addresses the structure and function of communal webs in a comparative way. In this chapter I present a comparative study of the webs of social spiders. As Witt (1975) indicates, for many spiders it is impossible to describe the behavior of the animal without taking into account the web-spider interaction. I shall therefore describe the adaptations and limitations of communal webs as related to the behavior of communal spiders.

Wilson (1975) argues for a loose definition of sociality that will allow examination of a variety of animal interactions from the point of view of sociobiology. I agree with Wilson, and include in this discussion species exhibiting vastly different levels of social organization. It is natural to compare the behavior of spider species with advanced levels of sociality to that of the eusocial insects. Michener (1969) has proposed a classification scheme to describe the stages of sociality among insects, and many arachnologists have attempted to adapt this scheme to social phenomena in spiders. However, for many species, not enough is known concerning the mating systems and general life-styles of the spiders to fit them comfortably into Michener's categories. For this reason, I have adopted Jackson's classification system for spiders (1978d) throughout this chapter. Jackson recognizes three general classes of spider sociality: (1) "solitary" spiders have nonoverlapping webs and exhibit aggressive interactions with other individuals throughout most of their lifetime; (2) "communal-territorial" spiders have interconnected web-support elements but each occupies a defended area, such as an orb; and (3) "communal-nonterritorial" spiders occupy a common web with little or no aggression. I shall be concerned mainly with the last two types of spiders.

Variation in Web Construction among Social Spiders

Burgess and Witt (1976) describe three general classes of spider webs:

1. Orb webs are two-dimensional "cartwheel" structures of geometric design. The circular orb serves as the catching surface. Orb webs may be vertical, as in solitary *Araneus* spp., or horizontal, as in *Cyrtophora* spp.

2. Sheet webs are two-dimensional webs with little or no symmetry. They may have horizontal or vertical orientations. The catching surface may be flat, convex, or concave. Sheet webs tend to be long-lasting structures and are resistant to environmental or prey-induced damage. Solitary *Agelenopsis* spp. build horizontal sheet webs with these characteristics.

3. Space webs are three-dimensional structures, and can be based on orb, sheet, or tangle webs. The communal spiders belonging to the genera *Agelena*, *Anelosimus*, *Mallos*, and *Stegodyphus* build sheet webs having such a three-dimensional structure. Burgess and Witt consider the orb webs of the communal spider *Metepeira spinipes* to belong in this class. Among the orb-building communal spiders, all webs exhibit common support lines arranged in a tangle in close proximity to individual orbs.

Webs of Orb-Building Communal Spiders

The most thoroughly studied of the orb-space webs is that of the araneid *Cyrtophora citricola*, a communal-territorial species (Blanke 1972; Burgess and Witt 1976; Lubin 1974, 1980b; Rypstra 1979). There is no evidence of active cooperation in this species. Individuals rest at their own hubs, and prey capture, web removal, web repair, and egg-sac construction are all solitary activities.

The web complex consists of numerous fine-meshed, nonadhesive horizontal orbs having either a dome or a bowl shape. Common supports for these webs include cacti, ornamental shrubs, and man-made structures such as fences. Although *C. citricola* is sometimes solitary, colonies of up to several hundred spiders have been described. These larger colonies may cover the entire upper surfaces of hedges.

The horizontal orb is semipermanent and is built and defended by the individual (Fig. 7.1). The hub is open, and irregular in both shape and size; the opening serves as the primary passage between the upper and lower web

Fig. 7.1. Orb web of *Cyrtophora citricola*. The horizontal orb (*O*) is suspended by a communal upper barrier web (*U*). A lower barrier web (*L*) contains a free space (*F*) beneath the orb. Debris (*D*) can be seen in both the upper and the lower barrier webs. Courtesy of Ann L. Rypstra.

surfaces. The orbs are arranged one above the other; an irregular, nonsticky, communal barrier web is located above and below the layered orbs. The space web of each orb serves as a knockdown trap for flying insects, as the primary site of intraspecific interactions, and as the structural support for the orb. The upper space web, which increases in density toward the center of the hub, is also where spiderlings build a tangled web before dispersal. Occasionally, an adult builds a retreat among the dried leaves in the upper barrier web. As the lower barrier web approaches the underside of the orb, its mesh becomes wider, creating a "free space" occupied by the web owner.

Whereas the orbs are constructed and defended by individuals, the space web that connects adjacent orbs is built communally (Fig. 7.2). Large web complexes often extend between trees. Most of the colony activity occurs at night, although prey capture and emergency web repair may occur during the day.

Cyrtophora molluccensis is a related communal-territorial species that builds a similar web complex. Colonies are durable and persistent. Lubin (1980a) described one colony known to occupy the same web site in Papua New Guinea for 12 years. The webs of this spider are found in forest clearings,

Fig. 7.2. Communal web of *Cyrtophora citricola*. Individual orbs (*arrows*) are suspended by a communal barrier web among the supporting cacti. Courtesy of Ann L. Rypstra.

and near human habitation on fences or ornamental trees and shrubs. *Cyrtophora monulfi* has similar habits, although it is usually solitary. *C. molluccensis* and *C. monulfi* build a conical silken retreat above the orb that serves as a diurnal resting place (Lubin 1973, 1974, 1980b).

The webs of the communal-territorial araneid *Metepeira spinipes* are similar in structure to those of the *Cyrtophora* species, the primary differences being that, in *M. spinipes*, viscid silk is used for the spiral threads and the orbs are vertical. This Mexican species has been extensively studied by Burgess and Witt (1976) and Uetz and Burgess (1979). Since Uetz (this volume) provides a detailed account of web building and predatory behavior in *M. spinipes*, I shall simply describe the major features of the communal web.

The spiders build among *Agave* plants, which provide a favorable microhabitat with sturdy support, water collection in the lower part of the plant, and high insect activity. Each individual web has five main features:

1. A nonadhesive space web serves both as a knockdown snare and as the primary support for the orb.

2. The retreat, located within the space web, provides a resting place for the web owner.

3. The orb is wide-meshed and has sticky spiral threads. Although insects may become entangled in the space web, the orb is the primary prey-capture site.

4. Signal threads run from the orb to the spider sitting in the retreat and serve to alert the spider to the presence of potential prey on the orb.

5. Egg sacs are located above the retreat.

Individuals construct their own retreats, orbs, signal threads, and egg sacs, but the tangled space webs are constructed communally, as in *C. citricola*. *Metepeira labyrinthea*, a related spider that is sometimes found in aggregations (but is usually solitary), builds a web like that of *M. spinipes* (Burgess 1978).

Group size in *M. spinipes* is related to the severity of the habitat and the availability of suitable prey. Spiders in severe habitats tend to occur solitarily or in small groups. Spiders in more favorable habitats occur in larger aggregations and exhibit smaller interindividual distances (Uetz, Kane, and Stratton 1982).

The araneid *Metabus gravidus* is a communal-territorial orb weaver that builds web complexes over narrow sections of running streams in Costa Rica (Buskirk 1975a, 1975b). Up to 70 spiders may be found in a colony, which is composed of individual orbs with viscid spirals, a common scaffolding supporting the orbs, and one or more communal retreats. Webs in the upstream half of the colony are more closely spaced than those in the downstream half, and tend to be occupied by juveniles. Younger spiders build vertical orbs near the bank, whereas adults build horizontal webs over the water. The distance between orbs is related to the age of the spiders, the adults exhibiting greater nearest-neighbor distances than the juveniles. Spiders spend the night under rocks or logs, which serve as communal retreats. Orb building and renewal are individual activities and occur at various times throughout the day. As a spider constructs its orb it strengthens and repairs the communal space web.

Eriophora bistriata (Araneidae) is unusual among the orb-space-web builders in that the web complex is not a long-lasting structure, the space web and each orb being renewed daily (Fowler and Diehl 1978). The web of this Paraguayan spider has two components: a diurnal retreat and a nocturnal capture web. The retreat, located on a branch near the construction site for the capture web, has very little silk, and thus spiders in the retreat are in close contact. As dusk approaches, spiders leave the retreat to begin construction of the capture web. The capture web consists of an irregular space web supporting individual orbs that are structurally similar to the orbs of *Araneus* spp. The silken support lines are spun first, by all the colony members. After constructing the communal space web, spiders begin building individual orbs, where they exhibit territorial behavior throughout the remainder of the night. At dawn the spiders eat the secondary support lines and individual orbs as they return to the diurnal retreat. Only the principal threads are left to aid in the construction of another orb-space web the following evening. These threads may extend 30 m or more from the diurnal retreat.

Other communal-territorial spiders build orb-space webs and exhibit territorial behavior similar to that described above. The uloborid *Philoponella republicana*, for example, constructs large communal webs containing hundreds of spiders, each defending an individual orb. Other uloborids, including *Uloborus arizonicus*, *Uloborus mundior*, and *Philoponella oweni*, also build communal webs, but the number of occupants is smaller. *Araneus bandelieri* is an unusual araneine species in which females build communal retreats for themselves and their egg sacs after living most of their lives on independent orbs. The reader is referred to the reviews of Shear (1970a), Burgess (1978), Gertsch (1949), Kaston (1965), and Wilson (1971) for further information on these spiders.

Numerous authors have commented on the absence of communal-nonterritorial behavior among gregarious orb weavers. Buskirk (1975a, 1975b) suggests that the limiting factor is the difficulty of enlarging an orb to allow simultaneous occupation by a number of individuals. In a similar vein, Rypstra (1979) argues that the precision required for orb construction precludes the evolution of communal weaving behavior in these species. In addition, the unitary nature of the orb may make advanced social behavior impossible, since webs must be spaced to maximize prey input (Lubin 1974). Krafft (1979) has suggested that nongeometric webs may be better adapted for colonial behavior because vibrational cues are damped by these webs and thus individual territories are harder to defend.

Webs of Other Communal Spiders

Communal-nonterritorial spiders that do not build orb-space webs are known in a number of families, including the Agelenidae, Amaurobiidae, Dictynidae, Eresidae, and Theridiidae (Burgess 1978). Communal-territorial spiders are found in these families and in the Dipluridae, Lycosidae, Oecobiidae, and Pholcidae (Burgess 1978). The webs vary in structure from loose tangles to tightly woven sheets resembling webworm nests.

The most extensively studied of the sheet-space-web weavers is the African *Agelena consociata*, examined by Darchen and Krafft and their coworkers since the mid-1960's. This agelenid species exhibits the most advanced level of social behavior among spiders, having many characteristics in common with the eusocial insects, including mutual tolerance, interattraction, a common nest, coordination of activities among colony members, a dominance hierarchy, permanence of the colony, and a "group effect" whereby individuals isolated from the group exhibit an increase in mortality and a decrease in metabolic rate (Krafft 1982; Darchen 1965). In contrast to the eusocial insects, *A. consociata* has no caste system (or its accompanying morphological specialization), a continued presence of males, and less reliance on chemical communication in the coordination of colony behavior (Krafft 1971).

A. consociata builds huge nests in the secondary forest, which are occupied by up to 1,000 individuals of both sexes and all stadia (Darchen 1965). The nest consists of three main sections: (1) one or more shelters, (2) a horizontal sheet web, and (3) a labyrinthine snare. All spiders cooperate in web construction, and no cannibalism has been observed. In the laboratory, *A. consociata* individuals placed at the base of plants climb up the foliage and begin to build retreats by enclosing the spaces among leaves with silk. Three or four spiders aggregate within the new retreats and elaborate the internal structure into a series of continuous, interconnected galleries. Spiders leave the retreats only at night, or when prey are available. During the day some spiders wait at the retreat entrances for prey. Mature colonies set aside some chambers as nurseries in which the young are fed and/or interconnected egg sacs are hung from the ceiling. The horizontal web is spun after the retreats. It is a densely woven, irregularly shaped sheet attached by silk lines to most of the galleries of the shelters. The horizontal web serves as the prey-capture sheet. The tangled snare is the last structure to be built, and is located above the prey-capture sheet. This vertical web serves as a knockdown trap and, like the other components of the nest, is composed of nonadhesive silk. In nature, the communal nest may be 3 to 4 m across at the base and 2 m high (Krafft 1969, 1970a, 1970b, 1971, 1979). A similar nest is built by the related species *Agelena republicana* (Darchen 1979).

Stegodyphus sarasinorum is an eresid exhibiting a level of sociality comparable to that of *A. consociata*. Shared characteristics include mutual tolerance, interattraction, a group effect, and cooperation in nest construction, prey capture, and care of the young (Kullmann 1972b). I have had the opportunity to work with both *S. sarasinorum* and the related *Stegodyphus africanus*. These species have similar habits and build similar nests. The following descriptions are based on my own preliminary observations and on the reports of Bradoo (1972a, 1972b, 1980), Jacson and Joseph (1973), Krafft and Roland (1979), and Kullmann (1972b).

The nest is composed of cribellate silk, and is organized into one or more communal retreats and a prey-capture sheet (Fig. 7.3). The communal retreat is an amorphous, spongy mass with irregularly shaped, nonreinforced openings leading to the web interior. Parts of the outer surface of the web have papery consistency, and the remains of a few prey items may be incorporated into the silk matrix (Fig. 7.4). Spiders spend most of the day within

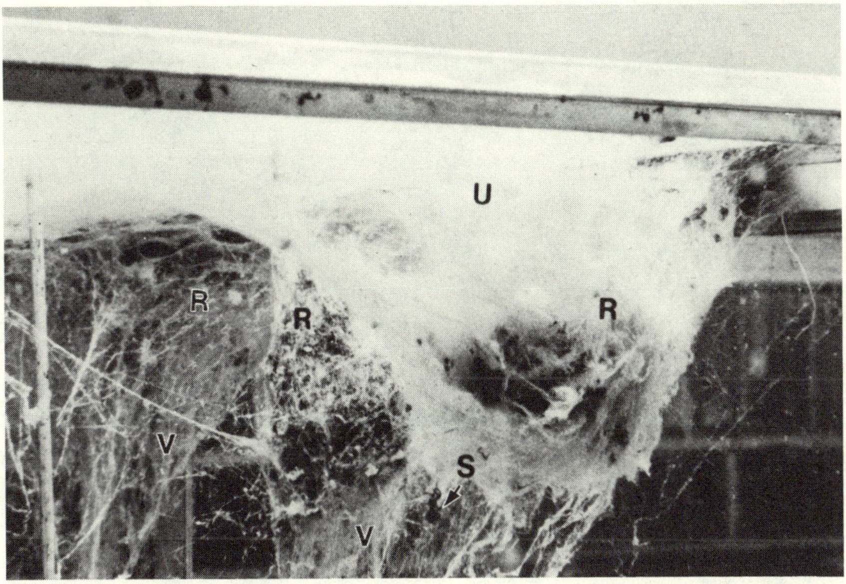

Fig. 7.3. Communal web of *Stegodyphus sarasinorum*. The web was built in the laboratory. Three communal retreats (*R*) are hung from the roof of the cage by an upper support web (*U*). The retreat at the center is the oldest, and had been abandoned by the time this photograph was taken. Animals moved freely between the other two retreats. Nest entrances leading to the web interior are visible on the surface of the retreat at the left. Vertical sheets (*V*) serve as prey-capture snares. A spider (*S*) is foraging on one of the sheets.

the retreat, leaving the retreat only when the web is disturbed by potential prey. The interior of the retreat is a loosely woven, spongy mass with interconnecting tunnels. Egg sacs have a papery appearance similar to that of parts of the web surface and are anchored within the web interior. Prey are captured on a two-dimensional sheet web extending from one side of the retreat. Two or more retreats may share a common prey-capture sheet, and spiders move freely from one retreat to another. Prey-capture sheets have a horizontal or nearly vertical orientation depending on the structure of the supporting vegetation. In the laboratory the retreat is built first, and then the prey-capture sheet.

Mallos gregalis is a dictynid spider found in the mountains of central Mexico (Fig. 7.5). This communal-nonterritorial species builds a sheetlike web having a superficial resemblance to the nests of tent caterpillars. Up to 100,000 individuals of both sexes and various stadia may occupy a large colony in a single tree (G. W. Uetz, pers. comm.). *M. gregalis* prefers to build in trees of the *Acacia* genus and has been reported to cover nearly three-quarters of an 18-m tree with webbing (Burgess 1979a).

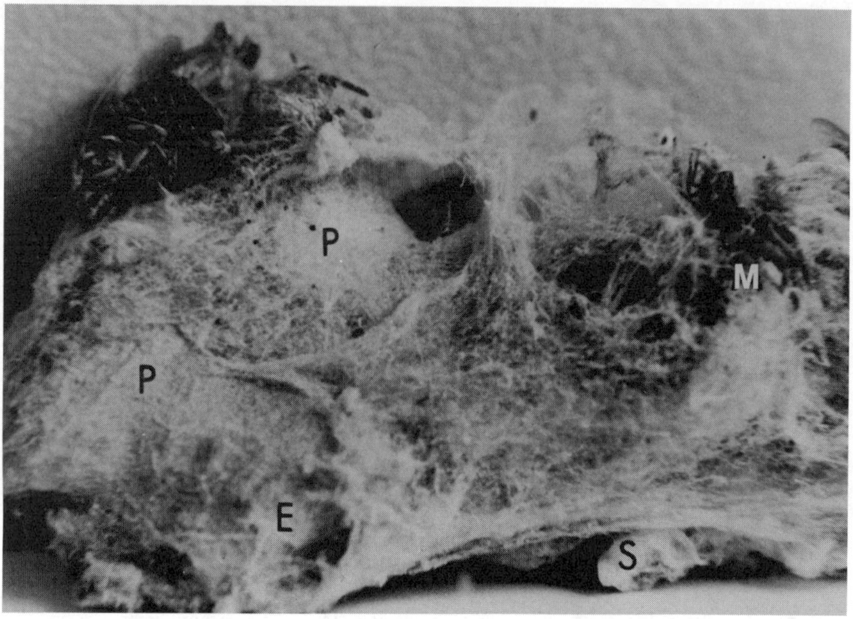

Fig. 7.4. Section of a web of *Stegodyphus sarasinorum*. Some areas of the web have a papery texture (*P*) when compared with the loosely woven areas characterizing much of the web surface. Entrances to the web interior (*E*) are not reinforced. The carcasses of a few prey items (a lepidopteran and flies), and the remains of a spider's molt (*M*) are visible on the outer web surface. At the bottom is an opened egg sac (*S*).

M. gregalis exhibits many of the social characteristics of both *S. sarasinorum* and *A. consociata*, including mutual tolerance and cooperative nest construction and prey capture (Burgess 1976; Diguet 1909a, 1909b, 1915). The tolerance exhibited among nest mates is such that *M. gregalis* will not feed on conspecifics, even if the alternative is starving to death (Jackson 1981b). An interesting group effect is observed in *M. gregalis*. Although females survive normally when isolated from nest mates (Jackson 1979a; Tietjen 1982), males housed with females tend to survive longer than isolated males. This sex-specific effect occurs because adult males do not take part in either prey capture or web building (Jackson 1981a, 1981b).

The following description of the web of *M. gregalis* is based on measurements of webs collected by J. W. Burgess and G. W. Uetz and on a report of Burgess (1979a). The three-dimensional web conforms closely to the geometry of the underlying plant, completely enswathing areas of vegetation with both cribellate silk and tough, noncribellate supporting threads (Fig. 7.6). Whereas the webs of the *Agelena* and *Stegodyphus* species suffocate the underlying vegetation, the leaves enclosed by the web of *M. gregalis* remain green and alive.

Fig. 7.5. Two adult females and a few immatures of *Mallos gregalis* beginning construction of a retreat on a philodendron in the laboratory. A captured housefly has been restrained in the nest matrix.

Fig. 7.6. Web of *Mallos gregalis*. The web is built around a tree branch; the leaves, although dead in the photograph, were green when the nest was collected. The upper surface of the web is shown. The carcasses of numerous prey items (mainly flies) cover the outer prey-capture sheet. Note how closely the web adheres to the underlying vegetation. This web was collected by G. W. Uetz.

Fig. 7.7. Outer surface of a web of *Mallos gregalis*. Prey items cover much of the surface. Areas of new web expansion (*N*) are relatively free of debris, although the remains of flies being incorporated into the nest matrix are visible (*arrows*). Some silk serves as support strands (*S*), and spiders often run along these lines when on the web surface. This web was collected by G. W. Uetz.

The outer surface is the prey-capture sheet. It is peppered with the remains of prey incorporated into the nest matrix (Fig. 7.7). Some sections of the outer surface are relatively clear of debris and represent areas of active silk deposition and web expansion. Other sections have a papery appearance. These papery areas, which run beneath the web surface, are water-resistant and probably serve to keep the web interior dry during rainfall; they may also aid in keeping the web interior free of windblown dust. Runways of thick dragline silk crisscross the upper surface of the web. Spiders run along these paths, and along similar trails in the web interior. These thickened lines probably also provide structural support. Tunnel entrances leading to the web interior are reinforced with silk (Fig. 7.8). The lower surface of the web has far fewer nest entrances than do the upper and lateral surfaces (Fig. 7.9). Leaves here are not as completely enswathed, and the silk is, for the most part, nonadhesive. The lower surface is not a primary prey-capture site (as evidenced by a carcass count), and spiders seldom wander into this area, preferring the upper and lateral surfaces.

Within the web interior, the silk is clean and elaborated into a series of interconnected tunnels and chambers (Fig. 7.10). As a result the interior of

Fig. 7.8. Outer surface of a web of *Mallos gregalis*. Tunnels (*T*) leading to the web interior are often associated with leaves near the web surface. Tunnel entrances are reinforced with non-cribellate silk (*arrow*). Silk with a papery appearance (*P*) extends into the web interior; this silk is largely nonsticky. This web was collected by G. W. Uetz.

Fig. 7.9. Lower surface of a web of *Mallos gregalis*. There are fewer nest entrances here, and the silk does not completely envelop the leaves (compare Fig. 7.6). In addition, fewer prey are captured on this surface. This web was collected by G. W. Uetz.

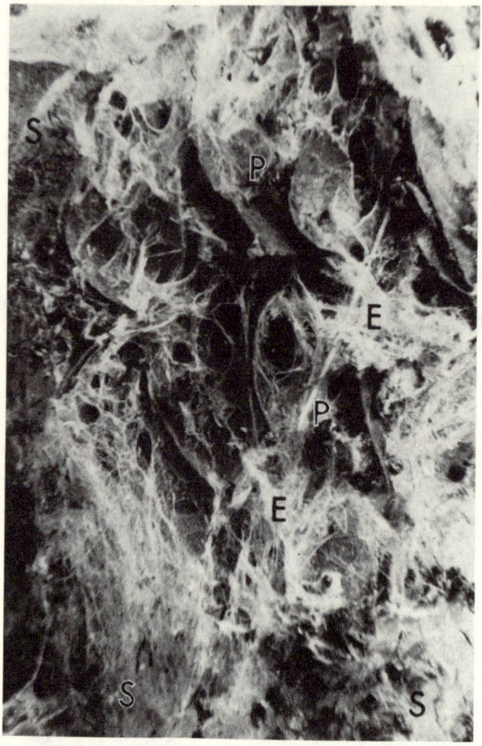

Fig. 7.10. Interior of a web of *Mallos gregalis*. Numerous interconnected tunnels and chambers give this area a spongy appearance. The silk is clean, except for the remains of a few prey that have been pulled into the web interior (*P*) and spider exuviae (*E*). Compare this area with the web surface (*S*), which has a dusty appearance and is dotted with the remains of many prey. This web was collected by G. W. Uetz.

the nest has a spongy appearance. Animals reinforce the tunnels and chambers with layers of noncribellate silk, much as they do the nest entrances on the web surface. The galleries deep in the center of the larger and older colonies are choked with excess silk, spider exuviae, the remains of egg masses, and dead spiders. This suggests that in the most mature nests the spiders inhabit only the outer layers of the web interior. Dissection of nests containing living spiders supports this hypothesis, since in dissected nests most of the spiders are found in the outer layers. The difficulties involved in dissecting such a structure, however, may affect the reliability of the findings. In addition, only a few large inhabited webs have been dissected, and all of these were built in the laboratory.

Between the outer leaves the silk is clean, and has a spongy appearance similar to that of the web interior. These between-leaf areas contain few prey remains, are among the first structures built, and are occupied by spiders during the day. These regions are also the primary sites of egg mass deposition. Unlike the eggs of *A. consociata* and *S. sarasinorum*, which are enclosed in a silken sac, the eggs of *M. gregalis* are largely nonenclosed, which permits observation of the developing spiderlings. The egg masses are suspended within chambers by "clouds" of finely spun silk produced by individual spiders. There are no special brood chambers set apart from other galleries as in *A. consociata*; rather, *M. gregalis* females appear to deposit egg masses within any suitable chamber.

M. gregalis does well in the laboratory and will construct webs on a variety of substrates, including living plants (e.g., *Philodendron* spp.) and artificial Plexiglas supports. In addition, animals build functional webs in containers of widely differing volumes—in petri dishes as well as in room-sized environmental chambers. The spiders can be raised through several generations in even the smallest of enclosures, and exhibit normal prey capture, courtship, copulation, and egg-mass deposition. Webs built in the laboratory show many of the features of natural nests, including a definite prey-capture area and a spongy interior. The advantage of having spiders build within transparent containers is that the behavior of spiders in the web interior is easily observed. Such behavior cannot be observed in natural colonies. Webs built in smaller cages also exhibit less intercolony variation in shape, allowing comparative analyses of web growth with reasonable sample sizes (Tietjen 1980).

Three members of the theridiid genus *Anelosimus* exhibit varying levels of sociality: *Anelosimus eximius* (from South America), *Anelosimus jucundus* (from Mexico), and *Anelosimus studiosus* (from North America). Each builds a nest consisting of an irregular sheet web spun over dead leaves, a labyrinthine tangle web above the sheet, and one or more retreats. The sheet web often resembles a webworm nest (Brach 1975, 1977; Burgess 1978; Kaston 1965; Shear 1970a). The level of social behavior of *A. eximius* is quite advanced—nearly as advanced as that of *Agelena consociata* (Wilson 1971). Neither *Anelosimus jucundus* nor *Anelosimus studiosus* has a level of social behavior approaching that of *Anelosimus eximius*; rather, in each there is a prolonged association of the spiderlings with the mother, which persists until the young have matured (Brach 1977; Burgess 1978). In the laboratory, *A. jucundus* adults can be maintained in groups with little or no cannibalism. Under such conditions, communal feeding is observed.

Amaurobius socialis (Amaurobiidae) and *Oecobius civitas* (Oecobiidae) are two unusual sheet-web weavers exhibiting varying degrees of sociality. The Australian *A. socialis* is a cave-dwelling species that spins a dense sheet web stretching 3.5 to 6 m along the roof of the cave. The web has openings whereby the spiders move between the web interior and the web surface (Kaston 1965). *O. civitas* is a small Mexican spider that builds a small sheet web beneath stones. The communally built prey-capture sheet is surrounded by 110 to 180 retreats. Although the retreats are inhabited by individuals, they are not defended territory. A spider entering the retreat of another will cause the resident animal to leave without any apparent aggression. The resident then moves to another retreat. If the second retreat is occupied, another displacement occurs. In addition, this species may be communal egg layers, the females cooperating in the construction of a single egg case (Burgess 1976; Shear 1970a).

Control of Web-Building Behavior

Web-building behavior among nonsocial spiders has been extensively studied. The research ranges from descriptive studies (Bentzien 1973; Eberhard 1973, 1975; Witt and Reed 1965) to ontogenetic studies (Benforado and Kistler 1973; Reed et al. 1970), studies of the effects of drugs (Witt, Reed, and Peakall 1968), and behavior-genetic or evolutionary analyses (Risch 1977). Kullmann (1972a) and Kaston (1964) provide excellent reviews of the literature on the evolution of web-building behavior among cribellate and ecribellate spiders, and these reviews are updated by Shear (this volume). The web-building behavior of social spiders, however, has not been as extensively studied. Most of the research is descriptive; there are only a few experimental studies, the exception being the work of Krafft and his co-workers on *Agelena consociata* (Krafft 1969).

Mass-Action Behavior

One characteristic of social groupings among spiders is cooperation in the construction of nests (Kullmann 1972b). The cooperative efforts of nest mates in spinning need not be a highly coordinated activity, but rather can proceed through the indirect interaction of colony members. Wilson (1971) calls this indirect cooperation "mass-action" behavior, and provides examples of such behavior within the eusocial insects. Among social bees, the seemingly coordinated activity of hive construction is controlled by mass-action behavior. Nest building in *Polistes* sp., for example, is the result of antagonistic behavior among individuals. While some workers are engaged in brood-cell construction, others are tearing down the newly made cells. This antagonistic interaction eventually determines the number of cells that are built and their location. Thus, the mass-action behavior controlling hive construction involves the uncoordinated, often conflicting actions of many workers largely oblivious of the behavior of sisters building nearby. At no time are individuals aware of the condition of the colony as a whole.

It is not surprising that the level of cooperation seen among social spiders is low. Krafft (1979) indicated that cooperative web-building behavior by *Agelena consociata* is the result of indirect interaction among individuals rather than complex communication. Each spider reacts in the same way to a given situation, and this predictability of behavior is the basis for cooperation in nest construction. Similarly, nest construction and maintenance in the communal-territorial *Metabus gravidus* results from the independent behavior of spiders, individuals modifying their activities according to their position on the web (Buskirk 1975b). Colony activity in *A. consociata* and *M. gravidus* is thus controlled by mass-action behavior as seen in the eusocial insects.

Web Building in Mallos gregalis

During the past 4 years I have investigated web construction and utilization of space in *Mallos gregalis* with the aid of a computer-controlled television camera (Fig. 7.11). Colonies of 1 to 20 individuals are maintained in 52.2-cm^3 petri dishes, and diffuse light is transmitted through the colonies to a solid-state television camera located above them. With this technique, spiders are seen by the computer as black dots on a white background. Animal location is recorded on a 32 × 32 array providing 1,024 possible positions within the camera's field of view. The positions of a number of spiders can be monitored simultaneously. Since the sensitivity of the camera is

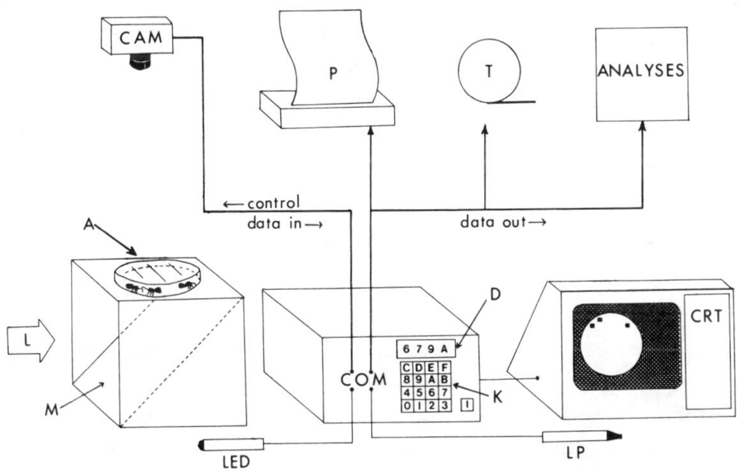

Fig. 7.11. Computer-camera apparatus used to monitor activity of *Mallos gregalis*. Light (*L*) is reflected from a mirror (*M*) through an area containing the spiders and their web (*A*). The camera (*CAM*), located above the arena, receives control information from, and sends data to, the computer (*COM*). A keypad (*K*) and display (*D*) allow the setting of exposure, number of frames per second, and mode of operation. The image can be displayed on a video monitor (*CRT*), where spiders are seen as black dots on a white background. Data can also be transmitted to a printer (*P*), magnetic media for storage (*T*), and/or a second computer for analyses. A light-emitting diode (*LED*) and light pen (*LP*) allow the user to label spiders or areas within the arena.

greatest in the infrared, the movements of the spiders can be recorded using a light source that is outside of their visual range (Crane 1949; DeVoe 1972; DeVoe and Zvargulis 1967). The growth of the web is recorded by programming the computer to construct a grey-level image of the nest. Since silk blocks out the transmitted light, grey level and silk density are proportional. It is thus possible to determine the relative silk density in 1,024 levels at each point on the 32 × 32 grid. A more complete description of these methods can be found in Tietjen 1980, 1982.

Preliminary analyses indicate that there is little initial activity as spiders begin construction. At this time animals aggregate in groups of two to six spiders depending on the size of the colony. Web building begins at each aggregation site, and spiders move among the groups. Within a few days, most animals settle at one aggregation site, and this becomes the area of maximal web growth. As this site expands, it eventually encompasses the abandoned web sites. Colonies in larger containers exhibit similar behavior (pers. obs.). Hollar and Witt (1979) have also recorded an initial fluctuation in the distribution of spiders followed by a stabilization of interanimal distance and utilization of space.

Observations indicate that *M. gregalis* is negatively geotactic, and colonies always begin construction on the highest structure within a container. In addition, *M. gregalis* exhibits positive phototaxis during the early phases of construction, by always building on the side of the arena facing a window, regardless of compass direction. Under diffuse overhead light, the spiders build without a preferred direction. Gravitational cues appear to direct web construction more than light cues, since colonies illuminated from below still build in the upper section of containers.

Self-Organizing Properties of the Web

Web construction in *M. gregalis* is influenced at first by the physical features of the environment (i.e., by gravity, light, and the qualities of an appropriate substrate). Once begun, however, the web is self-organizing, and *M. gregalis* individuals continue to add to existing silk despite changes in the environment. If, for example, spiders begin web construction on the side of the arena facing a window and on the following day the arena is rotated away from the window, animals continue to add silk to the original web site rather than reorienting toward the light.

Wilson (1971) describes similar self-organizing effects among the eusocial insects. Mound-building termites, for example, begin construction of archways to support the nest interior by building simple pillars. Chemical cues are present in the salivary secretions and fecal deposits that the workers use to cement the soil together. As the columns increase in height, pheromones accumulate and induce workers on adjacent pillars to begin building toward one another, thus creating the arch. This behavior does not depend on direct communication among the workers, but rather on the proximity and size of columns.

Jackson (1981a) has indicated that whereas adult females of *M. gregalis* are attracted to conspecific silk, silk-mediated interattraction is less pronounced in both males and immatures. In this context, it is interesting to note that measurements of interanimal distance show that *M. gregalis* males distribute themselves at greater distances from nest mates than do females (Jackson 1979a). In addition, when mixed-age colonies are transferred from small to large cages, adult females of *M. gregalis* tend to stay within the original nest whereas males and spiderlings tend to emigrate (pers. obs.). Data collected with the computer-camera apparatus indicate that when adult females are released into arenas containing previously deposited silk, they continue nest construction at the old web sites. This behavior occurs both when the old web sites are occupied by conspecifics and when the previously built nests have been cleared of spiders. These results suggest that the attractive properties of the web aid in organizing web elaboration and expansion by adult females. Similar instances of silk-mediated interattraction have been described in other species by Burgess and Witt (1976), Darchen (1965), Krafft (1970b), and Krafft and Roland (1979). (See Tietjen and Rovner 1982 for a review of the literature.) In many cases, the attraction depends not only on tactile cues but also on chemical cues provided by pheromones.

Among social bees, each hive has a unique colony-specific odor that is recognized by members of the group (Wilson 1965, 1971). This combination of pheromones prevents workers from moving from one hive to another, since they are immediately recognized as strangers by foreign colonies and are killed or driven from the nest. No such colony-specific odor has been found among communal spiders. Individuals of *A. consociata*, *M. gregalis*, and *Metepeira spinipes* can be transferred from one colony to another without apparent ill effects (Burgess 1976; Darchen 1965; Krafft 1970b). In some cases, the lack of intercolony recognition extends between species. *Agelena republicana* individuals transferred to *A. consociata* nests are not only accepted but also aid in nest construction (Darchen 1979; Krafft 1970c). Similar congeneric associations have been described for *Stegodyphus* spp. (Kullmann 1972b).

Circadian Rhythms

Another feature of communal organization is the circadian rhythms of the group, as measured by the synchrony of activity among group members and the temporal organization of individual rhythms (Harcourt 1978; McBride 1976). In his excellent review, Cloudsley-Thompson (1978) describes circadian rhythms in a variety of spiders. Recently, such rhythms have been described in noncommunal spiders by Ramousse and Davis (1976) and Riechert and Tracy (1975).

Among the orb-space-web weavers, diurnal rhythms in spinning behavior have been observed in the araneids *Cyrtophora citricola* (Blanke 1972; Rypstra 1979), *Eriophora bistriata* (Fowler and Diehl 1978), and *Metabus gravidus* (Buskirk 1975b) and the uloborid *Philoponella republicana* (Lubin 1980a). In general, these species exhibit peaks in activity during the night hours. At this time there is an increase in aggressive interactions, prey capture, and web-building behavior.

Similar rhythms have been observed among the sheet-space-web weavers. Krafft (1969, 1970c) recorded the activity of *A. consociata* in both laboratory and field, by means of actographic techniques utilizing a vibration pickup connected to the web. He found that at dusk spiders emerge from the retreats to begin the weaving of support structures and the repair or elaboration of the prey-capture sheet. This activity peak lasts only about a half hour; web-building activity declines rapidly thereafter and remains low for the remainder of the night. Web destruction and satiation affect only the number of spiders involved, not the duration of the activity period. Other general activities that occur throughout the night include cocoon construction, repair of retreats, and prey capture. The diurnal rhythms are cued by photoperiod.

Mallos gregalis, *Stegodyphus sarasinorum*, and *Stegodyphus africanus* exhibit rhythms in web-building behavior similar to that described for *A. consociata* (Bradoo 1980; Tietjen 1982). *M. gregalis* is unique among the communal-nonterritorial species in lacking a prominent group effect. This permits recording of the circadian rhythms both of colonies and of isolates. Using photoelectric recording methods, I monitored the circadian activity of *M. gregalis* in groups and in isolation (Tietjen 1982). Data indicated that *M. gregalis* individuals exhibited fluctuations in activity over several days such that a period of little or no movement was followed by a variable number of high-activity days. Such individual activity rhythms were out of phase among animals, and circadian patterns were superimposed on the individual cycles. Animals of high daily activity, for example, were highly active only during the night. Those of low daily activity showed little movement throughout the entire day.

In *M. gregalis*, *S. sarasinorum*, and *S. africanus*, circadian rhythms coordinate colony behavior so that nest construction occurs mostly in the evening. In *M. gregalis*, individual activity cycles determine whether spiders engage in self-maintenance or colony-maintenance activities. Animals of low daily activity remain in the interior of the nest, and are largely involved in self-maintenance behaviors such as courtship, copulation, egg-sac construction, and solitary feeding. Those of high daily activity move to the web surface and take part in colony-related behaviors, including expansion and repair of the web, and communal prey capture and feeding.

The factors controlling individual daily activity in *M. gregalis* were not determined. Data indicate, however, that degree of satiation, temperature, humidity, and photoperiod do not affect daily activity in isolates, at least over the relatively short time span of the experiments. During starvation periods of several weeks, animals do tend to exhibit an increase in activity and to spend more time on the prey-catching surface (Witt 1978). Over short periods of time, however, endogenous factors, such as sexual tone (Crane 1949), may be involved. Animals of high sexual tone, for example, would be expected to remain in the web interior and exhibit low daily activity.

Hundreds of spiders may occupy a single colony. Since individual activity cycles are randomly out of phase among nest mates, there is more or less constant daily activity in the colony as a whole. As one spider shifts to low daily activity, another spider replaces it. Buskirk (1975b) describes a similar situation in the communal-territorial *Metabus gravidus*. She suggests that the mixture of activity patterns seen among individuals in this species coordinates the maintenance of the colony scaffolding. In addition, individuals modify their behavior according to their location within the colony. An analogous situation exists in larvae of the tent caterpillar *Malacosoma pluviale* (Wellington 1957). High-activity larvae emigrate from the nest and become leaders of the infestation, and low-activity larvae are the followers. The ratio of active to inactive members in the colony affects the behavior of the entire group, controlling the frequency of future infestations and the propensity of the colony to defoliate new sections of the host tree.

One might expect that such a system of control would produce "cheaters"—animals that engage principally in self-maintenance rather than colony-maintenance activity (Maynard-Smith 1976). Physiological and/or behavioral restraints such as the need to empty silk glands may inhibit such behavior. In addition, in species with overlapping generations and long-term nest sites, one might expect a high degree of relatedness among colony members. In these species kin selection could inhibit such behavior (Eberhard 1975).

Budding

Small groups of *Mallos gregalis* sometimes leave the parent colony to establish secondary colonies (Diguet 1909b), a phenomenon known as budding. Emigration usually begins with a single individual's leaving the primary colony and beginning nest construction some distance away. This secondary colony is connected to the larger web by silk strands that other spiders follow to the new web site. The recruited individuals then aid in nest construction at the new site. The propensity of individuals to emigrate is related to the age of the parent colony: older colonies exhibit an increase in budding (P. N. Witt, pers. comm.). Apparently, as the interior of the web becomes choked with silk and/or the available substrate becomes fully occupied, spiders begin to emigrate. The succession of microbial flora may also affect budding (see below).

Darchen (1965) described similar budding behavior by *A. consociata*. Lubin (1980a) described budding and colony formation by *Cyrtophora moluccensis* and *P. republicana*. The social theridiid *Achaearanea wau* from Papua New Guinea exhibits massive dispersals of late-instar and adult female spiders (Lubin and Robinson 1982). Group emigration from the old web site is over silk highways up to 1 m wide and several meters long. The highways are constructed of loose parallel threads and are destroyed after several days' usage. I have twice observed *M. gregalis* building similar loosely woven sheets before dispersal in the laboratory.

Summary

Web construction among communal spiders has many characteristics in common with that seen among solitary spiders, including reliance on geo- and phototropisms and on factors related to substrate suitability. In all of the communal spiders studied there is no evidence of a sophisticated system of communication. Individuals act as independent units, changing their behavior in response to conditions in their immediate surroundings. Aside from mass-action behavior, the only other factors that affect cooperation in web construction are the self-organizing or attractive properties of the web and the activity rhythms of individuals.

Web-Related Adaptations for Communal Living

Various authors have described a number of advantages derived from group living (Buskirk 1975b; Jackson 1979b; Krafft 1969, 1979; Lubin 1974; Rypstra 1979; Wilson 1971, 1975). Advantages directly related to the communal web fall into three categories: (1) shelter, including protection from predators and environmental effects; (2) information transfer and coordination of colony behavior, via vibrational, chemical, and thigmotactic cues; and (3) exploitation of habitats, including prey capture and utilization of habitats not available to solitary spiders.

Shelter

Rypstra (1979) indicated that birds are attracted to the conspicuous webs of *Cyrtophora citricola*, where they attempt to steal prey from the outer sections of the colony. The web silk, however, obviously irritates the feathers and eyes of these kleptoparasites, and Rypstra suggested that long-lived avian predators learn to avoid such webs. Spiders in the interior areas of the colony gain additional protection from the outer barrier. Webs at the edge of the colony generally require the most repair of damage caused by the weather and predators. Horton (1980), working with the solitary species *Argiope aurantia* and *Argiope trifasciata*, reported similar results. He found that inexperienced blue jays attempt to steal prey from the webs of these spiders, but eventually learn web avoidance owing to a noxious stimulus provided by the silk.

Circadian rhythms in the occupation of web areas by spiders have been described in *Agelena consociata* (Krafft 1969, 1970c, 1971), *Cyrtophora citricola* and *Cyrtophora moluccensis* (Blanke 1972; Lubin 1974), *Eriophora bistriata* (Fowler and Diehl 1978), *Philoponella republicana* (Lubin 1980a), and *Mallos gregalis* (Tietjen 1982). In general, spiders remain in an inaccessible area of the web during the daylight hours and forage for prey in the evening. By restricting activity to the night, spiders may effectively avoid visual predators, as Rypstra (1979) has suggested for *C. citricola*.

Some of these species build opaque retreats that may provide protection from the noonday sun as the spiders move to the web interior. Diguet (1909a, 1909b) suggested that one of the major functions of the web of *M. gregalis* was to protect the occupants from rain and the hot sun. Temperature-dependent movements have been described in the solitary spiders *Agelenopsis aperta* (Riechert and Tracy 1975) and *Nephila clavipes* (Carrel 1978; Krakauer 1972; Robinson and Robinson 1973a). In the communal spider *C. citricola*, the positioning of animals on the web has been found to be affected by temperature, spiders preferring a temperature of about 27° C (Blanke 1972). Krafft (1970b) obtained similar results for *Agelena consociata*, which prefers a mean temperature of 23° C.

M. gregalis occupies a harsh environment, which in the late summer months is characterized by heavy rains, strong winds, and a mean daily temperature of 14 to 21° C (Burgess 1979a). Reba Williams at Georgia College tested the thermal-buffering efficiency of natural *M. gregalis* webs (pers. comm.). She positioned thermoelectric probes (with a sensitivity of 0.03° C) at various depths within nests to test how they respond to both conductive and radiant heat. Preliminary data indicate that the web of *M. gregalis* provides little insulation from conductive heat but that the nest buffers the effects of radiant heat. Short-term temperature variations of several degrees on the web exterior (such as occur when the sun passes behind a cloud) resulted in temperature variations in the web interior of only a few tenths of a degree. Similarly, Williams tested the relative transparency of the nest and found little attenuation of light attributable to the silk alone.

The webs of most communal spiders destroy the underlying vegetation.

The leaves enveloped by the webs of *Stegodyphus sarasinorum* and *Stegodyphus africanus* turn yellow as they die, which suggests that mortality is due to interference with light penetration, and therefore with photosynthesis. Silk webbing surrounding dried leaves has been observed in the nests of *Agelena consociata* and *Agelena republicana* (Krafft 1979), *Anelosimus eximius* (Robinson 1977a; Wilson 1971), *Anelosimus studiosus* (Brach 1977), and *C. citricola* (Lubin 1980b).

In contrast, the webs of *M. gregalis* do not destroy the underlying vegetation (Burgess 1979a). Williams did not have nests built on living vegetation available for analysis, but it is possible that leaf transpiration could affect temperature and/or humidity conditions within the web interior. The unusual construction and location of egg masses in *M. gregalis* may be related to the proposed transpiration effect. Enclosed egg masses suspended in the webbing of the mother's nest are common among other members of the Dictynidae (Chamberlin and Gertsch 1958; Kaston 1948). In *M. gregalis*, however, egg masses are not enclosed in a sac but are surrounded by a cloud of fine silk and are found between adjacent leaves or within the curls of single leaves. Such positions would allow the nonenclosed egg masses maximum benefit from leaf transpiration and reduce the likelihood of dessication.

Information Transfer

In addition to providing shelter for the animals within the nest, the webs of communal spiders are important channels for information transfer among colony members. Such webs extend the sensory perception of the spiders, which rely on a combination of chemical, vibrational, and tactile cues (Tietjen and Rovner 1982; Witt 1975).

Among the eusocial insects, chemical signals control much of colony behavior. Pheromones play important roles in, e.g., alarm, caste control, interindividual attraction, nomadic behavior, and feeding (Blum and Brand 1972; Shorey 1976; Wilson 1965, 1971).

Jackson (1981a) showed that a pheromone is responsible for the attractive properties of *M. gregalis* silk. Silk aged for 1 month and washed silk were found to be relatively unattractive when compared with fresh conspecific silk. In addition, males of related dictynid species such as *Mallos trivittatus* and *Dictyna calcarata* are induced to court in the presence of the females' silk. *M. gregalis* males, on the other hand, do not exhibit courtship behavior in the presence of female-produced silk. Jackson(1978b) suggested that, rather than relying on chemical communication with females, *M. gregalis* males incorporate a sexual advertisement routine into their daily time budget. He called this advertisement routine pluck walking (Jackson and Smith 1978).

Chemical communication has also been demonstrated in *Agelena con-*

sociata and *S. sarasinorum* (Krafft 1970b, 1971; Krafft and Roland 1979; Kullmann 1972b; Kullmann and Zimmerman 1971). In *A. consociata*, for example, silk, galleries, and both dead and living cocoons are attractive. Adult females prefer living egg sacs to dead egg sacs, however, which suggests that additional chemical cues are involved. In addition, spiders preferentially follow the silk trails of conspecifics. Similar behavior has been described for *S. sarasinorum*.

Among nonsocial spiders, contact chemoreceptors have been described as gently curving, blunt-tipped hairs found on the distal portions of appendages (Drews and Bernard 1976; Foelix 1970; Foelix and Chu-Wang 1973b; Harris and Mill 1977b). In contrast, mechanoreceptors are straight, sharp-pointed, and evenly distributed over the spider's body (Foelix and Chu-Wang 1973a; Harris and Mill 1977a). Similar receptors have been observed in *A. consociata, M. gregalis, S. africanus*, and *S. sarasinorum* (Krafft 1975; Kullmann 1972b; pers. obs.).

Signals transmitted by silk also include web-borne vibrations. The information transmitted by this means has been most thoroughly studied in the solitary orb-weaving spiders. Vibration signals are important in intraspecific communication, courtship behavior, and agonistic displays, as well as prey capture and recognition of potential parasites (Krafft 1978; Leborgne and Krafft 1979; Lubin 1980b; Suter 1978; Vollrath 1976; Witt 1975). In some cases it has been possible to record the response of the web to various signals. Masters and co-workers (1981, this volume) recorded transverse, longitudinal, and lateral vibration transmission in the orb of *Nuctenea sclopetaria*. They noted that longitudinal waves exhibited the least attenuation and suggested that these signals were the most important to the spider. Analyses of the spider's response to the signals range from electrophysiological readings (Barth and Bohnenberger 1978; Finck 1972) to behavioral descriptions (Finck and Reed 1979; Frings and Frings 1966). Most probably, the vibration receptor is the lyriform organ (Walcott 1969).

Communal-territorial species make extensive use of vibrational cues during agonistic encounters. Brach (1977) indicated that *Anelosimus studiosus* females recognize conspecifics by means of web-borne vibrations. Typically, males are tolerated by the resident female, but adult females are killed or driven off. In *C. citricola* and *C. moluccensis*, web jerking, tensioning, and shaking are seen during aggressive interactions (Lubin 1973, 1974, 1980b). Web jerking and tensioning are relatively low-intensity pulls on silk, differing mainly in their time sequencing. Web tensioning is a slow, long pull, whereas web jerking is performed much more quickly. Each is used to locate objects on the web. Web shaking involves large-amplitude oscillation of the horizontal web. Movement of the orb is effected by strong contractions of legs I or, in vigorous web shaking, legs I and II. Web shaking is also seen in response to theridiid kleptoparasites and sarcophagous egg parasites, and when potential prey become entangled in the upper space web. Sometimes these movements are transmitted to nearby orbs, which results in the arousal of neighboring spiders. From 50 to 75 percent of the colony may exhibit web shaking under these conditions (Rypstra 1979).

Among communal spiders, courtship behavior relies mainly on vibrational cues. In *M. gregalis*, for example, males exhibit pluck-walking behavior both as a component of courtship and as a sexual advertisement display. During pluck walking, the male holds the web with the claws of legs I, and then releases the silk while moving the legs dorsally. Such plucks are performed first with one leg I and then with the other. While performing an advertisement display males walk slowly through the nest plucking in a jerky, rapid manner. Females signal their readiness to mate by not moving away from displaying males (Jackson 1978a, 1979a; Jackson and Smith 1978).

Initiation of prey capture also depends upon vibrational cues. Lubin (1980b) has described prey capture in *C. citricola* and *C. moluccensis*. Vibrations from the prey entangled in the upper barrier web are transmitted to the spider waiting beneath the horizontal orb. Lubin suggests that the initial vibrations may permit discrimination of prey type, and if the prey is judged to be acceptable, it is located by means of web jerking and tensioning. If the insect falls onto the nonsticky orb, the spider runs along the undersurface of the web and touches the prey with legs I, perhaps for final prey discrimination via chemical and/or tactile cues. Data indicate that *Cyrtophora* spp. do discriminate among prey either early in the attack by vibrational cues or later by direct contact. Lepidopterans, for example, are immobilized by biting, whereas potentially dangerous or powerful prey, such as beetles, are first enswathed with silk and then bitten. Prey that do not fall from the upper barrier web are dislodged by means of web shaking. If web shaking fails to dislodge the prey, the spider uses web tensioning to locate the insect and climbs into the upper space web to capture it. The immobilized prey is then carried back to the underside of the orb, where feeding occurs. Other communal-territorial species such as *Anelosimus jucundus*, *A. studiosus*, *E. bistriata*, *Metabus gravidus*, and *Metepeira spinipes* exhibit similar prey-capture behavior based largely upon vibrational cues (Brach 1977; Burgess and Witt 1976; Buskirk 1975a, 1975b; Fowler and Diehl 1978).

Mallos gregalis exhibits prey-capture behavior that is typical of communal-nonterritorial species (Jackson 1977c). Spiders wait within the web interior until a fly lands and is entangled on the prey-capture sheet. As the fly attempts to escape, spiders are stimulated to run toward the source of vibration. If the fly stops buzzing, spiders stop advancing and wait. Location and biting of the fly results in increased struggle, and additional spiders are recruited by the stronger vibrational cues. One to ten spiders have been observed to participate in subduing prey, and nearly 30 animals may feed simultaneously following immobilization. Occasionally, prey may be transported into the web interior for storage (Burgess 1979a, 1979b).

When multiple prey items are entangled on the catching surface the spiders appear disoriented in their prey-capture behavior. Under these circumstances numerous spiders are recruited from the web interior. However, as the spiders attempt to orient themselves toward struggling insects, they are confused as they turn from one buzzing fly to another. Disorientation increases when one fly ceases to struggle and a second resumes struggling. Even with a single prey item spiders often overshoot the fly during their approach. I have observed spiders to search within a few millimeters of a struggling fly for so long that it escapes.

Agelena consociata, *A. republicana*, *Anelosimus eximius*, *A. jucundus*, *S. africanus*, and *S. sarasinorum* also rely on vibrational cues during prey capture (Brach 1975; Darchen 1965; Jacson and Joseph 1973; Krafft 1970b; Kullmann 1972b; Kullmann and Zimmerman 1971). Major differences in prey-capture behavior are related to the detaining properties of the web. In *Agelena consociata*, for example, the capture sheet is nonsticky and the upper space web serves mainly as a knockdown device. Prey capture in this species is much more vigorous than in *M. gregalis*, and involves an initial recruitment of many spiders rather than a few. The energetic attack is needed to subdue prey quickly before it can escape from the capture sheet. As the spiders swarm over the insect, vibrational and/or chemical cues reduce the likelihood of intraspecific aggression (Krafft 1971; Witt, Scarboro, and Peakall 1978).

Burgess (1979b) recorded vibration transduction by the web of *M. gregalis* in response to both the vibrations produced by natural prey and synthetic vibrations. Some of the prey were altered to affect their ability to produce vibrations. Prey included cockroaches, bees, anesthetized houseflies, flies with immobilized wings, and normal flies. Web vibrations were recorded using a phonograph cartridge attached to the web. The output from the cartridge was routed through an amplifier to an oscilloscope and magnetic recorder, and data were analyzed with a spectrograph. Burgess found that web vibrations of less than 30 Hz or greater than 700 Hz were attenuated by the web, and that resonances occurred between 100 and 400 Hz. Recordings of vibrations produced by houseflies and from artificial vibrations indicated that the web resonated strongly at the buzzing frequency of the flies. Behavioral observations showed that normally buzzing flies elicited mass recruitment behavior by the spiders, but that cockroaches, bees, and altered flies were ignored. Burgess suggested that walking insects or spiders do not elicit an attack because the vibrations that they produce are below the response window of 100 to 400 Hz, and that, similarly, ensnared bees producing vibrations of greater than 1 kHz are not attacked because such high frequencies are also attenuated. Thus, the web of *M. gregalis* is tuned to the vibration characteristics of the spider's natural prey, muscid flies (Burgess 1979a, 1979b).

W. D. Hollar (pers. comm.) used a more sensitive photo-optic transducer than did Burgess to record vibrations in the web of *M. gregalis*. He found that the vibrations produced by males during courtship and sexual advertisement are transmitted 11 to 12 cm through the web with little or no attenuation. In addition, the vibrations produced by normally walking females can be recorded over comparable distances. These results suggest that Burgess's findings are incomplete. It should also be noted that the behavioral sensitivity of the spider may differ from the observed vibration characteristics of the web, since sensitivity may be limited or accentuated by receptor structure (Barth 1982). In the solitary *Araneus sericatus*, for example, sensitivity of the vibration receptors can be altered by a change in leg position (Finck and Reed 1979). This spider may have both attentive and nonattentive postures on the web. The lyriform organs used by *M. gregalis* for vibration reception may have their sensitivity set in a similar manner (Burgess 1979b).

Witt (1975) concluded that although chemical cues provided by the web are important in the behavior of solitary spiders, information transfer by vibrational cues controls the activities of the animals to an even greater extent. The evidence presented here suggests that the same holds true for communal spiders.

Exploitation of Habitats

One function of communal organization is to allow members of the society to exploit habitats not available to solitary spiders. The cooperative spinning activities of *C. citricola* and *C. moluccensis* produce durable snares in open, grassy areas of forests. These areas are characterized by numerous flying insects and few competing spider species. Many noncommunal spiders are excluded from such habitats owing to the paucity of suitable substrates. Large *C. citricola* colonies may extend between trees and lie within established insect flight paths. Such structures could not be built by solitary individuals (Lubin 1974). Large prey-capture sheets built cooperatively require less individual labor. Isolated individuals of *Agelena consociata*, *Metabus gravidus*, and *Mallos gregalis*, for example, must work harder to build functional webs than individuals in social groupings (Buskirk 1975a, 1975b; Krafft 1979; pers. obs.).

Rypstra (1979) used the ratio of the number of prey captured to the number of prey striking the web as a measure of prey-capture efficiency in *C. citricola*. Five colonies with different numbers of inhabitants were observed. Rypstra found that efficiency was directly correlated with web size, and concluded that the greater complexity of the knockdown snare in the larger colonies was responsible for the observed effect. In addition, position within the colony affected prey-capture efficiency. Spiders near, but not directly at, the periphery of the colony exhibited the greatest success in prey capture. The increase in efficiency extended to other species building near the outer sections of the web. For example, *Leucauge* sp. individuals in the outer sections of the webs of *C. citricola* exhibited an increase in prey-capture efficiency when compared to nearby conspecifics not associated with the colony. These data indicate that prey-capture efficiency is increased in colonial groupings. Similar results have been described for other communal-territorial species, including *Anelosimus studiosus* and *Metabus gravidus* (Brach 1977; Buskirk 1975a, 1975b).

Colony size also affects prey-capture efficiency among communal-nonterritorial species. In *Mallos gregalis*, for example, fewer flies escape from large webs as compared to small ones (Jackson 1979a). The increased efficiency seen in the larger webs is due to the greater deposition of cribellate silk, which does not lose its stickiness over relatively long periods of time (Eberhard 1980b), thereby allowing a gradual buildup of the effective prey-capture surface. The web of *M. gregalis* serves as a trap, and not just as an alerting or detaining device as in many other communal-nonterritorial species. In *Agelena consociata*, for example, the nonsticky web only alerts the spiders to the presence of prey on the web (Krafft 1979).

Limitations of Communal Web Building

The major advantages of sharing a communal web include an increased efficiency in intraspecific communication and prey capture. There are disadvantages, however. For example, the communal web can attract predators and parasites to the population. In addition, living within a semipermanent web poses the secondary problem of nest sanitation.

Predation on Communal Spiders

The conspicuous webs of communal spiders attract a variety of visual predators. Although the nests may provide protection from some avian predators, other animals may take advantage of the high prey densities found in colonies. Vertebrate predators of *Cyrtophora citricola* and *Cyrtophora molluccensis*, for example, include geckos, chameleons, and passerine birds (Blanke 1972).

Ants, bees, and wasps are perhaps the most important predators of communal spiders. Predation by these insects has been described for *Agelena consociata*, *Anelosimus eximius*, *C. citricola*, *C. moluccensis*, and *Eriophora bistriata* (Blanke 1972; Fowler and Diehl 1978; Krafft 1970a; Vollrath 1978). Ants may also be important predators of *Mallos gregalis*. During the fall of 1980, one of my laboratory colonies of *M. gregalis* was invaded by ants (*Prenolepsis* sp.). The ants were apparently attracted by the fly carcasses on the web, since these were the first items that they removed from the nest. As they collected the flies, the ants began to steal egg masses and kill spiders. *M. gregalis* individuals did not appear to notice the presence of the ants until bitten, probably because the vibrations produced by the walking insects were damped by the web. Spiders were not observed to cooperate in protecting one another or the nest. In fact, unmolested spiders often moved away from the struggle between an ant and a nest mate. Although the ants occasionally attacked a spider in small groups, single ants never lost to a spider. More than 280 subadult and adult spiders were killed, and all 15 egg masses were removed from the nest, before I could bring the infestation under control. Apparently the overwhelming number of predators, coupled with the vibration-damping properties of the web, made the spiders easy prey for the ants.

In addition to insect predators, spider predators may be attracted to the nest. These heterospecific intruders may wander throughout the web, freely feeding on the inhabitants. Predatory araneids have been observed within the webs of *Agelena consociata*, *Anelosimus studiosus*, *C. citricola*, *C. mollucensis*, *M. gregalis*, and *Philoponella oweni* (Blanke 1972; Brach 1977; Burgess 1979a; Diguet 1909a, 1909b; Krafft 1970a; Lubin and Robinson 1982; Trail 1980).

Parasitism

Parasites may also be attracted to communal webs. Lubin (1974, 1980b) describes parasitism on the egg cases of *C. citricola* and *C. molluccensis* by wasps and sarcophagous flies. The parasitic dipterans are most prevalent at midday around the web complex. At this time female spiders hang by legs IV in the lower barrier web. Although this position is the most efficient for thermoregulatory behavior, it effectively decouples the spider from vibrational cues transmitted by the silk. Thus, the flies are able to move into the upper space web to parasitize the egg case without alerting the web owner. At other times of the day the female protects her brood by resting legs IV or legs III and IV on the egg case.

Spider kleptoparasites are often found in the webs of communal spiders. Such an association has been described between, e.g., theridiid kleptoparasites and the nests of communal-territorial *Cyrtophora* species (Lubin 1980b). Four *Argyrodes* species are kleptoparasitic on *Achaearanea wau* colonies (Lubin and Robinson 1982). The invading spiders not only steal prey but also attack egg sacs, young, and adults. The inhabitants of colonies with numerous kleptoparasites respond by removing the prey-capture sheet and moving to the retreats, where they wait for several days until the kleptoparasites abandon the colony.

Nest Sanitation

The elimination of wastes can affect the distribution of nutrients and parasites in the immediate environment surrounding a colony (McBride 1976). Among spiders, microorganisms such as the rickettsias and viruses are major disease-causing agents, and to a lesser extent fungal infections also affect spiders (Morel 1978). During an infection, the hepatopancreatic cells are usually affected by the organisms, which complete their life cycle when eliminated with the host's excreta. The wastes of the diseased spiders can then infect other members of the population.

Among social bees, behaviors associated with nest hygiene occupy much of the workers' efforts (Morse 1972; Visscher 1980). Dead and ill workers are carried from the hive to reduce the possibility of an epidemic. Mold and bacterial growth are inhibited by agents found in the combs, honey, and pollen stores. Propolis is used to entomb objects such as dead mice and insects that are too large to remove from the hive. The colony is thereby protected from microbial infections, since organisms covered with propolis dry out with no decay.

Many communal-nonterritorial araneids exhibit sanitary behavior analogous to that observed among the eusocial insects. Periodic bouts of nest-cleaning behavior have been observed in *Agelena consociata*, *Anelosimus eximius*, *Stegodyphus africanus*, and *Stegodyphus sarasinorum* (Krafft 1970a, 1971; Robinson 1977a). Dead nest mates, prey remains, morbid eggs, and other debris are thrown over the edge of the prey-capture sheet during nest cleaning. In *M. gregalis* the remains of prey are not removed from the web, but dead or dying spiders are transported to the exterior of the nest and morbid egg masses are destroyed (Tietjen 1980). In addition, the ring of excreta surrounding the nest of each of these species suggests that the spiders move to the exterior of the web to eliminate wastes (Bradoo 1980; Krafft 1969, 1970b; Tietjen 1980, 1981).

It is possible, however, that the ring of excreta surrounding the nest of *M. gregalis* does not result from sanitary behavior by web occupants (Fig. 7.12). Animals may deposit excreta uniformly on the web surface. Since animals tend to occupy only the upper and lateral surfaces of the web, few excreta are expected directly beneath the nest. Animals on the upper web surface may deposit excreta directly on the prey-capture sheet, but these wastes may be covered as the result of web-expansion activities. The only excreta that would be expected to fall to the ground are those produced by animals on the lateral web surfaces. Thus, the ring of excreta surrounding a web can be explained either by sanitary behavior or by the uniform deposition of wastes on the web surfaces spiders are observed to occupy.

In order to differentiate between sanitary behavior and the uniform waste deposition described above, I used a computer-camera apparatus to map the

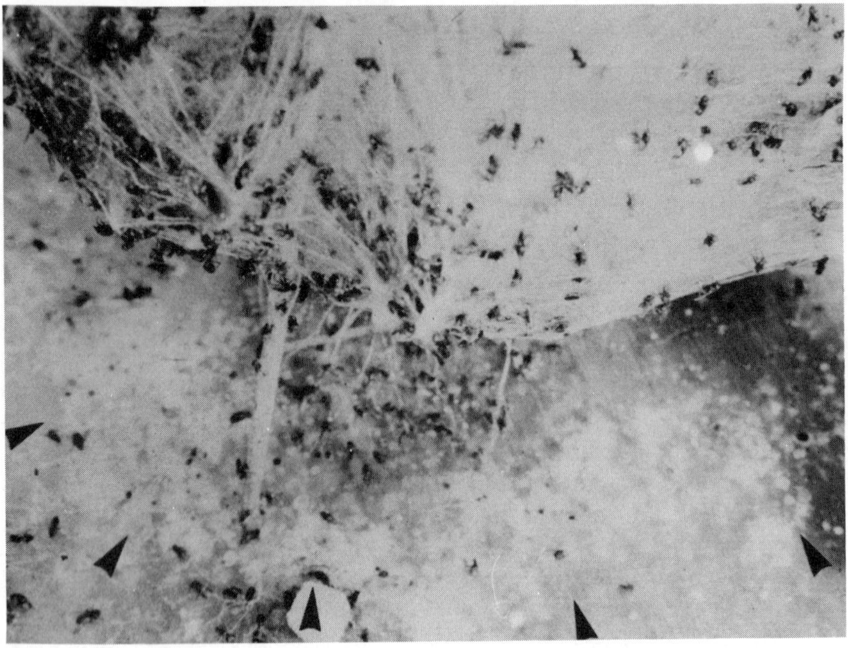

Fig. 7.12. Ring of excreta (*arrows*) surrounding a web of *Mallos gregalis* in the laboratory.

distribution of excreta as related to web density and spider movements (Tietjen 1980). Data indicated that the distribution of excreta was uniform with respect to web density. These data suggested that *M. gregalis* does not exhibit sanitary behavior, but rather deposits excreta uniformly throughout the nest. The computer was then programmed to record and analyze the movements of spiders over periods of several hours to determine which areas of the web showed the highest incidence of occupation. Data indicated that spiders occupied high-silk-density sites within the observed arenas for nearly 90 percent of the experimental period, despite the fact that such areas comprised less than 3 percent of the available sites. Comparing the observed distribution of excreta with that expected on the basis of the occupation of web areas, I found that spiders deposited excreta outside their normal areas of habitation. Thus, despite the uniform distribution of wastes with respect to web density, *M. gregalis* does exhibit sanitary behavior.

The consequences of fouling the web with excreta were tested by taking advantage of the tendency of older colonies to expand and bud. Unrestrained colonies usually exhibit lateral web expansion. Colonies confined within relatively small containers, however, can expand only by building below the primary colony. Under these conditions, the secondary colony is fouled by excreta from the nest above. As a result the lower colony exhibits both fungal and bacterial growth, whereas the upper colony shows little evidence of such infestation. Eventually, inhabitants of both colonies die, apparently from microbial infections. These spiders exhibit abdominal swelling that in some cases has enough internal pressure to rupture the integument. The liberated hemolymph has an atypical color; it is light brown, and has a greasy appearance. These symptoms are similar to those seen in other arthropods with microbial infections.

Other Nest Associates

Not all organisms associated with the communal nest are potentially dangerous to the inhabitants. Some have a negligible effect on the colony, such as the wasps that are sometimes associated with the nests of *A. eximius* (Vollrath 1978). Others may aid spiders by scavenging debris in the web, thereby reducing the chances that microbial growth will affect the colony. Various beetles, book lice, hemipterans, and moth larvae have been observed scavenging on the surfaces of social spider webs. The movements of the nest associates are such that predatory behavior by the spiders is not released. Arthropods are known to be associated with the webs of *Agelena consociata, Anelosimus eximius, A. studiosus, M. gregalis, and Stegodyphus* spp. (Brach 1977; Burgess 1979a; Diguet 1909a, 1909b; Krafft 1979; Robinson 1977a).

Laboratory-built and natural *M. gregalis* webs have a heavy, sweet odor when the colonies are healthy. When the colonies are unhealthy, they have a sour odor, and such odors often precede a population crash or mass emigration from the nest in the laboratory. In order to determine the origin of these odors, Edward Williams and I studied the microbial flora associated with the webs of *M. gregalis*. Data indicated that the silk supported little microbial growth. Flies fed upon by *M. gregalis* were found to have an altered microflora when compared both with flies fed upon by other spiders and with recently killed flies (more than 1 month old) upon which spiders had not fed. Broth cultures derived from flies removed from *M. gregalis* webs had the same sweet odor that was associated with the entire nest. Cultures grown from other flies, however, had a rancid odor. Flora from both types of cultures were used to inoculate plates for species comparison. Data indicated that two species of yeasts (based on colony morphology) were associated only with flies from *M. gregalis* webs. Mixed broth cultures of these yeasts were sweet-smelling, which suggests that the odor of *M. gregalis* webs is due to the presence of these organisms within the prey carcasses.

Diguet (1909a, 1909b) reported that in Michoacán *M. gregalis* webs are brought into homes and used as natural flypaper. He observed the high prey-capture efficiency of the webs, and suggested that the sweet-smelling nests may attract flies. To test this hypothesis, he compared the number of

Fig. 7.13. Fly pupae on the lateral surface of a web of *Mallos gregalis*. One pupa is clearly visible (*P*); the other pupae are partially hidden by flies (*arrows*) that soon after eclosion were captured by spiders. Note the entrances to the web interior (*E*).

flies landing on a web covered with paper to the number landing on a nearby paper structure. He concluded that the odor of *M. gregalis* webs attracts muscid flies. Similarly, Burgess (1979a) reported that *M. gregalis* webs in the field are surrounded by swarms of flies, and I have observed flies ovipositing on *M. gregalis* webs in the laboratory. The larvae scavenge among the carcasses and eventually pupate on the web (Fig. 7.13). On eclosion, the emerging flies are caught in the web and captured by the spiders. Prey attraction to webs has been described for a number of solitary araneids (Eberhard 1977b; Horton 1979; McKeown 1952).

In contrast, Jackson (1980a) recently reported that flies are not attracted to *M. gregalis* webs. I have since investigated possible attraction of flies by *M. gregalis* webs using a technique not available to Jackson, i.e., long-term automated recording. I recorded both the activity of a fly (its locomotion and grooming behavior) and its position within a two-choice olfactometer for a 1½-hour period. In addition, I used old *M. gregalis* webs as a stimulus, and recorded the sex of the flies. Data indicated that female flies are attracted to the webs of *M. gregalis*, which would account for the oviposition on webs described above. Most probably, the differences in the results are due to my using old web material as a stimulus and Jackson's using new web material.

Experiments are now in progress to determine if the attracting odor is due to the microorganisms mentioned earlier. If it is, then *M. gregalis* is unusual among the Araneae in using microbial symbionts for the attraction of prey.

Conclusion

The only communal spiders that have become the subjects of long-term study are *Agelena consociata*, *Cyrtophora citricola*, *Cyrtophora molluccensis*, and *Mallos gregalis*. The behavior and ecology of *A. consociata* are best understood, owing largely to the work of Krafft, and his integration of both field and laboratory research. *M. gregalis* and the *Cyrtophora* species have been studied mainly within a single setting: the former in the laboratory, the latter in the field. Studies of these species would benefit from an integrative approach. However, *M. gregalis* colonies, although locally abundant, have a patchy and unpredictable distribution, which would make it difficult for field researchers to plan a research program; and the *Cyrtophora* species would be hard to work with in the laboratory, owing to the spatial requirements of their large web complexes.

Comparative studies of communal spiders are few, and for the most part, restricted to comparisons of spiders in the same genus. Many research techniques, however, are sufficiently refined that they can be used to compare the behavior of more distantly related spiders. Perhaps such applications would produce additional unifying concepts for the study of social behavior in spiders.

As yet, the breeding structure of communal spiders is poorly understood. This is not the case among the eusocial insects, where much of the evolution of communal behavior can be explained by their unusual method of sex determination (Hamilton 1964; Lin and Michener 1972; Trivers and Hare 1976). Shear (1970a) suggested that research on the breeding structure of colonies may provide important insights into the social behavior of spiders. In addition, population studies should be carried out to determine the degree of relatedness within colonies. If intracolony relatedness is relatively high, then kin selection may be an important factor in the evolution of communal behavior among spiders.

Acknowledgments

I am most grateful to Peter N. Witt, who introduced me to research on communal spiders, and provided much encouragement during the early stages of my investigations. Thanks are also due George Uetz and J. Wesley

Burgess, who collected specimens and webs of *M. gregalis*, and Ann Rypstra, who generously allowed me to use her photographs. Alan B. Cady presented an earlier version of this chapter on my behalf at the American Arachnological Society Meeting in Knoxville, Tennessee in 1981. Allen Tubbs's comments were most helpful during the preparation of the manuscript. This research was supported by NSF Grant BNS 79–10186, and was conducted in the laboratories of the North Carolina Foundation for Mental Health Research, the Department of Biology at Georgia College, and the Department of Biology at Lindenwood College.

8

Web Building and Prey Capture in Communal Orb Weavers

GEORGE W. UETZ

COMMUNAL WEB BUILDING in spiders is a fascinating phenomenon, considering the kinds of solitary, aggressive predators that spiders are known to be. The number of recent papers on the subject attests to its interest to researchers in the field of sociobiology. (For reviews of the literature, see Shear 1970a, Burgess 1978, Jackson 1978a, Buskirk 1981, Krafft 1982, and Tietjen, this volume.) In this chapter I deal with communal web building in orb-weaving spiders of the family Araneidae, and in particular with communal web building in an orb weaver from Mexico, *Metepeira spinipes*.

Orb-weaving spiders are found in the families Araneidae and Uloboridae, and are known for their complex, intricate, and beautiful circular webs (Witt, Reed, and Peakall 1968). Several species in both of these families exhibit social aggregation and build communal webs. However, there are constraints on the sociality of orb weavers that make them an interesting subject of study within the "social" spiders. Unlike the sheet or tangle webs of communal spiders in other families (the Theridiidae, Hexathelidae, Agelenidae, Dictynidae, Eresidae, Amaurobiidae, Oecobiidae, and Pholcidae), orb webs cannot be built by more than one spider. The orb web is the result of a complex sequence of behaviors that is tightly controlled by the genetic "program" of individual spiders (Witt and Reed 1965; Witt, Reed, and Peakall 1968) and is not likely to be modified to allow participation by other individuals (Lubin 1974; Burgess and Uetz 1982). Thus, cooperative orb-web construction is precluded. Moreover, since the predatory behavior of orb weavers is tied to the use of their individual webs, orb weavers generally do not demonstrate communal prey capture and feeding. One species, *Eriophora bistriata*, a group-living orb weaver from Paraguay, does exhibit communal prey capture and feeding, but only when prey are large relative to the spider (Fowler and Diehl 1978). Prey as large as the spider or smaller are captured and fed upon solitarily. In general, it would appear that social organization among orb weavers is limited to aggregations of individuals building and occupying their own webs within a shared foundation web.

Within this set of constraints, orb weavers demonstrate several discernible types of group web building: (1) temporary and fortuitous aggregations; (2) permanent or semipermanent aggregations of two-dimensional webs built daily within a shared foundation web, off to one side of which are located individual or communal retreats; and (3) permanent or semipermanent aggregations of three-dimensional webs or web-retreat complexes in which the spiders reside.

Temporary and Fortuitous Aggregations

There are many otherwise solitary orb weavers that aggregate as juveniles or in response to particular environmental conditions, such as high concentrations of web supports or prey. All of the reviews cited above give many examples of species that form such aggregations, but a few species deserve mention here because they demonstrate a suite of characteristics that may be deemed preadaptive for social grouping of a more permanent nature.

Juvenile Webs

Many solitary orb weavers, as well as several communal orb weavers, form aggregations at some time (usually within several days) after emerging from the egg sac. This behavior is different from the maternal care patterns in other families (e.g., the building of nursery webs in the Pisauridae), since the mother is already dead and the juvenile aggregation web is spun by the spiderlings as they move about (McCook 1889). Juvenile webs have been studied in several species of orb weavers, including *Argiope trifasciata* and *Argiope aurantia* (Tolbert 1977), *Araneus diadematus* (Burch 1979), *Nephila clavipes* (Hill and Christenson 1981), and *Nephila edulis* (Austin and Anderson 1978).

Both development and environment are known to influence the nature, duration, and spacing of juvenile aggregations. For example, *A. diadematus* spiderlings, like those of other orb weavers, stay close to their siblings at first, but gradually move farther away as they mature (Burch 1979). Presumably, dispersal is stimulated by a decrease in yolk reserves and/or an increase in cannibalistic tendencies (Austin and Anderson 1978). However, environmental conditions such as temperature and humidity are also influential. Tolbert (1977) found that *Argiope trifasciata* spiderlings cluster close together in full sunlight and at higher relative humidity, and move farther apart in shade. Spiderlings eventually "balloon" by spinning silk and floating away on air currents, and it has been suggested that sensitivity to climatic variables affects the selection of the best moments for aerial dispersal (Richter 1970a, Greenstone 1981).

Hill and Christenson (1981) found that communal-web structure inhibits predatory responses in young *N. clavipes* spiderlings. While on the communal web, spiderlings feed on dead or wounded siblings and on immobile prey, but do not attack live prey. Third-instar spiderlings inhabit the communal web for some time after emergence; they then disperse, spin small orbs, and catch prey. When transferred back to a communal web, however, these spiderlings cease predatory behavior; and communal-web inhabitants transferred to orbs do attack prey. Occasionally, second- and third-instar *N. clavipes* spiderlings on communal webs were observed feeding in groups of two or more on the same prey item, although aggressive interactions over possession of prey were also observed. In my laboratory, spiderlings of *Metepeira spinipes*, a communal-territorial orb weaver, have been observed engaging in communal feeding, and possibly in cooperative prey capture of active *Drosophila*, on the communal web (M. J. Benton, pers. comm.).

Juvenile aggregations probably serve a variety of purposes; e.g., they provide protection from predators and adverse weather conditions, and facilitate development of the behavioral repertoire of the spiders. Burch (1979), working with *Araneus diadematus*, and Darchen (1965) and Krafft (1971, 1982), working with *Agelena consociata* (Agelenidae), all found that isolated spiderlings, deprived of interaction on the communal web, suffered greater mortality than communal spiderlings. These authors also found that communal life is not essential for growth or maturation, however. Possibly behaviors demonstrated by spiderlings while on the communal web serve the same function as play behavior in juvenile vertebrate predators: facilitation of development of behaviors used in adulthood in prey capture and feeding. Even though communal web structure appears to inhibit some predatory behaviors (Hill and Christenson 1981), the experiences gained on the communal web during this period of development may somehow ensure that some of the spiderlings survive. An important question arises at this point concerning the possibility that the behaviors of juveniles of solitary orb weavers differ from those of juveniles of communal orb weavers. Is it possible that, as Kullmann (1972b) has suggested, experience on communal webs influences tolerance of conspecifics or siblings in later life?

Web Clumping

McCook (1889) found that *Metepeira labyrinthea*, a common solitary orb weaver, occasionally joins webs together. This occurs in areas where microhabitat features (such as the presence of branches or leafless shrubs) permit large populations to inhabit small areas. Myers (1927) has reported occasional groupings of *Orsinome arborea*, a long-jawed orb weaver (Metinae) from New Zealand, along banks of streams and edges of ponds where spider density is high. Many other orb weavers, and many other web-building spiders, have been observed building webs together as the result of favorable concentrations of web attachments (Buskirk 1981; Burgess and Uetz 1982), and some authors have cautioned against interpreting this phenomenon as incipient sociality. For example, web clumping in *Nephila* sp., which has frequently been observed (Buskirk 1981; Struhsaker 1969), has been suggested by Farr (1976) to be a stochastic phenomenon, aggregations being likely to occur when population densities are high, or when several males are courting a single female (Christenson and Goist 1979). Recently, Farr's suggestions have been challenged on statistical grounds (Schoener and Toft 1983) and on the basis of further research into social behavior in *N. clavipes* (Brown et al. 1985; Rypstra 1985). Clusters of *Nephila* webs occur in areas of high prey density, and within these clusters aggressive interactions were not observed (Rypstra 1985). Experimental food deprivation resulted in increased aggression and dispersion of web aggregations, while supplementation of prey increased aggregation size (Rypstra 1985). Naturally occurring web aggregates may include from two to twenty individuals, and are most likely to occur in forest clearings where climatic factors and prey availability are favorable (Brown et al. 1985; Rypstra 1985; Uetz and Hodge, unpubl. observations). *Nephila* clearly demonstrates the capacity for some degree of social existence, given appropriate environmental conditions.

Web clumping associated with high prey density is commonly seen in other solitary spiders. A. L. Rypstra (1983) has reported web clumping of orb weavers in both intra- and interspecific groups in field enclosures maintained with a superabundance of prey. Honjo (1977) found that *Dictyna follicola* (Dictynidae) exhibited clumped dispersion patterns in man-made structures where fluorescent lights drew large quantities of insect prey. Aggregations of *Nuctenea sclopetaria* (Araneidae) occur on docks near water (McCook 1889), and at the Riverfront Coliseum Sports Arena in Cincinnati, Ohio (G. W. Uetz and W. Allen, unpub.), where the abundance of insect prey causes this species to exhibit the "aggregational response" to increased prey density (Riechert 1974b). During the summer, the population of *N. sclopetaria* at the Coliseum builds until it achieves high densities (100 spiders/m^3), and exhibits surprisingly small interindividual distances (10–

25 cm). Over several weeks, the daily web-building activities of these spiders result in what might appear to be a large communal mass of webbing occupied by hundreds of individuals. In actuality, this mass of webbing is composed of the degenerated remains of common attachment lines used by many spiders and a large amount of collapsed sticky thread from abandoned webs. These spiders are nocturnal, and spend the day in retreats in cracks and crevices in the building. The retreats become closely packed together, until spiders are less than 1 cm apart. Spiders capture insects, court, and mate on the "communal" web. Adult females defend their individual space off the web (or at the periphery of their orb, if they build one) while adult males wander about on the communal webbing or on the walls. Vigorous plucking of silk lines, web shaking, and grappling with intruders often occur, especially among females; these behaviors maintain interindividual spacing.

Interestingly, the behaviors exhibited by these fortuitously aggregating but solitary spiders are similar to those exhibited by tropical communal orb weavers like *Metabus gravidus* (Buskirk 1975a, 1975b, 1981) and *Eriophora bistriata* (Fowler and Diehl 1978). Both of these species emerge daily from a communal retreat and spin individual orb webs in a preexisting irregular silk scaffolding. Several authors have suggested that aspects of orb-weaver behavior (or, for that matter, of the behavior of most web-building spiders) such as tolerance of shared attachment lines, communication through web plucking, and temporary communal aggregation of juveniles may constitute preadaptations for communal web building (Shear 1970b; Kullmann 1972b; Burgess and Witt 1976; Burgess and Uetz 1982). It is tempting to speculate that the evolutionary pathway to the kinds of communal behavior exhibited by *M. gravidus* and *E. bistriata* might include a stage represented by the behaviors seen in, e.g., *N. sclopetaria* (Burgess and Uetz 1982).

Aggregations of Two-Dimensional Webs

Social organization in the Central and South American orb weavers *M. gravidus* and *E. bistriata* consists of permanent aggregations of two-dimensional orbs built daily within a communal silk scaffolding, off to one side of which is located a communal retreat.

Buskirk (1975a, 1975b) studied *M. gravidus* in Costa Rica, and found colonies of 5 to 75 individuals building webs attached to logs and tree branches over running water. Colonies remain populated year-round, but spiders are most numerous between May and October. Activity is diurnal, but spiders emerge from the retreat at different times to spin their individual webs, depending on hunger level and previous feeding time. Prey captured on individual sticky spirals are not shared. Buskirk (1975a) has shown that spacing within colonies is a result of social interactions. For example, individuals on their webs usually maintain a distance of 16 to 22 cm from their nearest neighbors. However, females allow males to approach significantly closer than other females. Moreover, intruding spiders release different levels of agonistic responses depending upon their distance from the web owner. Despite the agonism that takes place on individual webs during the day, there is no cannibalism, and, after they take down their orbs in the evening, spiders pass the night close together in the communal retreat (Buskirk 1975b).

Many early naturalists, including Darwin (1845), wrote about a large communal araneid (body length 2.5 cm) living in colonies of from 30 to 40 individuals on up to several thousand individuals in trees in the drier regions of Paraguay, Argentina, and Brazil (Fowler and Diehl 1978). It is not certain that all of these early reports were of the same species, but they probably were, the species in question being *E. bistriata*. Like *M. gravidus*, *E. bistriata* builds two-dimensional orbs in a communal silk scaffolding, and spends time off the web in a communal retreat. Unlike *M. gravidus*, *E. bistriata* is nocturnal, and spends the day in the retreat. A network of thick, communally spun silk lines radiates out from the retreat to nearby trees and shrubs. At night, spiders emerge from the retreat and spin individual orbs in the upper portions of these foundation lines; they return to the retreat at dawn (Fowler and Diehl 1978).

E. bistriata is the only communal orb weaver known to exhibit facultative communal prey capture and feeding. H. G. Fowler (pers. comm.) observed individual prey capture and defense of captured prey when prey were the same size as the spider or smaller. However, when prey were larger than the spider (1.5–2.25 × spider body length), two or more spiders were involved in capture and feeding (Fowler and Diehl 1978).

Aggregations of *E. bistriata* are not permanent in that, upon reaching sexual maturity, females and males disperse. Dispersal of adults occurs in early March, at which time females find sites away from the colony, lay eggs, and die. Spiderlings emerge from egg sacs in May, and establish communal groups that remain together throughout the year. Populations of *E. bistriata* vary in size and degree of developmental synchrony from generation to generation. The univoltine, developmental communality of this spider is similar to that of the theridiid *Anelosimus studiosus* (Brach 1977). In *E. bistriata*, the synchronous development of group-living siblings, facultative communal prey capture and feeding, and the dispersal of adults to lay eggs (and thus start new colonies) are all adaptive for colonizing suitable microhabitats in a semiarid environment where prey may show patchiness in space and/or time (Fowler and Diehl 1978). In any case, the social behavior of *E. bistriata* is highly advanced for orb weavers, and perhaps the developmental coloniality of this species represents another alternative route to sociality, as Kullmann (1972b) suggested for other families of spiders.

Aggregations of Three-Dimensional Web Complexes

There are solitary orb weavers that build three-dimensional web complexes combining a catching spiral and a barrier or space web. The three-dimensional space web may serve a prey-knockdown function, as in *Cyrtophora*, which build a nonadhesive horizontal orb (Lubin 1973, 1974). The space web may also provide a living site where a retreat may be built off the sticky spiral, as in *Metepeira* (McCook 1889; Burgess and Witt 1976; Levi 1977b). Lubin (1978) has suggested that three-dimensional barrier webs may also serve a moisture-collection function in arid environments.

Orb weavers building three-dimensional web complexes may represent a series of steps along another path to sociality, achieved by the joining together of separate living units to form a communal web (Lubin 1974). Individuals of *Cyrtophora citricola*, a species from Mediterranean and equatorial Africa (Kullmann 1958; Rypstra 1979), and *Cyrtophora molluccensis*, a species from New Guinea (Lubin 1974), build horizontal orbs surrounded by a buffer zone of space threads, which are attached to one another in a communal web. These species live in a communal-territorial manner (Jackson 1978a), and actively defend their individual webs by using agonistic behaviors such as leg jerking and web tensioning (Lubin 1974; Rypstra 1979). Individual spiders of these species accrue both benefits and disadvantages from living in permanent aggregations. Disadvantages include the conspicuousness of the web, which attracts predators and parasites, and causes prey to avoid the web (Lubin 1974). The disadvantages are offset somewhat by an increase in prey-capture efficiency at night (Rypstra 1979), and even during the day as a result of the combined knockdown effect of grouped webs (Lubin 1974). Moreover, permanent aggregation of individuals of all ages provides young spiders with protection from predators, and allows spiders to exploit and monopolize habitats that solitary spiders may not be able to use (see Tietjen, this volume, for a more complete discussion).

Within the genus *Cyrtophora*, there are several species that demonstrate different levels of social organization, ranging from solitary web building to facultative communal web building in favorable locations, on to permanent communal web building with overlapping generations (Lubin 1974). The large communal webs of *C. citricola* and *C. molluccensis* may contain several hundred individuals, and perhaps represent a logical conclusion to an evolutionary trend in this genus toward larger body size and web size (Lubin 1974). In *C. molluccensis* it is clear that this trend has facilitated aggregation,

Fig. 8.1. Communal web of *Metepeira spinipes*. Several catching spirals are visible.

and the invasion of treetops and gaps between trees, which contain abundant prey and few competing species. In both species, it is clear that increased efficiency in the capture of prey, particularly of larger prey and nocturnal prey, has played an important role in the evolution of communal web building (Rypstra 1979).

Metepeira spinipes, found in central Mexico, occurs solitarily, but more frequently occurs in aggregations of 5 to 150 or more individuals (Fig. 8.1). My colleagues and I have studied aspects of the behavior and ecology of this species during several trips to Mexico over the last ten years.

The web of individual *M. spinipes* is characteristic of the genus. It consists of a vertical, two-dimensional catching spiral and a three-dimensional space web with a retreat (Fig. 8.2). The retreat is a small cone of silk, which is often covered with insect remains or plant debris. Signal threads connect the hub of the spiral and the retreat, where the spider rests (McCook 1889; Burgess and Witt 1976). This three-dimensional spiral space-web composite is attached at its periphery to vegetation or other supporting structures by many reinforced guy lines. In the groups observed, individuals had their own retreats and spirals, but space webs were joined (Uetz and Burgess 1979). Although spirals are renewed frequently (usually daily), the communal space web persists and is elaborated by the web-building activities of numerous individuals. The resultant communal web is a mass of interconnected individual webs that support one another, attached at its periphery to the vegetation. Although the communal space web is built and maintained by the efforts of all colony members, and individuals interact while on the space web, each individual captures prey on its own spiral. The communal-territorial behavior exhibited by *M. spinipes* suggests that this species represents an intermediate stage in the evolution of social behavior in orb-weaving spiders, and thus makes it an interesting subject of study.

Group size in *M. spinipes* varies geographically, apparently in response to prey availability and environmental conditions. In severe habitats, where prey availability is low and environmental conditions are extreme (e.g., in high-altitude areas, or in desert grassland), spiders are predominantly solitary or occur in small groups. In intermediate habitats (e.g., agricultural areas with seasonal rainfall), spiders more frequently occur in aggregations. In favorable habitats, where climate is benign year-round and insects are abundant (e.g., in tropical areas), colonies are large.

In the following pages, I shall discuss web construction, web spacing, habitat and colony formation, aggressive interactions, and prey capture in *M. spinipes*. The behaviors discussed will serve to illustrate the biology of communal-territorial orb weavers; and it will be possible, following the discussion, to draw some conclusions regarding the evolution of communal web building among orb weavers.

Web Construction in Metepeira spinipes

The web-construction sequence in *M. spinipes* differs in some respects from the sequence in *Araneus diadematus* and other two-dimensional orb-web builders described by Witt, Reed, and Peakall (1968). Since the bulk of the space web persists from day to day in solitary and communal individuals, the sequence described here applies to both kinds of spiders. The supporting elements (the radii, frame, and signal threads) are constructed first (Figs. 8.2a,b). Radii are laid on approximately opposite sides of the hub in a manner that appears to balance tension within the structure. During this phase, the spider stops frequently to pluck the radial strands and reinforce the connections. Once all the radii are in place, a nonsticky spiral is constructed from the center out (Fig. 8.2c). After returning to the hub and plucking the web several times in all directions, the spider drops to the lower end of the frame and begins weaving the lower section of the sticky catching spiral (Fig. 8.2d). After completing this section of the spiral, the spider returns to the hub for more plucking and strengthening of hub connections. Then the central portion of the sticky spiral is constructed, from the outside in, in a continuous circling of the web (Fig. 8.2e). The distance between spiral turns is apparently measured with legs IV. The nonsticky spiral is taken up as the sticky spiral is laid, and ingested upon completion of the web. The

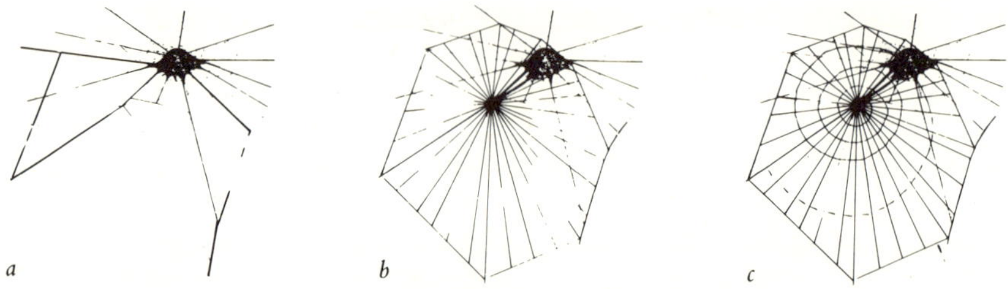

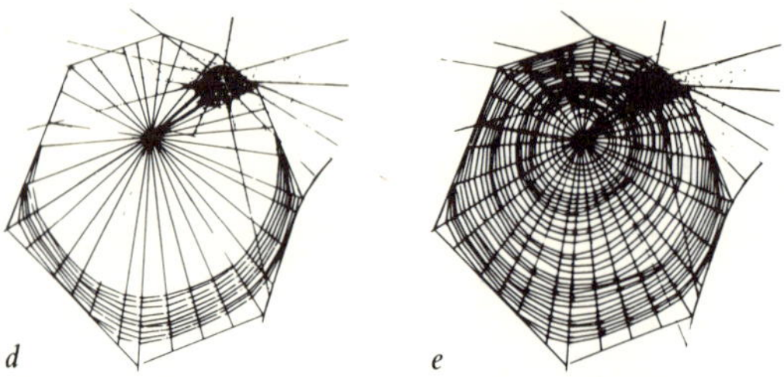

Fig. 8.2. Stages of web construction in *Metepeira spinipes*: *a*, retreat in space web; *b*, radii, frame, and signal threads; *c*, temporary (nonsticky) spiral; *d*, lower portion of sticky spiral; *e*, central portion of sticky spiral.

spider then returns first to the hub and finally to the retreat, where it rests with forelegs touching the signal threads. The entire web-construction process may take between ¾ and 3½ hours, depending on environmental conditions (heavy dew on webs inhibits web building) and on whether the spider experiences interference from other individuals. The construction process involves several successive stages and return loops, as indicated by a flow diagram (Fig. 8.3).

Spirals are renewed frequently, often daily. There is apparently a diurnal rhythm of web-building and other activities that is the same for solitary and communal individuals. The following description of the activity pattern is based on approximately 108 observer hours of study (686 observations of solitary spiders and 4,665 observations of communal spiders).

In the morning (0700–1200 CST), spiders are finishing their webs and returning to their retreats (Fig. 8.4a). Much of the daily prey capture occurs in this time period, when many pollen- and nectar-feeding insects are active. At midday, and for most of the afternoon (1200–1730), spiders are inactive in their retreats (Fig. 8.4b). However, since they sit in retreats with forelegs touching signal threads, they are in a capture-ready position. Vibrations of insects struggling on spirals are detected instantly through the signal threads, and spiders rush out onto the webs to capture the prey. Spiders in retreats also escape exposure to direct solar radiation, which may be intense in the afternoon hours. Later in the afternoon, a daily rainfall occurs, at least on most days during the rainy season (May–September). In the evening (1730–2400), as darkness falls, the ambient temperature drops from about 25 to

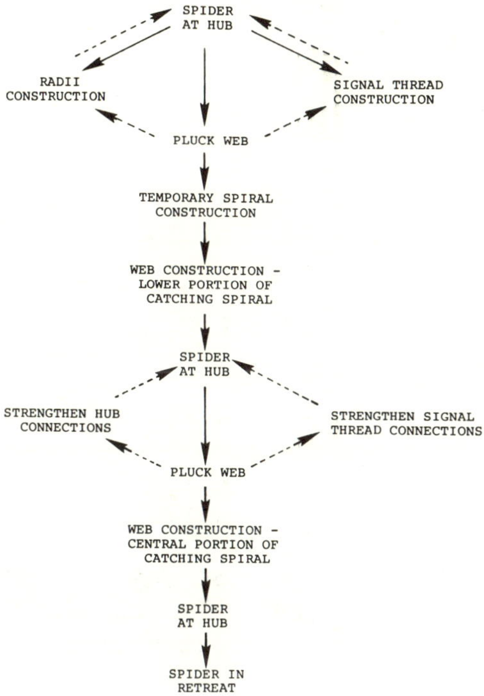

Fig. 8.3. Flow diagram of web construction in *Metepeira spinipes*.

16° C (Fig. 8.4c). Spiders become active at dusk, leaving their retreats to sit on the hub and capture prey (mosquitoes are active at this time), and to begin spiral destruction. Spiders are most active during the night (0000–0700), when spirals are taken down and renewed (Fig. 8.4d). (For a description of web removal in *Metepeira labyrinthea*, a solitary, temperate-zone species, see Carico, this volume.) During the most active period of web construction (0400–0900), agonistic interactions are frequently observed, and these often result in the stopping and restarting of construction. The frequency of agonistic encounters may be as high as one per spider per hour.

If all behavioral observations are grouped, there are no statistically significant differences between the activities of solitary individuals and those of communal individuals (*G*-test, $0.4 < p < 0.5$). However, if time periods are examined separately, communal individuals appear to spend more time in web construction and maintenance, and in agonistic interactions. Agonistic encounters involve web plucking and shaking, charging of intruders, and grappling with fangs open (but without biting). Among communal individuals, agonistic encounters are usually associated with reorientation or relocation of spirals within the communal space web, activities performed at

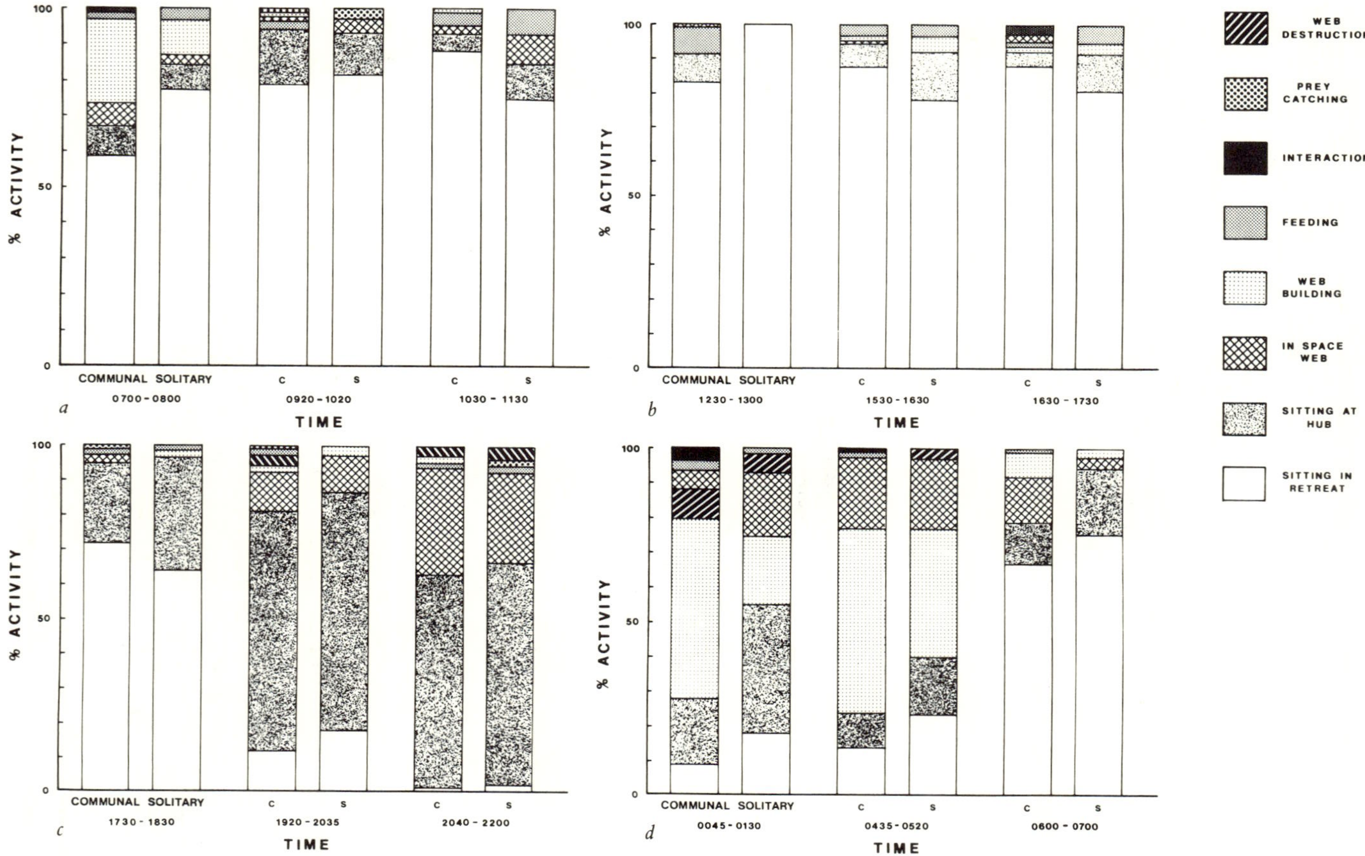

Fig. 8.4. Activities of communal and solitary individuals of *Metepeira spinipes* throughout the day: *a*, activities in the morning (0700–1200 CST); *b*, activities in the afternoon (1200–1730); *c*, activities in the evening (1730–2400); *d*, activities at night (0000–0700).

night. Agonistic encounters that occur during daylight hours are associated with prey-stealing attempts, web invasion, male defense of females against other males, and female responses to courting males. But contests associated with web-building activities are far more common, and influence the spatial arrangement of webs within colonies.

Web Spacing in Communal Metepeira spinipes

In communal groups of *M. spinipes*, the individual web is attached at its periphery to other webs—so that, as a result, it is impossible to determine an individual's contribution to the communal webbing that supports it. Uetz and Burgess (1979) measured polygons of communal webbing in order to estimate the volume of communal webs of *M. spinipes*. When the amounts of colony volume per individual are compared across a range of group sizes, it is obvious that the amount of space for each spider decreases as group size increases (Fig. 8.5). Individual space is compressed in larger groups of spiders, much more than if individual webs were merely attached to one another at their periphery. This compressibility indicates that these spiders tolerate conspecifics at closer distances than if webs were organized in "modular unit" fashion, as in a simple aggregation formed in an area affording a high concentration of web attachments or prey. Measured nearest-neighbor distances (i.e., distances between retreats) also follow this pattern (Fig. 8.6). These measurements suggest that although interindividual spacing is compressed in larger groups, there is nevertheless a lower limit to the distance maintained between neighbors.

Analysis of data pertaining to nearest-neighbor distances in colonies in different habitats reveals some interesting patterns (Fig. 8.7). Within-colony nearest-neighbor distances apparently vary with habitat (Uetz, Kane, and Stratton 1982), smallest distances being seen in habitats with benign climate and high insect activity, and greatest distances in severe habitats with low insect activity. In Córdoba, a tropical moist forest area, individuals are packed tightly together in large colonies of about 1,500 to 2,000 individuals

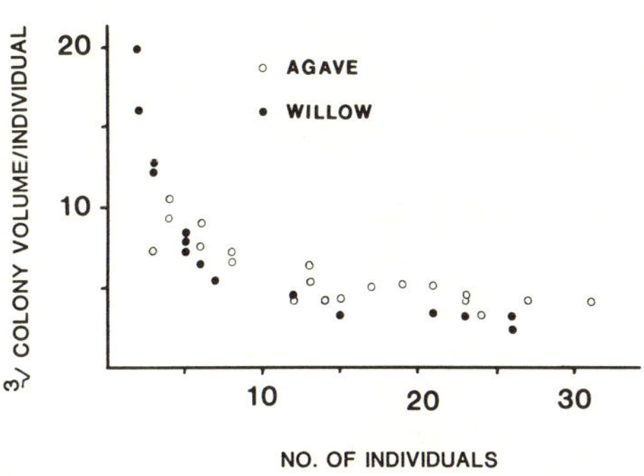

Fig. 8.5. Amount of colony volume per individual vs. colony size in *Metepeira spinipes*. Colony volume is expressed as a cube root for linearization.

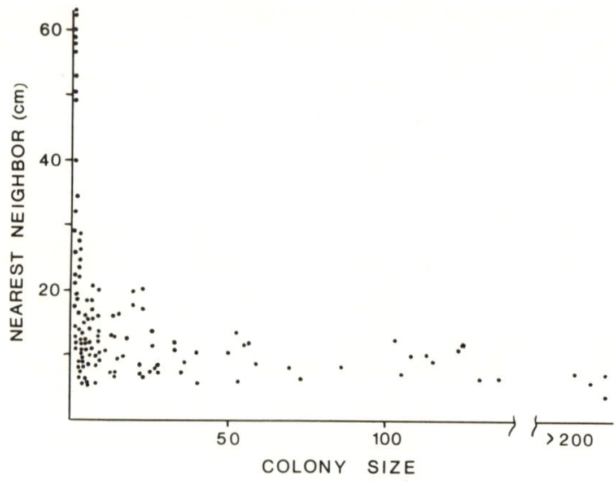

Fig. 8.6. Nearest-neighbor distance vs. colony size in *Metepeira spinipes*. Each dot represents a colony mean.

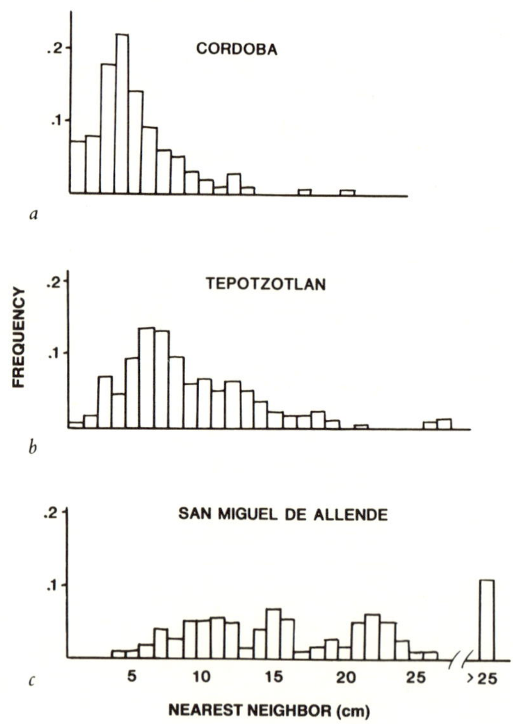

Fig. 8.7. Frequency distributions of within-colony nearest-neighbor distances in colonies of *Metepeira spinipes* at three sites: *a*, five colonies of approximately 1,500–2,000 individuals; *b*, 25 colonies of 25–150 individuals; *c*, 25 colonies of 3–15 individuals.

(Fig. 8.7a). The most frequently seen nearest-neighbor distance is 3 to 5 cm, which is very close, considering that these spiders have a body length of 6 to 10 mm. At the other extreme, in San Miguel de Allende, a desert grassland area, spiders are in small groups (3–15 individuals) and exhibit greater variance in spacing and greater nearest-neighbor distances (Fig. 8.7c). Tepotzotlán, an agricultural area with seasonal rainfall, is intermediate between the two extremes, and spiders there show intermediate group sizes (25–150 individuals) and intermediate nearest-neighbor distances (Fig. 8.7b). The influence of environment on availability of prey is reflected in the differences in nearest-neighbor distances. When a variety of sites are considered, there is a significant negative correlation between nearest-neighbor distance and prey availability (Fig. 8.8; $r = 0.98$, $p < 0.001$) (Uetz, Kane, and Stratton 1982). Field experiments in which colonies were moved and reestablished in areas of lower prey activity confirmed this finding: nearest-neighbor distance increased after transport, except where prey supply was supplemented by addition of attractants like cow dung or fruit (Uetz, Kane, and Stratton 1982).

Apparently, individuals of *M. spinipes* can tolerate conspecifics at closer distances in areas of higher prey availability. Riechert (1978a) found a similar reduction in interindividual distance over a latitudinal gradient in a desert funnel-web spider, *Agelenopsis aperta* (Agelenidae). This solitary spider is highly territorial, and exhibits a uniform dispersion pattern, which is maintained by aggressive interactions between individuals. The area that an individual spider defends may be larger than the spider's actual web; its size depends on the energy needs of the spider and the availability of prey. In lower latitudes or environments where climate is moderate and prey availability is

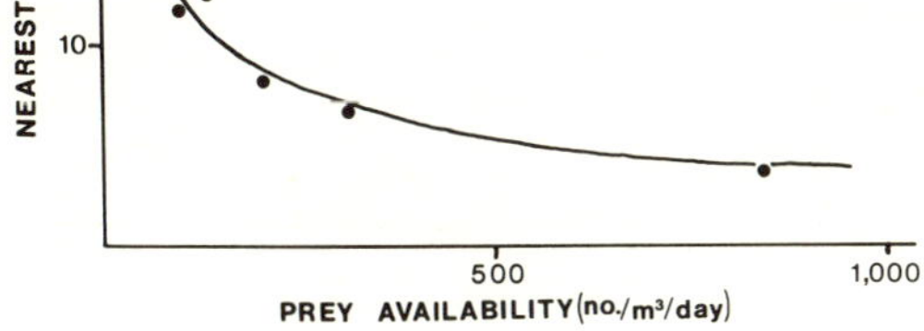

Fig. 8.8. Grand mean nearest-neighbor distance at a site vs. estimated prey availability (number of prey per cubic meter per day) in *Metepeira spinipes*.

high, territory size decreases. However, the fact that these distances can be changed experimentally by manipulation of prey availability suggests that the social spacing of this species is flexible, and not completely genetically fixed as in *A. aperta* (Riechert 1981). The rapid changes in spacing observed during the above-mentioned experiments (with nearest-neighbor distances changed within 10 days) is probably a function of the fact that *M. spinipes* individuals renew their catching spirals daily. Studies of foraging behavior in a variety of predatory animals have shown that activity level, aggression, and territory size increase over time with starvation, and decrease with satiation (Riechert 1978a, 1981). Thus, within-colony spacing patterns may reflect the nutritional state of the spiders, which in turn is influenced by the effect of environment on prey availability. These influences result in much variability in within-colony spacing. Recent studies have shown that in addition to environmental factors (e.g., food availability), there may also be genetic differences between populations that influence spacing and tolerance of other individuals (Uetz et al. in press; Uetz and Cangialosi, in press). These findings suggest that populations of *M. spinipes* represent either genetic subgroups within the species or newly evolved (or evolving) species with behaviors adapted to the differing selective pressures in their particular environments.

Burgess (1979a) and Burgess and Uetz (1982) have used a successive nearest-neighbor technique derived from plant-ecology sampling methods (Thompson 1956) to assess the dispersion patterns of spiders within communal webs. The method involves comparison of means of distances between first through twelfth nearest neighbors within a colony with expected values generated by a random-dispersion distribution model (Fig. 8.9)

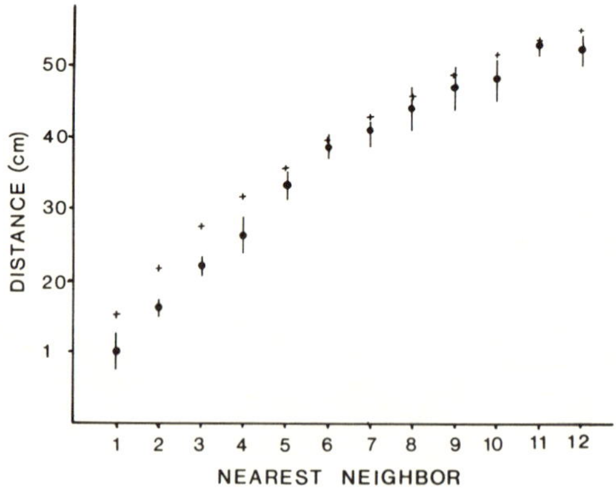

Fig. 8.9. Comparison of mean first through twelfth nearest-neighbor distances within a colony of *Metepeira spinipes* in Tepotzotlán with values generated by a random-dispersion distribution model. ●, actual values; +, expected values. The data indicate aggregated dispersion.

(Burgess and Uetz 1982). If measured values sufficiently exceed expected values, a uniform or regular dispersion of spiders is indicated; if measured values are sufficiently lower than expected values, an aggregated dispersion is indicated. Of 13 colonies in Tepotzotlán for which we had sufficient data to permit this type of analysis, three showed an aggregated dispersion, four a uniform dispersion, and six a random dispersion of spiders within the communal web. Moreover, in these colonies there was no apparent relationship between spacing pattern and body size. To confound matters further, orientation of spirals within colonies varied greatly and was not associated with any given dispersion pattern. In other words, a colony that showed an aggregated dispersion of spiders may have demonstrated aggregated, random, or uniform spiral orientations. Some colonies appeared to have all spirals oriented in the same plane, as if prey were coming from the same direction. Others had all spirals facing out from a central point where retreats were located, which may be a response to the radial symmetry of some plants (e.g., *Agave* and *Yucca*). Still others showed a variety of web orientations, having no discernible pattern.

It is apparent that a variety of factors must influence the spatial arrangement of webs within communal groups of *M. spinipes*. These factors probably include prey directionality, habitat architecture, size and age distribution of spiders, and silk density. The hierarchy of web-site quality within colonies, and accompanying hierarchy of individuals occupying sites, that have been seen in *C. citricola* (Rypstra 1979) have not been observed in *M. spinipes*, but such hierarchies may exist. At the Córdoba site, where generations overlap, spiderlings and small immatures occupy web sites at the periphery of colonies, whereas adults and subadults are distributed more in the center. Whether this is due to lifetime occupation of sites by individuals, or web architectural constraints based on differences in body size, or some other factor is not yet known. Obviously, the topic of within-colony spatial arrangements requires further research.

Habitat and Colony Formation in Metepeira spinipes

Studies of spider sociality suggest that a major selective factor leading to the development of communal web building is the potential for utilizing habitats or capturing prey that are not available to solitary individuals or other solitary spiders (Lubin 1973, 1974; Brach 1975, 1977; Buskirk 1975b, 1981; Rypstra 1979). For example, some communal orb weavers exploit large spaces between trees (Lubin 1974) or over streams (Buskirk 1975b) where no solitary spider could construct a web. The communal webs of *M. spinipes* may allow this species to occupy sites at the top of *Agave* or between large branches of *Opuntia*, to which solitary spiders may not have access. It is also possible that communal webs are architecturally more stable than solitary webs, and better able to withstand wind and rain encountered in open microhabitats. Spiders occupying these sites may also have access to kinds of flying prey not present in other microhabitats.

In general, both solitary and communal individuals of *M. spinipes* attach their webs to stout vegetation or other structures of some substance (Uetz and Burgess 1979); *Agave*, *Opuntia*, other cacti, mesquites, succulents, and shrubs, as well as signposts, power lines, and fences, are common web-attachment sites. Webs are found as low as 50 cm above the ground and as high as 100 m in the air on power lines.

Arachnologists traveling in Central America have noted enormous colonies of web-building spiders in power lines in mountain areas (J. W. Burgess, pers. comm.; P. N. Witt, pers. comm.; W. B. Peck, pers. comm.; W. J. Gertsch, pers. comm.). Until now, the identity of the "power-line spider" has been unknown, because these lines were high above the ground and inaccessible. In Córdoba, we found a midelevation area where these colonies had spilled over onto the ground, and discovered that they were made by *M. spinipes*. Obviously, the habitats occupied by this species are varied, but they all provide firm supports for web foundations and access to open space.

The communal space web may support dozens to thousands of individual webs. In larger colonies, the communal web is irregular in shape, and is draped over supporting vegetation, or suspended between structures by numerous reinforced guy lines. Two very large colonies in Córdoba and Orizaba were measured in an attempt to estimate colony size (Fig. 8.10). These colonies were found to contain approximately 5,300 and 6,500 individuals, respectively. Sixty-five hundred spiders, of approximately 70 to 100 mg live weight each, dispersed at densities of approximately 100 to 120 spiders/m^3, and suspended by fine silk scaffolding up to 10 m across, must require considerable support. The guy lines supporting colonies of this size are obviously permanent, in the sense that only natural disasters like falling trees or cyclonic storms, or human destruction, could dislodge them. They represent the construction efforts of hundreds or even thousands of individuals, and are capable of withstanding high winds and heavy rain.

As noted earlier, *M. spinipes*, though found solitarily, more frequently occurs in aggregations of 5 to 150 or more individuals. In most localities, these aggregations represent an annual cohort of individuals hatched at approximately the same time, or successive cohorts hatching from serial egg sacs. Mating and oviposition occur in the late fall. Individuals overwinter (pass the dry season) as eggs or spiderlings in the egg sac. Only in the moist tropics, where reproduction appears to occur year-round, is there overlap of a wide variety of age classes. Aggregations of *M. spinipes* remain together for long periods of time, and in most habitats usually consist of the offspring of the residents of the previous year.

Studies of marked individuals reveal that groups of adults and juveniles show almost no change in numbers for periods of up to 28 days. Spiders are

Fig. 8.10. Portion of a colony of *Metepeira spinipes* of approximately 6,500 spiders. Author on left, Michael Benton on right.

most likely to change web sites or leave colonies, when they molt, particularly in early instars. Although movement between colonies is not common, individuals from different colonies will build webs together, even when the colonies are in areas hundreds of miles apart (Burgess 1978). Observations of transplanted colonies in the field, as well as laboratory studies, indicate that individuals will build together even on open, unoccupied sites. When spiders are released onto unoccupied or cleared web sites in the field, or into containers in the laboratory, the process of web construction is always the same. Individuals climb to upper portions of the plant or container, laying silk as they go, and eventually develop bridge lines between supports. Once bridge lines are established, all spiders use them as web-attachment points, even if they did not take part in construction of the bridge lines. The rebuilding of a colony is accomplished by simultaneous (but not coordinated) web-building efforts of individuals. During such web-construction efforts, agonistic behavior is rarely observed. Spiders contacting each other merely detour around each other and continue the laying of framework lines. There appears to be a tendency to attach lines to other lines, as has been observed in other communal-web-building spiders (R. R. Jackson, pers. comm.). This may explain how spiders can establish webs over such large spaces without actual cooperative effort.

In nature, most web sites are repopulated annually by hatchlings from eggs laid by previous occupants of the site. Often the framework of a communal web, complete with egg sacs left behind by deceased females, persists throughout the dry season (November–April). Even when the framework itself is destroyed, egg sacs remain tethered to the vegetation by strong lines. When egg sacs are removed from a site, few if any spiders inhabit that site the following season, or even later. Thus persistence of a colony from year to year is dependent on successful reproduction of previous inhabitants, or on an influx of new colonists.

As spiderlings emerge from egg sacs at the onset of the rainy season (in May), they build webs on preexisting silk or vegetative supports. Laboratory observations of eclosion indicate that if silk is present, spiderlings remain aggregated and construct a communal web around the egg sac. If silk is not present, they disperse in various directions, and end up in the corners at the top of the cage. The first webs constructed are typical planar orb webs, lacking retreats and signal threads. After the next molt, spiders construct the typical spiral space-web composite that they will continue to build throughout their lives.

At the lower-elevation tropical forest site in Córdoba, the climate is moderate and moist year-round. Reproduction occurs throughout the year, and overlapping generations of individuals can maintain continuous occupation of a site over time. Spiderlings in some of these colonies appear to be distributed about the periphery of the communal web, or at the ends of guy lines attached to vegetation. Many of the large colonies in this area have persisted for years. Local residents told us that one colony, containing perhaps 10,000 individuals, had been there as long as they had lived there (about 12 years) and in this time had almost doubled in size.

Benefits and Costs of Communal Living

For group living to be selectively advantageous to solitary predators like spiders, the benefits accruing to the individual from communal living must exceed the costs. Benefit/cost ratios are probably expressed by survival rates, or by energetic expenditures associated with each aspect of group living. Since we have not measured either of these directly, we must rely on measurements of, e.g., time expanded in various activities, frequency of behaviors, and prey-capture rates as indirect indicators of cost or benefit.

A likely cost of communal web building is energy spent in aggressive encounters, or prey-catching time lost as the result of such encounters. Tolerance of conspecifics at close distances is at odds with the usual territorial, cannibalistic nature of spiders. Conspecifics constitute a significant portion of the diet of many species (Edgar 1969). We have observed very little evidence of cannibalism in *M. spinipes*, although it does occur occasionally. Aggressive encounters in *M. spinipes*, and in other communal orb weavers (Rypstra 1979; Buskirk 1975a), involve web shaking and plucking, which escalate to grappling, during which fangs are open but rarely used to bite. These findings are consistent with the finding of Riechert (1982) that territorial conflicts in spiders are more frequently resolved by communicative display interactions than by actual combat. Aggressive interactions in *M. spinipes* increase with colony size, and an individual in a colony of 50 is challenged for its web site approximately four times daily during the web-building period (Uetz, Kane, and Stratton 1982). Time spent in aggressive encounters is clearly a cost of group living, a cost that increases as groups grow larger.

Another potential cost of communal web building is increased visibility, and thus vulnerability, to predators (Rypstra 1979) or parasites (Lubin 1974). In a study comparing the reproductive success of solitary and communal individuals of *Philoponella oweni* (Uloboridae), Smith (1982a) found egg-sac parasitism to be an important cause of mortality. Communal individuals experience a significantly higher rate of parasitism by chalcid wasps than solitary individuals, presumably because communal groups are easier for the wasps to find. However, since communal individuals lay more eggs than solitary individuals, there is no significant difference between them with respect to the number of spiderlings produced. Smith suggests that because solitary and communal existence led to equivalent reproductive success, *P. oweni* individuals may exhibit either mode of existence. If this is so for other spiders as well, then perhaps other factors are more influential in tipping the benefit/cost ratio in favor of the evolution of social grouping.

Communal web building may in fact serve to protect individual spiders from predation, in a number of ways (Burgess and Uetz 1982). First, there is a geometric effect of clustering that can make predators' searching more difficult. Using a mathematical model, and the example of fish schooling, Brock and Riffenburgh (1960) demonstrated a decreasing probability of detection of individual prey by solitary predators as groups grow larger and cluster more tightly together. This effect would presumably hold for other group-living animals, including communal spiders.

Spiders may also gain protection from predators by securing a position on the inside of the web complex. Galton (1871), Williams (1964, 1966), and Hamilton (1971) have all noted the protective advantages of living within a "selfish herd." For spiders, individuals located on the outside of a colony are most likely to be preyed upon first; thus other colony inhabitants are safer from predatory attacks. The complexities of tangled layers of silk provide an extra buffer to spiders occupying inner web sites. Rypstra (1979) has shown that position within colonies of *C. citricola* is important. Spiders on the periphery of colonies are vulnerable to avian predators, whereas those well inside are protected. However, interior sites receive fewer prey. Intermediate sites (just inside the colony) provide the best combination of protection and prey, and are thus preferred. A significantly higher number of aggressive disputes over web sites occurs in these areas.

As mentioned earlier, a major factor leading to the evolution of communal web building is the potential for building webs in spaces to which solitary spiders may not have access (Lubin 1973, 1974; Brach 1975, 1977; Buskirk 1975b, 1981; Rypstra 1979). Communal webs may occupy open spaces or "flyways" between vegetation, and thus provide spiders with access to different kinds of flying prey. The architectural stability of communal webs may make them better able to withstand the effects of wind and rain encountered in open spaces, and allow spiders and their progeny to monopolize web sites for long periods of time. This is an important advantage, since competition for web sites providing optimum protection from the elements and positions near prey attractants may be considerable (Riechert 1981). In addition, spiders building their individual webs within preexisting, well-anchored silk scaffolding may capitalize on the silk and energy investments of other spiders and save on the energetic costs of web maintenance.

Communal spiders also have an advantage not shared by other group-living animals, in that the communal web serves both as a place to live and as a medium for transmission of vibrations. Spiders may use vibration signals as a means of communication (Witt 1975) or as a means of obtaining information about insect prey flying into, or predators encroaching upon, the web. My colleagues and I have seen several examples of the latter use of vibration signals in observing colonies of *M. spinipes*. On many occasions, when we attempted to capture spiders, antipredatory behaviors such as web flexing and vibrating (Tolbert 1975) and escape maneuvers were employed. During web disturbances many individuals are clearly aware of an intrusion, and often four to six individuals join in web shaking. The resultant mass response is impressive and startling, and would undoubtedly serve to confuse or even drive off some predators, such as hovering birds and wasps. The only predator that we observed, an unidentified orange wasp, never succeeded in capturing a spider despite numerous attacks on spiders in their retreats. As the wasp approached each retreat in succession, the spider dropped down onto the ground, apparently alerted by vibrations in the silk created by the wasp's wing beats. Perhaps communal webs perform an antipredatory function by serving as an "early warning system" (Burgess and Uetz 1982).

When an insect strikes the web of one individual, the vibrations of the ensuing struggle are transmitted through the silk to other individuals. Usually three or four spiders respond to these vibrations, either by running out to the hub of their own spiral and assuming a position of capture readiness

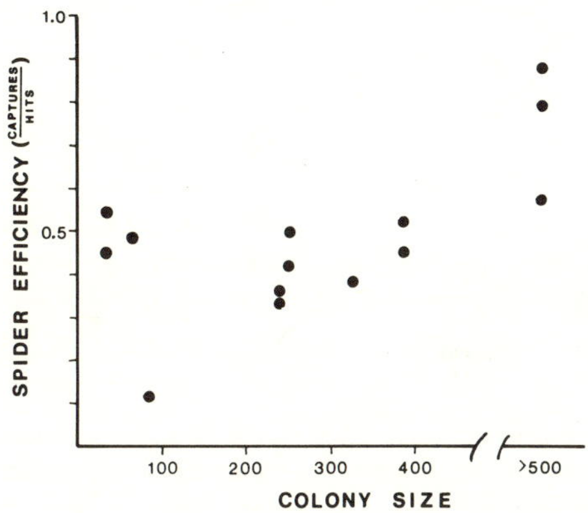

Fig. 8.11. Prey-capture efficiency (captures/hits) vs. colony size in *Metepeira spinipes*.

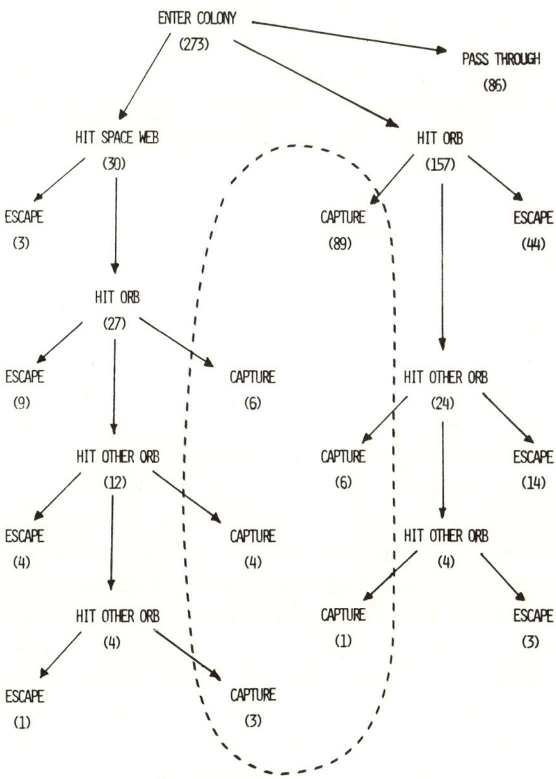

Fig. 8.12. Fate of insects flying into colonies of *Metepeira spinipes*.

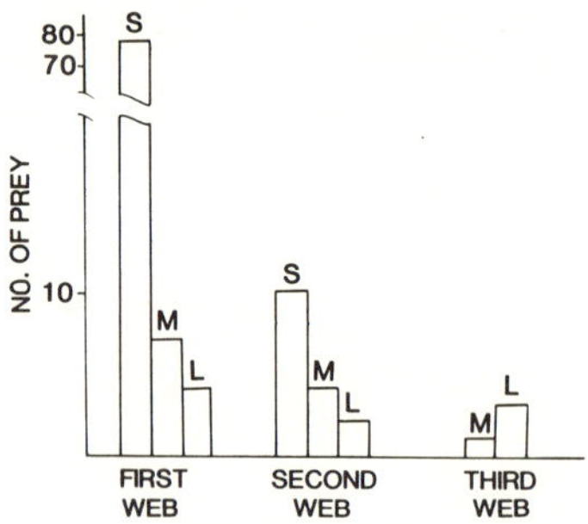

Fig. 8.13. Size distributions of insect prey captured on the first, second, or third web hit in colonies of *Metepeira spinipes*. *S*, small prey; *M*, medium-sized prey; *L*, large prey.

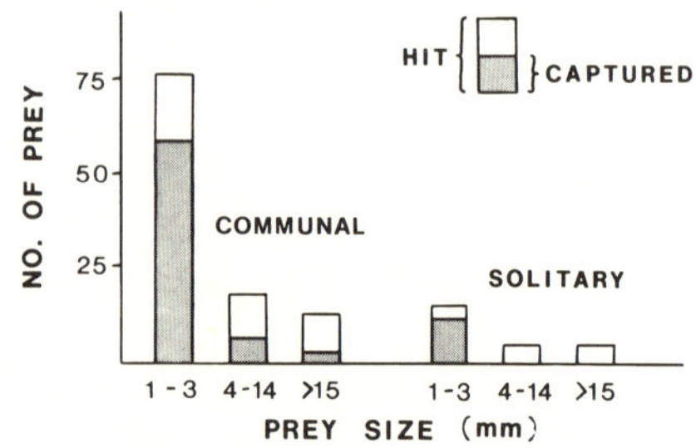

Fig. 8.14. Size distributions of insect prey captured by communal and solitary individuals of *Metepeira spinipes*.

or by approaching the web where the prey is being captured. Often one of these spiders is able to capture the insect in its own web if it escapes from the first web.

Wilson (1975) quotes an old Ethiopian proverb, "When spider webs unite, they can halt a lion." Although lions have never been observed as prey in spider webs, it is probable that the increased prey-knockdown effect of grouped webs allows communal spiders to capture both more prey and larger prey than solitary spiders. In communal *M. spinipes*, prey-capture efficiency (i.e., number of prey captured/number hitting the web) increases with colony size (Fig. 8.11), as Lubin (1974) and Rypstra (1979) have found for other communal orb weavers. This increase may be explained by the fact that a sizable portion of the prey captured in *M. spinipes* colonies consists of insects deflected by individual webs. In observing the fate of numerous insects flying into *M. spinipes* colonies (Fig. 8.12), we noted that the prey-knockdown effect of grouped webs allows communal spiders to capture 20 percent more prey than solitary spiders. We also noted that those prey captured after striking and escaping from several webs (Fig. 8.13) were large insects (e.g., beetles and bees). A comparison of size classes of prey captured by solitary and communal individuals reveals that there is no difference in the size distribution of insects hitting webs, but that a greater proportion of larger insects is captured by communal individuals (Fig. 8.14). Since insect biomass is a cubic function of body length, a slight increase in the number of larger insects captured may mean a sizable increase in the biomass available to a spider. The knockdown effect of grouped webs may thus make available to communal spiders more prey of greater biomass than are typically available to solitary spiders. We estimate (from insect weight and energy content/body length regressions) that a communal spider might obtain 33 percent more energy than a solitary spider in the same habitat.

Conclusion

It appears that communal web building in orb weavers like *M. spinipes* exists, not as the result of cooperative effort, but through long-term occupation of web sites, or sustained juvenile aggregations (Kullmann 1972b). Aggregations of *M. spinipes* are based on a degree of social interattraction, in that individuals tend to attach silk to preexisting silk. Spacing patterns suggest that these spiders tolerate conspecifics at closer distances than if individual web units were merely attached to one another. However, there is considerable temporal and geographic variation in the social-grouping tendency of *M. spinipes* that is clearly related to the availability of insect prey. We might speculate that the communal behavior of *M. spinipes* evolved as the result of an increased tolerance of conspecifics in habitats where prey were locally or seasonally abundant. In such habitats, territory size and interindividual distance would sometimes be reduced, and populations would be large. If web sites providing proper architectural support were patchily distributed (as *Agave* and *Opuntia* are) and of limited availability, selection would favor individuals capable of tolerating conspecifics and attaching webs together. In such situations, the advantages derived from the properties of the communal web (e.g., exploitation of habitats free of competing spiders, increased prey-capture efficiency, and increased protection from predators) would usually outweigh the advantages gained by maintaining maximal distance from conspecifics. Retention of an interindividual spacing mechanism like web defense, however, would allow flexibility and ensure survival if prey availability fluctuated. The end result would be a species that, like *M. spinipes* and other communal orb weavers, displays a combination of solitary and communal behavior.

9

Web Building, Predatory Versatility, and the Evolution of the Salticidae

ROBERT R. JACKSON

SPIDERS OFTEN are divided informally into two groups, the web builders and the cursorial hunters (Gertsch 1949). Web builders are usually thought of as sedentary animals with poorly developed vision. Cursorial spiders do not build capture webs and are subdivided into the short-sighted and long-sighted hunters, the jumping spiders (Salticidae) being considered the most clear-cut examples of the latter. In numbers of described species, the Salticidae are the largest family of spiders (Kaestner 1968; Prószyński 1971), and one of the major groups of animals in which highly developed vision has evolved (Land 1974; Forster 1982a, 1982b).

Salticids have a pair of large anterior median (or principal) eyes and six smaller lateral (or secondary) eyes. The posterior median eyes are typically much smaller than the other lateral eyes and their function has yet to be determined. With the other four lateral eyes the spider detects movement. Next the spider turns so that the image of the object falls on the retinas of the principal eyes. The principal eyes discriminate, by mechanisms that are not fully understood, between classes of objects such as prey, mates, and conspecifics of the same sex (Land 1972a).

In response to conspecifics, salticids perform courtship or threat displays (Peckham and Peckham 1889, 1890; Crane 1949; Plett 1962a; Jackson 1982a; Richman 1982). Prey, usually motile insects, are pursued by some combination of stalking, chasing, and leaping, the spider eventually inserting its fangs and holding on until its venom takes effect. (For surveys and examples of salticid predatory behavior, see Heil 1936, Drees 1952, Precht 1952, Precht and Freytag 1958, Plett 1962b, and Forster 1977a, 1977b.)

This portrayal is appropriate for most salticids, the "typical" jumping spiders, but not for those in the genus *Portia*. These spiders, which appear to be confined to northern Australia, New Guinea, Asia, and Africa, are of particular interest because they combine primitive morphological features (Wanless 1978) with unusual behavior. Their primitive features—especially their large posterior median eyes (Fig. 9.1), which function in movement detection (Jackson, in press)—are remarkable in salticids. Their behavior is perhaps unique: several species have been reported to build capture webs and invade alien webs (Gravely 1921; Sherriffs 1931; Bristowe 1941; Coleman 1978, 1980; Forster 1982a; Murphy and Murphy 1983; Y. D. Lubin, pers. comm.; W. A. Shear, pers. comm.). These are very unusual behaviors for salticids and raise interesting phylogenetic questions, since it has been suggested that the ancestors of the Salticidae were nonvisual web builders (Forster 1979; Shear, this volume).

In northern Queensland I located a population of *Portia fimbriata* and carried out both field and laboratory studies in collaboration with David Blest (Jackson 1981d, 1982d; Jackson and Blest 1982a, 1982b). These spiders were found to be unusually versatile predators that build large webs and use them

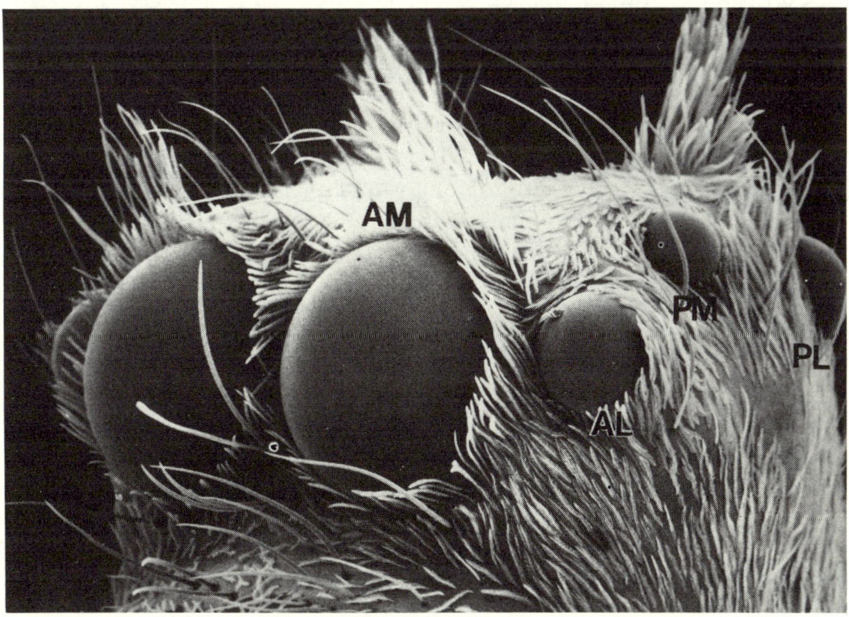

Fig. 9.1. Dorsal carapace of *Portia fimbriata*. The anterior median eye (*AM*) is larger than the other eyes. The posterior median eye (*PM*) is approximately the same size as the anterior lateral eye (*AL*) and the posterior lateral eye (*PL*). Courtesy of Rainer F. Foelix.

in various ways to capture both insects and other spiders. They also invade alien webs where they prey on the resident spiders, their eggs, and insects caught in the webs. Specialized movements of the legs and palps create web vibrations that seem to mimic the struggles of ensnared insects, and these often lure nonvisual spiders to the waiting *P. fimbriata*. Morphological specializations give the spiders the appearance of dead, tattered leaves, probably affording them protection from visually hunting predators. Their unusual appearance is also exploited in conjunction with specialized, slow stalking behavior to pursue prey with well-developed vision, namely other salticids, which fail to recognize the approaching *P. fimbriata* as a predator. The eyes of *P. fimbriata* have a mixture of primitive and advanced features (Blest et al. 1981; Blest, in press).

As a result of our studies, we proposed a hypothetical scheme for the evolution of typical salticids from nonvisual web-building ancestors, attempting also to account for the predatory strategy of *P. fimbriata*, and to predict the strategies of related species evolving in different types of environments, and those lacking the cryptic morphology of *P. fimbriata* (Jackson and Blest 1982a). We argued that the unique visual system of the salticids evolved in conjunction with predation by invasion of alien webs.

Comparative studies (in progress) of additional salticid species, especially other species of *Portia* (Subfamily Spartaeinae), species in other spartaeinine genera, and species in related subfamilies, especially the Old World lyssomanids, may clarify the evolutionary events that led to the adaptive radiation of the Salticidae, and account for the unusual predatory versatility of *P. fimbriata*.

In this chapter, I review the biology of *P. fimbriata* and the hypothesis concerning salticid evolution, and discuss relevant information about other species, especially other *Portia*. I also present a hypothesis concerning the origins of all araneomorph spiders.

Behavior in *Portia fimbriata*

Portia fimbriata first came to my attention by accident in 1979, when I was in northern Queensland casually looking for spiders in a tropical rain forest. Insects and spiders, especially typical web builders and cursorial salticids, were abundant in this dark, damp, warm habitat. Webs between the buttresses of large trees and on overhanging ledges and boulders were often cluttered with leaves, dirt, insect remains, and other debris. Pushing aside with my finger what I thought was detritus, I disturbed a resting *P. fimbriata* quite at home in an alien web. It was a startling introduction. The studies of this extraordinary spider that ensued (Jackson 1982d; Jackson and Blest 1982a) are the source of most of the information reviewed here.

Crypticity

P. fimbriata is a moderately large salticid (adult body length: male, 5–6 mm; female, 7–10 mm) colored in a mixture of the blacks, whites, browns, and pale yellows and oranges that characterize much of the debris that occurs in webs. Fringes and tufts of hairs on the legs and abdomen increase the resemblance to debris. When resting in webs, both their own and those of other species, all sex/age classes adopt a special posture, the "cryptic rest posture" (Fig. 9.2), which improves their concealment. Legs I, II, and III are

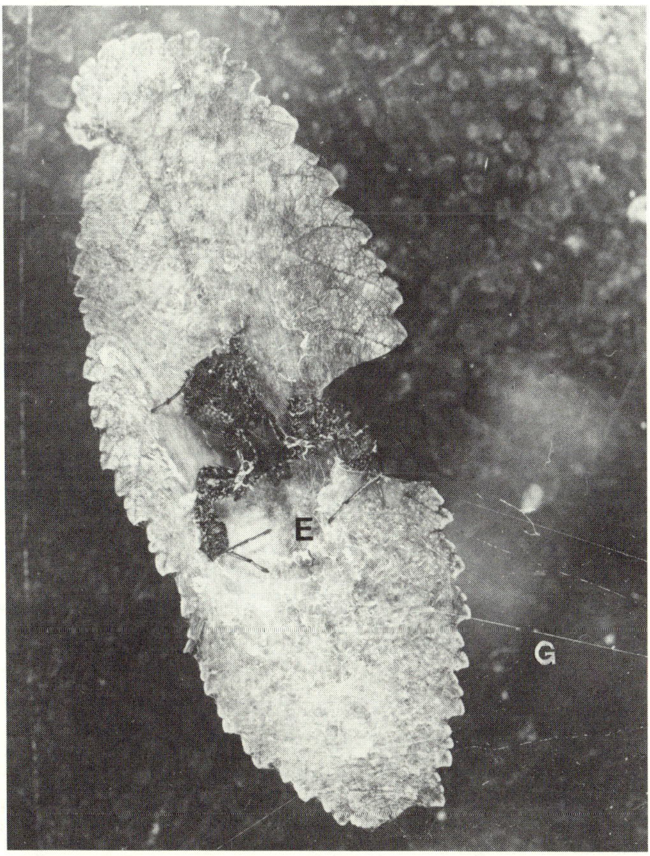

Fig. 9.2. Female *Portia fimbriata* in a cryptic rest posture. The spider is standing on a leaf suspended in a type 2 web. The leaf is oriented such that the long axis is vertical. The spider faces downward, palps retracted and legs highly flexed. Earlier, this female invaded the web, evicted the conspecific resident female, and ate the eggs on the leaf. Later she oviposited her own eggs on the leaf. *E*, remains of the previous female's egg case; *G*, guy line holding the leaf to the web.

held largely ventral to the body, the tarsi angling medially and often crossing over the midline of the sternum. *P. fimbriata* has a high carapace, steeply sloping anterodorsally from the abdomen. In the cryptic rest posture, the femurs of legs IV are angled anterodorsally, closely following the contour of the carapace, and medially so that the patellae almost meet. The remainder of each leg IV angles posterolaterally. The palps are retracted to the sides of the chelicerae, the tarsi angling ventrolaterally. The effect of this specialized posture is to obscure the outline of the body and appendages.

The unusual appearance and cryptic rest posture of *P. fimbriata* are apparently adaptations that conceal the spider from visually hunting predators and resemble the protective adaptations of many insects (Robinson 1969a). Since *P. fimbriata* is eucryptic (difficult to distinguish from its background) when standing on silk or debris in a web or on a tree trunk or a rock in the environment, it often goes undetected. The cryptic rest posture is significant because visually hunting predators have been shown to respond especially to the outlines of appendages and the body (Robinson 1973). Even when seen, *P. fimbriata* is often not recognizable as a spider because of its special resemblance to debris in the web. Although "special resemblance" (Poulton 1890) has sometimes been referred to as Batesian mimicry (Edmunds 1974; Forster 1982a), the latter term should be reserved for cases involving aposematic models, since the processes involved are likely to differ considerably (Wickler 1968; Vane-Wright 1980; Cloudsley-Thompson 1981; Robinson 1981).

Additional antipredatory adaptations seem to be associated with locomotion. *P. fimbriata* appears "mechanical" as it moves very slowly and jerkily across a web, a rock, or vegetation. A single step, advancing the spider only a few millimeters, might take as long as a full second; however, velocity is quite variable, even for successive steps within the same bout. The palps, the legs that step, and highly varied combinations of other legs wave in a choppy fashion as the spider advances, and continue to wave during the pauses of variable length between steps. There is little or no evident synchrony between the movements of simultaneously waving appendages, and abrupt, drastic changes occur in the velocity and amplitude of the movement of a given appendage during a given bout of waving.

Movement is known to be an important stimulus in prey detection for many visually hunting predators. As they walk, some leaf- and stick-mimicking phasmids and mantids rock forward and backward or from side to side in a manner suggestive of a leaf or stem being gently blown by a breeze in the forest (Robinson 1969a). The mechanical, jerky motion of *P. fimbriata* does not seem so readily interpretable in this fashion, but it may resemble flickering light penetrating the forest canopy and intermittently striking a dead leaf. It might be regarded as the locomotory equivalent of a "bizarre form" (Robinson 1969a) since the walking *P. fimbriata*, even if detected, simply does not resemble an animal.

Web Spinning and Web Use

P. fimbriata spins two types of webs. Type 1 webs, built by all sex/age classes, are horizontal platforms with diameters two to three times the spider's body length (Fig. 9.3). Adult females and juveniles also spin larger, three-dimensional type 2 webs of about 4,000 cm^3 (Fig. 9.4). Although their structure is variable and often difficult to discern, type 2 webs generally consist of three inclined sheets that converge at the bottom and surround a three-dimensional tangle of threads, often with one or more dead leaves suspended near the top.

To spin a type 1 web, *P. fimbriata* draws threads across a space, such as a horizontal gap in the vegetation, by detouring across neighboring objects (stems, tree trunks, rocks, etc.) while holding threads free of entanglement with one leg IV. Repeated crossings of the gap, initially by the same route

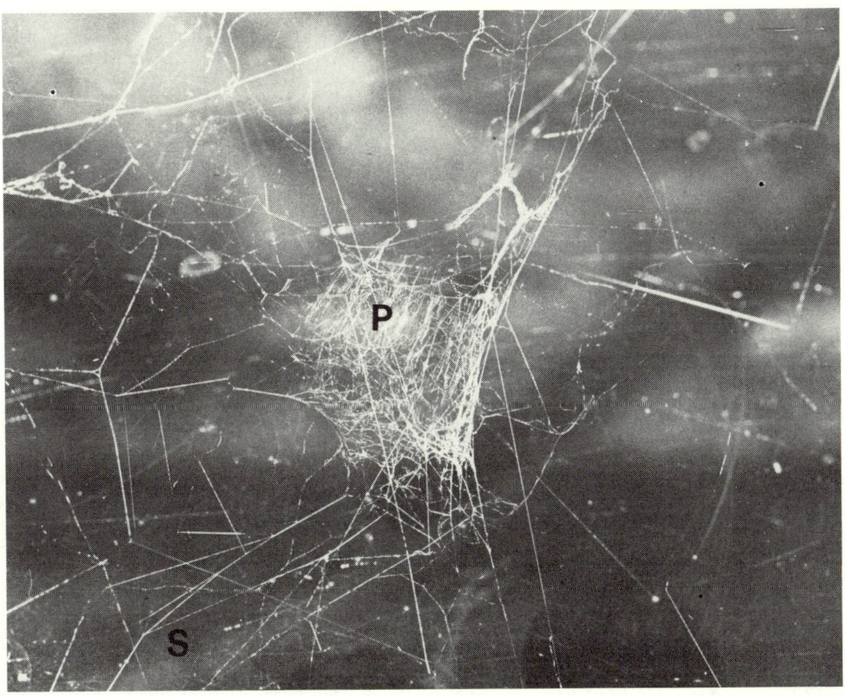

Fig. 9.3. Type 1 web of *Portia fimbriata*. *P*, horizontal platform, approximately 2 cm in diameter; *S*, scaffolding lines connecting the platform to rocks and vegetation (not shown). The spider normally rests under the platform.

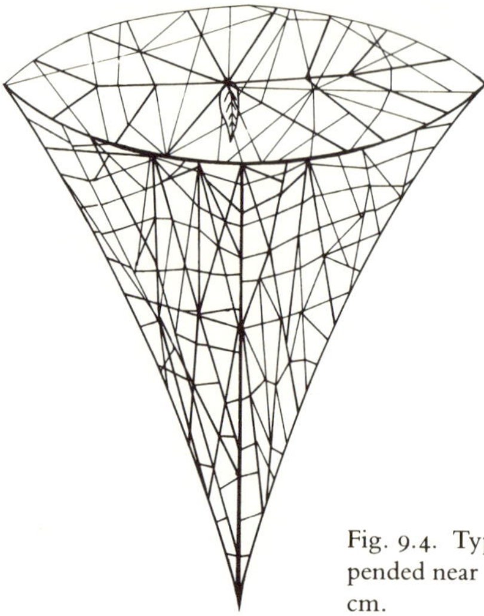

Fig. 9.4. Type 2 web of *Portia fimbriata*. A dead leaf is suspended near the top. The widest diameter is approximately 30 cm.

Fig. 9.5. Female *Portia fimbriata* spinning in her type 2 web. *R*, rock ledge to which the web is fastened; *L*, dead leaf suspended in the web.

and later on the threads themselves, are followed by excursions to the side that expand the bridge into a platform.

A type 2 web is spun by fastening lines to a leaf on the substrate and ascending stems, tree trunks, or other objects while trailing the silken lines. After attaching the lines, the *P. fimbriata* returns to the leaf by more or less the same route. This sequence is repeated, the spider ascending to successively greater heights and beginning to descend, first by dropping on draglines, and later by climbing down lines already laid. Eventually, the spider begins making attachments at multiple points on suspended lines. By repeatedly attaching and tightening new lines of silk while breaking old ones, the spider hoists the leaf upward. Fine manipulation and positioning of the leaf is achieved with the chelicerae by gripping edges and then leaning or walking forward or backward.

Once the leaf has been hoisted to nearly its final height, the spider spends much of its time making excursions in all directions to and from it. Next the spider concentrates its spinning to the side of and below the leaf, repeatedly covering existing lines with new silk, thereby creating the inclined sheets (Fig. 9.5). A recognizable type 2 web is usually built in about 2 hours, but in both the field and the laboratory spiders add to the webs daily, gradually making them more substantial.

As it walks on its web, *P. fimbriata* makes rotary probing movements similar to those of some araneids, theridiids, and other typical web-building spiders (Peters 1932; Jacobi-Kleemann 1953; Eberhard 1972b; Salzmann and Witt 1973). The legs rotate from the femurs, causing the tarsi to trace ellipses of 1–4 mm. Any leg might probe, but anterior legs probe more often. Probing seems to occur when the spider repositions legs but fails initially to make contact with silk, several rotations occurring before the tarsus contacts silk and comes to rest.

P. fimbriata actively selects leaves only a few centimeters in length and slightly concave in shape for suspension in its webs. Females place their eggs in the cavity of the leaf and cover them with silk. Subsequently, they spend much of their time standing over their eggs on the leaf while adopting the cryptic rest posture (Fig. 9.2).

By applying spots of paint to their legs and bodies, I labeled 59 specimens of *P. fimbriata* in the rain forest. Like many web builders in other families, the labeled spiders were remarkably sedentary, some remaining more than 48 days in the same webs, where they molted several times, matured, mated, and oviposited.

Cohabitation of adult males with subadult females, copulation taking place after the female matures, occurs in numerous species in both cursorial and web-building families (Bristowe 1941; Jackson 1977c, 1978c, 1982b, 1982c, 1982e; Jackson and Harding 1982; Pollard and Jackson 1982; Richman

1982; Krafft 1982). Cohabitation occurs in *P. fimbriata* also, but with some unusual variations. Sometimes it lasts more than a month, much longer than has been recorded in other salticids. Subadult males cohabit with subadult females, mating taking place after both molt, and there is indirect evidence of males cohabiting with females two molts from maturity.

Lengthy cohabitations and cohabitations during which both the male and the female molt, the female possibly molting more than once, seem to be adaptations permitted by the more sedentary habits of this web-building salticid as compared with cursorial species. More attention to cohabitation by web builders in other families might reveal additional examples of extended cohabitations. For example, in the web-building pisaurids *Inola amicabilis* and *Inola subtilis*, subadult males cohabit with subadult females and mating occurs after the pair molt (Davies 1982; R. R. Jackson, pers. obs.).

Predatory Behavior

P. fimbriata readily invades webs of agelenids, amaurobiids, araneids, diplurids, eresids, pisaurids, segestriids, stiphidiids, tetragnathids, theridiids, and uloborids and preys on the host spiders. Once on an alien web, *P. fimbriata* uses specialized movements of the legs and palps to generate vibratory stimuli to which the host spider responds in a manner more appropriate for potential prey, or perhaps a conspecific, in its web than for a potential predator. Sometimes the host spider is lured to the waiting *P. fimbriata* (Fig. 9.6). Sometimes the host spider responds with localized movements that assist *P. fimbriata* in maintaining its orientation as it stalks slowly across the web.

Several types of vibratory behavior occur, e.g., flutter, strike, and pluck, each of which may be performed by palps, legs, or both. To flutter, the two palps (Fig. 9.7) or the first two pairs of legs are positioned forward and rapidly oscillate against the silk. To strike, the two palps are moved together slowly up and back and then rapidly forward and downward against the silk. Any one or any combination of the first four legs sometimes strike in a similar fashion.

Fluttering and striking tend to occur soon after *P. fimbriata* contacts the alien web. Plucking is a highly variable behavior that predominates later in the encounter. A silk line is caught with the tarsus of a palp or leg and snapped up, down, or both (Figs. 9.8 and 9.9). The spider may pluck once or several times in succession, and the velocities and amplitudes of these movements are highly variable. For example, a plucking leg might move 1 mm or 7 mm. Usually only one palp plucks at a time. Although a leg sometimes plucks alone, more often a highly variable combination of legs pluck simultaneously.

P. fimbriata plucks both while stationary and while walking on alien webs. While stationary, it often faces its prey and performs various forms of pluck-

Fig. 9.6. Female *Portia fimbriata* (dorsal view) feeding on a juvenile amaurobiid (ventral view). *P. fimbriata* lured the amaurobiid to within a few millimeters before grasping it around the ventral cephalothorax. *A*, abdomen of the amaurobiid.

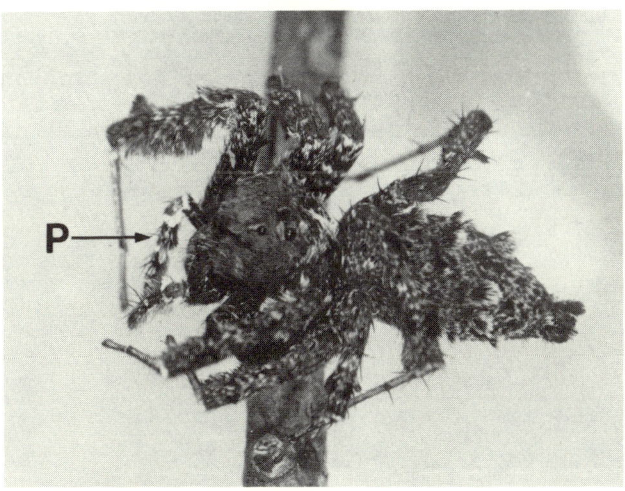

Fig. 9.7. Female *Portia fimbriata* (dorsal view) fluttering her palps on the silk of an uloborid web. *P*, palp.

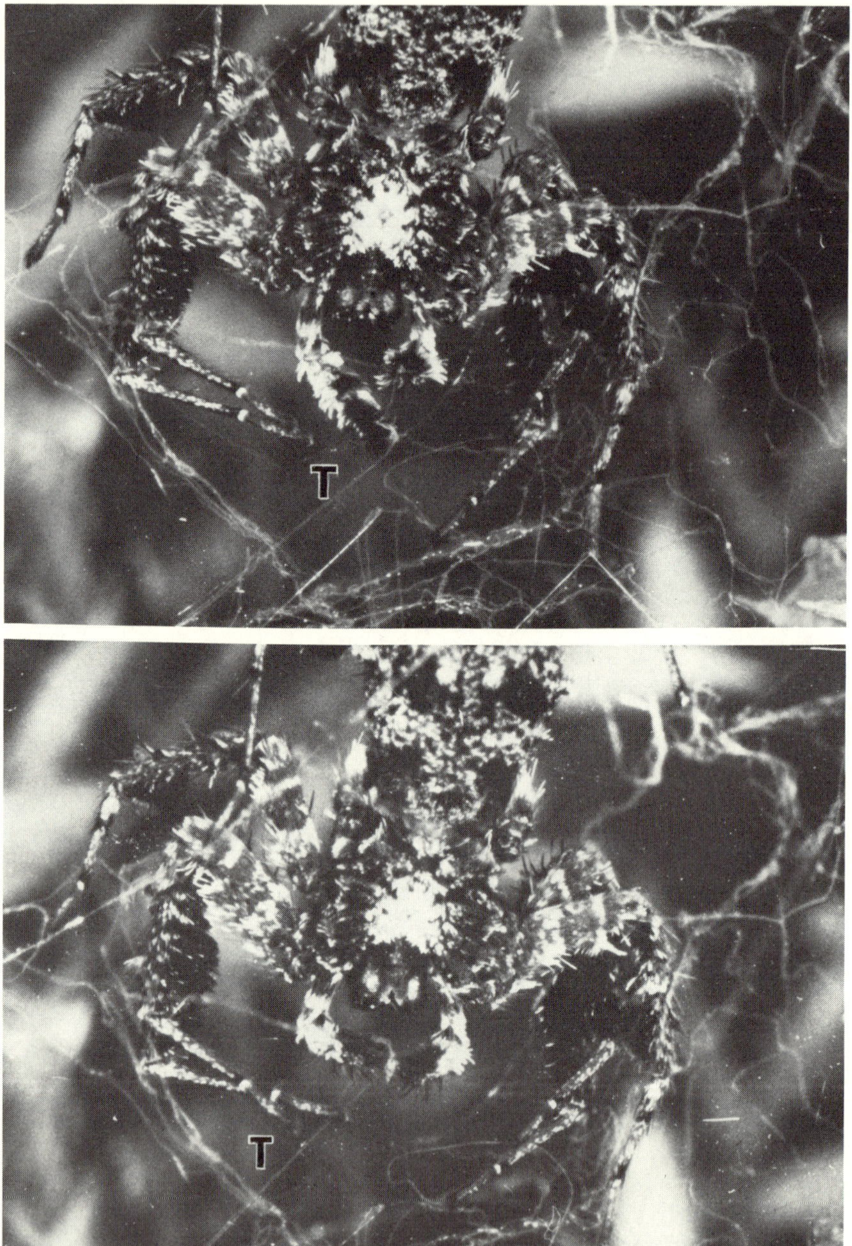

Fig. 9.8. Female *Portia fimbriata* (ventral view) plucking with a palp on the silk of an amaurobiid web. *a*, the left palp catches a thread (*T*). *b*, the palp retracts, causing the thread to move.

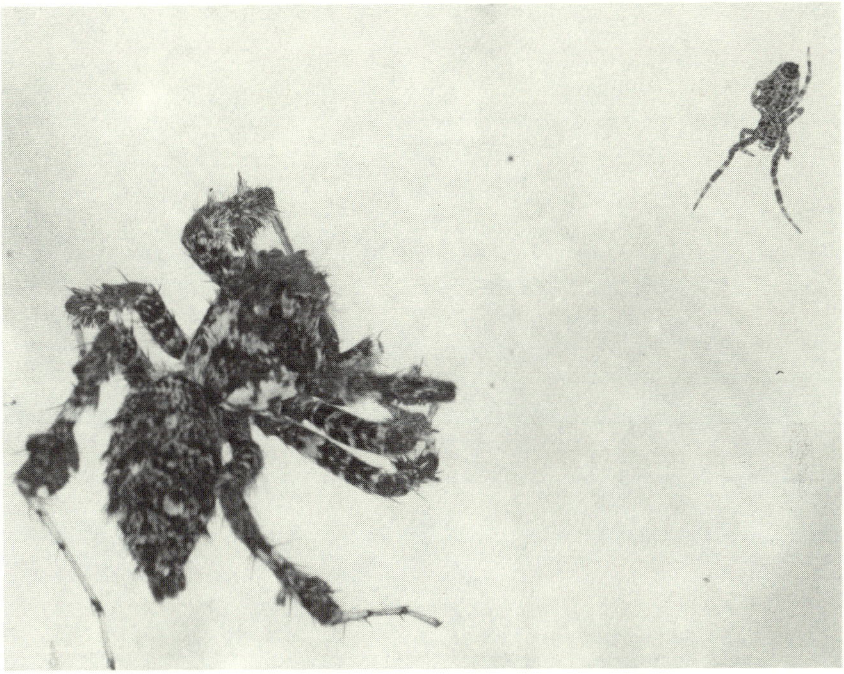

Fig. 9.9. Female *Portia fimbriata* (*left*, dorsal view) facing an uloborid (*right*, ventral view) and plucking by rapidly flexing her legs.

ing behavior; but if the prey begins to approach, the predator persists with a particular type of plucking, sustaining a more or less set velocity and amplitude and using specific appendages held in a stereotyped position. However, if the prey ceases to approach, *P. fimbriata* resumes vibrating in a more variable fashion.

Regardless of whether the prey approaches or is approached, *P. fimbriata* usually attacks by lunging from about 5 mm away. In preparation, the first two pairs of legs are slowly moved up and back, and the fangs are extended. *P. fimbriata* then attacks by moving rapidly forward and either grasping the prey with its chelicerae or simply stabbing it with its fangs. In the latter case, the prey runs away after being punctured, and after about a quarter of a minute goes into convulsions as the venom takes effect and paralyzes it. *P. fimbriata* watches the departing prey, then walks across the web and retrieves it.

Sometimes *P. fimbriata* enters web complexes of the social spiders *Badumna insignis* (Amaurobiidae: Lehtinen 1967; Desidae: Forster 1970) and *Philoponella congregabilis* (Uloboridae) and remains for many days at a time, evi-

Fig. 9.10. Female *Portia fimbriata* (dorsal view) feeding on an uloborid egg case (*E*).

dently harvesting the rich supply of food. *P. fimbriata* can feed not only on the social spiders but also on insects that land on the adhesive silk; and sometimes it takes insects directly from the chelicerae of the host spiders.

P. fimbriata finds yet another type of food in alien webs, food that is of particular interest because it is nonmotile. *P. fimbriata* bites through the silk around the egg cases of the host spiders and feeds on the contents (Fig. 9.10).

Away from webs yet another unusual behavior occurs: specialized stalking of other species of salticids. Salticid prey differ from other prey of *P. fimbriata* in having, like the predator, well-developed vision. However, the adaptations that apparently conceal *P. fimbriata* from its own visually hunting predators seem to have a dual function. They also conceal *P. fimbriata* from the salticids on which it preys.

The slow, mechanical aspects of its locomotion are especially pronounced when *P. fimbriata* stalks a salticid. Also *P. fimbriata* conceals the outlines of

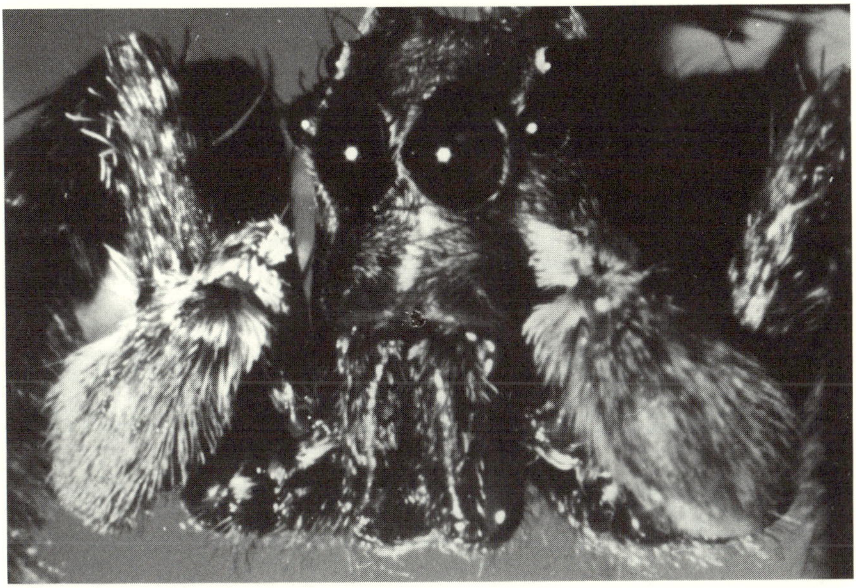

Fig. 9.11. Male *Portia fimbriata* (frontal view) with palps retracted to the sides of the chelicerae while stalking a cursorial salticid. Courtesy of D. Clyne and J. Frazier, Mantis Wildlife Films, Sydney.

the palps by holding them to the sides of the chelicerae, as in the cryptic rest posture (Fig. 9.11). During ordinary walking, however, and while stalking other types of prey, it customarily hold the palps in front of the chelicerae (Fig. 9.5).

Whenever the salticid prey turns and faces *P. fimbriata*, the latter ceases to approach, although it often continues to wave its legs and palps jerkily. While facing *P. fimbriata*, the salticid gives no indication of recognizing its peril. Often it grooms itself, especially its eyes. Sometimes it walks directly toward and even under or over the waiting predator. However, *P. fimbriata* does not resume its relentless approach until the salticid faces away (Fig. 9.12).

P. fimbriata usually attacks by slowly positioning itself over the salticid. When the salticid is facing away, *P. fimbriata* swoops downward to seize (Fig. 9.13) or stab it near the pedicel. A stabbed salticid runs away and goes into convulsions as the predator watches, then retrieves it. Like spiders in webs, salticids seem particularly susceptible to the venom of *P. fimbriata*.

P. fimbriata readily departs from its web to stalk and capture salticids on neighboring rocks and tree trunks, carrying them back onto its web to feed. Also, *P. fimbriata* often makes predatory forays across open ground to neighboring webs, and then returns to its own web.

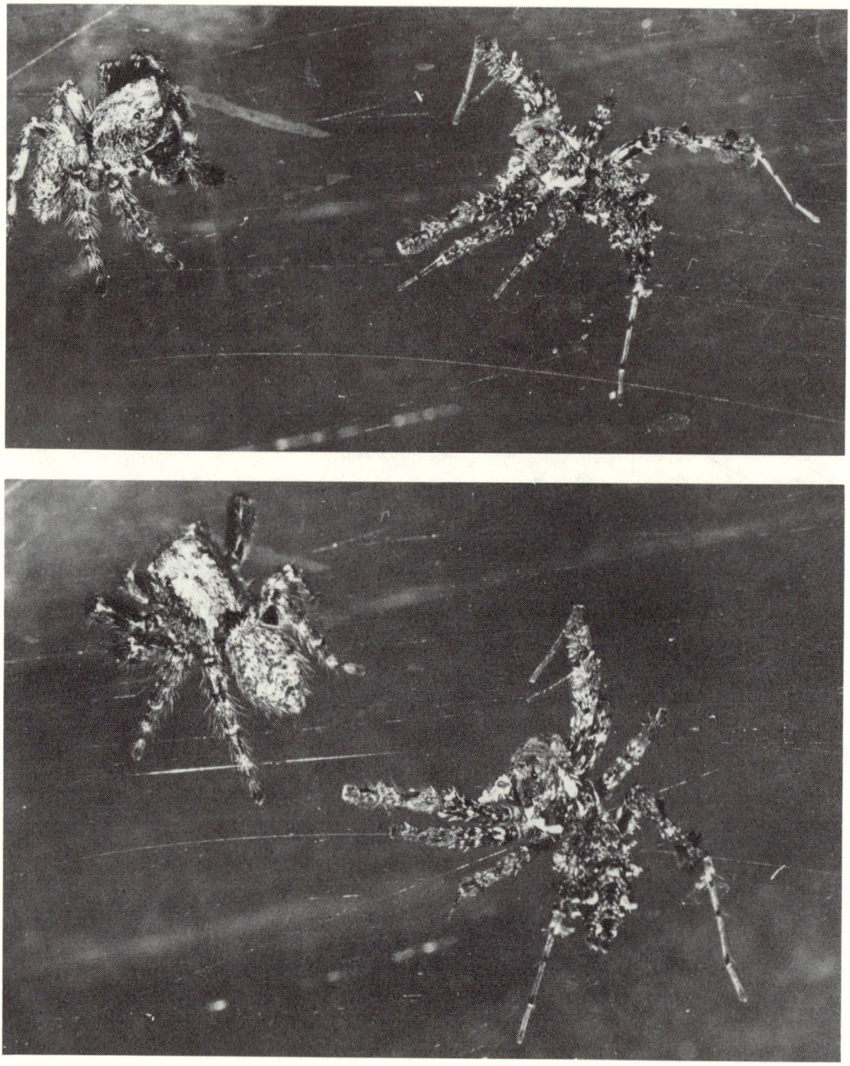

a

b

Fig. 9.12. Female *Portia fimbriata* (*right*) stalking a female cursorial salticid, *Plexippus validus* (*left*). *a*: *P. validus*, having detected movement behind her, turns and faces *Portia fimbriata*; *P. fimbriata* ceases to advance. *b*: *Plexippus validus* turns and walks away; *Portia fimbriata* resumes stalking.

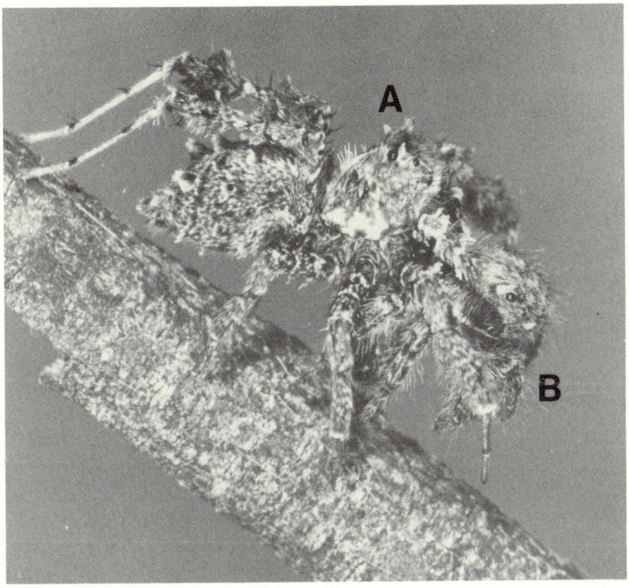

Fig. 9.13. Female *Portia fimbriata* on a stem just after grasping a female cursorial salticid, *Plexippus validus*, near the pedicel. Side view. *A*, carapace of *Portia fimbriata*; *B*, carapace of *Plexippus validus*.

Occasionally *P. fimbriata* feeds on insects encountered outside webs. However, in sharp contrast to typical salticids, *P. fimbriata* does not pursue insects very readily, persistently, or effectively. Sometimes it attempts to get within a few millimeters before lunging or simply picking up an insect. This tactic rarely works, and when it fails the spider rarely continues the pursuit. The agile stalking, chasing, and leaping behaviors used so effectively by typical salticids (Gardner 1964, 1965; Land 1971, 1972b; Dill 1975; Forster 1982a) are absent from the repertoire of *P. fimbriata*.

Nor do the webs built by *P. fimbriata* seem impressive as snares for insects. Dipterans and other insects sometimes adhere to the silk, but usually for only a few seconds; and the spider most often ignores them. On the other hand, the persistence with which *P. fimbriata* pursues other spiders is extraordinary. To stalk a salticid or vibrate on a web continuously for over an hour is not uncommon. A *P. fimbriata* observed in the web of a theridiid performed vibratory behavior for three successive days before capturing the host spider.

There is compelling evidence that *P. fimbriata* prefers spiders to insects as prey. For example, in one series of observations in the laboratory, 27 *P. fimbriata* were presented on successive days, in random order, with different types of potential spider and insect prey. All of the 135 spiders, but only 14 of the 54 insects, were captured and eaten. In another study, *P. fimbriata* readily left its web and walked over to transparent glass cages containing cursorial salticids, but this never occurred when the cages contained flies (Jackson and Blest 1982b). Dead, dried flies glued to corks and jiggled in front of *P. fimbriata* were usually ignored, but dead salticids presented in a similar way were usually stalked (Jackson, in press).

Insects seem to be incidental prey, taken by *P. fimbriata* only when easily obtained. However, insects landing on its web may be significant in assisting *P. fimbriata* to capture neighboring web spiders. It is well known that spiders of either the same or different species often fasten their webs together in nature (Brignoli 1966; Shear 1970a; Krafft 1970; Buskirk 1981, Uetz, this volume). This occurs with many species in the Queensland rain forest, including *P. fimbriata*. Also, *P. fimbriata* often spins its own silk over that of an adjacent alien web, sometimes eventually encompassing it and thus creating a compound web consisting of silk from two species. In nature, 60 percent of the webs of *P. fimbriata* were unequivocally contiguous with webs of other spiders.

When webs are interconnected, an insect on one web sends vibrations into neighboring webs. It is not uncommon for a spider to raid a neighbor's web to capture and eat an insect, and sometimes the other spider (Bristowe 1941; Buskirk 1975a; Jackson 1979a). When the raided neighbor is *P. fimbriata* this habit has fatal consequences, since *P. fimbriata* pursues the spider that pursues the insect. By fastening its web to a web complex of another species, *P. fimbriata* can potentially harvest numerous invading spiders, since vacated units in the complex are soon filled by other spiders.

Intraspecific Interactions

The Salticidae are one of the major groups in the animal kingdom in which visually mediated (and often complex) display behavior occurs during intraspecific interactions (Crane 1949; Jackson 1982a). Males court females with special postures and movements of their legs, palps, and abdomens. Sometimes there is specialized stepping behavior (dancing), in which one spider repeatedly approaches the other then backs away from it, along a zigzag path. The detailed characteristics of displays vary considerably from species to species. In male-male interactions, yet other postures and movements may occur.

In contrast, typical web-building spiders, with their poorly developed vision, rely mainly on specialized vibratory behavior during intraspecific communication. For example, males often pluck with their forelegs or drum with their palps in a specific fashion on the females' webs during courtship (Krafft and Leborgne 1979; Robinson and Robinson 1980; Robinson 1982a).

In fact, this simple distinction between the communicatory behavior of salticids and that of web-building spiders in other families does not always hold. Many, perhaps most, of the typical salticids build silken nests. These are usually tubular in shape and often located in places with little ambient light, such as under rocks on the ground, inside rolled-up leaves, and under the loose bark of trees (Kaston 1948; Jackson 1979b). The spiders remain in these when molting or ovipositing, and generally when inactive; sometimes they mate inside them (Plett 1962a; Jackson 1978c).

There is increasing evidence that vibratory courtship occurs frequently among Salticidae when males encounter females in nests (Bristowe 1941; Jackson 1977a, 1982b, 1982c, in press; Edwards 1981; Jackson and Harding 1982), the same males courting with typical visually mediated displays in the open when ambient light is adequate. For example, specialized pushing and pulling on the silk with legs may occur (Fig. 9.14), or the spider may insert his fangs and tug up and down by repeatedly raising and lowering his body.

Fig. 9.14. Male cursorial salticid, *Mopsus mormon*, performing vibratory courtship on the nest of a conspecific female. His abdomen (*A*) elevated and twitching, the spider probes with tarsi I (*T*) on the silk. The nest is built on a leaf. *D*, door, an elastic opening to the nest.

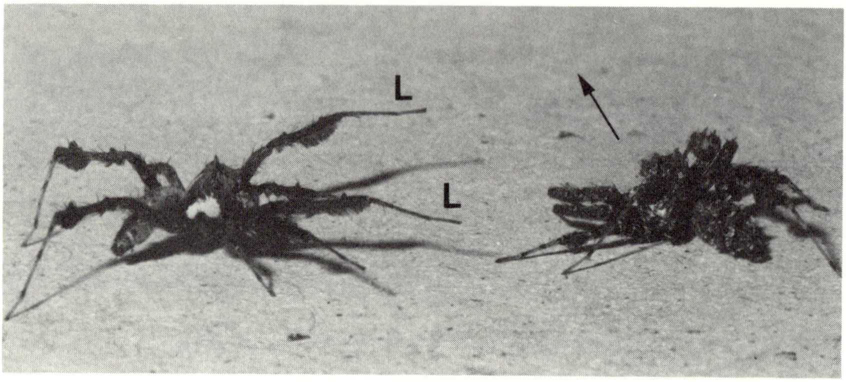

Fig. 9.15. Male *Portia fimbriata* (*left*) courting a conspecific female (*right*), his legs I (*L*) elevated and waving. The female walks away (*arrow*) to her right.

Fig. 9.16. Pair of *Portia fimbriata* copulating while suspended from the female's dragline. Ventral view of the female (*right*). The male (*left*, facing upward) applies his left palp. *S*, female's spinnerets; *A*, male's abdomen; *L*, male's right leg I.

Alternatively, with his legs on the silk, the male may rapidly vibrate his body up and down. Some recently obtained anatomical evidence suggests that these movements may be associated in some species with stridulation (W. P. Maddison, pers. comm.; G. B. Edwards, pers. comm.). Thus it is particularly interesting to consider the display behavior of *P. fimbriata*, a web-building jumping spider.

P. fimbriata males show no tendency to pluck or otherwise vibrate on the webs. Instead, they employ specialized postures and movements of the palps and legs (Fig. 9.15), regardless of whether the female is on or off her web, and these have been shown experimentally to be visually mediated behaviors (Jackson, in press). When distant from the female, the male erects his forelegs stiffly, angles them forward, and waves them up and down in an alternating fashion. When closer, he switches to a different display, in which the forelegs move rapidly from side to side, over a distance of only about 1 mm. The two legs repeatedly move medially toward, then laterally away from, each other. When still closer, the male walks over the female and begins to copulate after a period of tapping and scraping with his legs and palps.

Perhaps it is not surprising that vibratory courtship does not occur in *P. fimbriata*. Ambient light is more or less equivalent on and off the webs of *P. fimbriata*. The situations in which individuals of this species encounter each other are not as disparate as they are in species in which females may be either in nests in places with dim light or in the open in bright light.

However, some other features of the intraspecific behavior of *P. fimbriata* are unusual. Female salticids customarily keep all their legs on the substrate while mating, but *P. fimbriata* may copulate while suspended in the female's or an alien web. If they are away from webs, the pair may copulate after the male walks onto the female and the female lowers herself, with the male on board, by a line of silk (Fig. 9.16).

Although other salticids are not known routinely to mate hanging from a dragline, in a study of a cursorial salticid, *Phidippus johnsoni*, one out of about 400 observed copulations took place suspended from a line (Jackson 1980b). This apparently occurred when the female accidentally slipped from a stem after the pair had begun to copulate. *Portia fimbriata*, however, seems to be specifically adapted to this form of copulation.

Although *P. fimbriata* does not use vibratory displays during intraspecific interactions, it vibrates on the nests of another species of salticid. One of the most common of the cursorial salticids in the rain forest with *P. fimbriata* is a species currently known as *Lagnus kochi*, although it will be placed in a new genus in a taxonomic revision planned by F. R. Wanless (pers. comm.). *L. kochi* builds tubular nests and exhibits courtship versatility. Males use visually mediated displays when interacting with females outside, and vibratory courtship with females inside, nests. Most females are unreceptive and

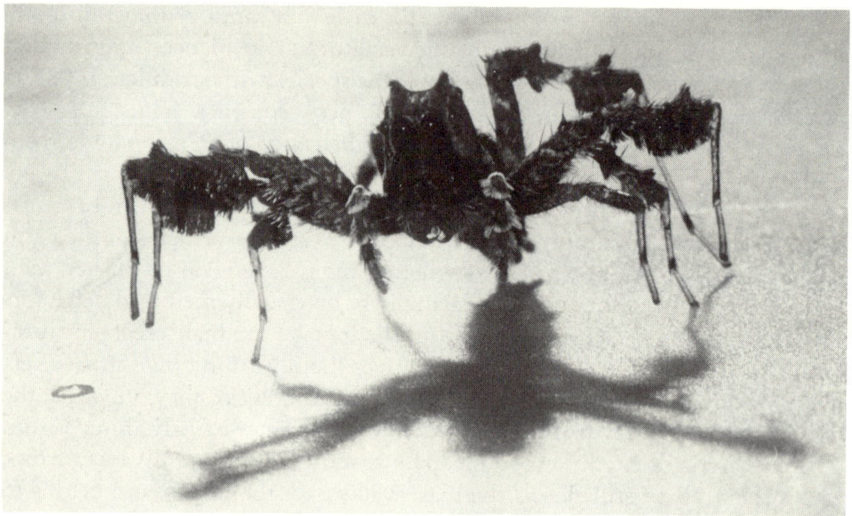

Fig. 9.17. Female *Portia fimbriata* (frontal view) performing a hunched-legs display and swaying to the left while facing a conspecific female (not shown).

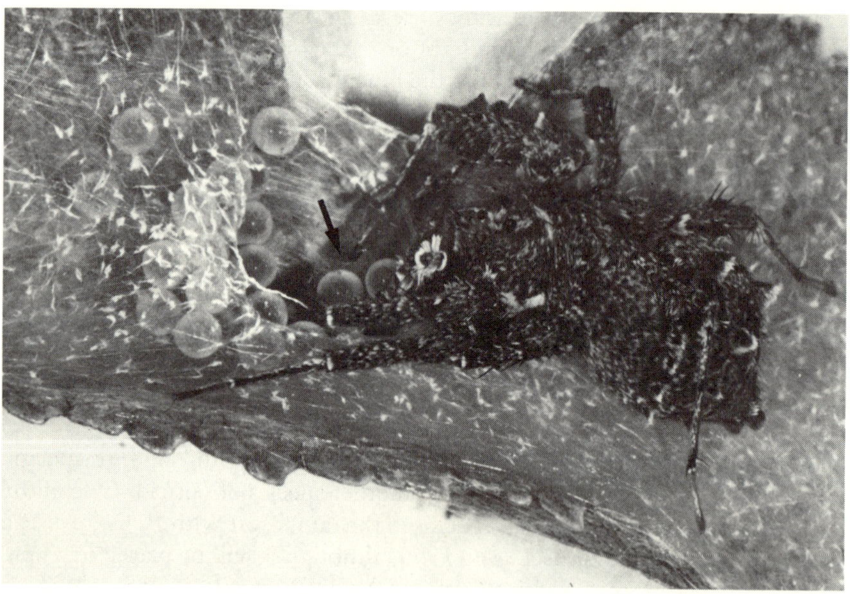

Fig. 9.18. Female *Portia fimbriata* (approximately dorsal view) feeding on the eggs of a conspecific. Having chased away the conspecific and opened her egg case, the spider rakes the eggs toward her mouth using legs I (*arrow*) and feeds on them one at a time. The spider faces toward the left.

eventually leave their nests to attack the males or decamp (R. R. Jackson, pers. obs.). When *P. fimbriata* walks onto an occupied nest of *L. kochi*, it plucks much as on an invaded web and captures the salticid that forsakes the safety of its silken tube.

Generally the displays of female salticids are less pronounced or at least have received less attention from zoologists than those of the males. However, display behavior in females is very pronounced in *P. fimbriata*. "Hunched-legs" displays are the major displays in both female-female and male-male interactions. The first three pairs of legs are highly flexed and held to the side of the body, the first two pairs usually off the substrate and slowly waving up and down. Often the spider sways slowly from side to side (Fig. 9.17), or the two spiders may repeatedly circle each other.

Female-female interactions frequently escalate. The two spiders may push back and forth after bringing their faces together, their fangs open and their legs extended to the side. Periodically, one of the spiders reaches over the other with one or more of its legs—both spiders may accomplish this—and a grappling match ensues, during which legs may be torn off. Eventually the spiders separate. If an intruding female evicts the resident from a web with eggs, she immediately goes to the eggs and eats them (Fig. 9.18). In some cases, she subsequently oviposits her own eggs on the suspended leaf in the captured web (Fig. 9.2).

Interactions between salticids of the same sex have traditionally been referred to as threat display, agonistic display, or ritualized fights (Crane 1949; Jackson 1982a; Richman 1982). However, it is rare that the contested resource is as evident as in these interactions involving *P. fimbriata*. For the resident the eggs are her progeny, but for the intruder they are potential food, and the web with its suspended leaf is a potential site for her own eggs.

After becoming accustomed to the slow, mechanical movements of *P. fimbriata* during predatory and other behaviors, one is quite startled to witness the rapid, smooth, and agile motion that characterizes intraspecific interactions, especially courtship. Females often charge across their webs toward courting males, make truncated leaps toward them whereby they advance only a few millimeters at a time, or leap directly onto them from several centimeters away.

Slow, jerky motion may be at a premium in other contexts, concealing *P. fimbriata* from both its prey and its predators. However, concealment is less relevant during intraspecific interactions, when the point seems to be to communicate. This is best accomplished by behaviors that make the spider more instead of less conspicuous (Smith 1977). Slow movement is, of course, also inappropriate when escaping from predators, and *P. fimbriata* moves rapidly when chased.

The slow movements used by *P. fimbriata* during normal locomotion are reminiscent of those of African chameleons. However, the chameleon, unlike *P. fimbriata*, is incapable of rapid movement (Guppy and Davison 1982).

Predatory Versatility

Some of the tactics in the predatory strategy of *P. fimbriata* are known to occur in other species of spiders. For example, various species of salticids feed occasionally on insect eggs (Hensley et al. 1961; Whitcomb and Tadic 1963; Whitcomb and Bell 1964; Hensley 1971; Jennings and Houseweart 1978; McDaniel, Sterling, and Dean 1981; McDaniel and Sterling 1982). Female spiders of numerous species sometimes eat their own inviable eggs (Kaston 1965). Recently hatched spiderlings of *Achaearanea tepidariorum* (Theridiidae) and *Misumenops* (Thomisidae) have been observed to feed on sibling eggs before leaving the egg case (Schick 1972; Valerio 1974a), and this is probably a widespread phenomenon in spiders (Kaston 1970; Polis 1981). *Clubiona cambridgei* (Clubionidae) is a short-sighted hunting spider that feeds voraciously on the eggs of salticids and those of conspecifics (Pollard, in press). Some other spiders, especially lycosids, feed on another type of nonmotile prey, dead insects (Bristowe 1958; Knost and Rovner 1975; Williams 1979).

Numerous species in a variety of families, but especially the Theridiidae, normally live in the webs of other species and feed variously as scavengers, kleptoparasites, and/or predators of the hosts (Wiehle 1928, 1931; Exline 1945; Archer 1947; Kraus 1955; Lamore 1957; Kullmann 1959, 1960a, 1960b; Legendre 1960; Exline and Levi 1962; Robinson and Robinson 1973a; Kirkendall 1979; Nyffeler and Benz 1980; Trail 1980; Wise 1982). The best-known kleptoparasites are certain *Argyrodes* (Theridiidae) that are specialists at taking insects from the webs of other species (Vollrath 1976, 1979a). *Curimagua bayano* (Symphytognathidae) is a specialist at taking food directly from the mouths of other spiders (Vollrath 1978). *Marpissa melanognathus* is a salticid that takes insects directly from the mandibles of ants (Bhattacharya 1936).

Some mimetid spiders have been observed to enter alien webs and feed on the host spiders after performing vibratory behavior that may mimic an insect or conspecific in the web (Gerhardt 1924; Archer 1942; Berland 1932; Bristowe 1941, 1958; Czajka 1963; Cutler 1972b; Lawrence 1981). *Archaea* (Archaeidae), *Lampona* (Gnaphosidae), and *Ariamnes* (Theridiidae) also seem to be specialists at preying on other spiders (Millot 1948; Legendre 1961; Millot and Legendre 1964; Clyne 1969, 1979; Forster and Blest 1979). Many other spiders, both cursorial species and web builders, are primarily predators of insects but take spiders as prey opportunistically, although not apparently using behaviors specialized for this type of prey (Bristowe 1941; Jackson 1977b; Mascord 1970; Polis 1981).

A few cursorial salticids have been observed to occupy alien webs and, in some cases, feed on the host spiders (Bristowe 1941; Lamore 1958; Enders 1974, 1975c; Tolbert 1975; Jackson 1977b). The manner of attack has been observed for two species, *Phiale* sp. (Robinson and Valerio 1977) and *Lagnus kochi* (Jackson, in press): the salticids leap into the victims' webs rather than stalking across the silk. *L. kochi* and some other cursorial salticids, however, walk across the silk of alien webs to reach leaves and other debris that they use as nest sites, but their locomotion is evidently impaired compared to that of the host spiders.

In contrast, there is no evidence that *Portia fimbriata* is hampered at all on any type of web. It walks with apparent ease on orb, sheet, and space webs, on sticky as well as nonsticky webs, and on both cribellate and ecribellate webs. It has no difficulty walking on the remarkably adhesive webs of social cribellate spiders that completely immobilize typical salticids. Also, no salticids other than *P. fimbriata* have been observed to use specialized vibratory behavior in prey capture.

Luring of insect prey by olfactory means has been credited to some araneid and dictynid spiders (Diguet 1909a, 1909b; McKeown 1963; Horton 1979; Tietjen, this volume). *Mastophora*, for example, attract moths apparently by emitting stimuli that mimic the pheromones of female moths (Eberhard 1977a; Stowe, this volume). These spiders, along with the mimetids and *P. fimbriata*, are arachnids that can join the better-known angler fish and *Photuris* fireflies in providing examples of aggressive or Peckhamian mimicry (Wickler 1968; Lloyd 1975).

Although aggressive mimicry, invasion of alien webs, use of fast-acting venom against other species, kleptoparasitism, and feeding on eggs each occur in other spiders, *Portia fimbriata* is apparently unique in performing all of these varied behaviors, along with specialized stalking of cursorial salticids that has no obvious parallels among other spiders.

The dietary ranges of predators are often envisaged as lying along a continuum from stenophagy to euryphagy, depending on whether few or many types of prey are taken. Stenophagous predators are expected to employ specialized behaviors adaptively tailored to the particular prey species in their narrow diets; euryphagous predators are expected to employ more generalized behaviors applicable to a wider variety of prey species and not so precisely tailored to particular ones (Levins 1968; Morse 1980). The jack of all trades is said to be master of none, but *P. fimbriata* seems to be a jack of all trades and the master of all. This highly versatile predator uses multiple tactics, each of which is adapted to, and very effective against, particular types of prey.

Predatory versatility occurs in other spiders. Many web-building species have prey-specific attack and handling behaviors for different types of insects that land on their webs (Peters 1931, 1933; Robinson, Mirick, and Turner 1969; Robinson, Robinson, and Graney 1972; Krafft 1970; Robinson and Olazarri 1971; Robinson and Robinson 1973a, 1974, 1976a; Harwood 1974; Valerio 1974b; Eisner and Dean 1976; Robinson and Lubin 1979a). The salticid *Stoidis aurata* is a specialist at preying on ants that uses different types of attack behaviors for different types of ants (Edwards, Carroll, and Whitcomb 1974). However, the behaviors used with different types of prey seem less disparate for *S. aurata* and the nonvisual web builders than for *P. fimbriata*. In fact, the predatory versatility of *P. fimbriata* seems more pronounced than any of the examples of versatility recently reviewed by Curio (1976).

P. fimbriata begins to seem like a bizarre arachnid aberration, to be added to some list of zoological curiosities. However, when we consider its taxonomic status, this spider takes on a different significance, and may shed light on the evolutionary origins of the Salticidae.

Phylogenetic Origins of the Salticidae

All salticids have functional spinnerets, but the ways in which they use silk vary considerably. Many, perhaps most, salticids spin tubular silken nests. Also, many species construct suspension structures on which they rest at night (Eberhard 1974; Carroll 1977; Robinson and Valerio 1977; Jackson 1982c). The spider may simply hang from the vegetation by a dragline, or may first build an intervening tangle of threads fastening the dragline to the vegetation. *Pellenes nigrociliatus* lives in snail shells that it suspends from the vegetation (Mikulska 1961).

Lyssomanes in the Neotropical region and various species in Australia, Asia, and Africa spin small, flimsy "webs" in the vegetation, especially over concavities in leaves, in which they molt, oviposit, and remain during periods of inactivity (Eberhard 1974; Wanless 1980a; Forster 1982a; Jackson, in press; G. B. Edwards and C. E. Griswold, pers. comm.). These webs, and the type 1 webs of *Portia fimbriata*, seem analogous to the nests and suspension structures of other salticids. However, type 2 webs are much more substantial, providing a base from which *P. fimbriata* invades adjacent alien webs or makes forays upon neighboring surfaces; and they have limited but effective capabilities of ensnaring prey, perhaps most importantly as bait for neighboring spiders.

Possibly *P. fimbriata* derived web building, along with web invasion and facility at walking on varied types of silk, from ancestors that were typical vagrant salticids. One might envisage suspension structures and resting "webs" as representing successive intermediate stages leading to web building. Leaping attacks on web builders, as practiced by *Phiale* sp. and *Lagnus kochi*, might be taken to represent a step toward the more specialized predatory invasion of webs practiced by *Portia fimbriata*.

However, it seems more likely that both *P. fimbriata* and typical salticids had ancestors that built and invaded webs, these characteristics being conserved by *P. fimbriata* but lost by the advanced species as they became vagrant hunters. Despite the large size of the family Salticidae (about 400 genera and 4,000 species: Prószyński 1971), spinning of large webs, specialized web invasion, and luring of prey have not been found to occur in any of the advanced species.

A satisfactory hypothetical scheme for the evolution of the salticids must account for the origin of the unique salticid eyes. The retinas of the lateral eyes of *P. fimbriata* resemble those of typical salticids in basic respects, but have less developed glial architecture (Blest 1983). In typical salticids, the retinas of the principal eyes have some unusual morphological features, including a peculiar boomerang shape when viewed in cross section, muscles attached to the retinas that move them in varied and complex fashions in a plane perpendicular to the visual axis, and segregation of the receptor cells into four layers (Scheuring 1914; Homann 1928; Dzimirski 1959; Land 1969a, 1969b; Eakin and Brandenburger 1971; Blest et al. 1981). The principal eyes of *P. fimbriata* are highly specialized, having the same basic features as those of typical salticids but also achieving greater acuity than those of other salticids as the result of a more pronounced telescopic construction (Williams and McIntyre 1980a, 1980b; Williams 1981).

One way to increase image size in an eye is to increase focal length. In the principal eyes of salticids, focal length has been increased by placement of the retinas at the rear of tubes that are surrounded by pigment and extend far back into the cephalothorax. However, there are limits to this method of increasing focal length in animals with small heads, such as the salticids.

In *P. fimbriata* there is a deep pit in front of the retina at the rear of each eye tube. This pit acts as a diverging lens and, in combination with the corneal lens at the front of the tube, provides the eye with a telescopic lens system. As a result, the resolving power of the retina is increased by about 50 percent. Similar diverging-lens systems occur in other salticid species, and this may be a universal component of salticid eyes. However, the telescopic effect is more pronounced in *P. fimbriata* than in any other salticid species investigated (Williams and McIntyre 1980a, 1980b). Diverging-lens systems like those found in salticids occur in some birds of prey that, owing to aerodynamic constraints, also have limited head space (Snyder and Miller 1978).

The biology of *P. fimbriata* suggests the hypothesis that the ancestors of the Salticidae were web-building spiders with poorly developed vision

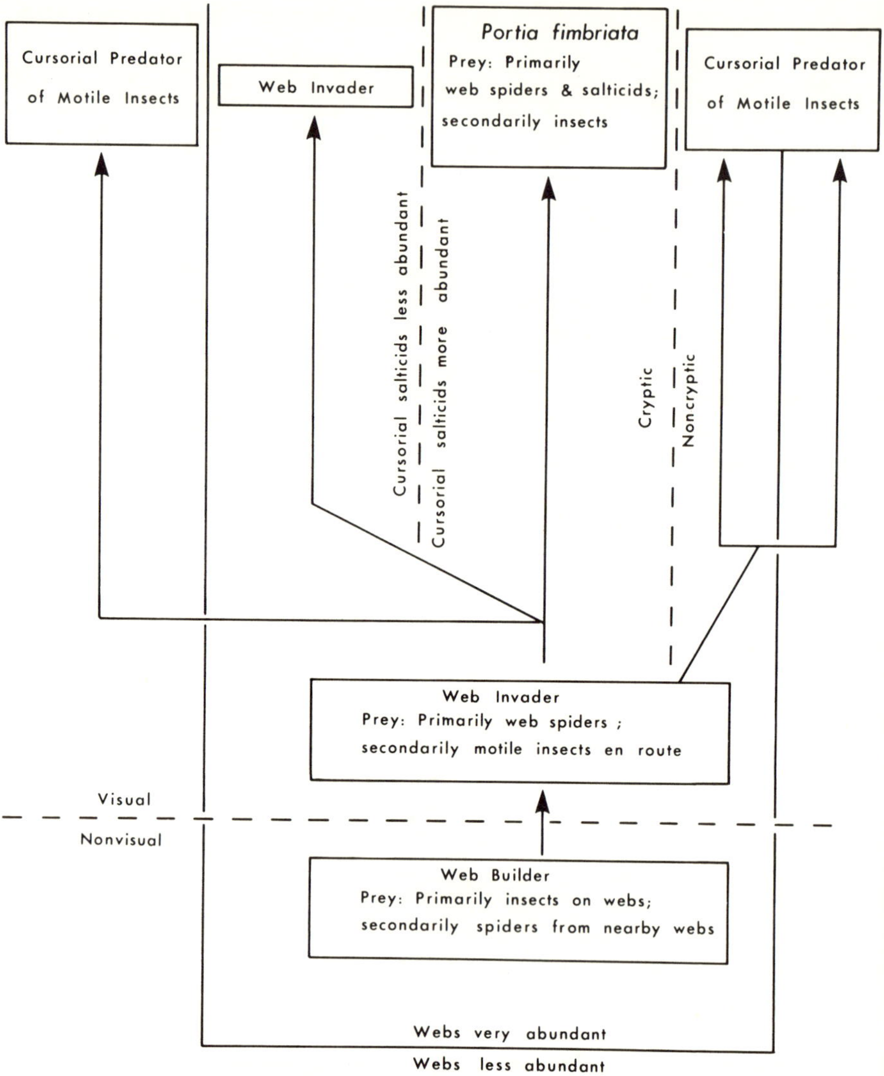

Fig. 9.19. Hypothetical scheme for the evolution of both *Portia fimbriata* and typical cursorial salticids from nonvisual (below horizontal dashed line) web-building ancestors. The arrows represent evolutionary transitions in behavior and habitat. The solid line distinguishes between habitats in which web spiders, as potential prey, are abundant (center of diagram) and habitats in which web spiders are less abundant (periphery of diagram). The vertical dashed line at the right side of the diagram distinguishes between salticids that are as highly cryptic as *P. fimbriata* (to the left of the line) and salticids that are noncryptic, or at least less so than *P. fimbriata* (to the right). The vertical dashed line near the center of the diagram distinguishes between habitats in which both web spiders and cursorial salticids are abundant (to the right of the line) and habitats in which web spiders are abundant but cursorial salticids are not abundant (to the left). Salticids specialized for web invasion alone are predicted in the latter case. Reprinted, by permission, from Jackson and Blest 1982a.

(Fig. 9.19). If we suppose that these ancestors lived in habitats in which webs of other species were numerous and often contiguous, as in the Queensland rain forest, then opportunities would arise for fortuitous capture of neighbors at web boundaries, and there would be scope for the evolution of a predator specializing as a web invader.

Improved detection and tracking of neighboring spiders from a distance would seem necessary. Web spiders customarily locate prey by means of vibratory stimuli (Barrows 1915; Peters 1931; Liesenfeld 1956; Suter 1978; Barth 1982). However, the patterning of vibratory stimuli is likely to vary significantly with the characteristics of the web (Langer 1969; Burgess 1979b; Frohlich and Buskirk 1982). If the hypothetical ancestor of the Salticidae shared with *P. fimbriata* the habit of invading diverse types of webs, use of a sensory modality independent of web dynamics would be advantageous. Vision is the major candidate (Marler 1968).

For a spider specialized in predation by web invasion, forays across open ground, rocks, or vegetation would seem necessary in order to locate new webs to invade. Locomotor adaptations, such as the ability to leap across barriers (Parry and Brown 1959) and the acquisition of claw tufts (Hill 1977), might evolve in this context. Also, the developed vision of this spider would create the potential for taking motile insects through visually mediated stalking, a predatory mode not being exploited by other spiders. If pre-salticids entered this new adaptive zone (Simpson 1953), the adaptive radiation of the typical Salticidae can be envisaged.

Pursuit times associated with predation on spiders on invaded webs were probably lengthy for these ancestral forms, as they are for *P. fimbriata*. If *P. fimbriata* were to use faster pursuits, it would probably be ineffective. A rapid approach across the web would probably alert the host spider. Leaping into webs might also be dangerous, especially when used against large web spiders. Since web spiders often attack rapidly (Barrows 1915; Robinson and Olazarri 1971), if *P. fimbriata* failed to land on its prey, it might be attacked before it regained its footing on the host spider's silk. On the other hand, faster pursuits are effectively used by cursorial salticids against insects (Land 1972a, 1972b; Hill 1979; Giulio 1980; Cutler 1980).

Current optimal-foraging theory (Pyke, Pulliam, and Charnov 1977) would imply that if insects were abundant, web spiders would be deleted from the diet of *P. fimbriata* because of the lengthy on-web pursuit times. However, understanding the unique predatory behaviors of *P. fimbriata* requires consideration of other factors besides search and pursuit times. The abundance of web-building spiders sympatric with *P. fimbriata* is an important factor, as are the adaptations by which *P. fimbriata* is concealed from its own visually hunting predators. These adaptations include not only the unusual morphology of the spider but also its slow and peculiar means of locomotion. If *P. fimbriata* were to move rapidly and smoothly to capture insects, the concealment provided by its morphology would probably be considerably reduced. If the potentially short pursuit time for predation on insects is weighed against the cost of sacrificed concealment, which is considerable for a highly cryptic spider, then predation on insects might not be favorable given the abundance of web spiders.

Unlike insects, salticids are appropriately pursued by *P. fimbriata* using slow, cryptic behaviors that conceal the predator from the acute vision of both its prey and its own predators. Natural selection can be envisaged as perfecting this type of predation in an environment such as the Queensland rain forest, where cursorial salticids are abundant.

To summarize, it is proposed that typical salticids evolved from web-building ancestors, and that the development of acute vision took place in the context of web invasion as a predatory tactic. Possession of acute vision, however, permitted vagrant predation on motile insects, the advantages of which led to the loss of web-associated behaviors and the adaptive radiation of typical salticids. The unique predatory strategy of *P. fimbriata* is the result of special circumstances: occupation of habitats in which web spiders and typical salticids are abundant, coupled with extreme crypticity as an anti-predatory adaptation.

Comparative Studies

The types of information needed for a critical examination of the hypothesis concerning salticid evolution proposed here are wide-ranging. One possible critical approach is to consider the natural history and behavior of other putatively primitive salticids.

The Lyssomaninae (Petrunkevitch 1928) are generally considered to be primitive. Establishment of a separate family, Lyssomanidae, for these spiders, although discussed perennially (Blackwall 1877; Homann 1950, 1971; Prószyński 1968; Platnick 1971; Esteban 1976; Galiano 1976, 1980; Wanless 1978, 1979, 1980a, 1981b), does not seem justified. The Old World lyssomanines, *Asemonea*, *Goleba*, *Macopaeus*, *Onomastus*, and *Pandisus* are apparently a polyphyletic group that will require the establishment of two new subfamilies. Their relationships with the New World *Chinoscopus* and *Lyssomanes* are uncertain, but in some cases they have probable affinities with *Portia* (Wanless 1980a, 1980b, 1980c, pers. comm.).

According to recent studies (Wanless 1978, 1981c; Murphy and Murphy 1983), there are about a dozen species of *Portia*. Along with *Brettus*, *Cyrba*, and ten other genera, *Portia* belongs in the subfamily Spartaeinae (Wanless 1979, 1981a, 1984). These genera occur predominantly in tropical Asia and

Africa, *Portia fimbriata* in tropical Australia and *Cyrba algerina* in the Mediterranean region being notable exceptions.

The Spartaeinae, the Old World lyssomanines, and various other plurident salticids from Africa and Asia, such as *Cocalodes*, seem to be broadly related primitive groups (Wanless 1980d, 1982, 1983). The majority of the Salticidae, however, belong to more than 20 subfamilies of advanced species (Petrunkevitch 1928).

It should be noted that subfamilial classification of the Salticidae presents formidable problems. Many of the subfamilies in Petrunkevitch's classification (1928) are artificial, as he realized, but his scheme nevertheless provides a starting point from which improvements can be made (Kaston 1948; Prószyński 1971).

Our hypothesis concerning salticid evolution predicts relationships between factors such as crypticity, habitat, and predatory behavior. Web building is not expected in morphologically advanced salticids. Even in primitive species, only those that are highly cryptic are expected to use the highly specialized predatory behaviors against other spiders described for *P. fimbriata*. As crypticity decreases, a tendency toward greater reliance on rapid, smooth locomotion, more appropriate for pursuit of motile insects outside webs, is expected, regardless of the abundance of webs to invade. Primitive species living in habitats in which web spiders are not numerous are expected to lack web-associated predation, regardless of their degree of crypticity. Specialized stalking of other salticids is expected to evolve only when typical salticids are exceptionally numerous and the predator is extremely cryptic. As we shall see, available evidence seems to corroborate these predictions.

Only one species of *Portia*, *P. fimbriata*, has been recorded from Australia; and, except for the study in Queensland, records of this species are rare. However, a second population of *P. fimbriata* was recently found in the Northern Territory, in a habitat strikingly different from the Queensland rain forest: the mouths of limestone caves in open woodland. Although web-building spiders from other families are numerous in these caves, cursorial salticids were judged to be much less abundant here than in the Queensland habitat. The two populations are morphologically similar, but they are behaviorally distinct in several respects. Most important, although the population of *P. fimbriata* from the Northern Territory builds webs and uses vibratory behavior to capture spiders on neighboring webs, it lacks, as predicted, the specialized behaviors by which the population from Queensland stalks and captures cursorial salticids.

Studies have recently begun on the natural history and behavior of populations of *P. fimbriata* from Malaysia and Sri Lanka and populations of four other *Portia* species: populations of *P. labiata* from Malaysia and Sri Lanka, a population of *P. albimana* from Sri Lanka, and populations of *P. schultzi* and

P. africanus from Kenya. Some of the preliminary results of these studies are particularly relevant.

All of the *Portia* studied share the unusual cryptic morphology of the Australian *Portia*. In addition, all of the Asian and African *Portia* build large webs and employ vibratory behavior similar to that of the Australian *Portia*. Sometimes they lure spiders to within lunging distance, sometimes they stalk slowly across webs while vibrating. On invaded webs, all *Portia* feed on insects and the hosts' eggs; *P. labiata* also feeds on kleptoparasitic spiders (*Argyrodes*), as does *P. schultzi* (Forster 1982a).

Web-building spiders are abundant in each of the Asian and African habitats. However, in these habitats, as in the limestone caves of the Northern Territory, cursorial salticids do not seem exceptionally numerous; and none of the Asian and African *Portia* uses the specialized stalking of cursorial salticids employed by the Queensland *Portia*. Like the *Portia* from the Northern Territory, the Asian and African *Portia* seem to occupy a category predicted by the model.

The Queensland *Portia* seem to be more extreme in the slowness of their locomotion, especially when stalking prey, and more consistent in their employment of vibratory behavior on alien webs. The other *Portia*, for example, are prone to leap into webs onto their victims from suitable vantage points outside the webs, something the Queensland *Portia* only rarely do.

In Queensland, all *Portia* were found within 100 m of permanent running water (a creek or river), on rugged and precipitous terrain. Because of the terrain and the enveloping canopy of the forest, ambient light levels in this habitat were consistently low. In contrast, the habitats of the other *Portia* were often on less sculptured terrain and/or in less dense forest, where ambient light levels tended to be higher.

The effectiveness of cryptic morphology and associated slow, mechanical locomotion in concealing *Portia* from visually hunting predators would seem to vary with ambient light levels. Certainly, to the human eye *Portia* are more easily recognized in brighter light; and a similar relationship between concealment and ambient light occurs in a New Zealand masking crab, *Notomithrax ursus* (Pack 1983), and in rodents (Dice 1947). Although all *Portia* are similar morphologically, the population from Queensland seems to be more effectively cryptic, as a result of living in a habitat with more subdued light. A more rapid pursuit, especially by leaping into a web, may be less successful on average, but rapid locomotion is advantageous, especially to a less cryptic *Portia*, in reducing the time the spider remains exposed to the view of a potential predator.

There seems to be a trend, as predicted, toward adoption of more rapid locomotion in the *Portia* populations that are not so extremely cryptic as the

Queensland populations. However, it should be emphasized that these rapid pursuits occur opportunistically. Very slow, patient pursuits, comparable to those of the *Portia* from Queensland, also occur frequently. The normal locomotion of these spiders is still much slower and more mechanical than that of typical salticids.

With one exception, of the numerous salticids observed in the same habitats as *Portia*, including salticids from allied genera, none is cryptic in the extreme fashion of *Portia*; and, as predicted, all seem to be efficient cursorial predators of insects. The exception is *Phyaces comosus* from Sri Lanka, a species probably not closely related to *Portia*. *Phyaces comosus*, like *Portia*, has a "special resemblance" to detritus. Like *Portia fimbriata* from Queensland, *Phyaces comosus* is highly proficient at capturing cursorial salticids; but, unlike *Portia*, *Phyaces comosus* does not spin or invade webs. Unlike *Portia*, the spartaeinines *Brettus adonis*, *Brettus cingulatus*, and *Cyrba algerina* are efficient cursorial predators of insects. However, like *Portia*, they are vibratory, web-invading predators of spiders.

Lyssomanes and other species that spin small weblike structures as resting and oviposition sites (see above) are from putatively primitive subfamilies. None is highly cryptic, and all seem to be efficient cursorial predators of insects (Forster 1982a; Richman and Whitcomb 1981). Their resting "webs" function as silken shelters analogous to the nests of many advanced species and in some cases ensnare insects. Since no advanced species seem to build them, I propose that resting webs are the result of an evolutionary reduction of the larger prey-capturing webs of ancestral spiders. The role of spinning behavior seems to have been reduced with the acquisition of a cursorial lifestyle.

Although some lyssomanines suspend their eggs on the silk of their "webs," the population of *Portia* from Queensland and the *Portia* populations from Asia and Africa oviposit on leaves or other debris suspended in their webs. The population of *P. fimbriata* from the Northern Territory, however, is more similar to the lyssomanines in that it oviposits directly onto type 1 silken platforms within its type 2 webs (Fig. 9.20).

Pheromonal communication may have phylogenetic implications. Males of many salticid species begin vibratory courtship or show other distinctive responses upon contacting vacant nests or draglines of conspecific females (Jackson 1976, 1981c, 1982b, 1982c; Edwards 1981; Yoshida and Suzuki 1981; Richman 1982). Similarly, males of some dictynid and theridiid species begin vibratory courtship when they contact vacant webs of conspecific females (Jackson 1978b; Ross and Smith 1979). Pheromones associated with silk and detected by contact chemoreception probably elicit these sexual responses by male spiders. Legendre and Llinares (1970) reported that males of

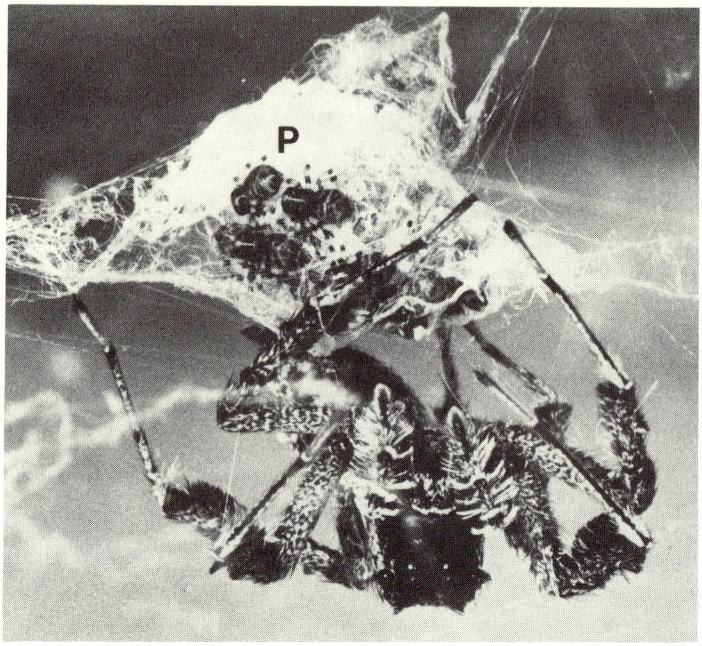

Fig. 9.20. Female *Portia fimbriata* from the Northern Territory hanging beneath her egg case on a type 1 platform in her type 2 web. Eggs have hatched and postembryos (*P*) are present.

the spartaeinine *Cyrba algerina* respond distinctively to substrates over which females have walked, although they did not state whether draglines were contacted.

Males of certain nonsalticid web-building species have been reported to locate conspecific females by means of olfactory pheromones (Tietjen and Rovner 1982; Olive 1982b). Crane (1949) reported that olfactory pheromones from females of certain salticid species lower the thresholds of conspecific males for sexual responses to visual stimuli. However, olfactory mate location has not been reported in the Salticidae; and because of the highly developed vision of the salticids, olfactory communication has generally been expected to have a minor role in courtship in this family (Kaston 1936; Platnick 1971; Robinson 1982a).

In recent studies olfactory mate location was demonstrated to occur in all tested lyssomanines and spartaeinines (*Lyssomanes viridis*, *Brettus cingulatus*, *Cyrba algerina*, *Portia fimbriata*, *P. labiata*, and *P. schultzi*) but not in 16 tested species from other subfamilies. Pheromonal communication might have been derived secondarily from cursorial salticid ancestors that relied solely on vision, but it seems more probable that it was conserved in certain species from nonvisual web-building ancestors.

The secondary genitalia of male spiders, their palpal organs, which have major significance in lower-level taxonomy (Levi 1982), may have implications concerning the origins of the Salticidae. In the Salticidae, as in many primarily cursorial families, palpal organs tend to be relatively simple (Shear 1967). In contrast, palpal organs in primarily web-building families are often complex, sometimes extremely so (Gerhardt and Kaestner 1938; Grasshoff 1974, 1975; van Helsdingen 1965, 1969). Thus it is of interest that unusually complex palpal organs occur in some of the putatively primitive salticid genera including genera closely allied to *Portia* (Wanless 1980a, 1980c, 1981a, 1981c, 1982). Even *Portia* have relatively complex palpal organs compared to other salticids.

Since it was argued that acute vision evolved in the Salticidae in conjunction with adoption of web invasion as a predatory tactic, and more specifically the invasion of a wide range of web types, specialists at web invasion in other families, lacking acute vision, are expected to invade a more restricted range of web types. Available evidence seems to bear out this prediction. Some mimetid species may be specialists at invading theridiid and araneid webs; however, more generalized feeding may occur in other mimetid species (Lawler 1972; Thompson 1972). *Taieria* and *Lampona* (Gnaphosidae) specialize at invading the webs of cribellate spiders (Jackson, in press). *Archaea* (Archaeidae) and *Ariamnes* (Theridiidae) do not seem to invade webs to capture their spider prey.

Origins of the Araneomorphae

Silk production is characteristic of all species of spiders, and plays a key role in the adaptations of both web-building and cursorial species (Savory 1928; Krafft 1982; Kullmann et al. 1975; Stern and Kullmann 1981; Shear, this volume). Although they do not spin webs, cursorial species typically make extensive use of silk in activities such as sperm induction, dispersal by ballooning, and construction of nests, egg cases, and draglines. In view of the possibility that salticids evolved from web-building ancestors and the pervasive use of silk by all spiders, questions arise about the ancestors of other primarily cursorial families.

The informal division of spiders into web builders and cursorial hunters is valuable only so long as it is not too closely examined. It is often difficult to define what is meant by a "web," and perhaps this is one of those occasions when being overly concerned about strict definition is counterproductive (Beer 1977). The prey-catching orbs of *Araneus diadematus* (Savory 1952), for example, are unequivocally webs and the term seems appropriate enough for type 2 webs of *Portia*. But the smaller type 1 webs of *Portia* and the resting webs of various salticids, which are generally not used in prey capture, create problems of definition. It would seem inappropriate to call the single line by which some salticids suspend themselves at night a web, yet prey-capturing devices consisting of few or even single lines, which are built by some araneids, theridiids, and uloborids (Akerman 1932; Holm 1939; Marples 1949, 1955, 1962; McKeown 1963; Hashimoto 1963; Clyne 1969, 1973; Forster and Forster 1973; Lubin, Eberhard, and Montgomery 1978; Eberhard 1979a), are often referred to as webs.

Dinopids, which cast silken nets over their prey (Akerman 1926; Baum 1938; Roberts 1955; Robinson and Robinson 1971; Austin and Blest 1979), do not fit comfortably into either the web builder or the cursorial hunter category. These nocturnal predators suspend themselves from a weblike array of threads while holding their nets, and detect their prey by means of highly specialized eyes (Blest and Land 1977; Blest 1978; Blest, Williams, and Kao 1980; Laughlin, Blest, and Stowe 1980). However, vision has no evident role in their courtship behavior, which seems basically similar to that of more typical web builders (Clyne 1967).

Scytodes are cursorial spiders that squirt a gummy substance from their chelicerae to immobilize their prey (Monterosso 1927; Bristowe 1931; McAlister 1961). Related genera, however, build prey-catching webs (Valerio 1974b). Female *Scytodes* build webs when they oviposit, but do not use these webs in prey capture (Dabelow 1958).

The Araneidae, Tetragnathidae, Linyphiidae, and Theridiidae are primarily web builders. However, bolas spiders, which tether male moths to droplets of glue at the ends of silk lines they hold by their legs (Stowe, this volume), are araneids, as are *Celaenia*, which spin typical araneid orb webs as juveniles (McKeown 1963) but make no apparent use of silk in prey catching as adults. Similarly, juvenile, but not adult, *Pachygnatha* (Tetragnathidae) spin webs (Balogh 1934). *Arcys* and *Archemorus* are araneids that make no apparent use of silk at any stage (Main 1976), and *Drapetisca* seem to be cursorial linyphiids (Comstock 1948; Kaston 1948; but see Kullmann 1961). *Euryopis* (Theridiidae) do not build webs; instead they trail bands of silk as they encircle and entangle their prey (Carico 1978; Porter and Eastmond 1982). These are all examples of something that seems to have happened repeatedly in primarily web-building families: reduction and sometimes complete abandonment of web building (Forster 1970; Forster and Wilton 1973; Main 1976; Opell 1979; Levi 1980b; Stowe, this volume).

Prey-catching methods similar to those of *Euryopis* occur in certain gnaphosids, hersiliids, oecobiids, and urocteids (Bristowe 1930b, 1941; Crome 1957; Glatz 1967; Clyne 1969; Burgess 1976; Jackson 1976; Main 1976). The gnaphosids and hersiliids in question are cursorial, although some other hersiliids are web builders. Oecobiids and urocteids build small weblike structures the functions of which seem largely sensory and from which the spiders run out to attack their prey (Kullmann et al. 1975).

The Lycosidae, Pisauridae, Clubionidae, and Oxyopidae are primarily cursorial families. However, *Aulonia* and *Sosippus* (Lycosidae) spin large prey-catching sheet webs (Brady 1962; Job 1974; Brach 1976). Females of many pisaurid species spin large nursery "webs" in which they place their eggs shortly before hatching (Bristowe 1958; Carico 1973). These webs have no evident role in prey catching. However, *Pisaura* juveniles build webs that aid in prey capture, although the adults are cursorial (Lenler-Eriksen 1969; Carico 1985), and a number of other pisaurid species build large prey-catching sheet webs (Bristowe 1941; Davies 1982; Nentwig 1985). Some clubionids build rudimentary webs (Savory 1928) or weblike scaffolding around their nests (Peck and Whitcomb 1970), but the functions of these are poorly understood.

The Oxyopidae are of particular interest, since Simon (1901) considered this family to be closely related to the Salticidae, and Lehtinen (1975) supported this view by noting similarities in hair ultrastructure. Females of some species build tentlike "webs" around their eggs (Whitcomb 1962; Main 1976), and these are somewhat similar to the resting webs of *Lyssomanes* and other salticids. Also, "webs" of uncertain function are built by juveniles of *Peucetia* (Kaston 1972b), and adults suspend themselves from a weblike scaffold when ambushing prey (Rovner 1980a). Some oxyopids, like *Portia fimbriata*, routinely mate suspended from the female's dragline (Gerhardt 1927, 1933; Whitcomb and Eason 1965, 1967). Moreover, like certain salticids, many oxyopids rest at night suspended from draglines attached to vegetation (Lowrie 1971; Cutler, Jennings, and Moodie 1977). Rovner (1980a, 1980b) argued on the basis of morphology and behavior that the Oxyopidae evolved from web-building ancestors, and recently a web-building oxyopid was discovered in Costa Rica (Griswold, in press). Perhaps the Oxyopidae and Salticidae had a common nonvisual web-building ancestor.

Admittedly, attention has been focused here on the Araneomorphae, the largest suborder of spiders. The Mygalomorphae and Mesothelae, however, are divergent lineages that would warrant separate consideration (Kaestner 1968). The findings concerning *P. fimbriata*, the phylogenetic hypothesis they instigated, and the observations on other species suggest the more encompassing hypothesis that all araneomorph spiders had web-building ancestors. This hypothesis would seem to be implied by Forster and Wilton's (1973) and Platnick's (1977) arguments that all araneomorph spiders passed through a cribellate phase, since the cribellum, a specialized spinning plate responsible for the adhesive component of hackled webs, seems to be fundamentally an adaptation of web builders. Extant cribellate spiders are primarily web builders. Both Simon (1864) and Lehtinen (1967) came close to proposing this hypothesis.

Web building was probably independently lost many times, as the result of various factors. Carico (1978) suggested two possible factors: olfactory attraction of prey, as occurs in, e.g., the bolas spiders (see also: Eberhard 1980a; Stowe, this volume), and specialized selection of microhabitats with high prey densities, as occurs in, e.g., *Euryopis* species. Acute vision is not predicted to evolve as a result of these two factors. The biology of *P. fimbriata* may allow us to glimpse the peculiar set of factors that led to the evolution of cursorial salticids, with their unique eyes and visually mediated behavior.

Acknowledgments

Field and laboratory studies were carried out in Australia during the tenure of a visiting fellowship in the Department of Neurobiology at the Research School of Biological Sciences of the Australian National University. I thank Professor G. A. Horridge, F.R.S., for the hospitality of the Department. I thank David Blest for his friendly collaboration on some of the studies in Australia and for many stimulating discussions about spiders, eyes, and evolution. Financial support for field studies in Asia and Africa and laboratory studies in New Zealand was provided by National Geographic Society Grant 2330-81, a grant-in-aid from the Erskine Foundation of the University of Canterbury, and grants from the University Grants Committee of New Zealand and the Academic Staffing Committee of the University of Canterbury. I thank the Conservation Commission of the Northern Territory in Australia, the Unit Penyelidikan Sosio Ekonomi, the Univesiti Pertanian Malaysia (Faculty of Veterinary Medicine and Animal Science), and the Institiut Penyelidikan Perhutanan in Malaysia, the Department of National Museums and the Department of Wildlife Conservation in Sri Lanka, and the Ministry of Regional Development, Science, and Technology, the Department of Wildlife Conservation, and the University of Nairobi (Department of Zoology) in Kenya for their generous assistance. I also thank Brian Walsh in Australia, Bertha Allison, Lloyd Whitten, and Wendy Corley in Malaysia, Dr. Thelma Gunawardane and Lakshman Weeratunga in Sri Lanka, and Professor George Kinoti in Kenya for assistance with field work. I thank Mary Catharine Vick for assistance with field work. I thank Susan Hallas, Simon Pollard, and M. C. Vick for comments on the manuscript. I am indebted to Terry Williams (Christchurch) for photographic assistance and to Densey Clyne and Jim Frazier (Mantis Wildlife Films, Sydney) for permission to use one of their photographs of *P. fimbriata*. I am grateful to Fred Wanless at the British Museum of Natural History in London and Valerie Davies and Robert Raven at the Queensland Museum in Brisbane for assistance in identifying spiders and for valuable discussions.

10

The Role of Silk in Prey Capture by Nonaraneomorph Spiders

FREDERICK A. COYLE

PROGRESS IN UNDERSTANDING the prey-capture behavior of nonaraneomorph spiders (Liphistiidae and Mygalomorphae) lags far behind the advances being made by workers studying such behavior in araneomorph spiders. Much of the information on prey capture non-araneomorph spiders is fragmentary and widely scattered throughout literature on the taxonomy and natural history of these spiders. Although Buchli (1969) published a partial review of the literature on nonaraneomorph prey capture, he did not examine the relationship between silk constructs and prey capture as deeply or as broadly as the topic merits.

The role of silk in nonaraneomorph prey capture has not been sufficiently appreciated. My own observations of the Antrodiaetidae, Atypidae, Dipluridae, and Ctenizidae, and a review of the literature on nonaraneomorph silk constructs and prey-capture behavior, reveal that virtually all nonaraneomorph spiders capture their prey at or close to the entrance of a tubular silk retreat, and that parts or extensions of this retreat, although frequently performing protective and other nontrophic functions, often help the spiders to detect and locate (and in some genera even entrap) prey. In this chapter I shall describe these silk constructs and the evidence that bears upon their roles in prey capture. I hope that the observations and hypotheses presented here will stimulate the increased research effort needed to better evaluate the multiple selection pressures acting on the constructs of these fascinating spiders. Such research should also help us test phylogenies, understand the origin of entrapment webs, and assess the trophic impact of nonaraneomorph spiders upon their communities.

For important new research on mygalomorphs, see note on p. 305.

The Sensory Mechanisms Involved in Prey Capture

In order to evaluate the role that nonaraneomorph silk constructs play in the detection and localization of prey, it is necessary to determine what kinds of prey-generated stimuli these spiders can perceive and to understand the sensory mechanisms involved. Although these goals are far from being attained, some helpful information is available.

The great majority of nonaraneomorph spiders do not use vision for prey detection or localization. Most nonaraneomorphs are crepuscular or nocturnal, have small eyes, and/or detect prey from within hidden retreats. Den Otter (1974) showed that covering the eyes of a theraphosid, *Sericopelma rubronitens*, did not alter the form or effectiveness of its prey-capture behavior. Buchli (1969) similarly manipulated the ctenizid *Nemesia caementaria* and achieved the same results. Petrunkevitch (1911) and Coyle (1971) demonstrated the noninvolvement of vision in prey capture by *Dugesiella hentzi* and *Antrodiaetus unicolor*, respectively. Although Gertsch (1979) implies that vision is important in diplurid prey capture, my observations of *Euagrus* indicate otherwise (F. A. Coyle, in prep.). Without providing convincing supporting evidence, Main (1957) states that vision probably plays an important role in prey capture in some aganippine ctenizids.

Much evidence indicates that olfaction, which is apparently the function of the tarsal organ (Foelix 1970; Dumpert 1978), is not used by nonaraneomorphs to locate prey (Petrunkevitch 1911; Coyle 1971; Marples and Marples 1972). I have found it possible to trigger prey-capture behavior in antrodiaetids *Sphodros*, *Ummidia*, and *Euagrus* by creating substrate vibrations with a toothpick. Klingel (1966) found that freshly killed insects placed on or between *Liphistius batuensis* signal lines did not trigger prey-capture behavior. Den Otter's (1974) experiments with *Sericopelma rubronitens* indicate that olfaction does not play a part in this tarantula's decision to capture prey but that after the capture a contact chemical sense affects the decision to eat or reject the prey. Presumably the contact chemoreceptors involved are the morphologically distinctive chemoreceptor hairs observed on the tarsi of many araneomorph spiders (Foelix 1970; Harris and Mill 1977b) and the mygalomorphs *D. hentzi* (Foelix 1970) and *Euagrus* spp. (F. A. Coyle, in prep.). Observations that hungry *N. caementaria* (Buchli 1969) and *Cyrtocarenum cunicularium* (Decae, Caranhac, and Thomas 1982) do not try to capture certain insects that appear to use defensive odors suggest that olfaction may occasionally play a role in nonaraneomorph prey capture.

It is unlikely that nonaraneomorph spiders use sound detectors to sense prey, for it is doubtful that many of their prey generate airborne sounds. Furthermore, Moggridge (1874) demonstrated that a capture-ready *Bothriocyrtum californicum* was unresponsive to a great variety of sounds. Petrunkevitch (1911) demonstrated that *D. hentzi* is insensitive to sound. *Sericopelma* trichobothria generate action potentials only at unnaturally high sound intensities (Den Otter 1974). An auditory slit sensillum like the tarsal slit of *Cupiennius salei* (Barth 1967) has not yet been demonstrated in a non-araneomorph spider.

Most of the observations of prey-capture behavior in nonaraneomorph spiders support the hypothesis that prey are detected by substrate-vibration receptors in the pedipalps and legs of these spiders. Some of the slit sensilla (i.e., the lyriform organs and single-slit sensilla) on the legs of araneomorph spiders are important substrate-vibration detectors (Walcott and Van der Kloot 1959; Liesenfeld 1961; Walcott 1969; Barth 1978, 1982). Barth and Seyfarth (1979) and Rovner and Barth (1981) have demonstrated that the wandering araneomorph *C. salei* relies on these substrate-vibration detectors for locating prey and mates at a range of up to 1 m on banana plants. Brownell (1977) demonstrated that a sand scorpion's basitarsal compound slit organs, which are sensitive to substrate vibrations (specifically Rayleigh waves), enable the scorpion to detect and localize insects moving on the sand surface up to 50 cm away. Den Otter (1974) studied another type of sensillum that may be an important detector of prey-generated substrate vibrations in several nonaraneomorph taxa. He demonstrated that the distinctive club-shaped sensilla present on the dorsal surfaces of all *S. rubronitens* tarsi trigger action potentials in the presence of substrate vibrations. Sensilla having a similar form and position are found in other theraphosids and some diplurids (Lehtinen 1980), *Liphistius* (Murphy and Platnick 1981), and ctenizids of the genera *Ummidia* (Buchli 1969) and *Cataxia* (Main 1969), but they are not present in several genera of ctenizids, antrodiaetids, atypids, and diplurids that I have examined. The leg setae and macrosetae that Den Otter (1974) demonstrated to be touch receptors in *S. rubronitens* are found on all spiders (Foelix 1982) but probably do not become involved in prey sensing until the prey is contacted. Main (1957) suggested that aganippine cephalic setae, which are probably in contact with the trap-door undersurface when the spider is in the foraging posture, may be important in prey detection.

The detection of prey-generated air currents may be an important prey-sensing process in some nonaraneomorph spiders. Several workers have shown that spider trichobothria are nicely adapted for detecting the presence and direction of prey-generated air currents (Goerner 1965; Goerner and Andrews 1969; Harris and Mill 1977a). The araneomorph *C. salei* can locate a crawling fly up to 20 cm away by orienting itself towards the fly-generated air current, presumably by means of its trichobothria (Barth 1982). Den Otter (1974) demonstrated that *S. rubronitens* trichobothria located on the dorsal surfaces of the tibia and more distal articles of the legs and pedipalps, are sensitive to air currents such as those generated by the movements of a cricket leg 2 cm away. Although all other nonaraneomorph spiders probably possess trichobothria of the type studied by Den Otter, they may not be important prey detectors in many of these species since retreats, burrow walls, and entrance structures probably shield the pedipalps and legs of many capture-ready spiders from prey-generated air currents.

The Silk Constructs and Their Relationship to Prey Capture

In this section I shall describe for each nonaraneomorph spider family the retreats and associated silk structures that may play a role in prey capture, and I shall review the evidence that bears upon such a prey-capture role.

Liphistiidae

The liphistiids construct a silk-lined burrow and cover the entrance with a thin "wafer-type" trap door. Some individuals of *Liphistius malayanus* and cave-dwelling *Liphistius batuensis* construct a short, camouflaged silk tube on hard soil or rock surfaces with a second, smaller trap door, apparently serv-

Fig. 10.1. Burrow entrance of *Liphistius desultor* on Penang Island, Malaysia. Signal lines extend from the entrance rim. Reprinted, by permission, from Murphy and Platnick 1981.

ing as an escape hatch (Bristowe 1932, 1952; Klingel 1966), at one end. *Liphistius* species attach to the silk entrance rim from six to eight long, multifilament silk lines that radiate out over the substrate and greatly extend the spider's prey-sensing area (Fig. 10.1) (Bristowe 1932, 1952, 1975, 1976; Savory 1960a; Klingel 1966; Murphy and Platnick 1981). In its normal foraging posture, the spider sits under the slightly opened trap door with the tarsi of its pedipalps and first two or three pairs of legs in contact with the proximal ends of the signal lines. Observations involving both real prey and mechanically generated vibrations (20–400 Hz) indicate that vibrations transmitted from the prey over any one of these threads inform the spider of its presence and location (Klingel 1966). There is no evidence that these lines serve to slow the prey. The spiders emerge quickly from the entrance to seize the prey, and then, if they have emerged completely, return to the entrance, reopen the door, and enter the burrow. Mechanical vibration of the trap door also triggers the prey-capture response (Klingel 1966).

Species of *Heptathela*, the other liphistiid genus, do not construct signal lines (Bristowe 1932; Shimojana 1965; Haupt 1977), and consequently have a proportionately smaller prey-sensing area than *Liphistius* (Haupt 1979). Haupt (1979) concluded that *Heptathela nishihirae* detects prey with substrate-vibration receptors on the pedipalps and first two pairs of legs. However, the manner in which the capture-ready spider extends the tarsi of these six appendages out from under the trap door and beyond the entrance rim so that the tarsal trichobothria are exposed suggests the possibility that the spider may be able to detect prey-generated air currents.

Dipluridae

Although some diplurids excavate burrows for retreats, apparently the majority do not. *Kiama lachrymoides* (Main and Mascord 1969), *Teyl luculentus* (Main 1975), *Xamiatus* (Raven 1981), *Aname* (Main 1976, 1982b; Raven 1981), and some species of *Stanwellia* (Main 1972), *Aparua* (Forster and Wilton 1968), *Cethegus* (Main 1960; Raven, in press), and *Ixamatus* (Main 1968; Raven 1980a) excavate their burrows, and usually line them with silk; but most diplurids instead construct tubular or funnel-shaped silk retreats in soil crevices or under stones, logs, or moss mats, or above ground in protected spots such as at the bases of plants, under loose bark, or in grooves in rough bark. These include some *Aparua* species (Forster and Wilton 1968), some *Cethagus* species (Raven, in press), some *Ixamatus* species (Main 1968; Raven 1980a), most *Australothele* species (Raven, in press), and the genera *Euagrus* (Stevenson 1908; Chamberlin 1924; Chamberlin and Ivie 1945; Gertsch 1973), *Ischnothele* (Petrunkevitch 1911), *Stenogrocercus* (Raven, in press), *Diplura* (Vollrath 1978; Kirkendall 1979), and *Microhexura* (Coyle 1981a). The presence of *Aparua*, *Cethegus*, and *Ixamatus* species in both categories, and the finding that *Calisoga longitarsus* individuals can excavate burrows, occupy preformed burrows, or, in more humid conditions, live in thin-walled silk tubes in spaces under logs or rocks (Bentzien 1975), indicate that in the Dipluridae the shift from a burrowing to a nonburrowing lifestyle, or vice versa, does not require elaborate genetic reconstruction.

A great variety of silk entrance structures have evolved in this family. The burrow entrances of three *Stanwellia* species and some other diplurids are collar- or turretlike extensions of the silk burrow lining (Main 1972, 1976) that, like the better-known antrodiaetid collars and turrets (Coyle 1971), may play a role in prey detection. Two undescribed burrowing diplurid genera from western Australia construct trap doors (Main 1968, 1976), which are not built by any other diplurids. Unlike other trap doors, these remarkable diplurid doors swing open 180° and lie flat on the ground. "The spider sits poised across the hinge line with legs spread out over the upturned underside of the opened door" (Main 1968). This foraging posture suggests that, in addition to performing functions such as flood protection (Main 1976), these doors may aid in prey detection by extending the substrate-vibration detection range of the spider's tarsi. Another remarkable observation by Main (1976) is that some diplurids (she fails to name the taxa) may capture aerial insects "such as moths, chafer beetles, and flying ants" without using entrapment silk: "The light-bodied spiders jump upwards from their funnel-like holes and grab prey in flight."

Many diplurid taxa rely upon funnellike extensions of their retreat silk to detect and capture prey. The tiny diplurids of the Nearctic genus *Microhexura* live within irregular, flattened, thin-walled tubular webs under moss mats or other materials lying on the ground or on rock surfaces (Coyle 1981a). These spiders appear to locate prey by means of silk-conducted vibrations generated by the prey when it contacts the tube or the poorly defined, shallow, funnel-shaped openings at the tube edge.

Recent field and laboratory observations of web design and prey capture in several species of *Euagrus* (F. A. Coyle, in prep.) extend our knowledge well beyond that presented in the brief notes of Stevenson (1908), Chamberlin (1924), Chamberlin and Ivie (1945), and Gertsch (1973, 1979). The *Euagrus* retreat is a silk tube, often irregular and sometimes branching, that lies hidden under a rock, log, or dead *Agave* leaf, in a crevice in an earth bank or rock outcrop, within a decomposing log, or in some other similarly well-protected microhabitat (Fig. 10.2). This retreat tube opens outward into irregular interconnected funnel-, curtain-, and sheetlike extensions that serve as the primary sites of prey capture (Figs. 10.3–10.5). In the smaller species these capture portions of the web tend to be small, confined to spaces under or within materials on the ground surface, and therefore inconspicuous. But in the larger species the capture webs are usually extensive and

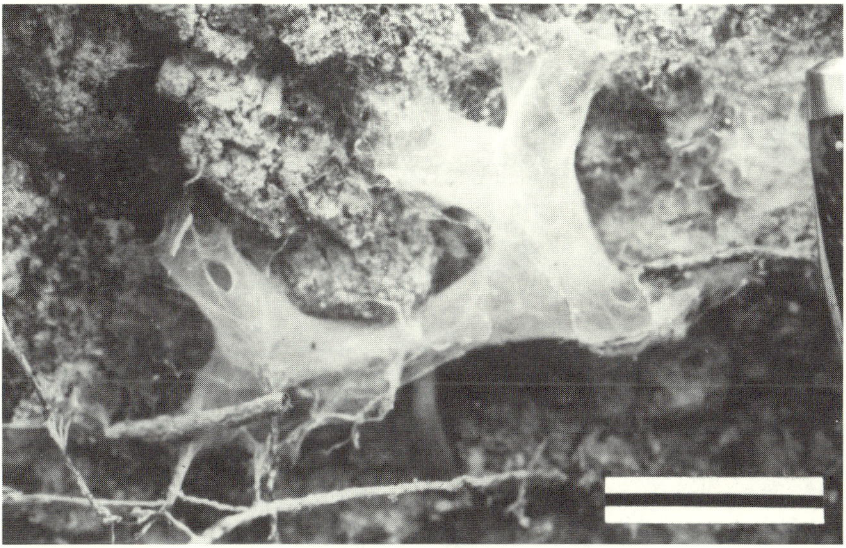

Fig. 10.2. Portion of a retreat of *Euagrus* sp. near El Punto, Oaxaca, Mexico. The capture web and part of the retreat were destroyed while excavating to expose the retreat, which was in a crevice in an earth road bank. The scale line represents 5 cm.

Fig. 10.3. Capture web of *Euagrus* sp. near El Punto, Oaxaca, Mexico. The retreat lies hidden within a crevice in the earth road bank. The scale line represents 5 cm.

Fig. 10.4. Capture web of *Euagrus* sp. near Ajijic, Jalisco, Mexico. The scale line represents 5 cm.

Fig. 10.5. Capture web of *Euagrus* sp. 24 km southeast of Acatlán, Puebla, Mexico. The retreat (not visible) penetrates 15 cm into soil at the base of a shrub. The scale line represents 10 cm.

conspicuous, extending well out from the retreat to cover an area of up to 600 cm^2. These larger capture webs are partly attached to vertical surfaces nearby and consequently may extend upward from the primary substrate for several centimeters; they thus constitute true space webs that are penetrated by irregular tunnellike access routes.

Both web design and prey remains indicate that these *Euagrus* webs serve primarily to help capture ground-dwelling arthropods such as beetles and ants. They are entrapment webs; at least several kinds of ants and beetles readily become and remain entangled in them. The irregular mesh of silk is composed of extremely fine anastomosing filaments (about 0.1 μm in diameter) that have been found to adhere to glass; but exactly what causes this adhesive quality is not yet clear. The spiders, which capture prey both day and night, rely heavily, perhaps exclusively, upon prey-generated vibrations of the silk to detect and locate prey. Web vibrations generated with a hand-held transparent glass rod 15 cm long and 1 mm in diameter trigger the prey-capture response. The trip from the retreat out to the prey frequently consists of a series of a few short dashes separated by pauses. A pause is typically not followed by another dash until prey movement occurs. Insects (dead or alive) that remain motionless after initial contact with the web are usually not located by the spider.

At least some species of the Australian genus *Cethegus* extend out from their retreat entrance a space web composed of a dense labyrinthine mass of threads partly suspended from nearby log, rock, tree-trunk, shrub, or grass surfaces (Main 1960, 1976; Raven, in press). A few to several funnellike passageways fan out from the retreat tube and penetrate and provide access to the entire web. Raven's (in press) laboratory observations and lists of field-collected prey remains indicate that the webs of *Cethegus daemeli* and *Cethegus palipes* are true entrapment webs that snare large numbers of land snails and at least some beetles and millipeds. *Australothele*, *Stenogrocercus*, and two other undescribed genera of Australian diplurids in the same subfamily as *Cethegus* construct webs that, although similar to those of *Cethegus*, are proportionately smaller and often less exposed (Raven, in press).

Some species of *Ischnothele*, a tropical diplurid genus, construct impressive aerial funnel webs that Petrunkevitch (1911) found to be abundant in low plants near Veracruz, Mexico, and that he initially mistook for *Agelenopsis* webs. These aerial webs, which I have also observed in southern Mexico, consist of a roughly horizontal single- or multilayered capture sheet that fans out from one to three funnels that emerge from the entrance of a tubular silk retreat hidden in a crevice in a rock outcrop, earth bank, or tree trunk (Figs. 10.6, 10.7). Owing to the irregular mesh of silk that surrounds the funnels, these webs are in part space webs. Two large *Ischnothele* capture sheets each extended out about 18 cm from the retreat entrance and were about 30 cm

Fig. 10.6. Web of *Ischnothele* sp. 26 km east of Cintalpa, Chiapas, Mexico. The retreat (not visible) extends back behind roots of a tree clinging to a vertical rock road bank. The scale line represents 10 cm.

Fig. 10.7. *Ischnothele karschi* female from Kilifi, Kenya, at the entrance of her retreat. Courtesy of F. M. Murphy.

wide. One small-bodied species of *Ischnothele* that I have observed near San José, Costa Rica, constructs small exposed capture webs of interconnected funnels and sheets on the surface of the forest floor litter. *Ischnothele* webs are similar to *Euagrus* webs in texture and ability to entrap prey.

Other diplurid genera construct funnel-and-sheet capture webs at least as impressive as those of *Ischnothele*. Vollrath (1978) described the web of a Panamanian *Diplura* species as a sheet or funnel covering an area of 50 × 50 cm. "The spider rests in the entrance of its retreat, a silken tube, waiting for prey to fall onto or crawl over the sheet. The prey is bitten in a sudden lunge and dragged into the retreat for consumption." I have observed in Colombia and Panama *Diplura* webs consisting of a horizontal aerial sheet up to 0.5 m^2 in area with an aerial mesh extending up to 2 m above the sheet. Whether this mesh serves to stop airborne insects or to support the sheet, or serves both functions, remains to be investigated.

These diplurid funnel-and-sheet webs, which are common and often conspicuous in many parts of the tropics, are the largest and probably the most effective prey-capture constructs evolved by extant nonaraneomorph spider taxa. Apparently the behavioral and ecological analogues of agelenid and *Sosippus* webs (Simon 1889; Petrunkevitch 1911; Gertsch 1979), these webs merit more extensive study than they have received to date. The effectiveness of the diplurid funnel-and-sheet-web prey-capture strategy is evidenced by the fact that some tropical American species support "a wide range of arthropod commensals, kleptoparasites, and other symbionts" (Robinson 1977a), including dipteran and lepidopteran larvae, beetles, and *Argyrodes* (Vollrath 1978) and *Philoponella* (W. G. Eberhard, pers. comm.) spiders. Five species of *Mysmenopsis* (Mysmenidae) have also been found to live as kleptoparasites on the webs of *Diplura* and *Ischnothele* species (Platnick and Shadab 1978b; Vollrath 1978; Kirkendall 1979). A remarkable association between a Panamanian *Diplura* species and the first known ectoparasitic spider, *Curimagua bayano* (Symphytognathidae), has been described by Vollrath (1978). F. W. Murphy (pers. comm.) has reported that the webs of *Ischnothele karschi* in Africa are inhabited by an abundance of spider and insect symbionts.

Hexathelidae

The Hexathelidae, recently erected by Raven (1980b), is the sister group of the Dipluridae, which it closely resembles in silk architecture. Some *Hexathele* species normally construct tube webs under rocks, logs, or loose bark; others often construct their tubes in crevices in the ground; and others (*Hexathele petriei*, *Hexathele huka*, and *Hexathele waipa*) consistently excavate burrows and line them with silk (Forster and Wilton 1968; Forster and Forster 1973; Raven 1976). Silk threads, which probably extend the prey-sensing area, radiate out from the tube-web entrance of *Hexathele hochstetteri* (Todd 1945) and from the burrow-entrance rim of *H. huka* (Forster and Forster 1973). *H. petriei*'s prey-sensing range is probably extended by a weblike multitude of anastomosing threads that extend several centimeters from the burrow-entrance rim over the surrounding ground surface and ground vegetation (Forster and Wilton 1968).

Most species of *Bymainiella* live in tube webs under rocks or logs. In at least one of these species, *Bymainiella lugubris*, the tube entrance is a funnel opening into a sheet (Raven 1978). One rain-forest-dwelling species, *Bymainiella boycei*, has become arboreal, constructing its tube web in crevices in tree bark and extending from the tube-web entrance a short silk funnel up to 5 cm in diameter (Raven 1978) that probably serves a prey-detection function. Prey remains indicate that this species captures at least some aerial insects (namely, Diptera and Lepidoptera).

Porrhothele species usually construct tube webs in cavities beneath stones, logs, or loose bark, or in crevices in earth banks or trees (Todd 1945; Forster and Forster 1973). Todd (1945) observed that threads usually radiate out from the entrances of these tubes. Laing (1973, 1975, 1978, 1979, in press) has learned much about prey capture in *Porrhothele antipodiana*. Although this species usually constructs a single tube that opens out into a fan-shaped sheet, the tube may branch and have two or more entrances. Sheet size is highly variable; some tubes have no sheets (or poorly developed sheets), other tubes have sheets up to 40 cm wide. Laing (1973) observed that spiders living in exposed microhabitats (on banks or in trees) have significantly larger capture sheets than those living under logs or stones, and suggests that this increase in sheet size is related to the lower prey densities found in the exposed microhabitats; i.e., somehow this species adjusts its prey-sensing area to prey density. Prey are detected from within the tube and located on the basis of web vibrations and perhaps other kinds of tactile information such as web tension. Although the spider sometimes initially misjudges the direction or distance required to reach the prey, search duration is short. (The mean time from prey contact with web to spider contact with prey is 2.0 seconds for 12 nighttime trials.) Most prey are not entangled in the web, but simply move more slowly and with more turns than when on the ground, so a speedy attack is important. (Spiders may move as fast as 0.2 m/sec.) Although beetles, isopods, and millipeds form the great bulk of *P. antipodiana*'s prey, some predominantly aerial insects are captured (wasps, bumblebees, and moths). These nocturnal spiders were reluctant to capture prey during the day; eight of 14 that did move up to the tube entrance in response to daytime web vibrations waited there for many seconds before dashing to the prey. (The mean time from prey contact with web to spider contact with prey increased to 29.0 seconds.) Laing (1979) suggests that this slow reaction time to daytime web vibrations may be an adaptation to avoid wasp predation.

Spiders of the dangerous Australian genus *Atrax* burrow into the rotten wood or crevice of a tree or into the soil under a rock or log, and usually extend the burrow-lining silk out from the entrance to form one or a few funnels (McKeown 1952; Levitt 1961; Hickman 1964; Mascord 1970; Main 1976). Hickman (1964) observed that silk strands that radiate out from the lower margin of each funnel "tend to entangle the feet of ground-roving insects." McKeown (1952) describes how *Atrax formidabilis* sometimes extends 45-cm-long parallel silk strands out from its funnel across a tree trunk. He claims that frogs stick to these strands and that the vibrations transmitted by the strands elicit a prey-capture response and guide the spider. These putative adhesive traplines deserve careful study.

The African hexathelid *Macrothele darcheni* is of considerable interest because it is the only communal nonaraneomorph spider known (Darchen 1967). This spider lives in low communal webs (up to 1 m^2 in area) composed of interlocking horizontal sheets and tubular retreats. In one such web 20 adult and subadult females lived together. Limited laboratory observations revealed that although two or three spiders oriented themselves towards the same prey, each prey was attacked by only one spider, which then moved to its retreat to feed. Agonistic behavior was never directed towards web mates.

Mecicobothriidae

The Mecicobothriidae, a small family of New World mygalomorphs that may be closely related to the Dipluridae (Platnick 1979; Gertsch and Platnick 1979), construct funnel-and-sheet webs similar to those of the diplurids and hexathelids. Most species construct a tubular or funnel-shaped retreat that opens out onto a flat prey-capture sheet, although additional tubes and sheets may be laid down beside, above, or below the primary sheet (Gertsch and Platnick 1979). The webs of *Megahexura fulva* resemble agelenid webs; normally the retreat is constructed in ravine bank crevices and the funnel and capture sheet are exposed. The webs of *Hexura* and the tiny *Hexurella* are usually built under leaf or conifer-needle litter or rocks. Wendel Icenogle (pers. comm.) reports that roughly half of a *Hexurella* web is a tube or tubes, and the rest is composed of sheets or strands of silk that are probably the principal prey-capture areas. While feeding *Hexurella* in captivity, he observed that the spiders oriented themselves towards the prey only if the prey were moving in the web, which indicates that *Hexurella* probably rely upon silk-transmitted vibrations for prey localization.

Fig. 10.8. Burrow entrance of *Antrodiaetus unicolor* near Tacoma, Virginia. *Left*, collar closed; *right*, collar open. The scale line represents 10 mm. Reprinted, by permission, from Coyle 1971.

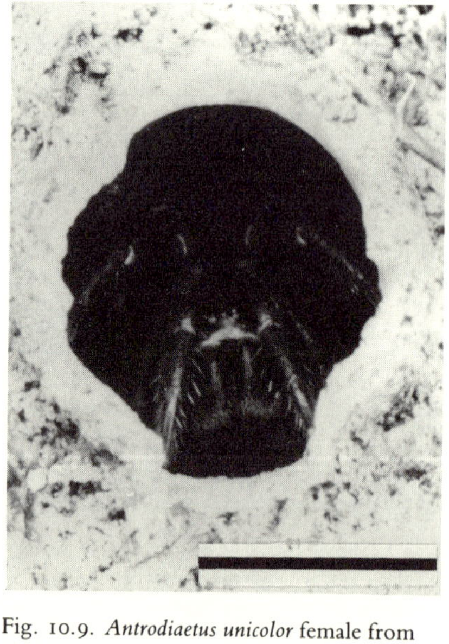

Fig. 10.9. *Antrodiaetus unicolor* female from Cullowhee, North Carolina, in a foraging posture at the entrance of her burrow. The scale line represents 10 mm.

Antrodiaetidae

The Antrodiaetidae, a small mygalomorph family restricted to North America and Japan, exhibit an interesting variety of silk burrow-entrance structures even though all 27 known species (in three genera) construct tubular silk-lined burrows in the ground (Coyle 1971). *Antrodiaetus* species construct collapsible collars, *Atypoides* species construct either collapsible collars or rigid turrets, and *Aliatypus* species construct trap doors.

All 13 species of *Antrodiaetus* extend the burrow-lining silk slightly beyond the entrance and add soil and/or other ground-surface materials to this silk extension (Fig. 10.8). All but young spiders collapse this short, flexible collar inward during the day to close off and camouflage the entrance. At night the collar is open, and the spiders assume their foraging posture inside the entrance (Fig. 10.9). Observations of over 100 large *Antrodiaetus unicolor* individuals in the foraging posture indicate that the spider is positioned just below the collar base (where the collar emerges from ground-surface materials), facing upward and out of the burrow. The spider's venter rests close to the lowest sector of the nonvertical burrow wall. The pedipalps, first legs, and second legs extend forward, and their tarsi touch the silken inner surface of the collar near its base to create six sensing points around the circumference of the entrance. Close examination of photographs of 28 spiders indicates that this positioning is stereotyped and bilaterally symmetrical (Fig. 10.10).

Considerable evidence indicates that prey-induced substrate vibrations transmitted to these tarsi through the collar silk and sensed by mechanoreceptors (probably slit sensilla) provide all the information used by the spider to detect the presence and location of prey. Since, in both the field and the laboratory, accurate strikes were frequently elicited by gently vibrating either the collar or the adjacent ground substrate with an artist's brush, a leaf petiole, or a similar object, chemoreception is ruled out. That such vibrations can elicit an accurate prey-capture response even when the entrance collar is completely closed demonstrates that vision is not necessary for prey sensing. Three other observations indicate that vision is not involved in *A. unicolor* prey detection and localization: (1) prey capture is normally performed at night, (2) *A. unicolor* eyes are very small, and (3) the field of vision of *A. unicolor* in the foraging posture is, at best, an inverted cone of space projecting out of the burrow entrance, leaving a blind region around the outside of the collar where prey are normally detected. The ability of *A. unicolor* to detect prey position accurately, an ability demonstrated by strike-accuracy test results to be presented later, must depend in large part upon its ability to compare the vibration intensities at all six tarsus-collar contact points simultaneously.

Fig. 10.10. Tarsus-collar contact points of 28 *Antrodiaetus unicolor* adult or near-adult females in a capture-ready posture. 0° is the dorsalmost point of the entrance opening. Bars indicate the range of positions for each tarsus; triangles indicate the mean position. *RP*, right pedipalp (N = 28); *LP*, left pedipalp (N = 28); RL_1, right leg I (N = 20); LL_1, left leg I (N = 17); RL_2, right leg II (N = 11); LL_2, left leg II (N = 8).

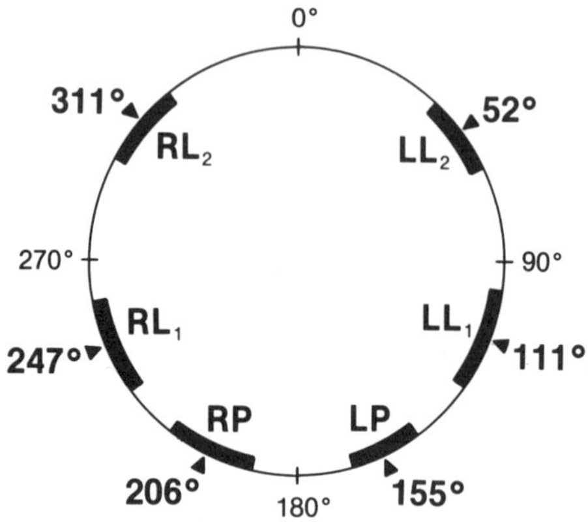

Fig. 10.11. Burrow entrance of *Atypoides gertschi* in Lassen Volcanic National Park, California. The collar is open. The scale line represents 10 mm. Reprinted, by permission, from Coyle 1971.

Many observations involving substrate vibrations generated by both live prey and probes indicate that the prey-detection range of *A. unicolor* is small. For example, during strike-accuracy tests performed with adults and near-adults (entrance diameter ranged from 12 to 19 mm) by moving a vibrating artist's brush slowly along the ground surface towards the collar until a retreat or prey-capture response occurred, the stimulus-to-entrance-rim distance ranged from 0 to 18 mm, with a mean of 5 mm (N = 397). It was difficult to measure this distance with greater accuracy than ±2 mm, and it should be noted that the spider may actually detect the vibrations at greater distances but respond only after the amplitude of these vibrations at the receptors increases. Since in many cases the response did not occur until the flared-out collar edge was touched, it appears that the collar itself is a part of the spider's prey-sensing apparatus and that it may increase the spider's prey-sensing radius and/or strike accuracy.

This limited knowledge of prey-capture behavior in *A. unicolor* better enables us to appreciate the functional significance of entrance designs found in two species of the closely related genus *Atypoides*. Unlike the *Antrodiaetus* collar, the collar of *Atypoides gertschi* does not remain erect when opened, but collapses outward and lies upon the surrounding ground surface (Fig. 10.11) (Coyle 1971). This significantly increases the collar circumference and the area of contact between the collar and the ground surface and thus probably increases the spider's prey-sensing area. *Atypoides riversi*, the conspicuous "California turret spider," constructs a rigid elongate collar (or turret) that is smooth white silk on its inner surface but is camouflaged and kept erect by soil particles and pieces of plant material attached to its outer surface (Figs. 10.12, 10.13) (Coyle 1971). The capture-ready spider is positioned just inside the turret opening, not at the turret base. Field observa-

Fig. 10.12. Turret of *Atypoides riversi* near Felton, California. Side view. The scale line represents 10 mm.

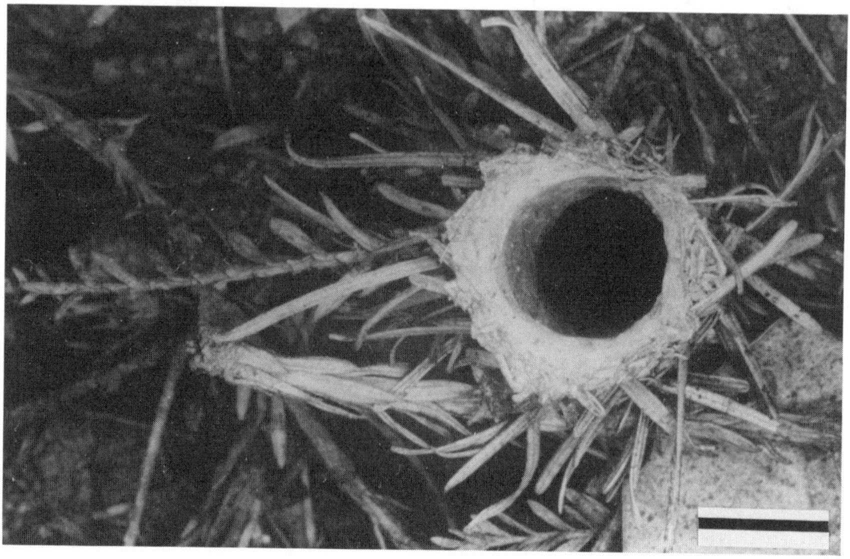

Fig. 10.13. Turret of *Atypoides riversi* shown in Fig. 10.12, seen from above. The scale line represents 10 mm.

Fig. 10.14. Turret of *Atypoides riversi* near Lake Tahoe, California. Attached pieces of litter form a radial pattern. The scale line represents 20 mm. Reprinted, by permission, from Coyle 1971.

tions indicate that, as in *Antrodiaetus unicolor*, prey are detected by sensing prey-generated vibrations conducted to the tarsi by the turret. Since I was able to trigger prey-capture responses by gently vibrating the ends of linear litter attached to the turret silk, it appears that this litter extends the spider's prey-sensing area. The strongly radial arrangement of the linear litter attached to some turrets in the Sierra Nevada populations (Fig. 10.14) resembles the pattern of linear litter used in the classic "twig-lining" strategy of some Australian Ctenizidae (Main 1957). The especially tall turrets of the coastal populations extend the prey sensing area vertically, and may also, by providing attractive launching or landing sites, increase the number of flying insects available for capture. This hypothesis is supported by Vincent's (1980) observation of winged carpenter ants and termites among the prey rejectamenta of these spiders and his observations of spiders striking at tachinid flies resting on the turrets.

The genus *Aliatypus* has evolved a trap-door entrance structure closely resembling that of many ctenizids. All 11 species construct a broadly hinged trap door of silk plus soil and/or plant material from the ground litter (Figs. 10.15–10.17) (F. A. Coyle and W. Icenogle, in prep.). The burrow lining flares out at the entrance to form a narrow lip against which the relatively thin door can fit snugly. The *Aliatypus* foraging posture appears to be similar to that of *Antrodiaetus unicolor*: the spider's venter rests against the lowest sector of the burrow wall and the tarsi of the pedipalps and legs I and II are positioned on the entrance rim just inside the slightly opened door. Detailed observations of appendage placement are lacking, however. Successful prey-capture responses, similar in form to those of *A. unicolor*, are triggered by substrate vibrations applied close to (not more than 5 mm from) or in contact with the entrance rim or door edge at night; vision and olfaction are not involved in prey detection and location. The burrow entrances of *Aliatypus aquilonius* are unique in possessing tough tabs of silk plus soil (and sometimes pieces of moss) that are extensions of, and radiate out along the ground surface from, the ventral and lateral sectors of the entrance rim (Figs. 10.15,10.16). The burrow entrances of 90 to 100 percent of all adults have such tabs, which vary in number from three to six and in length from 1 to 5 mm. Touching the tip of a tab triggers an accurate prey-capture response; apparently the tabs serve to extend the spider's prey-sensing area. The elongate pieces of litter that are occasionally attached to the entrance rims of other *Aliatypus* species may serve the same function (Fig. 10.17).

Atypidae

The atypids possess an interesting mixture of primitive and specialized characteristics, many of their specializations being part of a unique strategy of capturing prey through the walls of a tough, aboveground silk tube. Al-

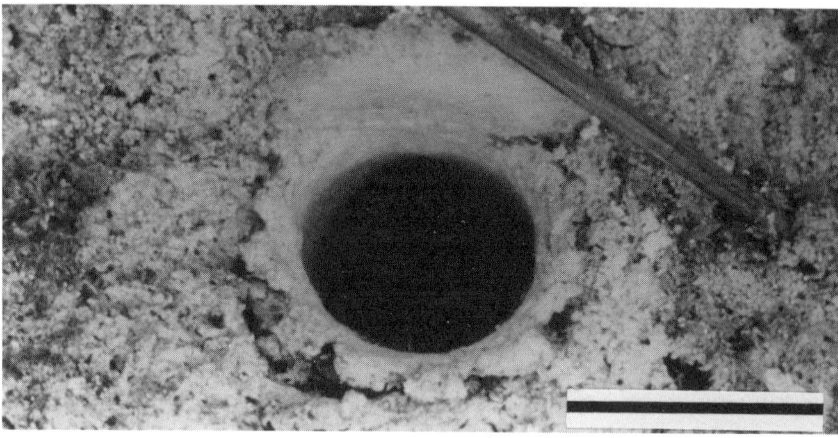

Fig. 10.15. Open burrow entrance of *Aliatypus aquilonius* in Grizzly Creek Redwoods State Park, California. Silk-plus-soil tabs are attached to the entrance rim. The scale line represents 10 mm.

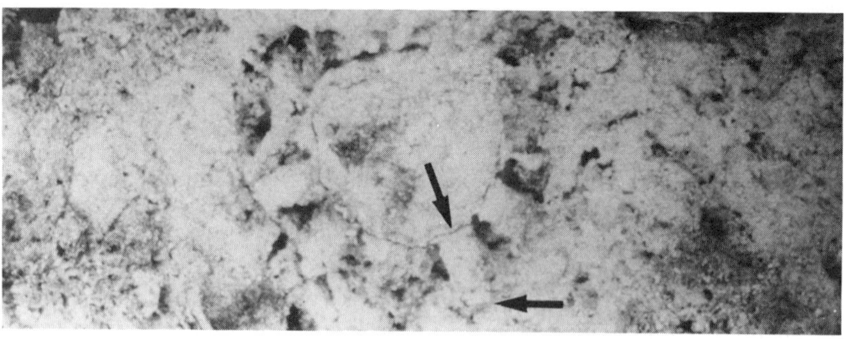

Fig. 10.16. Closed burrow entrance of *Aliatypus aquilonius* near Redway, California. One arrow rests on the trap door and points to the base of one of the silk-plus-soil tabs attached to entrance rim; the other arrow points to the tip of the same tab. Same scale as for Fig. 10.15.

though little is known about *Calommata* (Van Dam and Roberts 1917; Blandin 1971), the web architecture and prey-capture behavior of *Atypus* and *Sphodros*, the other two atypid genera, are well known (Enock 1885; Broen and Moritz 1964; McCook 1888; Poteat 1890; Gertsch and Platnick 1980; Coyle and Shear 1981). The silk lining of the burrow is extended above ground to form a long, tough, opaque silk tube (the "purse web") that either lies roughly parallel and close to the ground surface (*Atypus*) or extends vertically against a tree trunk or other support (*Sphodros*) (Figs. 10.18, 10.19).

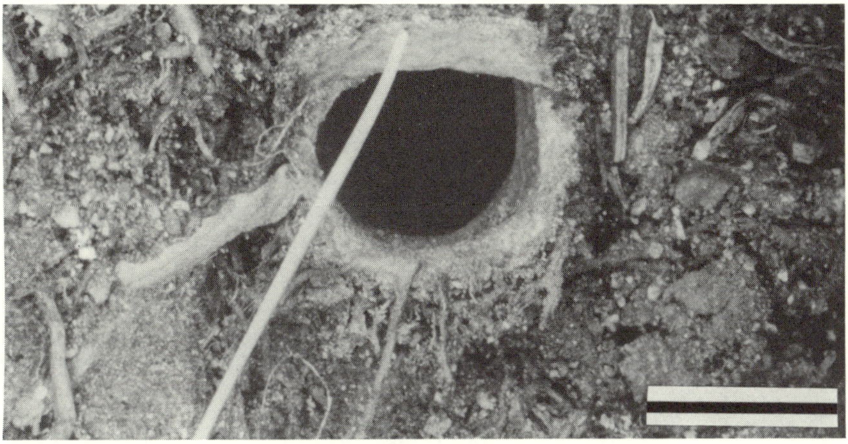

Fig. 10.17. Open burrow entrance of *Aliatypus janus* in Sequoia National Park, California. Linear litter is attached to the entrance rim. The scale line represents 10 mm.

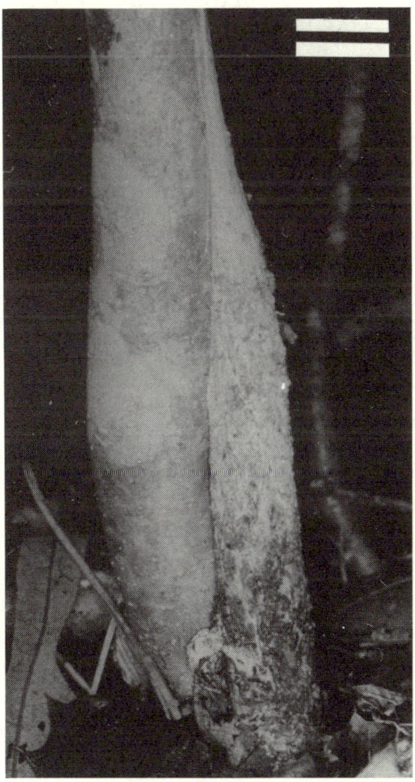

Fig. 10.18. Aerial tube of *Sphodros atlanticus* near Webster, North Carolina. The scale line represents 20 mm.

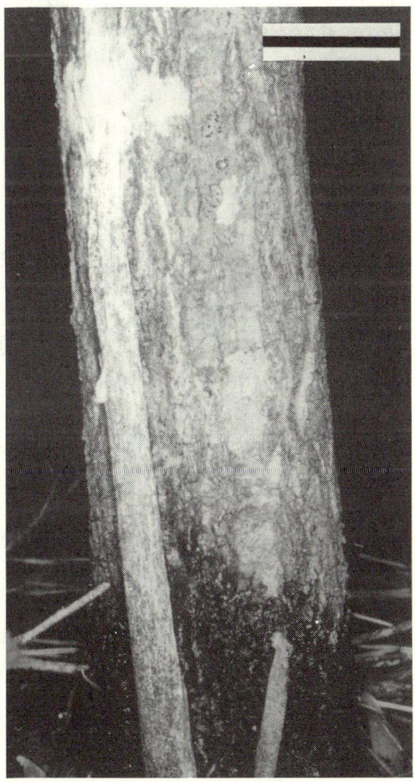

Fig. 10.19. Aerial tubes of *Sphodros abboti* in Suwannee River State Park, Florida. The scale line represents 50 mm.

When prey walk on the aerial tube, they generate vibrations that are transmitted down the tube to the spider, which relies on these vibrations to locate the prey. The spider strikes through the tube wall with its unusually long fangs to pin down the prey. Then the spider cuts through the tube beside the prey with one fang, and pulls the prey through the slit into the tube. The aerial tube thus aids in prey detection and location, but does not entangle prey. Clearly the prey-sensing area of a purse web is much greater than that of an antrodiaetid collar or trap-door entrance. There is some evidence that the vertical aerial tubes of *Sphodros* may allow these spiders to harvest slightly more aerial insects than antrodiaetid spiders can capture (Coyle and Shear 1981).

Theraphosidae

In spite of the large number of species and large size of theraphosids, they are not well known taxonomically or behaviorally. The contributions of Baerg (1958) and Minch (1978a, 1978b, 1979a, 1979b) to our knowledge of tarantula behavior are noteworthy. A search of the scattered literature reveals that the view that tarantulas are "wandering hunters" (Buchli 1969; Gertsch 1979) is a misconception. It appears that most, perhaps nearly all, tarantulas construct relatively permanent tubular silk retreats, and capture prey at or near the retreat entrance. They are predominantly sit-and-wait hunters.

The ground-living tarantulas occupy silk-lined retreats in burrows that they have excavated, or in abandoned burrows, or in natural cavities under rocks or logs. Commonly the silk lining is extended out of the entrance to form a silk rim or low collar that sometimes incorporates pieces of leaves (Fig. 10.20) (Petrunkevitch 1952; Yates 1968; Minch 1978a), a mat or sheet of silk on the adjacent ground surface (Baerg 1958; Cazier and Mortenson 1964; Clyne 1969; Mascord 1970; Main 1976; Minch 1978a; Gertsch 1979), or a funnel of silk attached to ground-surface litter or vegetation (Petrunkevitch 1911; Baerg 1958; Uyemura 1966, 1967). Nocturnal field observations by Minch (1978a, pers. comm.) of the prey-capture behavior of *Aphonopelma chalcodes* suggest that some of these silk structures may aid in prey capture.

The burrow entrance of *A. chalcodes* has a low collar of silk that is connected to a network of webbing covering the surrounding ground surface for a radius of several centimeters. Minch (1978a) reported that, throughout the observation nights, "tarantulas awaited the approach of prey within the entrance of their burrows. The tarsal claws of the first and second pairs of legs and the pedipalps maintained contact with the collar silk encircling the burrow entrance." Rarely did a spider emerge from the entrance and then only for brief periods. "Occasionally the spiders partially emerged from the burrow during the night and cleared away any living or dead plant material contacting the spider silk, using the pedipalps and front legs." More than

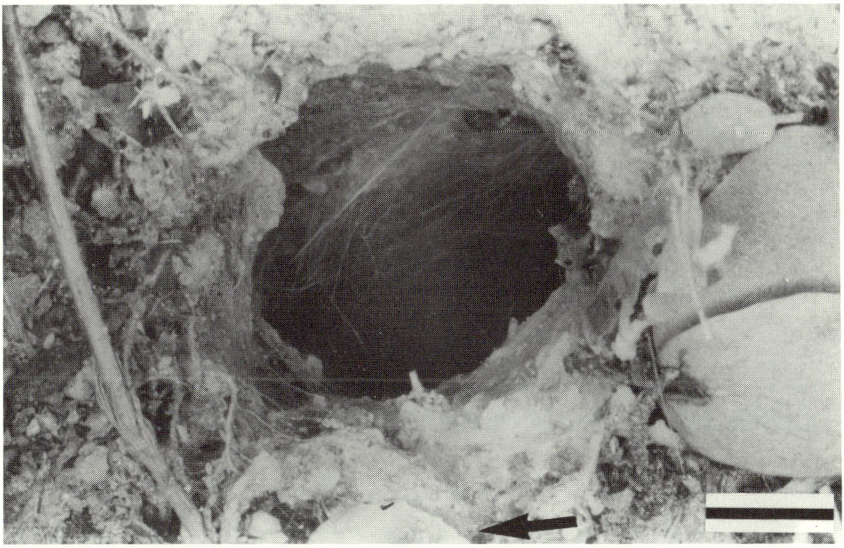

Fig. 10.20. Burrow entrance of *Aphonopelma* sp. near Pinehurst, California. The silk lining extends several millimeters out from the entrance over the substrate. The arrow points to the end of a leaf to which are attached numerous silk strands. The scale line represents 10 mm.

once Minch (pers. comm.) has seen a tarantula rush out of the burrow in response to an insect walking several centimeters away on the ground webbing and then pause if the prey ceased to move; as soon as the prey resumed motion, the tarantula moved directly to it and captured it. The spider is presumably relying upon one or both of two different detection mechanisms: trichobothria that detect prey-generated air currents, and silk- or soil-vibration detectors such as the slit sensilla or club-shaped bothria.

Both Petrunkevitch (1911) and Baerg (1958) observed that some tropical burrowing tarantulas spend nearly the entire night on the ground surface beside their burrow entrance. Whether the rosettes of dried leaves attached to the entrance rims of some California tarantulas (Gabel 1972) aid in prey detection or in preventing debris from falling into the burrow, or serve some other function, needs to be investigated. Similarly, the claims that the silk sheet around the burrow entrance of *Selenocosmia crassipes* in Australia serves to entangle prey (Mascord 1970; Main 1976) need to be tested. McCook (1887) observed the capture of a locust that had become entangled in the sheet of silk spun by a captive *Dugesiella hentzi*.

Even less is known about the arboreal tarantulas. The fact that they possess adhesive tarsal hair pads (scopulas) that allow them to cling to smooth vertical leaf surfaces does not necessarily mean that these spiders wander freely, as some authors suggest (Buchli 1969; Gertsch 1979). Many of these arboreal species construct densely woven tubular silk retreats on trees, smaller plants, or buildings (Bates 1863; Petrunkevitch 1929; Baerg 1958; Cooke 1972; Stradling 1981). Usually these are hidden in crevices or cavities, or camouflaged by attaching organic debris or tiny plants to the silk. The large amounts of silk (and energy) that must be used in constructing these retreats, and some long-term observations by Baerg (1958), Cooke (1972), and Stradling (1981), indicate that a retreat may be occupied by the same spider for as long as 3 years.

Stradling's (1981) observations of a large number of *Avicularia avicularia* living in flattened tubular silk retreats attached to tree trunks are especially informative. At night the spider rests on the trunk within 1 m of its retreat with its head down and legs spread wide. Approaching prey (typically katydids and small gekkonid lizards) are detected at a range of 20 cm or more apparently by substrate-vibration receptors that are sensitive to vibrations transmitted through the bark of the tree (no silk was observed on the bark surface), and are "seized in a sudden rapid dash of great precision." However, during the day the spider rests in its silk retreat, and will dash out and capture prey that touch the retreat. Baerg (1958) observed thin sheets of silk extending from the arboreal retreats of two *Avicularia* species, sheets that he postulated might play a role in prey capture. H. W. Levi (pers. comm.) has observed arboreal tarantulas in the Caribbean that sit at night in capture-ready postures just inside the entrances of their tubular retreats.

Ctenizidae

Much has been written about the fascinating burrow and entrance structures of these widespread, and in many localities surprisingly abundant, spiders. The great majority of ctenizids close their entrances with trap doors constructed of silk plus soil and/or pieces of plant material (Fig. 10.21). Attached to a broad sector of the entrance rim with a tough silk hinge, these doors range from thin, flexible "wafer-type" flaps that usually close down against a relatively broad and flat silk lip forming the entrance rim, to thick and rigid "cork-type" doors with beveled edges that fit snugly into a tough entrance rim with a complementary bevel. Although these doors and entrance rims serve primarily for protection, they are also part of the spiders' prey-detection system.

Surprisingly few observations of ctenizid prey-capture behavior have been published. Incomplete observations of foraging posture in *Ummidia* (Buchli 1969; Coyle 1981b) and the aganippines (Main 1957) indicate that the tarsi of the pedipalps and anterior legs rest on the entrance rim and that the slightly opened door may rest on the tarsi and/or the cephalothorax. There seems to be general agreement that substrate-vibration sensors are probably the key receptors used by ctenizids in sensing prey (Moggridge 1874; Main 1957;

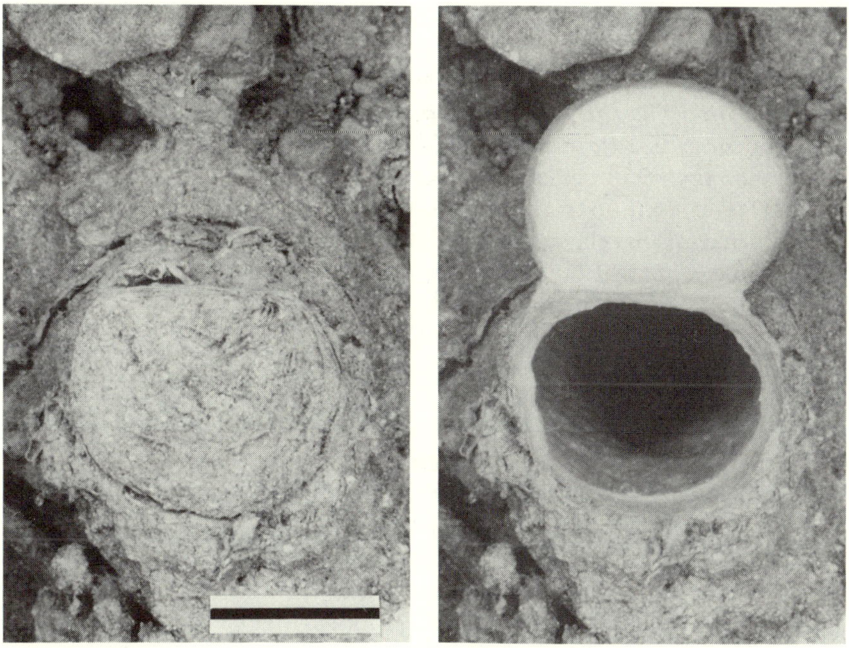

Fig. 10.21. Burrow entrance of *Hebestatis theveneti* in Santa Barbara County, California. *Left*, trap door closed; *right*, trap door open. The scale line represents 10 mm.

Buchli 1969; Coyle 1981b). Observations of some ctenizids (Todd 1945; Marples and Marples 1972; Coyle 1981b) indicate that prey-generated door vibrations can be detected and that sensitivity to prey-generated vibrations does not extend much beyond the edge of the trap door and entrance rim; usually the prey must touch the trap door or entrance rim before the spider will strike. However, Buchli (1969) suggested that *Nemesia caementaria*, which holds its door open fairly wide and extends its pedipalps and first two pairs of legs a bit in front of the entrance rim, often holding the first pair of legs fully extended in the air, has a much greater prey-sensing area. Such a posture suggests that this spider may be relying on the detection by trichobothria of prey-generated air currents.

Some ctenizids have evolved modifications of the standard trap-door-entrance design that serve to extend the prey-sensing area. Spiders such as *Actinoxia versicolor*, *Idiops colletti*, and some *Nemesia* and *Cantuaria* species characteristically incorporate relatively large pieces of plant material into their trap doors and/or entrance rims (Moggridge 1873; Pickard-Cambridge 1889; Smith 1908; Forster 1967; Forster and Wilton 1968). This practice, in addition to camouflaging the entrance, may increase the prey-sensing area, since some of these pieces extend out beyond the entrance rim. A photograph of *Cantuaria vellosa* burrow entrances in Forster and Forster (1973) reveals sizable irregular silk-plus-soil extensions of the entrance rim that may serve the same purpose. I have observed linear litter attached to the rims of several *Cyclocosmia torreya* burrow entrances (Fig. 10.22). Main (1957, 1976, 1978) documents a clear-cut case of prey-sensing-area extension in several aganippine trap-door spiders that use silk to attach the ends of linear litter (twigs, leaves, etc., up to 25 cm long) to the entrance rim in a radiating pattern, a behavior that she calls twig lining.

Some ctenizids that build trap-door entrances extend their prey-sensing area by running silk lines out from the entrance rim. Buchli (1969) suggests that such silk lines constructed by some individuals of both *Cteniza sauvagesi* and *Nemesia dubia* may enhance prey detection. Bacelar (1933) describes the numerous silk lines that radiate out from the burrow entrance of *Nemesia meridionalis*, which "dashes out to catch an insect whenever one touches one of the taut threads" (Buchli 1969). Decae, Caranhac, and Thomas (1982) have recently investigated the behavior of the Greek ctenizid *Cyrtocarenum cunicularium* and have concluded that Erber's (1868) remarkable report that this spider snares insects with a sheet web that it extends out from its trap-door entrance each night is totally fallacious.

Some ctenizids do not construct trap doors. Members of the *Cantuaria huttoni* species group have no door over their burrow entrance, which is rimmed by a relatively broad band of silk that may extend the prey-sensing radius (Forster 1967; Forster and Wilton 1968; Forster and Forster 1973). Most of the species of *Dyarcyops* from Australia (Mascord 1970; Main 1976) and *Damarchus cavernicolus*, a cave spider from Malaysia (Bristowe 1952), do not build trap doors. Blandin and Célérier (1977) describe an interesting West African ctenizid, *Aporoptychus lamottei*, which constructs no trap door. Instead, the burrow-lining silk is extended out of the entrance to form a collar a few millimeters wide that collapses outward to lie flat on the ground surface. Several silk lines extend from this collar out over the ground surface and, with the collar, probably serve to enlarge the prey-sensing area. The spider turns the collar inward to close the burrow entrance.

Species in at least three ctenizid genera extend their burrow lining above the ground surface to form a funnel, turret, or aerial tube that may extend their prey-sensing area and, in addition, increase their harvest of non-ground-dwelling arthropods. Three species of *Cyrtauchenius* from North Africa construct rigid turrets camouflaged and strengthened with soil and plant material (McCook 1890). The turret of one of these species is covered by a trap door, another is not, and the third has what appears to be silk signal lines radiating from its entrance. *Hermachastes schreineri* from South Africa constructs a rigid turret of many pieces of plant debris bound together

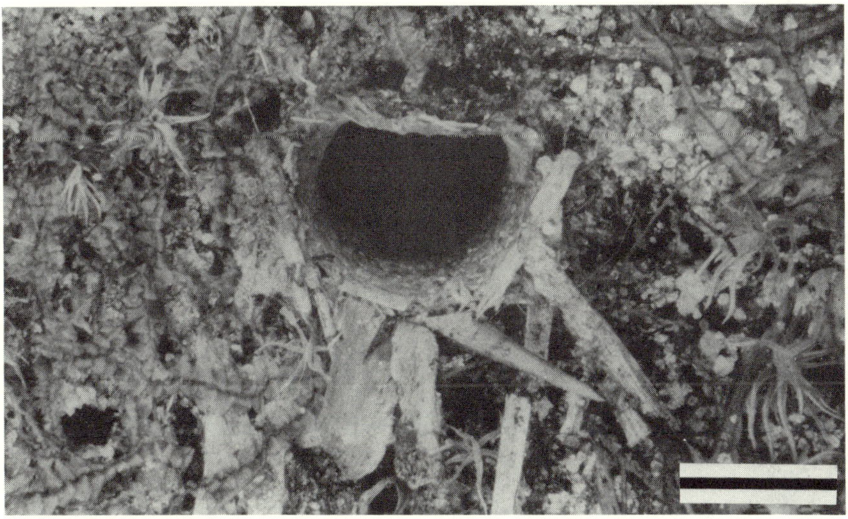

Fig. 10.22. Burrow entrance of *Cyclocosmia torreya* in Torreya State Park, Florida. The trap door is open. Linear litter is attached to the entrance rim. The scale line represents 5 mm.

by silk (Purcell 1903). *Dyarcyops robertsi* of Australia constructs a tough, camouflaged silk tube that extends from its burrow up to 30 cm above the ground surface (Mascord 1970; Main and Mascord 1974; Main 1976). These tubes are attached at their upper ends to tree trunks or other objects and are believed by Main (1976) to function as antiflooding devices. The resemblance of these tubes to the purse webs of the atypid genus *Sphodros* suggests that they may provide a large prey-detection surface and that the spider may run down the tube's outer surface to capture prey that are out of reach from the entrance. However, Mascord (1970) claims that "all prey are captured at the mouth of the tube." The fact that moths, in addition to beetles, are captured (Mascord 1970) indicates that this tube increases the opportunity for capture of flying insects. Two other *Dyarcyops* species, *Dyarcyops fuscipes* and *Dyarcyops pulchellus*, extend their silk burrow lining out from the entrance and attach it to grass or litter to form a low funnel (Mascord 1970; Main 1976).

Some ctenizids, such as *Conothele arboricola* (McKeown 1952; Main 1957; Crome 1962) and *Pseudidiops opifex* (McCook 1890), are truly arboreal and live in short, exposed silk tubes that are camouflaged with bark fragments and provided with trap door entrances. I would not be surprised if it is found that, as in the Atypidae, the entire exposed surface of the tube plays a role in prey detection.

Migidae

Migids construct tubular silk retreats either within excavated burrows in the ground or on the surfaces of tree trunks or rocks. Nearly all species close the entrances of these retreats with trap doors (Berland 1932), although *Calathotarsus coronatus* lives in a burrow that opens into a funnel web (Buecherl 1962), and one species of *Homogona* constructs a turret without a door (Main 1976). The aerial retreats are usually much shorter than the burrows, and their exposed surfaces are camouflaged and strengthened with pieces of plant material (lichen, moss, bark, etc.). In the African genus *Moggridgea*, some species, such as *Moggridgea mordax*, construct burrows in the ground (Purcell 1903), but others, such as *Moggridgea dyeri*, *Moggridgea abrahami*, and *Moggridgea microps*, construct tubular retreats on tree bark (Pickard-Cambridge 1889; Van Dam and Roberts 1917); and in some species of *Migas* some individuals construct ground burrows and others construct aerial tubes (Forster 1967; Forster and Wilton 1968). This indicates that the evolutionary shift from burrow construction to the construction of exposed tubular retreats, a shift that has occurred also in the Ctenizidae and Barychelidae, does not require gene-frequency changes at a large number of loci.

I suspect that, as in the Atypidae, the entire exposed surface of a migid's aerial tube is involved in prey detection. If this hypothesis is correct, the individuals in aerial tubes should enjoy a greater prey-sensing area than those in burrows, and, if they are able to run over the outer surface of the tube, should be able to take full advantage of the increased prey-sensing area. Tubes with trap-door entrances at both ends, which are occasionally found in *Migas* species (Todd 1945) and in *Moggridgea dyeri* and *Moggridgea abrahami* (Pickard-Cambridge 1889), might allow more effective access to the entire tube surface during prey capture, although the second door may serve primarily as an escape hatch. Another possible advantage of occupying an aerial tube rather than a burrow is that tree trunks and branches may be sites of especially high prey density, for they are the only access routes available for a multitude of nonflying arthropods, such as worker ants, that forage in the canopy.

Actinopodidae

The natural history of the Actinopodidae, a small family that is apparently the sister group of the Migidae (Platnick and Shadab 1976), is poorly known. Only the genus *Missulena* has been fairly well studied (Main 1956; Mascord 1970). *Missulena* species construct an expanded burrow entrance and cover it with two adjacent trap doors that face away from each other. The hinge lines of the two doors form a V, the area within the V being covered by a thick horizontal silk sheet (Main 1956). If, as I suspect, the spider simultaneously monitors both doors and the silk-supported "roof" between them, this complex entrance structure probably provides a significantly larger prey-sensing area than an unexpanded entrance with a single trap door. The prey-sensing area is further enlarged by strands of silk radiating from the entrance rims out over the ground surface for 2 to 3 cm (Main 1956).

Barychelidae

Apparently, most species of barychelids construct silk-lined burrows or shallow chambers in the ground, but, as in the Migidae, some build arboreal tubular silk retreats. Nearly all species construct trap-door entrances; quite a few construct two such entrances per burrow or tube. For example, *Cyphonesia obesa* constructs a shallow oblong chamber in the soil with two trap-door entrances (Blandin and Célérier 1977). Only several millimeters separate the hinges of the doors, which open in exactly opposite directions. Main (1956) reports finding barychelids in western Australia with similar double-door burrows. Like the double-door entrance of the actinopodid *Missulena* (Main 1956), these constructs may provide a significantly larger prey-sensing area. Blandin and Célérier (1977) hypothesize that the barychelid double-door constructs are designed primarily for escape from predators; they suggest that when, e.g., a pompilid wasp forces its way into one entrance, the spider escapes through the other. Much earlier, McCook (1890) suggested the same behavior for the barychelid *Stothis astuta*, which has a trap-door entrance at each end of its peculiar U-shaped burrow.

The arboreal barychelids of the genus *Sason*, from Sri Lanka, construct on bark a short tubular silk retreat with a trap door at each end (Berland 1932). Feio (1952) describes a similar double-door retreat built by *Paracenobiopelma gerecormophila*, a Brazilian arboreal barychelid. As I have suggested for the arboreal double-door tubes of the Migidae, perhaps the exposed tube surface functions as a relatively large prey-sensing area to which the extra door allows more effective access.

Two barychelids that do not construct trap doors, *Lampropodus iridescens* from Australia (Main 1976) and *Psalistops melanopygia* from Venezuela (Simon 1889), attach a collar of leaves or other materials to their entrance rims, and may thereby extend their prey-sensing ranges. In *P. melanopygia* these materials radiate well out from the entrance rim. Another doorless species, the North African *Leptopelma elongata*, extends its burrow lining above ground to form a bright white, dense, vertical funnel-shaped web about 7 cm tall supported by surrounding plants (Simon 1873; Moggridge 1874; Buchli 1969). The spider dashes up from the burrow to capture insects that enter this remarkable funnel web (Simon 1873). Alfred Russel Wallace, co-author of the concept of natural selection, suggested that this web may attract insects by virtue of its resemblance to a flower (McCook 1890).

Comparison of the Prey-Capture Effectiveness of Collar-Door and Trap-Door Spiders

In order to understand why a diverse array of entrance constructs has evolved among nonaraneomorph spider taxa, it is necessary to understand not only the multiple functions that each type of entrance performs but also how effectively each function is performed. Much evidence indicates that trap-door entrances can be effective in protecting their builders from dessication, temperature extremes, fires, floods, and perhaps predators (Moggridge 1873, 1874; Pocock 1895; Bristowe 1930a; Main 1957, 1978; Buchli 1969; Coyle 1971; Gertsch 1979). But many successful nonaraneomorph spiders do not build trap doors, and there is evidence that some of these spiders have evolved from trap-door-building ancestors. Consequently, it is likely that other types of entrances can perform some functions more effectively than trap-door entrances.

Accordingly, I decided to compare the prey-capture effectiveness of a collar-door spider, *Antrodiaetus unicolor*, with that of the related trap-door spider taxon *Aliatypus* and, if differences were found, to see if they are related to the differences in entrance structure. The entrances, foraging postures, and prey-capture responses of these two taxa, both of which are in the family Antrodiaetidae, are described above, and in more detail elsewhere (Coyle 1971; F. A. Coyle and W. Icenogle, in prep.). Thirty-one individuals of *Antrodiaetus unicolor* were observed in the laboratory and 52 in the field; all of these were from near Cullowhee, North Carolina. Twenty-three individuals of *Aliatypus* (from the species *Aliatypus californicus*, *Aliatypus isolatus*, and *Aliatypus thompsoni*) from California and Arizona were observed in the laboratory. Every spider was an adult female or a large subadult. The laboratory-tested spiders were kept in a springtime photoperiod regime in a humid basement room, where they lived in burrows that they constructed in large soil containers positioned so that the soil surface sloped 30°. In nature the burrow entrances of these spiders are usually on surfaces sloping 20° to 90°.

Stimuli consisting of artificially generated substrate vibrations meant to represent those normally generated by prey were presented to these spiders. The stimulus tool was a number 1 artist's brush attached to an electric-toothbrush motor. The stimulus approach radius was preselected from one of eight radii extending from the entrance center at 45° intervals. The 0° radius extended through the dorsalmost point on the entrance rim, the 45° radius was 45° clockwise from this, etc. The tip of the vibrating brush was applied to the substrate about 4 cm from the entrance and then moved slowly along the radius towards the entrance rim until the spider struck or the brush reached the rim. The following data were recorded: the radius tested; whether the spider retreated, struck, or showed no response; and whether the strike was directionally accurate, i.e., whether the stimulus approach radius passed between at least two of the spider's outstretched appendages. All stimulus tests were administered at night, with low-intensity red illumination, and only to individuals in the foraging posture. No spider was tested more than once each night.

The results of the stimulus presentations are given in Tables 10.1 and 10.2. Determinations of the statistical significance of result differences were based upon 2 × 2 tests of independence using the *G*-test (Sokal and Rohlf 1973) and a probability level of $p < 0.01$. The results for the laboratory-tested and field-tested *Antrodiaetus unicolor* samples were combined because they were not significantly different; the results for the three *Aliatypus* species were combined for the same reason. The high frequency of retreats, even at those radii where the highest strike frequencies were observed, may be partly due to differences between brush-generated vibrations and those generated by actual prey; it is not due to learning during the course of the experiment, since the probability of retreat did not increase with increased testing for individuals that were tested more than twice.

Tables 10.1 and 10.2 show that for both taxa the propensity to strike (the percentage of stimuli resulting in strikes), the strike accuracy (the percentage of strikes that are accurate), and the percentage of stimuli that result in accurate strikes (a value that is determined from the previous two and should be an indicator of prey-capture effectiveness) vary in a regular pattern around the entrance rim. For both taxa all three of these values are significantly higher at the ventralmost point on the entrance rim (the 180° test radius) than at the uppermost point (0° test radius), and the values decrease regularly as one proceeds upward from the 180° radius on either side of the entrance rim. Very probably, a key reason for this more effective response to prey approaching the ventral sector of the entrance is that four of the six sensing appendages (the pedipalps and legs I and II) are positioned on the ventral sector of the entrance lining, an arrangement that should result in greater vibration sensitivity and localization accuracy there than at the dorsal sector. Dorsal-sector stimuli are sensed, since the great majority of no-strike responses there were retreats, but they are probably sensed poorly. Another problem for the spider is that effective execution of a strike at a dorsal-sector stimulus is probably more difficult because of the greater amount of turning and antigravity movement required during such a strike. All of these factors probably decrease the chance that a strike at prey approaching the dorsal sector of the entrance will be accurate, and help explain why the spider is programmed to be more reluctant to strike at such stimuli.

The prey-capture effectiveness of *Antrodiaetus unicolor*, as measured by the percentage of stimuli resulting in accurate strikes, is significantly greater (at

TABLE 10.1
Results of Vibration-Stimulus Presentations to Antrodiaetus unicolor

Test radius (°)	Stimuli	No responses	Retreats	Strikes	Inaccurate strikes	Percent stimuli resulting in strikes	Percent accurate strikes	Percent stimuli resulting in accurate strikes
0	54	5	42	7	2	13	71	9
45	51	3	37	12	3	24	75	18
90	51	3	34	14	2	27	86	24
135	56	0	25	31	1	55	97	54
180	51	3	15	33	0	65	100	65
225	52	3	21	28	1	53	96	52
270	49	2	35	15	1	24	92	22
315	52	4	34	14	2	27	86	23
TOTAL	416	23	243	154	12			
Mean percent for all test radii						37	92	34

TABLE 10.2
Results of Vibration-Stimulus Presentations to Aliatypus Species

Test radius (°)	Stimuli	No responses	Retreats	Strikes	Inaccurate strikes	Percent stimuli resulting in strikes	Percent accurate strikes	Percent stimuli resulting in accurate strikes
0	24	1	23	0	0	0	0	0
45	25	1	22	1	1	4	0	0
90	21	0	15	6	0	29	100	29
135	24	0	13	11	0	46	100	46
180	22	0	9	13	0	59	100	59
225	23	1	13	9	0	39	100	39
270	22	0	13	7	0	32	100	32
315	24	0	22	2	1	8	50	4
TOTAL	185	3	127	49	2			
Mean percent for all test radii						26	96	25

a level of $p < 0.05$) than that of *Aliatypus*, owing primarily to the poor capture effectiveness of *Aliatypus* at the dorsal sector of the entrance, where the trap-door hinge is located. The propensity to strike, strike accuracy, and percentage of stimuli resulting in accurate strikes at the dorsal sector (i.e., the 315°, 0°, and 45° test radii) are significantly lower for *Aliatypus* than for *Antrodiaetus unicolor*. There is no significant difference between the two taxa for these same parameters at the ventral sector of the entrance (i.e., the 135°, 180°, and 225° test radii). Since the ratio of retreats to no responses was as high at the dorsal test points as at the ventral ones, the door hinge of *Alia-*

typus does not prevent sensory information from reaching the spider. However, because of the hinge structure and/or the foraging-posture pattern of tarsal placement, there may be less sensory information available to *Aliatypus* than to *Antrodiaetus unicolor* about dorsal-sector stimuli, a reduction that makes it less likely that a strike will be on target. There are other, perhaps more important reasons that a strike directed around the ends of the hinge is more likely to be unsuccessful or even hazardous: execution of an accurate turn during the strike may be difficult; a longer reach is required; there may be more friction with the door; and there may be a risk of hinge damage. Consequently, it seems reasonable to conclude that, because of the trap door, *Aliatypus* are genetically programmed to be more reluctant than *Antrodiaetus unicolor* to strike at prey-generated substrate vibrations at the dorsal sector of the entrance. If trap-door entrances reduce foraging efficiency, it is easier to understand why trap doors might be lost during the evolution of some mygalomorph taxa and why many trap-door spiders have evolved prey-sensing attachments and other constructs that increase their prey-capture effectiveness.

Evolution of Entrance Constructs

As information accumulates on the form, functions, and intra- and interspecific variation of nonaraneomorph spider constructs and on the construction behaviors involved, we shall be able to generate and test hypotheses about the direction of their evolution. Such hypotheses can then help in testing phylogenies. We should proceed cautiously, however, since evidence indicates that the phylogeny-constructing utility of some of these behavioral characters may be low because they are prone to rapid evolution: (1) both underground and aboveground or arboreal retreats are found within some species (*Liphistius batuensis*, *Liphistius malayanus*, and *Calisoga longitarsus*) and genera (*Aparua*, *Ixamatus*, *Hexathele*, *Bymainiella*, *Moggridgea*, and *Migas*); (2) some genera (*Cantuaria*, *Dyarcyops*, and *Cyrtauchenius*) contain both trap-door builders and species that do not build trap doors; (3) there is much variation in entrance number and sheet-web development in *Porrhothele antipodiana*; (4) some migid species construct both single- and double-door retreats; and (5) some populations of *Aganippe* are obligatory twig liners, others are facultative twig liners, and others are not twig liners (Main 1975).

As shown in Fig. 10.23, the trap door is a common entrance-closure structure in nonaraneomorph spiders. Although outgroup comparison with the amblypygids indicates that a trap door was not built by the first spiders, it is possible that it originated early in spider evolution, that it was lost in the

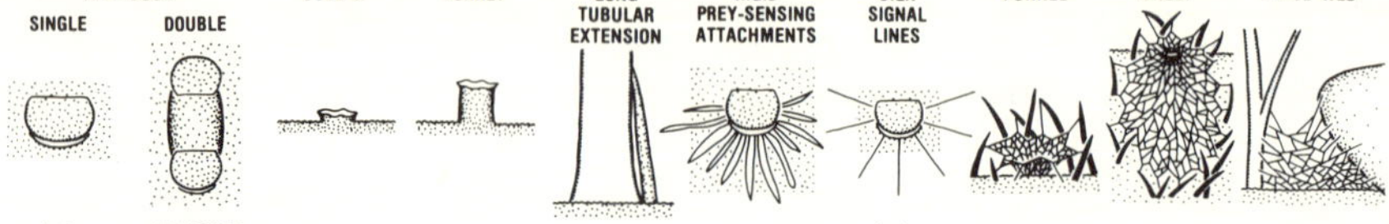

	TRAP DOOR SINGLE	TRAP DOOR DOUBLE	COLLAR	TURRET	LONG TUBULAR EXTENSION	RIGID PREY-SENSING ATTACHMENTS	SILK SIGNAL LINES	FUNNEL	SHEET	SPACE WEB
LIPHISTIIDAE	+ +	LIPHISTIUS					+ +			
DIPLURIDAE	UNDESCRIBED GENERA		STANWELLIA					+ +	+ +	EUAGRUS CETHEGUS
HEXATHELIDAE							HEXATHELE ATRAX	+ +	+ +	
MECICOBOTHRIIDAE								+ +	+ +	
ANTRODIAETIDAE	+ +		+ +	ATYPOIDES		ATYPOIDES ALIATYPUS				
ATYPIDAE					+ +					
THERAPHOSIDAE			+ +					DUGESIELLA CHILOBRACHYS EURYPELMA	AVICULARIA SELENOCOSMIA	
CTENIZIDAE	+ +		APOROPTYCHUS CANTUARIA DYARCYOPS	CYRTAUCHENIUS HERMACHASTES	DYARCYOPS	+ +	CTENIZA NEMESIA APOROPTYCHUS CYRTAUCHENIUS	DYARCYOPS		
MIGIDAE	+ +	MIGAS MOGGRIDGEA		HOMOGONA				CALATHOTARSUS		
ACTINOPODIDAE		MISSULENA					MISSULENA			
BARYCHELIDAE		+ +	LAMPROPODUS PSALISTOPS			PSALISTOPS		LEPTOPELMA		

Fig. 10.23. Distribution of entrance constructs among nonaraneomorph families. Plus signs indicate that the construct is common within the family; a genus name indicates that the construct is built by at least one species in that genus.

ancestral araneomorphs, and that mygalomorph trap doors are homologous to those of the liphistiids. It is also possible that, as Bristowe (1930a), Main (1976), and Gertsch (1979) maintain, trap doors have originated several times independently in the Mygalomorphae. Main supports her position by describing three different trap-door-construction behavior programs found in mygalomorphs, although she fails to identify the taxa. The occurrence of only two small trap-door-building genera in the diplurid-hexathelid group, and the unusual way in which these doors function, strongly indicate that trap doors are apomorphic in that group. The frequent occurrence of trap doors in the Liphistiidae, Ctenizidae, Migidae, and Barychelidae indicates that these structures are plesiomorphic within each of these families. Clearly, we need to learn much more about door-construction behavior, the details of trap-door structure, and the inter- and intrafamilial relationships of mygalomorph spiders before we can develop a sound hypothesis of trap-door evolution.

Trap doors probably have been lost a number of different times during the evolution of the Ctenizidae, Migidae, and Barychelidae, as evidenced by the fact that at least some species in each of the following genera lack trap doors: *Cantuaria*, *Dyarcyops*, *Damarchus*, *Aporoptychus*, *Cyrtauchenius*, *Hermachastes*, *Calathotarsus*, *Homogona*, *Lampropodus*, *Psalistops*, and *Leptopelma*. The tendency of trap doors to reduce prey-capture effectiveness, as indicated by the study comparing antrodiaetid collar-door and trap-door spiders, may help explain selection against trap-door construction, especially in non-arid environments where the primary protective functions of trap doors are not so important for survival. This may well be the case in the New Zealand genus *Cantuaria*. Although most species of *Cantuaria* construct trap doors, those in the *Cantuaria huttoni* species group have a doorless entrance (Forster 1967; Forster and Wilton 1968; Forster and Forster 1973). The trap-door species of *Cantuaria* live in relatively dry and open habitats, and the doorless species live in more humid forest habitats.

Numerous other nonaraneomorph entrance structures have evolved (Fig. 10.23), and many of these appear to function primarily to increase prey-capture effectiveness. Although they sometimes provide entrance closure, collars also serve to extend the spider's prey-detection area. These collars, which may be erect or extend outward to lie flat upon the ground surface, might have been derived either from the rims of open (doorless) entrances or from wafer-type trap doors. The turret of the antrodiaetid *Atypoides riversi* presumably has evolved from a collapsible collar like that possessed by all close relatives; it has probably extended the ancestral prey-sensing area. It is likely that the long aerial capture tubes of the Atypidae and the ctenizid genus *Dyarcyops* are derived from either collars or turrets. Some species of *Dyarcyops* construct collapsible collars (Main 1976).

Many ctenizids and one antrodiaetid species consistently attach pieces of plant material to parts of the entrance (Fig. 10.23), usually the entrance rim, thereby extending their prey-sensing area. This practice has clearly evolved independently several different times. Tablike silk-plus-soil entrance-rim extensions like those of *Aliatypus aquilonius* also serve to extend the prey-sensing area. Signal lines of silk that extend out from the entrance rim and likewise increase the prey-detection area occur in four families and have probably evolved *de novo* in each of these and several times independently in the Ctenizidae.

It is probably from such a simple system of signal lines that the funnel, sheet, and space webs of the diplurids, hexathelids, and mecicobothriids evolved. Within the Hexathelidae can be seen examples of all three of these probable evolutionary stages: (1) radiating signal lines (in *Hexathele hochstetteri* and *Hexathele huka*), (2) funnel (in *Hexathele petriei* and *Bymainiella boycei*), and (3) funnel with sheet (in many species in several genera). The funnel-and-sheet-web entrances in the Theraphosidae, Ctenizidae, Migidae, and Barychelidae have probably evolved independently, perhaps by a similar process. The question of the independent origin in several taxa of similar silk constructs is addressed by Shear (this volume).

These observations and the universal nonaraneomorph practice of living in a tubular silk retreat support the hypothesis of McCook (1889) and Pocock (1895) that the prototype spider web was simply a tubular silk retreat and that trapping webs evolved from silk signal lines extending out from the entrance rim of this retreat. Bristowe (1930a) supports this hypothesis; however, his suggestion that the earliest spiders were wanderers with daytime retreats conflicts with the observation that virtually all nonaraneomorphs wait within or close to their tubular retreats to ambush passing prey. McCook (1889) and Pocock (1895), and later Savory (1960b) and Kaston (1964, 1966), claim that the first silk lines extending out from the archetypal tube entrance were preadapted for prey detection but initially served other functions; e.g., they supported the entrance opening, discouraged predators, or provided orientation cues for the returning spider. Evidence presented here strongly indicates that even the shortest extension of an entrance rim may improve foraging success; consequently, the preadaptation step need not be included in the hypothesis.

The genetic-developmental program possessed by most extant nonaraneomorph spiders suits them well for the capture of nonflying arthropods, and, as Bristowe (1954, 1975) suggests, has probably not changed much since it originated back before the origin of winged insects. The only nonaraneomorph group to experiment seriously with strategies for capturing aerial insects is the Dipluridae, yet even these spiders may be substantially restrained from capitalizing on this rich resource by inherited constraints, such as a book-lung respiratory system not well designed for water conservation (Anderson 1970) or paraxial chelicerae that may work best when prey are on a firm substrate (Bristowe 1954; Laing 1975). In spite of the frequent evolution of prey-sensing extensions, most nonaraneomorphs probably experience lower food intake rates than the majority of araneomorph spiders. These low consumption rates are dictated by their sit-and-wait strategy, relatively small prey-detection area, and nonadhesive silk. However, because of their relatively low metabolic rates (Anderson 1970; Anderson and Prestwich 1982), long life cycles, and well-protected retreats, these spiders are especially well equipped for surviving in environments with inconsistent food supplies (Anderson and Prestwich 1982). Accordingly, in such environments, most notably arid ones, they sometimes occur in densities unmatched by araneomorphs of equal size (Main 1957, 1968, 1978; Coyle 1971, unpub.; Marples and Marples 1972; Vincent 1980). Consequently, in spite of their failure fully to enter the aerial-web adaptive zone, these primitive spiders are important members of the predator component of some communities, at least in terms of persistence and biomass.

Summary

A review of the literature on the prey-capture behavior and silk constructs of nonaraneomorph spiders, and some new observations, reveal that nearly all the prey of these spiders are captured at or close to the entrance of a tubular silk retreat and that portions or extensions of this retreat entrance often help the spiders to detect and locate prey by extending the range of substrate-vibration receptors located on the pedipalps and legs. These entrance constructs, which may take the form of radiating signal lines, attachments of litter, collars, turrets, funnels, or sheets, have evolved many times independently in several nonaraneomorph families. Probably the diplurid, hexathelid, and mecicobothriid funnel-and-sheet webs, some of which are large aerial trapping webs, have evolved from a system of silk signal lines radiating out from a retreat entrance.

Trap doors, which serve primarily for protection, are apparently plesiomorphic within the Liphistiidae, Ctenizidae, Migidae, and Barychelidae. The loss of trap doors by several taxa within some of these groups and the attachment of prey-detection devices to the entrance are more easily understood in the light of experimental evidence that trap doors can limit prey-capture effectiveness. The presentation of substrate-vibration stimuli to capture-ready antrodiaetid spiders revealed a significantly greater ($p < 0.05$) prey-capture effectiveness (as measured by the percentage of stimuli resulting in accurate strikes) for the collar-door spider *Antrodiaetus unicolor* than for *Aliatypus* trap-door spiders. The propensity to strike and the strike accuracy are significantly lower ($p < 0.01$) for *Aliatypus* than for *Antrodiaetus unicolor* at the dorsal sector of the entrance, where the trap-door is hinged, but not at the ventral sector. Apparently *Aliatypus* are programmed to be reluctant to strike at substrate vibrations near the trap-door hinge because the hinge decreases the chance that a strike will be successful.

Acknowledgments

I thank Susan Monroe for her assistance in the study of prey-capture effectiveness in the antrodiaetids. Bill Eberhard provided a helpful critique of an earlier version of this chapter. My observations of antrodiaetid prey capture and *Euagrus* and *Ischnothele* webs were made possible by NSF Grants GB-34128 and DEB-8104682.

Note added in proof: A revolutionary paper on mygalomorph spider classification by Robert J. Raven (1985: The spider infraorder Mygalomorphae (Araneae): cladistics and systematics. Bull. Amer. Mus. Nat. Hist., vol. 182, pp. 1–180) appeared too late to be taken into account in the final version of this chapter. Many significant taxonomic innovations, including new generic, subfamilial and familial names, reassignments of genera to families, and, most important, the dismemberment of the traditional Ctenizidae into three families placed at some distance from one another in the system, are proposed and convincingly argued. The reader of this chapter is referred to Raven's paper for information on the new system, and comparisons with the traditional one used here and in the Taxonomic Glossary.—Ed.

11

Web Removal Patterns in Orb-Weaving Spiders

JAMES EDWIN CARICO

MOST STUDENTS of spider biology are familiar with how orb webs are built and know that they are usually replaced each day. Several descriptions of orb-web construction can be found in the popular and technical literature spanning more than a half century. In contrast, however, it seems that the majority of araneologists are unaware of how webs are removed. Even the most detailed descriptions of orb-web construction in the modern literature do not mention the removal phase. Indeed, Nielsen (1932) may have been the only one to make more than a passing reference to how orb webs are removed, in his description of the behavior of *Araneus diadematus*. Certainly, until now there has been no published study devoted to the patterns of orb-web removal.

My observations of web construction in several araneid genera have indicated that the removal phase of this behavior involves fixed action patterns that rival the better-known patterns of the construction phase in complexity. Also, the important interrelationship between construction and removal in the daily web-building cycle has not been sufficiently appreciated.

My objectives in this chapter are, first, to describe two basic patterns of web removal used by certain orb weavers, and, second, to discuss and emphasize how these patterns relate to the daily cycle of web-building activities of these spiders.

Methods

My observations of miscellaneous orb weavers began in a casual way during the summer of 1976, but it was not until the summer of 1979 that I started to collect data more systematically. Most observations were made in my backyard and garden in Lynchburg, Virginia. An ordinary headlamp was used to observe nocturnal activity in the field and laboratory. Because the light from such a lamp might disrupt normal activity, confirmation of the critical sequences was made with the use of a safelight filter. No differences in the activities were observed in the filtered light. Series of photographs of web-removal sequences were taken in the field and laboratory with a Nikon F2 camera and a lens-mounted electronic ring flash. The drawings of removal patterns are based on photographs and observations, and are stylized to demonstrate more clearly the broad aspects of these activities. A more detailed analysis of these patterns, which is beyond the scope of this study, would require the use of cinematographic devices. Data presented here pertain only to adult females, the exact ages of which were not known.

Several species were observed as the opportunity arose, but data were collected more systematically for only three species: *Acacesia hamata*, *Cyclosa turbinata*, and *Neoscona hentzii*. These species were selected because they are relatively abundant locally and demonstrate the two characteristic types of web removal. I brought some specimens of *A. hamata* and *C. turbinata* into the laboratory to permit observations under more controlled conditions. The bridge and attached stabilimentum of webs of *C. turbinata* adult females along with the web occupants, were removed from their positions on objects in the field, affixed to square wooden frames, and brought inside. Adult females of *A. hamata* were induced to build webs in square wooden frames by placing an individual near an upper corner, rotating the frame 90°, and tapping the frame so that the spider dropped on its dragline to the other upper corner. The frame was then returned to the upright position, and the now-horizontal dragline was later used by the spider as a bridge line in the construction of a new web.

Descriptions of Web-Removal Patterns

In the course of this study, two distinct patterns of web removal became clear. For simplicity, I shall refer to these as the "slow-removal" and "fast-removal" patterns, names that reflect one of the essential differences between them. Colleagues have suggested that calling these, respectively, the "snow-plow" and "window-shade" patterns would be even more descriptive of the removal behaviors involved.

Slow-Removal Pattern: Cyclosa turbinata

C. turbinata is found throughout the United States and as far south as Central America (Levi 1977a). It is quite common in the Lynchburg vicinity, and places its vertical web on relatively stable structures such as woody shrubs, small trees, and fences (Fig. 11.1c). The horizontal diameter of the

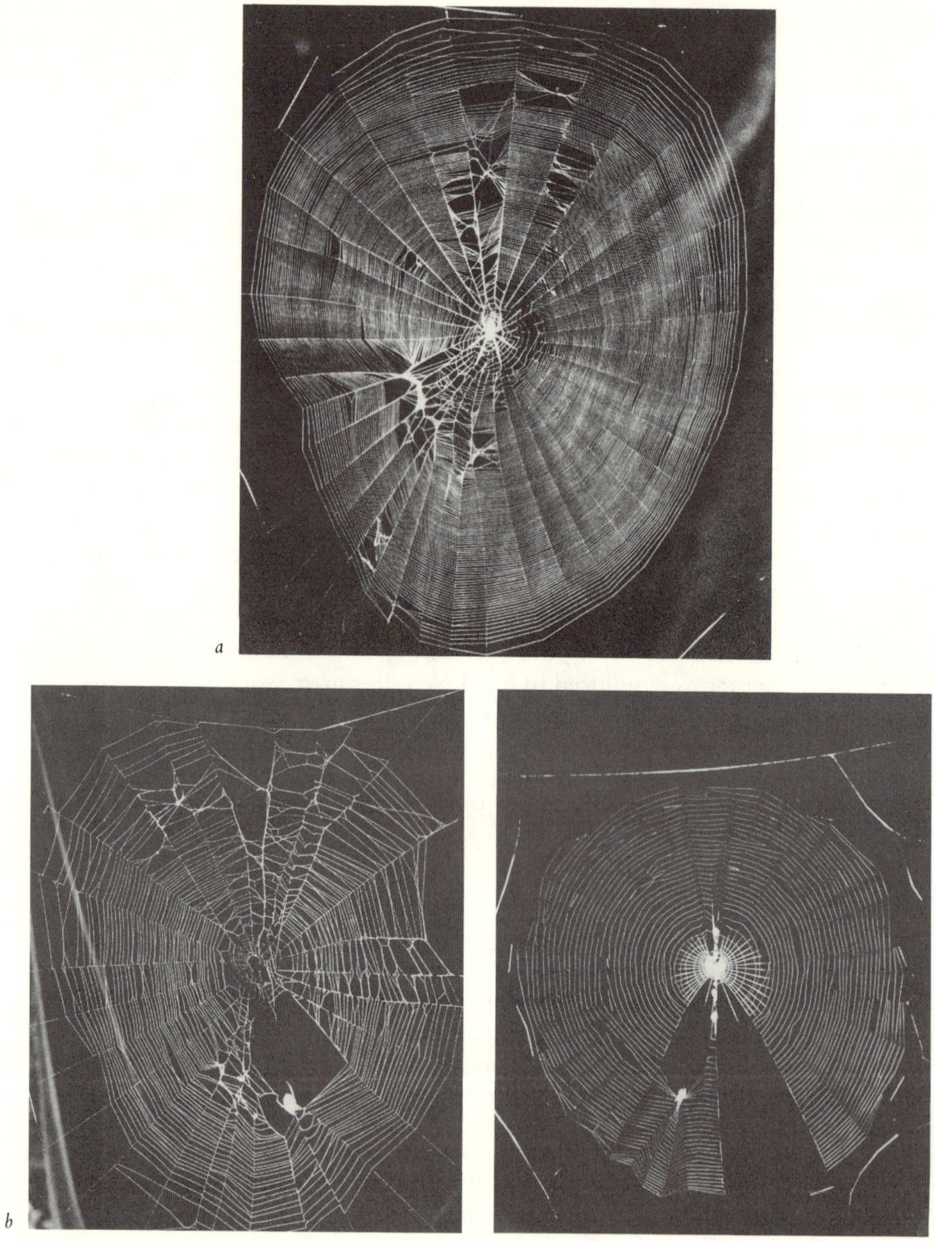

Fig. 11.1. Orb webs used in web-removal studies: *a*, web of *Acacesia hamata* (spider at hub); *b*, web of *Neoscona hentzii* (spider during first removal sequence); *c*, web of *Cyclosa turbinata* (spider during second removal sequence).

prey-catching area averages 12.8 cm. Characteristically, there is a thickened bridge line and a vertical stabilimentum through the center that contains prey remains. Egg sacs are placed in the upper part of the stabilimentum (Levi 1977a). The spider remains at the hub day and night, and is camouflaged by debris. Most prey capture occurs during the day. The outer frame is roughly square and is affixed by four or more attachments to the mooring points. The frame lines and stabilimentum are not removed, and only the prey-capturing area is renewed daily.

In response to the appropriate stimuli (see below), the spider moves radially towards the frame, gathering up three to five radii and associated sticky spiral threads. The process roughly resembles the making of a furrow by a snowplow. During this step, a dragline is paid out that later becomes a radius. By the time the spider reaches the frame line, a "pie-wedge" area of the catching web has been removed. The spider attaches the new radius to the frame and uses it to return to the hub. Destruction of the old web involves a series of behavioral steps, the "removal sequence" (Fig. 11.2), which are repeated until the entire catching area is removed. This sequence begins with the spider at the hub.

Typically, the first removal sequences occur on either side of the stabilimentum in the lower part of the web (Fig. 11.1c), as was reported for *Cyclosa conica* (Nielsen 1932). As web removal continues, the spider removes areas on alternate sides of the web and moves upwards, so that the last area of the web removed is at the top. After all the web is removed, a "skeleton"

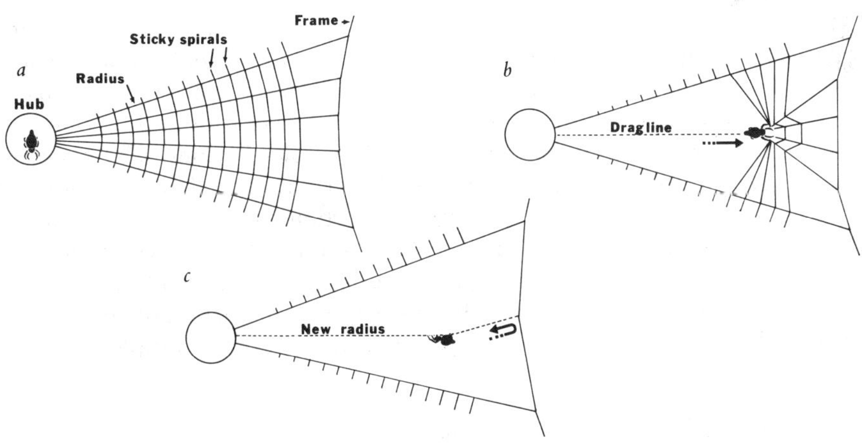

Fig. 11.2. Steps in the removal sequence of *Cyclosa turbinata*: *a*, wedge-shaped segment of a complete web before removal; *b*, spider removes sectors while paying out dragline; *c*, spider returns to the hub after attaching the dragline to the frame (the dragline becomes a radius in the new web).

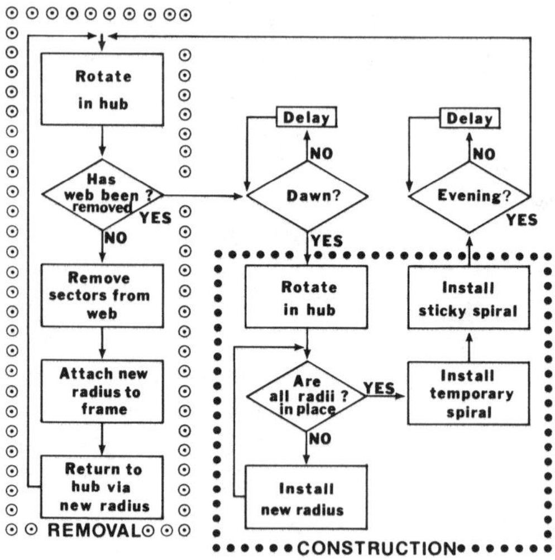

Fig. 11.3. Slow-removal pattern of *Cyclosa turbinata* and its relationship to the steps in the construction phase. Each rectangle represents an identifiable act; each diamond represents a break in the sequence that presumably requires a stimulus to proceed to the next step. Dotted circles represent the construction of the initial radii during removal.

web remains composed of the old frame lines, the stabilimentum, and six to ten new radii. The spider remains inactive at the hub for a variable time interval between the completion of web removal and the initiation of construction.

The initiation of web removal apparently requires a combination of extrinsic and intrinsic stimuli. Since web removal occurs only at night, darkness is clearly an important component. However, since removal may begin any time between dusk and dawn, an intrinsic factor or an additional, undetected extrinsic factor is apparently also involved. Controlled experiments may be necessary to determine the stimuli required to initiate removal.

The stimulus required to begin web construction must be the first light of dawn, since construction always begins at this time, and requires about one hour to complete. Six to ten radii are already in place as a result of removal of the previous web; construction begins with the installation of additional radii, followed by the familiar steps culminating in the completed web (Fig. 11.3).

Fast-Removal Pattern: Acacesia hamata

Detailed descriptions of the web of *A. hamata* are found elsewhere (Levi 1977c). Briefly, it is a closely meshed, delicate web with a triangular frame that is placed in weeds and shrubs (Fig. 11.1a; see also Fig. 5.4a).

Web removal in this species involves a single, unrepeated series of steps, and differs greatly from web removal in *C. turbinata* (Figs. 11.4, 11.5). Time taken to remove the web ranged from 60 to 75 seconds at temperatures of 21° to 24° C.

In response to the appropriate stimulus, the spider moves downward from the hub, trailing a dragline attached to the hub. As it moves down, the spider pulls the web under greater tension, so that the hub and bridge are seen to sag (Fig. 11.4b). The spider proceeds down the catching area, crosses a secondary frame line, and cuts one of the outer frame lines (Fig. 11.4c). The resulting sudden release of tension causes the web to become deformed, especially on one side. The spider then proceeds across to the frame line on the opposite side and crawls farther downward to a point near where the two side frame lines join, turns, attaches the dragline to the lower mooring thread, and cuts the remaining outer frame line (Fig. 11.4d). When this line is cut, the lower part of the web loses most of its tension, springs upwards, and partially collapses. The only tension remaining is that created by the connection of the dragline from the hub to the mooring line. Often the connection between the dragline and mooring line does not hold, so that when the frame line is cut, the only tension remaining in the entire lower part of the hub is that produced by the weight of the spider suspended on its dragline. In any case, the spider proceeds upwards on the dragline to the hub, and past the hub on a vertical radius to the bridge line, gathering up the orb as it goes (Fig. 11.4e). By the time the spider reaches the bridge line, the catching area of the web is completely collapsed upwards against it, and the configuration formed by the web is a Y if the dragline remains attached (Fig. 11.4f). The procedure resembles, in a crude way, the raising of a roller type of window shade. The spider attaches the dragline at its point of contact with the bridge line, and then cuts the bridge line near the attachment point. The spider proceeds laterally while reeling in the bridge line and the collapsed orb. At this time, tension is significantly reduced, the bridge line sags, and the spider spans the gap between the bridge line it is reeling in and the dragline it is paying out. When the spider reaches the limit of the collapsed orb, it turns, attaches the horizontal dragline to the frame bridge line, cuts the dragline, and returns to the attachment of the vertical dragline, reeling in the horizontal dragline as it goes (Fig. 11.4g). It crosses the point of attachment, often without reattaching the vertical dragline, and continues to reel in the remaining part of the bridge line and collapsed orb until it reaches the mooring point, whereupon it attaches the dragline to the support. The spider then proceeds to the sheltered area on the support, usually the axil of a leaf less than 10 cm away, and assumes a resting position until the beginning of the next building sequence at dusk.

At least two different stimuli are able, independently, to initiate removal.

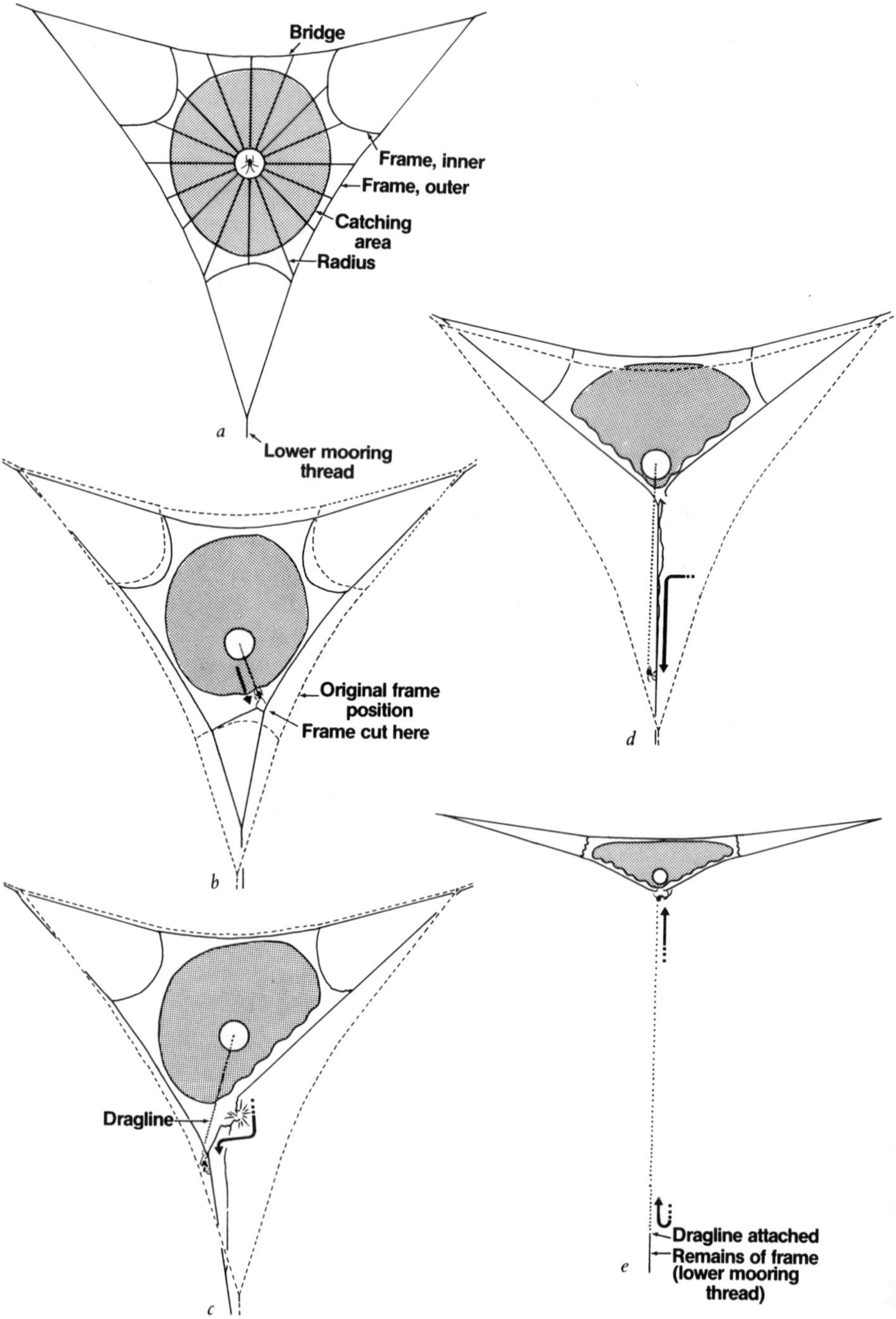
Bridge
Frame, inner
Frame, outer
Catching area
Radius
a
Lower mooring thread
Original frame position
Frame cut here
b
Dragline
c
d
e
Dragline attached
Remains of frame (lower mooring thread)

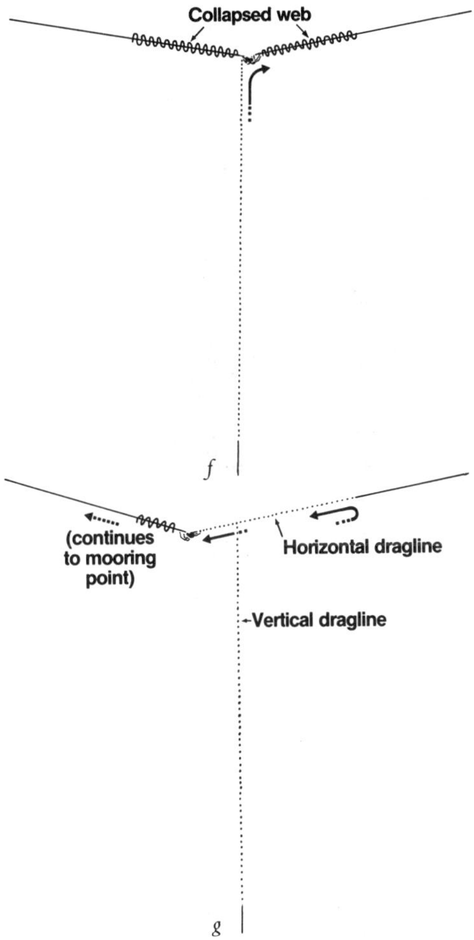

Fig. 11.4. Steps in the fast-removal pattern of *Acacesia hamata*: *a*, complete web before removal (for simplicity the radii in the following steps are not drawn but are presumed to be present); *b*, spider moves down from the hub to the frame while increasing tension on the web; *c*, spider cuts the frame, which partly collapses the web, and goes to the other side of the web and down; *d*, spider continues down and releases tension on the frame, which permits further collapse of the web; *e*, spider attaches the dragline to the mooring line and proceeds upwards, causing even more relaxation of tension and further collapse of the web; *f*, spider continues upwards, causing complete collapse of the web against the bridge, attaches the dragline to the bridge, and reels in the bridge and collapsed web to one side; *g*, spider turns and reels in the dragline on the bridge and the remainder of the old web.

All webs were taken down at early dawn; therefore daylight must stimulate removal. Light from a headlamp, however, never initiated this behavior. Rain often initiates removal in most individuals. Sprinkling webs in the field with a hand-operated mister initiated removal of one-third of the webs tried. Commonly, in case of destruction by rain or some other force, two webs might be produced in a single night, and at least three were made by one individual.

Alternation of Patterns: Neoscona hentzii

N. hentzii is a common, large orb weaver that reaches its full prominence from early September to mid-October. Its distribution in eastern North America extends from Massachusetts and Ontario southward to Florida and

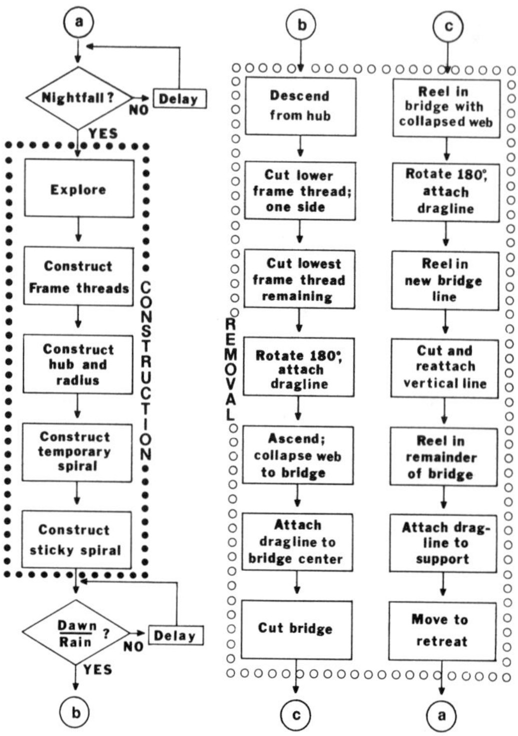

Fig. 11.5. Fast-removal pattern of *Acacesia hamata* and its relationship to the steps in the construction phase. Each rectangle represents an identifiable act; each diamond represents a break in the sequence that presumably requires a stimulus to proceed to the next step.

central Mexico and westward to eastern Kansas and Arizona (Berman and Levi 1971). Its web is vertical, and has an open hub; a retreat is built near one end of the bridge line (Fig. 11.1b). The web is placed in trees and shrubs and on buildings.

Web-construction activities in this nocturnal spider are quite variable. In an attempt to find a pattern in the daily cycle, I kept careful records on 23 individuals. Of the 14 individuals that were observed to remove webs, seven used the slow-removal pattern and six the fast-removal pattern; only one used both patterns during the survey period.

When the slow-removal pattern was used, the time taken to remove the web ranged from 35 to 55 minutes; web removal was followed immediately by web construction, which required about one to one and a half hours. Initiation of removal typically occurred at dusk, but happened at any time before dawn. Webs stayed up all day, but with few exceptions spiders went to their retreats at dawn.

When the fast-removal pattern was used, removal occurred during early dawn and took three minutes and thirty seconds to five minutes and forty-five seconds. One reason more time was needed, as compared with the web-removal time of *A. hamata*, is that the web of *N. hentzii* typically had more than three mooring points to detach. During the daylight hours, only the thin lines forming a Y, which may become the first radii, remained and were frequently broken. Spiders began construction of the new web at dusk, utilizing the Y lines if they were intact.

Observations of Other Species

During the course of this study, I made observations of several other species whenever the opportunity arose. For some species only a single observation was made, and none of the species was studied systematically.

The slow-removal pattern was observed in *Micrathena gracilis* (Walckenaer), *Micrathena sagittata* (Walckenaer), *Argiope trifasciata* (Forskål), *Argiope aurantia* (Lucas), *Metepeira labyrinthea* (Hentz), *Nuctenea cornutus* (Clerck), and *Mangora* sp.

The fast-removal pattern occurred in two additional species. *Verrucosa arenata* typically builds its web between midnight and dawn and leaves it up no later than noon. Numerous observations were made of web removal in this species, and in each case the fast-removal pattern was used. This species was observed to remove its web during rain. The fast-removal pattern was also observed in three individuals of *Eustala anastera*. This species usually does not leave the web up during the day (Levi 1977a), and I have observed its removal at dawn. Sprinkling water on the webs did not elicit removal.

The webs of these two species are similar, i.e., vertical, large, and open-meshed. Also, there are usually more than three mooring attachments of the frame to the substrate, and therefore more time is required for web removal in these species than in *Acacesia hamata*.

The Place of Web Removal in the Cycle of Web-Building Activity

Working with *Araneus diadematus* and summarizing the work of others, including Reed (1968), Peters (1970) listed three phases of construction: (1) frame and radii construction, (2) auxiliary spiral construction, and (3) catching spiral construction. Eberhard (1972b), in a careful study of *Uloborus diversus*, developed a more detailed list of activities: (1a) exploration, (1b) frame thread construction, and (1c) hub and radius construction; (2) temporary spiral construction; and (3) sticky spiral construction. Neither author discussed web removal. However, it is clear from my studies that removal of

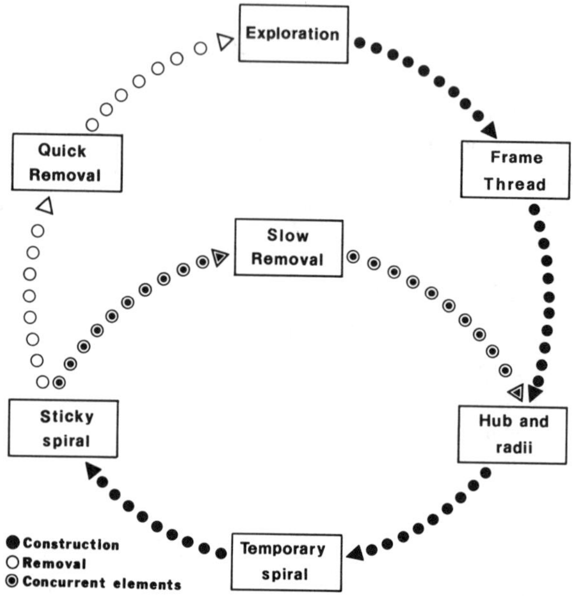

Fig. 11.6. Interrelationship of orb-web removal and construction. The construction steps are after Eberhard (1972b). The slow-removal pattern bypasses the steps prior to hub and radii construction and overlaps with the installation of the first few radii. The fast-removal pattern usually involves destruction of the entire web and thus requires the spider to rebuild the entire web.

the web is an important component of the daily web-building activity of the orb-weaving spider (Fig. 11.6). Furthermore, when spiders use the slow-removal pattern, destruction of the web is an important, overlapping component of web construction because the first few radii of the new web are put into place as the old web is removed. The exploration and frame-construction phases, cited as basic building steps by both Peters and Eberhard, are bypassed entirely, and only the catching area is renewed (Fig. 11.6). Therefore, in the typical daily web-building cycle of spiders using the slow-removal pattern (perhaps the majority of orb weavers), all the construction steps listed by these authors are not included. Obviously the exploration and frame-construction steps are included if the original web is destroyed or the spider changes web sites.

Spiders using the fast-removal pattern, however, typically must follow all these steps for each construction (Fig. 11.6). If nothing of the web remains at the end of the day, as is most often the case, then construction must begin with exploration. On the other hand, if the Y or bridge line survives, then construction begins with the surviving structure as the starting point. In *Acacesia hamata*, if only the bridge line survives, it is pulled into a Y. Whether a

Y survives or one is formed from the bridge, its intersecting lines serve as the "protohub" around which a frame is built.

Whichever pattern is used, there is merit to considering the removal pattern an important phase in the cycle of web-building activities, along with the construction phases. In Figs. 11.3 and 11.5 I have shown each pattern as part of a cyclical process, and I have demonstrated the interrelationships of these patterns within the daily web-building cycle in Fig. 11.6.

There is a tendency for spiders that use the slow-removal pattern (1) to have the web up during the daylight hours and to be diurnal feeders, (2) to have specialized structures associated with the catching area (e.g., a stabilimentum, as in *C. turbinata*, or a signal line to the hub, as in *Metepeira labyrinthea*), and (3) to build a permanent retreat and keep the same web site for several days. Correspondingly, there is a tendency for spiders that use the fast-removal pattern (1) to be nocturnal feeders and remove the web at dawn or in the morning, or with rain, (2) to lack specialized structures associated with the capture area, (3) to change web sites frequently, and (4) to make vertical webs. Perhaps one advantage of using the fast-removal pattern in the dawn hours is that the increasing light might make a spider that is moving about for a long period of time more attractive to a sighted predator. This is a particularly critical factor for *V. arenata*, which often removes its web at midday. This species also tends to make its rather delicate web in vulnerable, open areas, using frame lines that commonly exceed 2 m in length.

It seems safe to predict that either or both of these removal patterns are employed by the majority of orb weavers, and especially by those that do not have highly specialized or unusual webs. Very limited and casual observations of juveniles indicate that they, too, utilize these two patterns.

Atypical Webs

What about the methods of web removal in those orb-weaving species that have atypical webs? Some examples of atypical orb webs are discussed by Stowe (this volume). These modified webs, which are more common in the tropics, are thought by Stowe to be the result of specialization on certain kinds of prey. *Wixia ectypa*, a nocturnal spider, makes an unusual, reduced, "asterisk" web (Stowe 1978) that has only radii and a hub and lacks a frame and spirals. During a single observation of a juvenile, which was being photographed, the spider removed the web by climbing each radius and cutting it near its attachment to the twig, leaving only a single, horizontal thread. The entire web was removed in less than 10 seconds. The procedure followed resembled the slow-removal pattern, except that the time involved was extremely short.

Mecynogea lemniscata produces a complex web composed of a "labyrinth" surrounding an intricate, horizontal, dome-shaped orb. Juveniles of this species remove the web by first cutting the radii of the orb at its perimeter, thus causing it to collapse or roll up towards the hub. Other suspension lines for the interior of the orb are then cut, and the collapsed orb, resembling a ball of cotton, falls down into the lower labyrinth, where it remains. After several days a few collapsed orbs may accumulate in this region.

The unusual "inverted ladder" of the nocturnal spider *Scoloderus cordatus* is a vertical orb with the upper part greatly elongated above the hub (Eberhard 1972a; Stowe 1978). According to Stowe (1978, pers. comm.), this species removes the web using the fast-removal pattern.

Conclusion

I hope that the foregoing descriptions and discussion of web removal patterns have demonstrated that web removal is an important component of the behavior of orb-weaving spiders. I believe, further, that web construction is a cyclical process, and has not previously been appreciated as such. I therefore suggest that forthcoming descriptive studies of orb-web-building behavior would be significantly enriched if they included descriptions not only of the process of removal but also of the complete web-building cycle.

Acknowledgments

I wish to thank Dr. Herbert W. Levi, Shirley K. Whitt, and Nelle A. Carico for their suggestions and advice.

12

The Monophyletic Origin of the Orb Web

JONATHAN CODDINGTON

TWO VENERABLE groups of spiders spin orb webs: the cribellate Uloboridae, weavers of calamistrated sticky silk, and the ecribellate Araneidae, weavers of true viscid silk (Figs. 12.2, 12.1). "Cribellate" spiders possess a cribellum and a calamistrum. The cribellum is a flat, complex spinning plate from which the cribellate silk issues; the calamistrum is a comb of bristles on the fourth metatarsi used to comb the cribellate silk from the cribellum. "Ecribellate" spiders lack these structures. Other aspects of the morphology of the two families of spiders also differ considerably, yet both groups spin very similar orb webs. When juxtaposed to the morphological disparity between the spiders themselves, the similarity in the form of their webs and in the manner in which they are constructed has inspired a classic evolutionary controversy—the single versus the dual origin of the orb web. Behavioral evidence has defined one group, "orb weavers," including some but not all cribellate spiders, but morphological evidence has defined another, "Cribellatae," including some but not all orb weavers.

The problem can be viewed as two contingent dilemmas. First, are all orb webs homologous as orb webs, and/or all cribella homologous as cribella? Second, if the answer to both of these questions is yes, then which homologue is primitive and which derived, or do they both define the same group? There are several possible answers to these two dilemmas, and all have been suggested at one time or another in the history of the controversy. An additional problem contributing to the controversy over the orb web is secondary loss or complete modification of features in only some species of a monophyletic group. Occasionally, the species retaining the features are split off as a separate taxon defined by the retained primitive features. That taxon is not then synonymous with any real evolutionary lineage, and is termed "paraphyletic" because it includes only some of the species legitimately belonging to the lineage.

History of the Controversy over the Orb Web

The controversy over the orb web dates from the response of Thorell (1886) to some innovations in spider classification proposed by Bertkau (1878, 1882). Thorell pointed out that nearly all arachnologists before Bertkau had grouped spiders "not on differences in their *organization*, but on certain peculiarities in their *habits*, especially on their mode of locomotion and the form of their webs" (p. 301). In fact, those arachnologists recognized some version of the taxon "orb weavers" and placed all orb-weaving spiders in it. The taxon had various names: Orbiculariae (Walckenaer 1802), Orbitelae (Latreille 1825), Orbitèles (Walckenaer 1837–47, Vol. 1), Epeiriformes (Simon 1864), Radspinnen (Menge 1866), and Orbitelariae (Thorell 1869, Dahl 1883). Lister (1678), Clerck (1757), Sundevall (1833), Westring (1861), and Keyserling (1882) also grouped all orb weavers known to them together (information from Bonnet 1959: 5017–34). (Although the taxon "orb weaver" therefore dates conceptually from 1678, cribellate orb weavers were routinely included in it only after 1789, when Olivier first described a cribellate orb weaver, *Zosis geniculatus*, placing it in Linnaeus's genus *Aranea*.)

Bertkau had studied the anatomy of spiders extensively, and he used anatomical details rather than "habits" to define spider taxa. Bertkau thought the practice more scientific, an opinion with which Thorell agreed, at least in principle (1886: 301). Bertkau's Cribellatae was widely accepted at the time and afterward, possibly in part because of the general shift from the study of natural history in the early part of the nineteenth century to comparative anatomy in the latter half. Actually, Bertkau was not the first to propose a taxon defined by the possession of a functional cribellum. Blackwall, who had discovered the cribellum in 1833, grouped all cribellate spiders in the "family" Ciniflonidae (Blackwall 1841). Blackwall's suggestion had not convinced his colleagues, but Bertkau's did, perhaps because of his reputation as an anatomist and because he offered a comprehensive classification of spiders based on anatomical details.

In effect, Bertkau chose one of the solutions to the two dilemmas outlined above. He said that all cribella were homologous as such, but that all orbs were not. He said that the orb was convergent in cribellates and ecribellates, and that the cribellum had never been lost. Hence the Cribellatae were a monophyletic group. (Bertkau included Zoropsidae, Miagrammopidae, Filistatidae, Oecobiidae, Dinopidae, Uloboridae, Dictynidae, Eresidae, and Amaurobiidae.) The homology of all cribella as such, and of the colulus with the cribellum, was soon demonstrated embryologically by Dahl (1901) and Montgomery (1909).

Most araneologists have since agreed with Bertkau's innovations, e.g., Simon (1892), Pocock (1900), Dahl (1904), Comstock (1948), Berland (1932), Caporiacco (1938), Bristowe (1938), Gerhardt and Kaestner (1938), Millot (1949), Bonnet (1959), Kaston (1972a), and Gertsch (1979). The geometric regularity of orb webs makes appealing the idea that orb webs represent the most derived stage in a transformation series of web architectures. Many authors have commented that the orb web is a superb solution to the problems that a sessile, myopic insectivore must face in catching prey (Thorell 1886; Pocock 1895; Comstock 1912; Hingston 1920; Tilquin 1942; Savory 1952; Kaston 1964; Witt 1965, 1975; Langer 1969; Eberhard 1972a, 1981a; Chacón and Eberhard 1980; Kullmann 1972a; Denny 1976; Robinson 1977b; Gertsch 1979). For example, in a book aimed at a lay audience Gertsch (1979: 164) described the orb web as follows:

> The two-dimensional snare known as the orb web is a crowning achievement of the aerial spiders. . . . To the evolutionist it is only the last step of a series that has resulted in a circular design—an inevitable shape; and the spider has no more to do with spinning such a symmetrical web than a "crystal has to do with being regular." The web, among all objects produced by lesser creatures an unrivaled masterpiece, is above everything else a superb snare. . . . That a similar trap, produced by a like series of instinctive actions, should have evolved among a separate line of spiders might well seem an impossibility. Nevertheless, the cribellate uloborids have fashioned a web that, except for the substitution of the hackled band for the beaded spiral lines, is a faithful reproduction of the snare of the [araneoid] orb weavers. . . . This most highly evolved of all aerial webs is the result of the random activities of aerial prototypes, which finally established order among the irregular lines in the horizontal platform.

Implicit in that viewpoint may be a disinclination to believe that orb webs, once evolved, would have been lost except in isolated cases by further adaptation to even more specialized life-styles. If orbs are adaptively superior to other web architectures, the notion of convergent evolution on the orb-web style of architecture is more plausible. Not surprisingly, authors who favor arguments for the superiority of the orb as an architectural design are usually sympathetic to the dual-origin theory of the evolution of the orb web.

There are of course other answers to the dilemmas outlined above. For example, Pickard-Cambridge (1897–1905: 45) argued that not all cribella were homologous as cribella:

> With regard to the cribellate forms there appears to be no reason why the cribellum and correlated calamistrum should not have become independently specialized in several groups, each of these groups falling under different surrounding influences as they developed.

Lehtinen (1978) expressed a similar opinion regarding the Filistatidae.

Almost immediately after Bertkau published (1882), the obvious alternative occurred to Thorell (1886). He suggested that all cribella and all orb webs were homologous as such, but that the cribellum was a primitive feature for a large group of spiders, perhaps even for all true spiders:

> And this again [that most adult male cribellate spiders lack the cribellum] appears to me to prove that the cribellum and calamistrum are organs that have originally belonged to the order of spiders in general, and have in the course of time been reduced and lost in a part of them, those namely that no longer wanted them; and this independently of their greater or less affinity. (p. 322)

Thorell also argued that the orb web was uniquely derived:

> Even if we do not assign, in general, any great weight in the phylogeny and classification of the spiders to the form of their webs, it must be admitted that it is at least probable that spiders that fabricate regular or so-called geometrical webs have a common origin and belong to one and the same higher group, or, in other terms, that this industry cannot have arisen spontaneously and independently in two or more different and natural groups. (p. 324)

Finally, Thorell pointed out what to him seemed a conceivable but absurd possibility:

> Only on the supposition that all spiders have originally constructed such [geometric] webs, but that most of them have in the course of time lost this talent, could it be admitted that spiders belonging to radically different groups can give the webs such an artificially finished and almost identical form. But for such a supposition there is no reasonable ground. (p. 324)

To anticipate my own argument, Thorell's supposition may not have been as unreasonable as he thought, at least for an unexpectedly large fraction of "all" spiders.

Of course, Petrunkevitch (1928, 1933) thought that the cribellum was a primitive feature of all true spiders, and he agreed with Thorell (1886) and Wiehle (1927, 1928, 1931) that the uloborid and araneid orb webs were homologous as orb webs (Petrunkevitch 1926: 428). They therefore disagreed with most of their colleagues on those issues, but araneologists were apparently unanimous that, wherever orb webs occurred, they indicated monophyletic, and not paraphyletic, taxa. Specifically, neither the taxonomists cited above nor Petrunkevitch nor Wiehle nor Thorell believed that the non-orb-weaving araneoid families might have evolved from orb-weaving ancestors. For the taxonomists who followed Bertkau, the refusal to accept that linyphiids or theridiids had lost the orb-web was logically consistent. However, for Petrunkevitch, Wiehle, and Thorell it was illogical, since they held that Araneoidea were monophyletic, yet that both the orb web and the cribellum were homologous as such.

Lehtinen was the first modern araneologist to argue that Cribellatae as defined were paraphyletic. At first he also held that orb weavers were a monophyletic group (1967: 393–94). He later decided that the orb was convergent in uloborids and araneoids (1978; pers. comm.). Finally, Platnick (1977), in a review foretelling the "final nail in the coffin of the Cribellatae," summarized the available morphological evidence and argued that the distribution of characters among Mesothelae, Mygalomorphae, hypochiloids, and the remainder of the old Araneomorphae is such that the cribellum must be a primitive feature for all true spiders. Specifically, he showed that Crome's (1955) contention that the occurrence of four dorsoventral pairs of abdominal muscles in cribellates was synapomorphic was in fact a misinterpretation of a symplesiomorphic character. The paraphyly of the Cribellatae has become widely accepted (Brignoli 1979; Kaestner 1980; Kullmann and Zimmerman 1976; Forster 1970; Levi 1980a, 1982; Foelix 1982). The morphological argument surrounding the cribellum and the categorical level at which it defines a monophyletic group seems to have been settled. But what is the effect of the reinterpretation of the cribellum on the presumed relationship between uloborids and araneoids?

The collapse of the old Cribellatae has removed the single best reason for not regarding the orb web as a synapomorphy of a group that includes at least uloborids and araneoid orb weavers. No compelling reason has been suggested to ally uloborids with any other group, whereas the orb web itself may be a substantial reason to ally them with other orb-weaving spiders. However, the reinterpretation of the cribellum is irrelevant to the question of the homology of orb webs as such and the group that they properly define—all that has been accomplished is the removal of arguments that for a century have effectively prevented consideration of what is, after all, an independent issue. The same two interpretative dilemmas still persist: are all orb webs homologues, and, if so, what does the monophyletic group defined by orb webs include?

Both Kovoor (1977a) and Opell (1979) suggested that Uloboridae and Araneidae might be closely related, Kovoor on evidence of similar silk glands, Opell on evidence of similar web form and the femoral trichobothria shared by uloborids and tetragnathines. Lehtinen (1978), Brignoli (1979), and Levi (1980a) wondered whether orb webs might not be the primitive web architecture for the entire superfamily Araneoidea, and possibly homologous to the similar web spun by the Uloboridae. Although the impetus behind many of these speculations has been the similarity in web form, none of these authors actually discussed web form or construction behavior in any detail. Kullmann (1972a) considered web form, but he assumed that the Cribellatae were a monophyletic group. He supported what may be called the classical point of view—that the uloborid and araneid orbs were convergently evolved. Eberhard (1972b, 1981a, 1982) presents a summary of over 10 years of research on the building behavior of orb-weaving spiders. He gives data on most major groups of orb weavers (uloborids, araneids, metines, tetragnathines, nephilines, anapids, symphytognathids, theridiosomatids, and mysmenids). Although emphasizing the paucity of information to support generalizations, he suggests that, thus far, the evidence favors the convergent evolution of uloborid and araneoid orbs. Eberhard (1982) also suggests that the absence of an orb web in theridiids is probably original and does not represent the loss or modification of an orb web, thus agreeing with Thorell (1886), Kullmann (1972a), and most araneologists in between.

In my view, the evidence points the other way. The orb webs of uloborids and araneids appear to be homologous. If Araneoidea are monophyletic, then the primitive web architecture of the superfamily is indeed the orb web. Therefore araneoids that do not spin obvious orbs, such as linyphiids, theridiids, nesticids, and mimetids, have either lost the ability or modified the form of the orb beyond recognition.

An assessment of the positive evidence for the single origin of the orb web, and consequently for the monophyly of orb weavers, seems overdue. First the monophyly of the relevant taxa must be considered to ensure that the groups under discussion are real; then evidence pertaining to silks, silk glands, construction behavior, and web form can be used to test the idea that orb webs, and orb weavers, evolved only once.

Monophyly of Uloboridae, Dinopidae, Araneoidea, and Araneoid Families

Table 12.1 lists the taxa relevant to this discussion and their presumed synapomorphies.

Uloboridae

Opell (1979: 465) identified three synapomorphies for the Uloboridae: loss of poison glands, a ventral row of distinctive macrosetae on the fourth metatarsi, and two apical cymbial setae on the male palp. However, Opell (1982a) also found similar macrosetae on the fourth tarsi of Dinopidae, though he did not mention in which species. Kovoor (1977a) suggested other synapomorphies. The minor ampullate glands of uloborids are in a unique inverted position, compared with their position in all other spiders. Uloborids have a unique pair of glands serving the posterior lateral spinnerets, which Kovoor named the "pseudoflagelliform" glands (examined only in *Uloborus*, *Zosis*, and *Hyptiotes*). Also, the second-instar spiderlings build a distinctive orb web (Szlep 1961). Finally, uloborid web-building behavior is unique in at least four ways (discussed below, along with the justi-

TABLE 12.1
Synapomorphies of Uloborids, Dinopids, Araneoids, and Araneoid Families

Taxon	Character(s)
Uloborids	inverted minor ampullate glands, pseudoflagelliform glands, apical cymbial setae, loss of poison glands, behavioral traits
Dinopids	web structure, eyes in three rows
Araneoids	aggregate glands, flagelliform glands, serrate hairs, iL_4 push during sticky-silk attachment, paracymbium (?)
Theridiids	lobed aggregate glands
Nesticids	distinctive basal plate of trichobothria
Linyphids	palp conformation, route of ejaculatory duct
Nephilines	radii doubly attached to frame, oL_4 downward slide to locate inner sticky-silk loop, temporary spiral not removed
Araneids	rotation of palpal bulb, spin-wrap attack, reduced tapeta in posterior median eyes
Metine-tetragnathines	modified paracymbium, elongate chelicerae
Theridiosomatids	sternal pit organs, palp conformation, route of ejaculatory duct, temporary spiral as circles
Symphytognathids	fused chelicerae
Mysmenids	male metatarsal clasping spurs, femoral tubercle
Anapids	labral spur between chelicerae

NOTE: For references and more detailed discussion, see text.

fication for viewing the behaviors as derived). Uloboridae appear to be monophyletic by these nine characters.

Dinopidae

Dinopidae are also a family important to the consideration of the evolution of the orb web, but little is known of their morphology or behavior. Their monophyly is supported by their unique web form (Fig. 12.13) and their eye arrangement in three rows. A careful reading of Theuer's thesis (1954) on the biology of *Dinopis spinosus* makes quite reasonable the inference that the species spins a modified orb. The sequences in its construction behavior, for example, are similar to frame, radius, temporary spiral, and sticky spiral construction in orb weavers. But Theuer was not specifically evaluating the dinopid web as an orb web when he described the behavior, and, in the light of Eberhard's (1982) identification of salient characters in orb-web construction, many critical observations are missing. If indeed the web of *Dinopis* (and *Menneus*: Akerman 1926) is a modified orb,

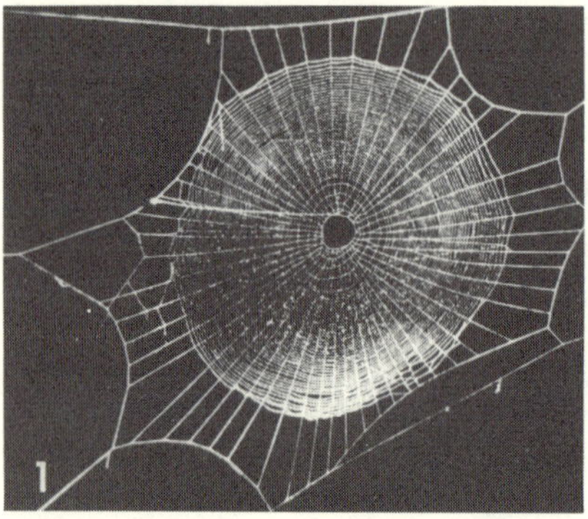
1

2

3
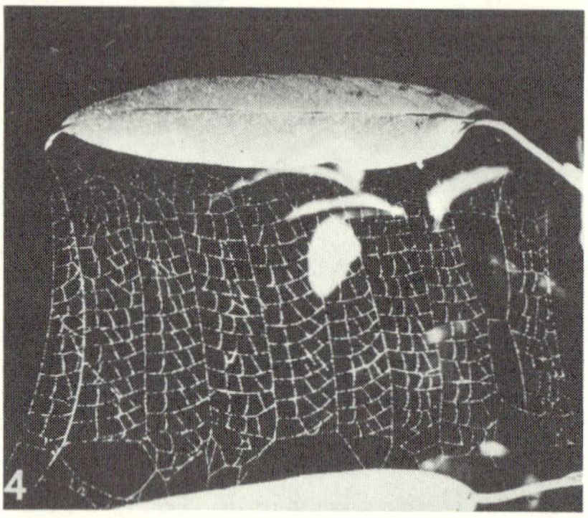
4
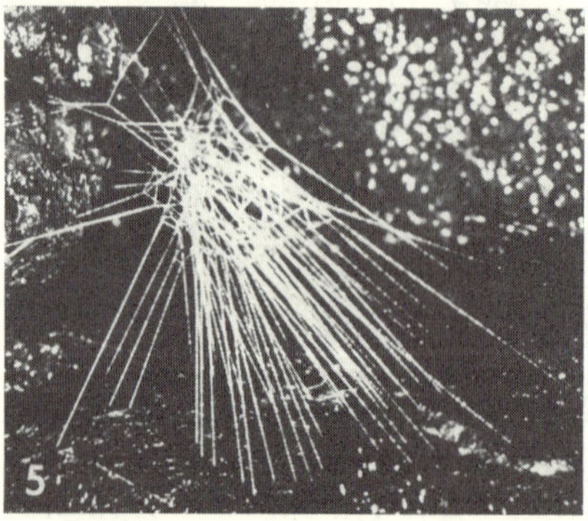
5

6

and all orb webs are homologous, the family would have to be the sister group of Uloboridae, Araneoidea, or their inclusive taxon. Both uloborids and dinopids possess pseudoserrate plumose hairs, which are lacking in araneoids (Green 1970: 8). But loxoscelids and thomisids have similar hairs, and Green noted that in general many intermediate hair types exist. Pseudoserrate plumose hairs, however, are weak evidence that dinopids may be the sister group of uloborids rather than araneoids. Shear (this volume), citing Kullmann 1975 and Opell 1979, points out that the cribellate silk of dinopids and uloborids is "puffed" along its length (see the photographs in Kovoor 1977a, and Kullmann 1972a), and states that the cribellate silk of other cribellates, e.g., *Hypochilus* (Hypochilidae), *Callobius* (Amaurobiidae), *Filistata* (Filistatidae), *Dictyna* (Dictynidae), *Stegodyphus* (Eresidae), lacks this feature. At present the feature appears to be a synapomorphy for uloborids and dinopids. Because araneoids are ecribellate, it can never be known whether or not their cribellate ancestor spun "puffed" cribellate silk. In other words, the feature could also be a synapomorphy for all orb weavers. Similarly, the pseudoserrate plumose hairs may also be an orb-weaver synapomorphy.

Uloborids and dinopids also use a wrap attack to subdue prey, and face away from the prey when wrapping (Eberhard 1967; Robinson and Robinson 1971). However, a wrap attack occurs in various other spider taxa, e.g., metine-tetragnathines (Eberhard 1982), Oecobiidae and Hersiliidae (Eberhard 1967), Pholcidae (pers. obs.), Nesticidae, Theridiidae, and Araneidae. Although more comparative work is needed on wrap attacks to separate convergent from homologous similarity, the literature suggests that the orientation of dinopids and uloborids towards prey is distinctive. This feature may be a synapomorphy of the two groups. Dinopids possess ventral macrosetae on their fourth tarsi.

Because both uloborids and araneoids possess a pair of specialized glands

---

Figs. 12.1–12.6: Webs of araneoid spiders. 12.1. Web of a typical araneid (*Micrathena gracilis*). Note the open hub (bitten out after sticky spiral construction) and the adjacent nonsticky spiral. 0.2×. (Voucher No. VII.3.1981:nph18). 12.2. Web of a typical uloborid (*Uloborus penicillatus*). The reticulate pattern of the hub indicates that the radii are laid early in nonsticky spiral construction and that hub modification is absent. 0.5×. (Voucher No. VIII.6.1980:6ph32.) 12.3. Representative linyphiid sheet web (unknown tropical genus). The spider is just visible beneath the sheet at the center of the web. 0.4×. (Voucher No. VII.12.1980:12ph30.) 12.4. Web of *Synotaxus* sp. (Theridiidae). The continuous vertical lines delimit the modules of the web. The numerous short vertical line segments are sticky silk. The spider can vary the number of modules spun from night to night. Approximately 0.2×. Courtesy of M. K. Stowe. 12.5. Web of *Achaearanea* sp. (Theridiidae). The distal ends of the straight lines are sticky. The central network is the retreat. 0.5×. (Voucher No. VIII.30.1980:16ph7.) 12.6. Web of *Eidmanella pallida* (Nesticidae). The distal ends of the branched vertical lines are sticky. The spider hangs upside down in the upper part of the web. 0.4×. (Voucher No. VIII.4.1981:7ph16.)

7
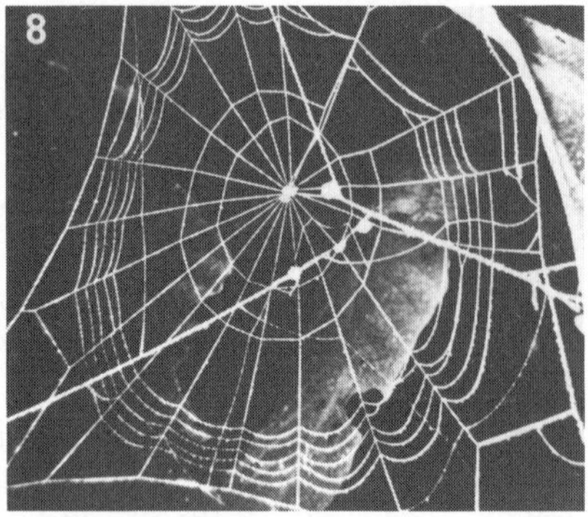
8

9
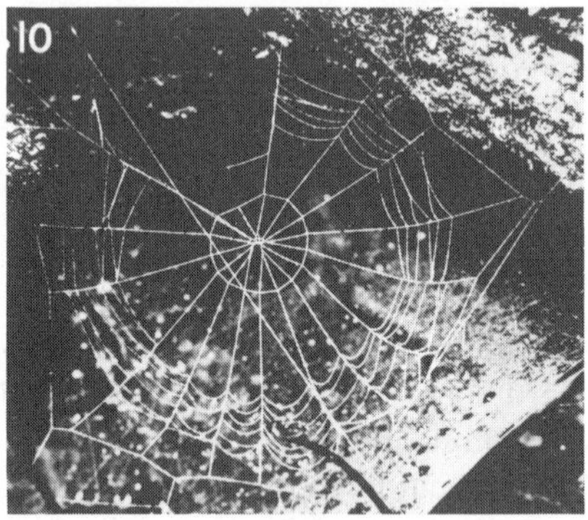
10
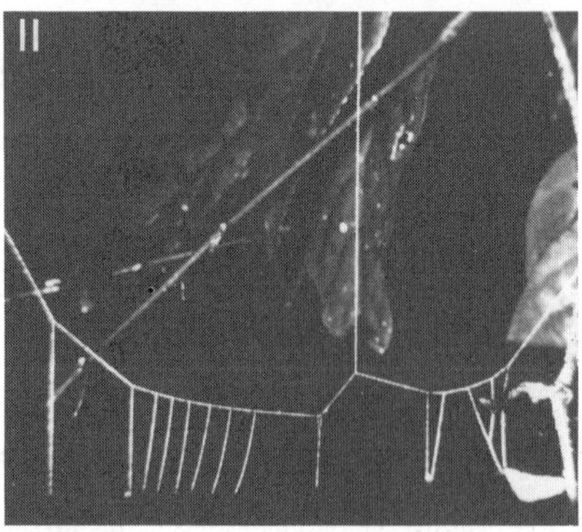
11
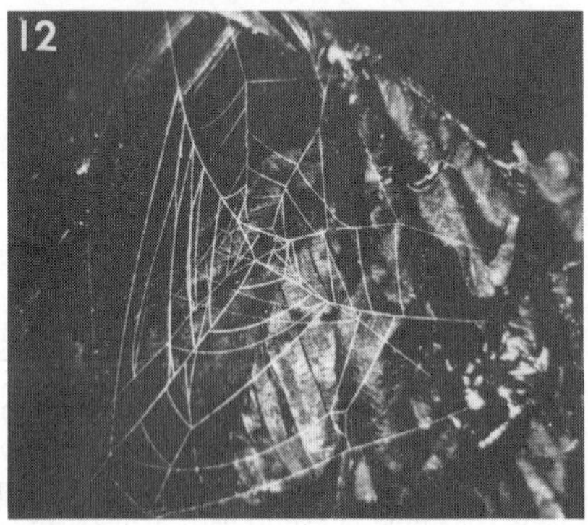
12

exiting on the posterior lateral spinnerets (either pseudoflagelliform or flagelliform; see below), it may be predicted that dinopids also have a similar pair of glands. If the dinopid glands are recognizably distinct from pseudoflagelliform glands and flagelliform glands, one must infer that the common ancestor of the three taxa merely had "specialized" glands. On the other hand, the predicted dinopid glands may share features with either taxon. Because the condition in dinopids is unknown, the former position is taken here. On the whole, little is known about dinopids, and consistent inclusion of the group in the following discussion is impractical.

Araneoidea

Araneoidea have been called one of the most distinctive and easily defined groups of spiders (Lehtinen 1967, 1975). Defining features that have been proposed are (1) a body covering of serrate hairs, (2) a fairly consistent trichobothrial pattern, especially the lack of tarsal trichobothria, (3) a web-spinning habit, (4) a globose abdomen, (5) complicated palps, and (6) spinneret arrangement (features 1–4: Lehtinen 1967; features 5 and 6: Kaestner 1980). Is any of these characters a convincing synapomorphy for the superfamily Araneoidea? Green (1970) pointed out that serrate hairs do occur in other families, and Lehtinen (1975) confirmed it. Only if the immediate outgroup of araneoids, and preferably several immediate outgroups, lacked serrate hairs could the araneoid serrate hair be considered a synapomorphy for the superfamily. No recent detailed arguments identifying a sister group to the Araneoidea other than the Uloboridae or Dinopidae exist, and both of these taxa reportedly lack true serrate hairs. Thus, the araneoid hair type could be a true synapomorphy, given favorable cladistic structure in the outgroup relationships.

Figs. 12.7–12.12. Webs of theridiosomatid spiders. 12.7. Web of *Epeirotypus* sp. (Theridiosomatidae). Note the similarity to a conventional orb: two-dimensional frame, radii, sticky spiral, and hub loops. 0.3×. (Voucher No. IV.13.1981:15ph31.) 12.8. Partially constructed web of *Epeirotypus* sp. (Theridiosomatidae). Note the nonsticky "circles" instead of a spiral and the incomplete sticky spiral. 0.4×. (Voucher No. VII.30.1981:nvph31.) 12.9. Web of *Theridiosoma gemmosum* (Theridiosomatidae). Note the radial anastomosis. 0.5×. (Voucher No. VI.30.1981:2ph12.) 12.10. Incomplete web of *Theridiosoma gemmosum* (Theridiosomatidae). The single nonsticky circle is characteristic of the genus. Compare this hub (before radial anastomosis) with that in Fig. 12.9. 0.3×. (Voucher No. VII.2.1981:gen.ph15.) 12.11. Web of *Wendilgarda* sp. (Theridiosomatidae). The lower vertical lines are sticky and are normally attached to the surface film of moving tropical streams. Note the lack of a frame and the dissimilarity to a conventional orb web. 0.5×. (Voucher No. VIII.11.1980:nvph1.) 12.12. Web of *Ogulnius* sp. (Theridiosomatidae). A frame is lacking and the radii are attached directly to the substrate. Note the irregular course of the sticky-silk line. 0.3×. (Voucher No. VII.6.1979:1ph1.)

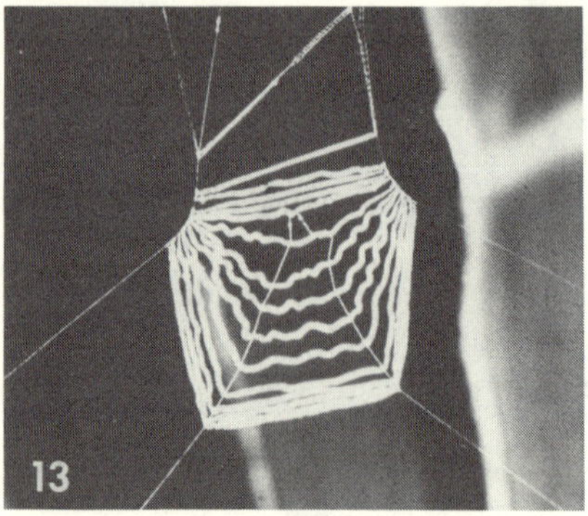
13

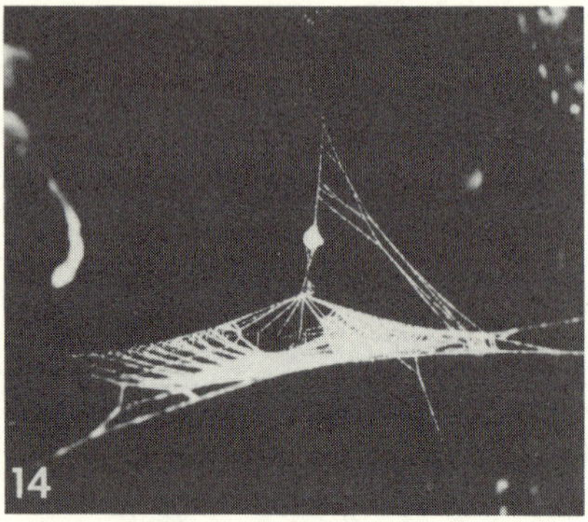
14

15

16

17

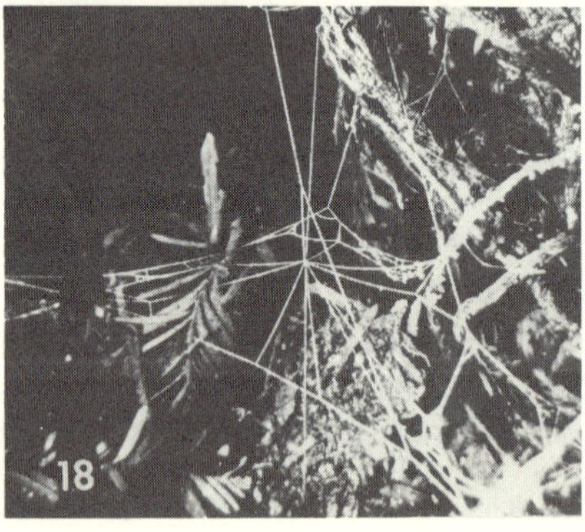
18

With regard to the second feature, Lehtinen (1980) showed that the trichobothrial pattern as such is not unique to Araneoidea. The third character is a "web-spinning habit," but what sort of web? If the orb is taken as the primitive web architecture of araneoids, the uloborids are a likely sister group. Although the orb might define the inclusive taxon, it could not then be synapomorphic for the Araneoidea. If the primitive araneoid web form is not the orb, but something like the web spun by theridiids (Fig. 12.5), or linyphiids (Fig. 12.3), as suggested by Pocock (1895), Comstock (1912), Kaston (1964), and Kullmann (1972a), then many other taxa spin similar webs, e.g., amaurobiids, pholcids, ochyroceratids, agelenids, psechrids, and tengellids. Without a more precise description of the "web-spinning habit," the character is not unique to Araneoidea.

The fourth character, a globose abdomen, occurs in many araneoid families, but, like the "web-spinning habit," the character is indistinctly defined and characterizes other spider families as well. Likewise, the complicated palps and spinneret arrangement have yet to be shown to be unique to Araneoidea.

One discrete character, not widely cited in the taxonomic literature, can be emphasized as a synapomorphy of the superfamily: aggregate silk glands. Kovoor (1977c) summarized the conclusions of Apstein (1889), Hopfmann (1935), and Anatasiu-Dumitrescu (1941, 1942) and suggested that aggregate glands were unique to Araneoidea. These glands provide the viscid coating of the sticky-silk lines, and have been found in Theridiidae (*Steatoda*, *Theridion*, *Teutana*, and *Latrodectus*), Linyphiidae (*Linyphia*), Araneidae (*Araneus*, *Argiope*, *Cyclosa*, and *Nemoscolus*), metine-tetragnathines (*Pachygnatha*, *Tetragnatha*, *Meta*, and *Zygiella*), and nephilines (*Nephila*) (Kovoor 1977c; Kovoor and Lopez 1980). Aggregate glands have been found in all araneoids

Figs. 12.13–12.18. Webs of dinopid and anapid spiders. 12.13. Web of *Dinopis spinosus* (Dinopidae). Approximately 3×. Courtesy of M. K. Stowe. 12.14. Web of *Anapis* sp. (Anapidae). The egg sac is positioned above the hub. The straight vertical line is constructed simultaneously with the radii. The diagonal lines at right are part of the sticky spiral. 0.9×. (Voucher No. VII.11.1980:16ph23.) 12.15. Web of *Anapisona simoni* (Anapidae). Infrequent accessory radii are visible. The catenary lines attached to the central vertical line are part of the sticky spiral. Note the egg sac just above the hub. 0.8×. (Voucher No. VII.11.1980:10ph38.) 12.16. Web of *Anapis heredia* (Anapidae). Note the numerous accessory radii, radial anastomosis, out-of-plane radii, and infrequent sticky-silk lines attached to the central vertical line. 1×. (Voucher No. VIII.10.1980: 7ph20.) 12.17. Web of *Chasmocephalon shantzi* (Anapidae). Note the accessory radii, radial anastomosis, out-of-plane radii, and out-of-plane sticky-silk lines. 1×. (Voucher No. IV.4.1980:11ph2.) 12.18. Web of *Chasmocephalon shantzi* (Anapidae) before sticky spiral construction. The out-of-plane radii are constructed simultaneously with the planar radii. 0.5×. (Voucher No. IV.5.1980:nvph28.)

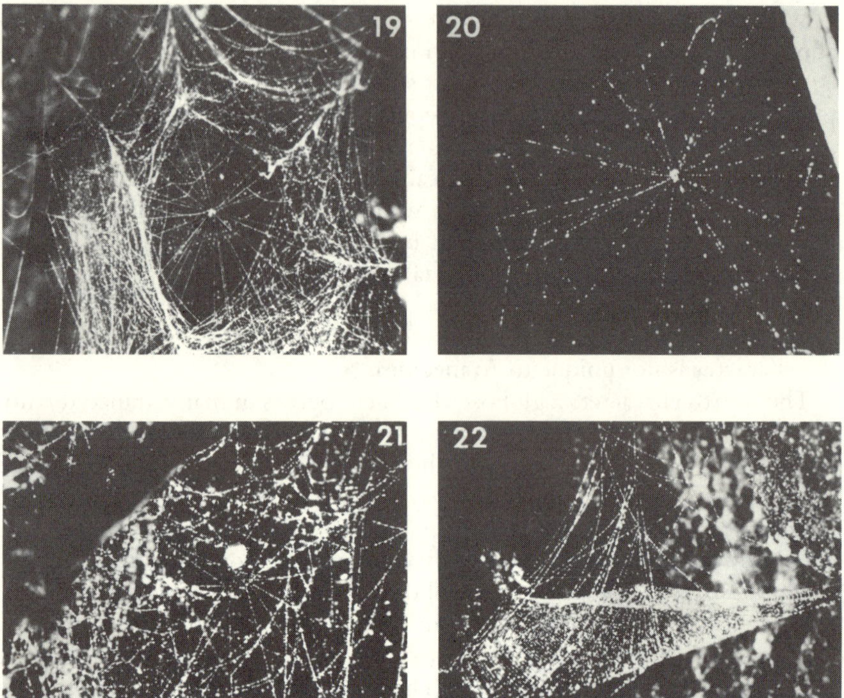
19
20
21
22
23
24

investigated thus far, and never in nonaraneoid species (45 nonaraneoid genera are cited in Kovoor 1977c). Although few araneoid taxa have been studied histologically, the webs of theridiosomatids, anapids, symphytognathids, mysmenids, and nesticids possess sticky silk lines more or less identical to those characteristic of the five lineages mentioned above. Because of the excellent correlation between the complex gland type and the appearance of viscid-silk lines in the web, it seems safe to presume that species having that kind of sticky silk also have aggregate glands. On the basis of that provisional inference, aggregate silk glands may stand as a synapomorphy of the superfamily.

Kovoor (1977c) found that flagelliform glands occurred in all araneoids studied (Table 12.2 below). She found similar glands in uloborids, the pseudoflagelliform glands. Although the glands are indeed similar, they can be distinguished by histological characters (Kovoor 1977a, 1978; Kovoor and Zylberberg 1979), so that true flagelliform glands become an additional synapomorphy for araneoids.

Shear (this volume) argues that the paracymbium of the male palp in Araneoidea is unique to the group, a possible fourth araneoid synapomorphy. Neither uloborids nor dinopids possess a paracymbium (Archer 1948). Heimer (1982) investigated the function of the paracymbium and stated that together with the median apophysis it served to arrest the rotation of the palpal bulb during expansion. He stated that the cymbial setae and median apophysis of uloborids acted in the same way, implying homology of the uloborid cymbial setae and araneoid paracymbium. But he also stated that the details of palpal rotation in uloborids and araneoids differed considerably, although he did not make the nature of the difference clear. However,

---

Figs. 12.19–12.24. Webs of mysmenid and symphytognathid spiders. 12.19. Web of *Mysmena* sp. (Mysmenidae). The outer envelope of sticky silk has been torn away to reveal the three-dimensional pattern of radii inside the web. Entire webs are roughly spherical or egg-shaped. The spider is at the center of the web. 0.8×. (Voucher No. VII.29.1980:11ph8.) 12.20. Web of *Mysmena* sp. (Mysmenidae) with radii completed, but early in sticky spiral construction. Note the few sticky-silk lines connecting the radii, and the frame lines supporting the radii. 1×. (Voucher No. VII.4.1980:nvph19.) 12.21. Section of a web of *Mysmena guttata* (Mysmenidae). The spider is barely visible beneath the egg sac positioned just above the hub. Radial anastomosis is present. 2.0×. (Voucher No. VIII.8.1980:8ph11.) 12.22. Web of *Maymena ambita* (Mysmenidae): The architecture is similar in all respects to that of anapid webs (compare Figs. 12.14–12.18). 1.8×. (Voucher No. VII.4.1981:3ph5.) 12.23. Web of an unidentified symphytognathid genus from Puerto Rico. The web displays numerous accessory radii and radial anastomosis, and is strictly two-dimensional and bounded by a frame (not shown). 1.2×. (Voucher No. VIII.27.1980:5ph23.) 12.24. Web of a *Patu* sp. immature (Symphytognathidae). Nearly all the radii visible are structural, although a few accessory radii are present. 0.7×. (Voucher No. VII.12.1980:5ph7.)

Opell (1979) hypothesized that the cymbial setae were part of a stridulation mechanism in the cladistically primitive uloborid genera *Waitkera*, *Tangaroa*, and *Polenecia*, so that the function of the cymbial setae is unclear. The different hypothesized functions do not necessarily conflict, because the stridulatory function may be original and the function in arresting the palpal bulb derived. The evidence therefore suggests homology, but also needs much more study.

Eberhard (1982) has found two behavioral features correlated with the attachment of viscid-silk lines to nonsticky-silk (NS) lines. All araneoids that he observed use the outside third and fourth legs to grip the NS line as the sticky-silk (SS) line is attached (Eberhard 1982, character B1). However, uloborids use the same grip in frame, radius, and temporary spiral construction, and an elaboration of it in SS construction (pers. obs.; see below). The second character also involves SS construction. During SS attachment, the inside fourth leg of the animal holds the SS line and extends in a characteristic fashion (Eberhard 1982, character C1). The latter character is absent in uloborids and dinopids (pers. obs.), but is present in all araneoid taxa examined thus far and appears to be a fifth synapomorphy. Eberhard (1982, character E1) also states that all araneoid orb weavers attach the sticky spiral to all radii encountered, whereas uloborids skip some attachments. But to judge from photographs and drawings of dinopid webs (*Dinopis subrufus*: Baum 1938; Roberts 1955; Clyne 1967; *Dinopis longipes*: Robinson and Robinson 1971; *Dinopis spinosus*: Theuer 1954; *Menneus camelus*: Akerman 1926) dinopids also attach SS lines to all radii, so that uloborids are probably autapomorphic in this respect. However, as far as is known, the five characters discussed above are unique to the superfamily Araneoidea.

Araneoid Families

Platnick and Shadab redefined the families Anapidae (1978a) and Mysmenidae (1978b), and Forster and Platnick (1977) redefined the family Symphytognathidae. Their synapomorphies for each taxon are listed in Table 12.1. Millidge (1977, 1980) argued that Linyphiidae (including the erigonine genera) are monophyletic on the basis of a conglomerate of characters of the male palp. Although the linyphiid palp is relatively well known as the result of the work of Millidge, Merrett (1963), van Helsdingen (1969), and others, those characters now believed to be unique to Linyphiidae may also appear in other araneoid families as their palp morphology becomes better known. At present, however, Linyphiidae as defined by Millidge seem to be monophyletic.

The Theridiidae are usually considered a poorly defined group, diagnosed mainly by the comb setae of the fourth tarsi (Levi and Levi 1962). Two characters correlated with the comb setae deserve more emphasis in any diagnosis of the theridiids: the possession of "lobed" aggregate glands and the use of sticky silk in the wrap attack to subdue prey. Kovoor (1977b) showed that these distinctive lobed glands are modified aggregate glands. As noted, the silk glands of few araneoid families have been investigated, but thus far these modified glands are unique to the theridiids. Eberhard (1967), Robinson (1969b, 1975), Robinson and Robinson (1971), and Robinson, Mirick, and Turner (1969) have documented the diversity of wrap attacks in spiders. The wrap attack of theridiids, however, is distinctive because the silk early in the attack is wet viscid silk, not dry silk. The spider throws viscid silk at the prey from some distance, and uses the comb setae of the fourth legs to handle the silk. I have seen viscid-silk wrap attacks in *Latrodectus*, *Achaearanea*, *Enoplognatha*, *Theridion*, *Steatoda*, *Argyrodes*, *Synotaxus*, and *Chrysso*. Apparently it also occurs in *Coleosoma* (Cutler 1972a), *Euryopis* (Carico 1978), *Anelosimus* (T. E. Christenson, pers. comm.; Eberhard 1979a), and other theridiid genera (Bristowe 1958). The lobing of the gland and its distinctive short duct (Kovoor 1977b) probably enable the spider to deliver large volumes of viscid silk at the beginning of the wrap attack. Araneids, uloborids, and dinopids, of course, use a wrap attack, but their swathing silk is said to be dry rather than viscid, and reportedly issues from a different gland (Kovoor 1977c).

In the past, Nesticidae have been defined by the form of the paracymbium in the male palp, but Lehtinen and Saaristo (1980) stated that the nesticid paracymbium is not diagnosable apart from that of linyphiids (Millidge 1977), tetragnathines (Levi 1981), mimetids (Shear 1981), or the other araneoids that possess a paracymbium. However, Lehtinen and Saaristo did state that the form of the basal plate of nesticid trichobothria is unique to the family (p. 58). The comb setae on nesticid fourth tarsi are similar to those of theridiids, but the form of their aggregate glands is unknown. Bristowe (1958) mentioned that *Nesticus cellulanus* uses a wrap attack "similar to Theridiidae," and Kirchner and Kullmann (1972) confirmed that this species does use viscid-silk to wrap prey. The viscid-silk wrap attack and comb setae suggest that nesticids are the sister group of theridiids. If nesticids also have lobed aggregate glands, they may yet be a monophyletic lineage (by the basal plate of the trichobothrium) within theridiids. Should that synapomorphy fail, no differential diagnosis of nesticids apart from theridiids is available. In any case, references below to "theridiids" are to the lineage including both groups.

Wunderlich (1980) proposed that the sternal pit organs, first described by Archer (1953), are a synapomorphy for Theridiosomatidae if that taxon is relimited to include only *Epeirotypus*, *Ogulnius*, *Theridiosoma*, and *Wendilgarda*. In all theridiosomatid palps the conformation and topology are consistent (a relatively huge tegulum, a hood-shaped conductor, and a bladelike median apophysis occur in similar orientations), and the route followed by the ejaculatory duct from the reservoir to the embolus is nearly identical in all genera. In *Epeirotypus* and *Theridiosoma* the temporary spiral appears as one or two circles (Eberhard 1982, character H2; pers. obs.). *Wendilgarda* and *Ogulnius* lack temporary spirals (pers. obs.).

Araneidae are a large and diverse family. This group may need to be re-limited before it is truly monophyletic. Its overall limits have certainly been constrained by the presence or absence of an orb web in otherwise doubtful taxa. For example, when theridiosomatids (as defined above) were considered a subfamily of the Araneidae, every type species of a new genus was first considered a theridiid. The morphology of these species obviously suggested theridiids to taxonomists such as Pickard-Cambridge and Keyserling. When these workers discovered the orblike webs of the species, they were transferred to Araneidae. In a like manner, the web form seems to be the only rationale for retaining nephilines, tetragnathines, and araneids in the same taxon. However, the existence of orb webs in other families means that this feature cannot be synapomorphic for Araneidae unless the form of the web or the construction behavior is unique.

Levi (1983) states that, in contrast to other araneoids, all araneids except the tetragnathines, metines, and nephilines have the bulb of the palp twisted in the cymbium so that the sclerites face medially; he also states that the same group has reduced tapeta in the posterior median eyes. Eberhard (1982) points out that during their wrap attack these araneids spin the prey and wind silk on it as on a bobbin ("spin-wrap attack"). When referring to "Araneidae" or "araneids," therefore, I mean the lineage comprising Araneinae (including gasteracanthines), Mastophorinae, Argiopinae, and Cyrtophorinae, as those groups are defined by Levi (1983). The group appears to be monophyletic by the above three characters. Eberhard (1982, character J2), relying mainly on the work of Robinson and Robinson (1980), indicated that araneid males perform a distinctive behavior termed "tarsal rubbing" during courtship. But tarsal rubbing occurs in metine-tetragnathines (*Leucauge papuana*: M. H. Robinson, pers. comm.), and a similar behavior occurs in theridiosomatids (*Wendilgarda* and *Epeirotypus* spp.: pers. obs.). Araneids also use a lateral tap of the outside first leg to locate the inner loop of sticky spiral during SS construction (Eberhard 1982, character A2), but that feature occurs in uloborids.

Nephilines use a downward slide of the fourth tarsus to locate the inner loop of sticky spiral (Eberhard 1982, character A3), but *Nephila clavipes* also taps laterally with the outside second leg (Eberhard 1982; pers. obs.). Nephiline radius construction is unique in one detail (Eberhard 1982, character F2), and nephilines also do not remove the temporary spiral during SS construction. The last-named character also occurs in *Cyrtophora* and *Mecy-*

nogea, but those genera are araneids, thus implying parallel losses. Nephilines, including *Nephila*, *Herennia*, and *Nephilengys*, appear to be monophyletic by these three characters.

Levi (1983) states that tetragnathines and metines tend to have a modified paracymbium and distinctive, elongate endites and chelicerae. How the metine-tetragnathine paracymbium is to be distinguished from that of nesticids or linyphiids is not clear—the character may be a symplesiomorphy (see Shear, this volume). The occurrence of elongate endites or chelicerae is also not an especially well defined character. The monophyly or paraphyly of the metine-tetragnathine lineage is especially critical because in many ways it exemplifies the "primitive araneoid" (Levi 1981a). To facilitate discussion the metines and tetragnathines will be treated here as if they were a monophyletic lineage, but that monophyly is tenuous.

Families such as Archaeidae, Hadrotarsidae, Mimetidae, Micropholcommatidae, Textricellidae, and Nicodamidae have often been included in the Araneoidea because of their general appearance (Levi 1982). The inclusion of Nicodamidae is especially problematic if the cribellate *Megadictyna* are retained in that family. But so little is known about the behavior and webs (if any) of these families that they cannot be discussed in this context.

Comparison of Uloborid and Araneoid Orb Weavers

I shall now examine in detail the similarities and differences between uloborid and araneoid orb weavers. First I compare silks and silk glands (and other morphological traits), then web-building behavior, and finally the web form itself.

Silk Glands and Morphology

Again, because nothing is known of their silk glands, dinopids cannot be included in the comparison of silks and silk glands, even though it is likely that they are orb weavers (J. Coddington, in prep.). However, the excellent recent work of Kovoor (1972, 1977a, 1977b, 1977c, 1978), Kovoor and Zylberberg (1979), and Kovoor and Lopez (1980) allows a general comparison of the silk glands of uloborids and araneoids (Table 12.2). Some of the similarities are clearly symplesiomorphies. For example, Apstein (1889) believed that ampullate, piriform, and aciniform glands were present in the ancestor of all true spiders. Kovoor (1977c) thought, however, that the category "aciniform" was so indistinct that only the first two gland types could be confidently assigned to the common ancestor of all true spiders. On the other hand, Millot (1949: 661) believed that the aciniform gland was probably the plesiomorphic gland type in spiders. Cylindrical glands also occur

TABLE 12.2
Occurrence of Silk Glands and Spinnerets in Uloboridae and Araneoidea

Gland	Spinnerets: Anterior median[a]	Anterior lateral	Posterior median	Posterior lateral
Uloboridae (*Uloborus* sp., *Zosis geniculatus, Hyptiotes paradoxus*)				
Major ampullate[b]	—	2	—	—
Minor ampullate[b]	—	—	2	—
Pseudoflagelliform	—	—	—	2
Cylindrical[b]	—	—	6 or more	8 or more
Aciniform[b]	—	—	many	many
Piriform[b]	—	many	—	—
Cribellate	many	—	—	—
Araneoidea (14 genera in six lineages; see text)				
Major ampullate[b]	—	2	—	—
Minor ampullate[b]	—	—	2	—
Flagelliform	—	—	—	2
Cylindrical[b]	—	—	2	4
Aciniform[b]	—	—	many	many
Piriform[b]	—	many	—	—
Aggregate	—	—	—	4

NOTE: For references and more detailed discussion, see text.
[a]Functional anterior median spinnerets are absent in araneoids.
[b]This type of gland occurs in nearly all araneomorph spiders.

in almost all true spiders (they are absent in some groups: Kovoor 1977c). Hence the commonality of ampullate, piriform, aciniform, and cylindrical glands in uloborids and araneoids is probably symplesiomorphic. Araneoids therefore differ from uloborids in three gland types: cribellate, flagelliform, and aggregate.

The first difference is easily explained. If the thesis that all true spiders are descended from a cribellate ancestor is correct (Platnick 1977), the presence of cribellate glands in uloborids is symplesiomorphic and their absence in araneoids derived, so that the absence could be considered an additional synapomorphy of the superfamily (although a "loss" character). The difference is consonant with the hypothesis that uloborids are an immediate outgroup to the araneoids.

The flagelliform glands of araneoids are presumed to be the source of the core fibers of the SS line (Sekiguchi 1952; Peters 1955b; Anderson 1970; Work 1981). As Kavanaugh and Tillinghast (1979) point out, Sekiguchi, who discovered the gland and claimed to have demonstrated its function, did not actually prove that the silk from the flagelliform gland formed the core fibers of the araneoid viscid-silk line. Nevertheless, the circumstantial evidence accumulated since Sekiguchi's work is compelling (Witt, Reed, and Peakall 1968; S. O. Anderson 1970), and most workers have taken the association for granted. In her study of uloborid silk glands (in *Uloborus* sp., *Zosis geniculatus*, and *Hyptiotes paradoxus*), Kovoor (1977a, 1978) named a pair of glands similar to the araneoid flagelliform glands the "pseudoflagelliform" glands. She showed that the two types of glands were similar in placement, shape, and staining reactions and in their absence in adult males. On the basis of this evidence for homology she suggested that the pseudoflagelliform glands served a similar function in uloborids, i.e., provided the core fibers of the SS line. To my knowledge, no one to date has confirmed her suggestion, but the morphological evidence for homology seems compelling, and an analogous function may be predicted.

The SS core fibers in uloborid and araneid orbs do differ substantially in their physical properties. Araneid SS core fibers are highly elastic, nonbirefringent elastomers (Work 1981); they can stretch up to six times their resting length. Ampullate-gland silk will stretch only about one and one third its resting length (Denny 1976; Work 1981). In *Steatoda triangulosa* (Theridiidae) the SS lines are not as elastic as the SS lines of araneids (pers. obs.). Uloborid SS lines are also not as elastic as araneid SS lines (Eberhard 1976), although the SS lines of uloborids are nevertheless more elastic than their own NS scaffolding lines (Eberhard 1972b: 437). Eberhard's observations are apparently the first comparison of araneid and uloborid silk properties, and more extensive quantitative comparisons would be valuable. For example, it is unknown whether the differences in orb-weaver SS lines are differences in degree, perhaps due to varying proportions of amino acids, or whether the differences preclude homology of the glands. However, comments by Lucas, Shaw, and Robinson (1960), S. O. Anderson (1970), and especially Denny (1976) suggest that simple quantitative changes in the amino-acid composition of silks can affect their physical properties considerably. On the whole, Kovoor's hypothesis that the uloborid pseudoflagelliform glands and the araneoid flagelliform glands are homologous seems reasonable.

The hypothesis is also corroborated by outgroup comparison. If orb weavers are monophyletic, the sister group of the orb weavers must have been cribellate because uloborids are both orb weavers and cribellate, and the cribellum is a primitive trait of all true spiders. Of course, most non-orb-weaving cribellate spiders use cribellate silk. Kovoor (1977c: 151) states that the core fibers of the SS lines in the Filistatidae, Eresidae, and Amaurobiidae are certainly ampullate silk. Morphologically, uloborids and araneoids are unique in having specialized glands serving the posterior lateral spinnerets.

At least some araneoids are specialized in one other respect. Both Work (1981) and Kavanaugh and Tillinghast (1979) report the presence in araneoid webs of a distinctive nonbirefringent cement at all SS-line and radius junctions ("SS-R" cement). This type of cement, which is found only at these junctions, is universal in both araneoid lineages studied thus far (*Araneus*, *Neoscona*, *Eriophora*, *Argiope*, and *Micrathena*; *Nephila* and *Nephilengys*). Work (1981) comments that the source of the cement is unknown, and that the mechanism by which it operates is wholly mysterious. Its occurrence in the above two lineages suggests that it may be an additional synapomorphy for araneoids, or a subset of araneoids. It appears to be absent in both uloborids (*Uloborus penicillatus*) and dinopids (*Dinopis* sp.) (Opell 1979; Kullmann 1975). If its occurrence is restricted to SS-R junctions, the cement would be a useful diagnostic for establishing homologies between elements in araneoid webs that do not resemble orbs.

The final difference in the silk glands of uloborids and araneoids is the unique presence of aggregate glands in araneoids. In all araneoid species studied thus far there are two pairs of aggregate glands opening on the posterior lateral spinnerets (Kovoor 1977c: 147–48). The hypothesis that uloborids are an immediate outgroup of araneoids would therefore imply that the araneoid aggregate glands are the homologues of some gland type in uloborids, unless that homologue has been lost or transformed in uloborids. Outgroup comparison makes this event unlikely, because araneomorph spiders in general have only cylindrical and aciniform glands serving the posterior lateral spinnerets. Both of these gland types are still present in uloborids. The cylindrical glands are exclusively concerned with cocoon silk; they are usually rudimentary in juveniles and absent in males. Because juvenile araneoids of both sexes use aggregate-gland silk, ontogenetic comparison indicates that cylindrical glands are not the homologues of aggregate glands. Either aggregate glands arose de novo or they are homologues of the aciniform glands.

Orb-weaving spiders usually have dozens or hundreds of aciniform glands per spinneret (Kovoor 1972, 1977a), meaning that with respect to one another the glands can only be homonoms, not homologues (Riedl 1979). (Homonoms, as contrasted with homologues, are standardized morphological substructures that occur in large numbers in an individual, e.g., setae or nephrocytes in spiders.) Homonoms cannot be distinguished individually and thus hypotheses about strict homology (always about morphological singulars) do not apply. However, homology between homonoms as unique sets can be hypothesized. The relative number of aciniform glands in uloborids is thus uninformative. No reason exists to suppose that araneoid aggregate glands arose de novo, and therefore it is simplest to presume that aggregate glands, as a group, are homologous to aciniform glands, as a group. Kovoor (1977a) considered the homology of uloborid aciniform and araneid aggregate glands quite plausible. She suggested that a number of aciniform glands might have coalesced to form the enormous aggregate glands unique to Araneoidea (p. 100).

As Levi (1980b) pointed out, there are few morphological similarities between uloborids and araneoids, and many differences. Most of the differences (Lehtinen 1967, 1975, 1978) are probably araneoid autapomorphies. However, the work of Wilson (1962a, 1962b, 1969) documents another potential synapomorphy for uloborids and araneoids. These two groups possess a particular kind of valve in the anterior lateral spinnerets that controls the spinning of the ampullate-gland dragline and the scaffolding of orb webs. In addition, they have a unique muscle, the duct-levator muscle. Wilson (1969) stated that, among 12 families and 18 genera studied, these two characters occurred only in *Uloborus*, *Steatoda*, *Theridion*, *Meta*, *Zygiella*, *Tetragnatha*, *Pachygnatha*, and *Araneus* (representing four orb-weaver lineages). Wilson (1969) also examined *Linyphia* and found neither the valve nor the muscle, but he offered no details concerning the spinneret morphology of linyphiids. Despite this negative evidence, it can still be predicted that the features will characterize all araneoid higher taxa. Wilson (1969: 107) stated that the control valves and muscles showed "differences of detail which prevent one from assuming that they are strictly homologous," but he did not explain his concept of homology, and he was interpreting the morphology under the explicit assumption that, regardless of his evidence, the two groups were convergent. Significantly, the control valve and muscle were absent in *Amaurobius ferox* and *Amaurobius terrestris*, cribellate species in a family (Amaurobiidae) that is a potential sister taxon to the orb weavers. *Amaurobius* also lack specialized glands on the posterior lateral spinnerets (Kovoor 1977c). Comparison with one potential outgroup to the orb weavers therefore supports this feature as a synapomorphy of orb weavers.

In summary, the evidence of the specialized glands on the posterior lateral spinnerets (either pseudoflagelliform or flagelliform), the ampullate-gland control valve, and the duct-levator muscle supports the monophyly of the orb weavers and none of the differences between uloborids and araneoids refutes it. Of course few taxa have been studied. The monophyletic hypothesis would indeed be refuted if, for example, synapomorphies that precluded the relationship with araneoids were found linking uloborids with other non-orb-weaving taxa. However, no such evidence has been found, and in fact the monophyly of orb weavers seems to be the only hypothesis supported by any substantial evidence at all.

Web-Building Behavior

Even though the comparison of orb-weaver silk glands and spinneret morphology suggests that orb weavers are monophyletic, the monophyly could be refuted if a detailed comparison of web-building behavior in uloborid and araneoid orb weavers showed incompatible differences between them. The assertion that the orb webs of the two groups are convergently evolved usually has rested on arguments about the adaptive value of web form. One would not necessarily expect details of web construction to be similar. In fact, if many possible ways to construct and attach silk lines exist, the probability of identical or closely similar behaviors should be small in any one instance, and minute for several independent instances combined. Of course, one could also claim that the details of behaviors are themselves adaptive, and that any similarity in them is also due to convergence under selective pressures. But is there any similarity for which this claim cannot be made? Clearly the only imaginable limit to that argument is imagination itself. There is no way to defeat the position. The best that one can do is to show that similarities are improbable by chance alone, and that no or very little evidence contradicts the simpler hypothesis of similarity due to common ancestry. Common ancestry, it must be admitted, is a simpler explanation than convergent evolution in otherwise equivalent cases. How then do uloborids and araneoids construct orb webs?

Web construction can be represented conveniently by a flowchart. The use of flowcharts to represent stereotyped sequences of behavior follows Robinson and Olazarri (1971). The overall sequences of web construction in uloborids and araneoids are compared in Fig. 12.25. The uloborid sequence is summarized from Eberhard's (1972b) study of *Uloborus diversus* and the araneoid sequence from Peters (1937), Tilquin (1942), and Savory (1952). Each sequence has been verified by personal observations. The overall sequences are obviously similar, but some of the similarity is probably due to what Seilacher (1973) has called "fabricational constraints," i.e., constraints imposed by the properties of the materials used, or by physical laws (such as geometry, gravity, mechanics, or temporal sequence). Fabricational constraints produce structural similarities due neither to heredity nor to natural selection—they are artifacts of limited possibilities. For example, because silk lines can support no bending or compressive force, constructions made of them must be in tension, and that limits the possible geometries of simple structures with few peripheral anchor points (to, e.g., planes, catenaries, or hyperbolic paraboloids, composed of straight-line segments). It also means that if all lines are to be taut in webs, only certain combinations of angles at the nodes of the web are possible. The basic sequence primary radius construction–NS construction–SS spiral construction is also inevitable. These similarities might still be due to common ancestry, but the possibility of independent origin is hard to exclude because no reasonable alternatives exist. Just as common ancestry is a simpler explanation than selection, fabricational constraint is a simpler explanation than common ancestry.

A more detailed look at construction behaviors, however, shows that they are more often similar than different, and that the differences do not refute the hypothesis that araneoids are derived from a cribellate orb-weaving ancestor. Primary radius construction (including exploration behavior) has not

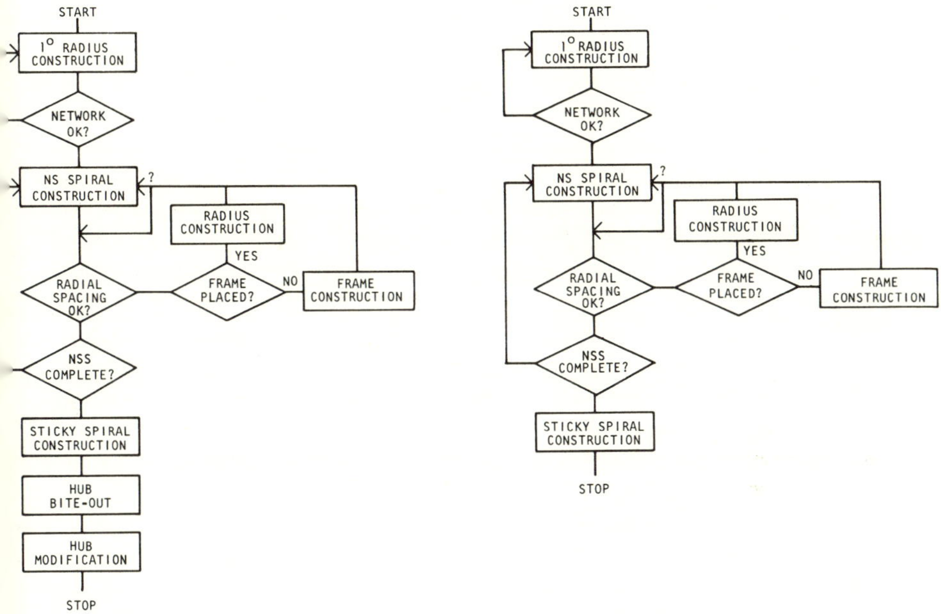

Fig. 12.25. General sequences of web construction of araneoids except nephilines (*left*) and of uloborids and nephilines (*right*). The araneoid sequence is summarized from Peters 1937, Tilquin 1942, and Savory 1952, the uloborid sequence from Eberhard 1972b. The inclusion of nephilines with uloborids is based on my own observations of *Nephila clavipes* and data from Eberhard 1982. NSS = nonsticky spiral.

been studied in detail in any group, but the process of frame construction in both groups is strikingly similar. *Uloborus diversus* and *Araneus diadematus* both construct a radius each time they construct a frame line. Tilquin (1942) has termed these "secondary" radii. By the time all frame lines are complete, a more or less equal number of secondary radii are in place, thus defining a "protohub," and construction of the "tertiary" radii continues. (Often "secondary" frames are added in the corners of the web by both uloborids and araneoids.) Uloborids use all the movement patterns seen in araneoids, but they also intercalate an additional movement pattern not seen in araneoids (Eberhard 1972b: 425): they cut the second half of the frame loose and move out farther on the first half before reattaching the second half and making the final secondary radius. Araneoids do not move the second half of the frame line, but make the secondary radius directly (Peters 1937; pers. obs.). In effect, uloborids add a behavior not seen in araneoids. Because all movement patterns in araneoid frame behavior occur in uloborid frame behavior, the additional movements of uloborids are probably autapomorphies for the group (Nelson 1978).

But are there other conceivable ways to make frame lines? Yes, because several perfectly feasible alternatives are given in descriptions of the process that are in fact incorrect although perfectly plausible (e.g., McCook 1889; Comstock 1912; Hingston 1920; Levi and Levi 1968; Dugdale 1969). I have watched frame construction in *Philoponella*, *Nephila*, *Chrysometa*, *Leucauge*, *Tetragnatha*, *Azilia*, *Argiope*, *Eriophora*, *Micrathena*, *Gasteracantha*, *Neoscona*, *Cyclosa*, *Salassina*, *Theridiosoma*, *Epeirotypus*, *Anapis*, *Anapisona*, and *Maymena*. The Uloborid genus *Philoponella* behaved essentially as described for *Uloborus* by Eberhard (1972b), and the araneoid genera as described by Peters (1937), Savory (1952), and Foelix (1982). With the difference noted above, frame lines and secondary radii are made the same way by both uloborids and araneoids.

Both araneoids and uloborids construct frames and radii as a subroutine within hub construction (Fig. 12.25) (Hingston 1920; Peters 1937; Tilquin 1942; Eberhard 1972b). *Uloborus diversus* completes the primary radii (and therefore the protohub), and then begins laying hub loops. Hub construction is interrupted almost immediately to make frame lines (and therefore secondary radii) and/or tertiary radii, then resumed, and the cycle repeated. This "nested" behavioral sequence continues until all radii are constructed, whereupon hub loop construction passes smoothly into temporary spiral construction. Eberhard (1972b: 429) stated that

> There was no abrupt transition between radius and hub construction and temporary spiral construction. The spider continued making hub attachments after laying the last radius, and after circling the hub one or more times, it began to move away from the hub as it continued its progress in a circular direction attaching its line to each radius it crossed.

Some araneoids, such as *Neoscona*, *Eriophora*, *Tetragnatha*, *Azilia*, *Theridiosoma*, *Chrysometa*, and *Epeirotypus*, space the loops of the temporary spiral abruptly wider than those of the hub loops (pers. obs.), but the patterns of leg movements appear indistinguishable from those of uloborids. Other araneoids, such as *Nephilengys*, *Herennia*, *Nephila*, *Salassina*, *Micrathena*, *Gasteracantha*, *Cyclosa*, *Zygiella*, and *Metepeira* make the transition between hub loops and temporary spiral gradually, but again using the same leg movements. In both groups, therefore, there is no qualitative distinction between hub loops and the temporary spiral, and the movement patterns involved appear identical. The hub loops and temporary spiral together can thus simply be called the nonsticky spiral. A similar point was made by Tilquin (1942). In most araneoids, frame and radius construction is completed earlier in nonsticky spiral construction than it is in uloborids, so that it often appears to occur before the latter behavior.

Another obvious, but unnecessary, similarity between uloborids and araneoids concerns the direction in which the temporary and sticky spirals are laid. In both groups the temporary spiral, being part of the nonsticky spiral, is laid from the hub outward, and the sticky spiral from the periphery inward. As Eberhard (1972b), Levi (1980a), and other authors have noted, this particular sequence seems to be arbitrary. However, only two possibilities exist (another example of fabricational constraints), and therefore the odds are even that two groups would coincide in sequence by chance alone.

Uloborids also differ from araneoids in making "doubled" radii and in biting out the protohub formed by the primary radii before commencing hub construction (Eberhard 1982, character F4; 1972b: 426–27). To make a doubled radius, the uloborid attaches its line to the hub, exits along a preexisting radius, attaches the trailing line to a point on the frame, and returns to the hub, thus forming a doubled radius. (Because of the bilateral symmetry of the spinnerets, all single lines in a web consist of two strands, but these strands can be seen only with a microscope, and when ethologists speak of a "doubled" line they generally mean that the animal has spun two lines, each made of two strands.) Araneoids follow the same sequence, but when returning to the hub they cut and reel in with the first legs the line laid on the way out, while laying a new single line behind them with the fourth legs. This line becomes the definitive radius (Eberhard 1982, character F1). Because uloborids "cut and reel" during primary and secondary radius construction, the absence of cutting and reeling during tertiary radius construction is probably another autapomorphy for the group. (In all araneoids except nephilines, radii are always cut and reeled.)

Other, apparently derived features of most araneoids are biting out the center of the hub after completion of the sticky spiral ("hub bite-out" in Fig. 12.25) and either adding lines to the hub or leaving it open ("hub modification"). (For a description of these behaviors see Eberhard 1981a: 195.) Neither uloborids nor nephilines modify their hubs after they spin the SS line (Eberhard 1982, character G1).

If uloborids are an immediate outgroup of araneoids, one might expect these behavioral traits (i.e., hub loops as part of NS spiral construction, frame and radius construction persisting throughout NS spiral construction, and lack of hub bite-out and hub modification) to persist in some araneoid group. In fact, all these traits occur in nephilines, a group also considered primitive with respect to other araneoids on morphological grounds (Levi 1983). Eberhard (1982, character F2) states that, like uloborids, *Nephila*, *Nephilengys*, and *Herennia* do not cut and reel radii. *Nephila clavipes* does not cut and reel any lines at all in its web (pers. obs.), but this difference is probably an autapomorphy of the nephilines (if it also occurs in other nephiline taxa). There is also no clear distinction between hub loop construction and temporary spiral construction in these genera. Nephilines are unique in attaching the second line of the doubled radius to a point slightly displaced from the attachment of the first line on the frame. As Eberhard (1982) points out, this peculiarity is apparently another autapomorphy of the group. The basic pattern of frame construction and radius construction, and their occurrence during NS spiral construction, can be taken as synapomorphies of all orb weavers, and as primitive behavioral traits for the araneoids. Certainly such similarities cannot be ascribed to fabricational constraints.

Another similarity between uloborids and araneids is the use of a lateral tap of outside leg I to locate the innermost loop of the SS line (Eberhard 1982, character A2). Nephilines also use a lateral tap, but of the second leg rather than the first. Thus all three taxa use a lateral tap of an outside leg, in contrast to other araneoid orb weavers, so the trait is presumably primitive (although the use of oL_2 in *Nephila* may be autapomorphic). Nephilines also use a downward slide of the fourth leg holding on to the radius to detect the innermost SS loop as the SS line is attached (Eberhard 1982, character A3). This trait also appears to be an autapomorphy of the nephiline genera.

All araneoids use the outside third and fourth legs to brace the radius when the SS line is attached; they also use the inside fourth leg to pull or push the SS line from the spinnerets as SS attachment occurs (Eberhard 1982, characters B1 and C1, respectively). Uloborids also use the oL_3L_4 grip to brace the radius (*Philoponella*: pers. obs.), but they use the inside fourth leg as well to brace the radius; they do not manipulate the SS line at all (Eberhard 1982, characters B2 and C2, respectively). However, *Uloborus diversus* uses an oL_3L_4 grip on radii to attach the NS spiral while the iL_4 holds the NS spiral line (Eberhard 1972b: 429; pers. obs. of *Philoponella*). The identical behavior by araneoids in a different context (the handling of SS lines) is quite plausible if araneoids and uloborids are sister groups. The use of the L_3L_4 NS-line grip appears to be a synapomorphy for orb weavers, and the iL_4 push of the SS line during SS construction a synapomorphy for Araneoidea. The addition of the iL_4 leg to the oL_3L_4 grip during SS construction in uloborids is apparently another autapomorphy.

Finally, uloborids and araneoids differ markedly in how they produce SS lines. In araneoids they are formed almost as a by-product of movement. The juxtaposition of aggregate and flagelliform spigots on the mesal surface of the posterior lateral spinnerets apparently allows the formation of viscid-silk lines without the intervention of legs. All cribellates have to brace both fourth legs against each other to comb cribellate silk from the cribellum onto the core fibers. But this difference in behavior is clearly due to anatomy, a fabricational constraint imposed by the nature of cribellate silk. The lack of SS-combing behavior might be claimed as a further synapomorphy of the araneoids, but the character is so obviously correlated with anatomy that it is not independent evidence apart from aggregate glands of the monophyly of Araneoidea. Araneoids often do perform "L_4 pulls" (Eberhard 1981a) when combing sticky silk that are reminiscent of the combing of uloborids, but the resemblance seems too superficial to claim homology.

One real difficulty is the nearly total lack of information on spider groups that might be sister groups of the orb weavers, such as dictynids, tengellids, amaurobiids, eresids, or filistatids. Should such a behavior as a L_3L_4 NS-line grip be found in those groups, the character would become a symplesiomorphy for the orb weavers. (However, Eberhard (1982) did not find it in cribellate-silk production in *Filistata*, *Tengella*, or *Mallos*.) It may be invalid in a specific case, and it certainly is invalid in general, to assume that because a web does not look like an orb it is not derived from one or is not a primitive precursor of one. Orb-weaver monophyly is not finally corroborated, however, until those behaviors basic to orb weavers (e.g., L_3L_4 NS-line grip, frame behavior, radius behavior, oL_1 lateral tap to locate lines, NS spiral construction, and frame and radius behavior as a subroutine in NS spiral construction) are known to be absent in the potential outgroups.

In summary, comparison of the construction behavior of uloborids and araneoids indicates that many similarities could be synapomorphies insofar as most of the similarities are not due to fabricational constraints. The simplest interpretation of the differences between the two groups is usually as autapomorphies of uloborids or as derivations of behaviors present in uloborids, and therefore as synapomorphies for the Araneoidea. The available behavioral evidence supports the monophyly of all orb weavers and the sister relationship of the uloborids (and/or the dinopids) to the superfamily Araneoidea. Eberhard's work (1972b, 1981a, 1982) has shown that construction behavior provides many diverse behavioral details that corroborate the monophyletic hypothesis.

Web Form

The final comparison is of web form, perhaps the feature of orb webs that has most convinced workers that the orb-web architecture is highly evolved and highly adaptive, and therefore likely to be the result of convergent evolution. It was the similarity of uloborid and araneoid orb-web architecture that probably provoked the hypothesis of convergent evolution of the orb web (Figs. 12.1, 12.2). After all, if the webs did not look similar the whole controversy would not have arisen. For Thorell (1886), and those who agreed with him, that similarity was the troubling flaw in Bertkau's argument for the monophyly of the cribellates (see above). Orb webs have always inspired admiration for their design. In view of Bertkau's evidence that they were convergently evolved, that admiration led to an easy rationale for the convergence—the great adaptive value of the orb architecture explained the troubling similarity. But is that reputed adaptive value even relevant to the issue of convergence?

Seilacher (1973) suggests that similarity of form has three basic causes: fabricational constraint (discussed below), common ancestry (advocated here), or function. "Function" includes convergence, because to invoke convergent evolution as an explanation of form one has to postulate that a particular solution to a problem in natural design has such adaptive value that natural selection renders convergence on that solution likely. But that solution need not be the best by independent design criteria to claim convergence. It only need be good enough so that within the smaller set of "good enough" designs it is likely that two similarly endowed groups might happen on the same solution. If the hypothesis of convergence does not claim optimality, then the only way it can be conclusively disproved is by demonstrating that orb webs have no adaptive value at all—the negation of optimality. That position is contradicted by the existence of spiders relying on orb webs to catch prey. Even though many authors have speculated that orbs are superbly designed to do what they do (see, e.g., Witt 1965, 1975, Burgess and Witt 1976, Eberhard 1981a, and works cited therein) few have claimed that orb webs are the best imaginable design. Indeed, Witt (1965) argued that although they work well, they are not optimal in any one sense, but rather are a compromise between many conflicting requirements. It seems clear that all that can be demonstrated experimentally is that orb webs are functional ways to do whatever it is the observer imagines they do.

Could convergence still be used as an explanation if orbs could be shown to be better than any existing web architecture for specific purposes, such as the entrapment of flying insects in their flyways, or the rapid transmission and localization of vibrations, or the efficient use of materials in relation to area covered? One could still not claim to have proved convergence, because any number of equally good and equally likely web architectures might exist if only the observer could imagine them, build models, and demonstrate their adaptive equivalence to orb webs. For example, the web of *Synotaxus* (Fig. 12.4) is also a nearly invisible vertical snare suspended from a few points, and it is effective at trapping flying insects (pers. obs.). Yet it is built on a modular, rectilinear plan (Eberhard 1977b). The webs of *Scoloderus* spp. (Eberhard 1975; Stowe 1978) and the metine genus in New Guinea (Robinson and Robinson 1972) are architecturally (but not in mode of construction) more similar to the web of *Synotaxus* than to the webs of other orb weavers. Numerous other orb weavers have also modified the basic architectural style of an orb web; alternative designs can and do exist (Stowe, this volume). It may also seem likely that signal transmission would be less efficient in a sheet web than in an orb web (Witt 1965). However, the metatarsal lyriform organs of web spiders are extremely sensitive vibration receptors (Walcott and Van der Kloot 1959; Liesenfeld 1961; Barth 1972). The critical variable is the total travel time of the spider to the capture site; the ease of moving on a dry sheet may compensate for the presumed attenuation of the signal. Thus, in the absence of the extreme claim of optimality (or its opposite), a claim of convergence based on adaptive value again depends on unspecifiable probabilities and unmeasurable approaches to adaptive perfection. To claim convergence, therefore, it is not enough to show that orb webs are good at what they do, because one cannot show that they are perfect, nor can one show that they are perfectly not perfect, and the middle ground is the purview of fruitless contention. Indeed, their comparison to perfection is itself limited by the observer's ability to imagine perfection. Therefore orb webs cannot be demonstrated to be convergent (or homologous) by arguments involving their adaptive value alone.

A claim of convergence can be justified only by evidence showing that one group of orb weavers is more closely related to non-orb weavers than to other orb weavers. It is never great adaptive value that substantiates convergence (although it may make it more plausible); it is always and only evidence that shows that two taxa similar in some feature cannot be so closely related that the similarity is due to the simpler explanations of fabricational constraint or common ancestry. Such evidence is what Bertkau (1882) and the workers who agreed with him thought they had in the fact of the cribellum. With the demonstration that Bertkau's Cribellatae are paraphyletic (Lehtinen 1967, Platnick 1977), the reason for believing that orb webs are convergent evaporates, because there is no other evidence based on shared derived characters that uloborids are more closely related to any other group except araneoids. Evidence of relatedness that refutes the interpretation of similarity as homology is a prerequisite for a claim of convergence, simply because convergence presupposes some genealogical hypothesis. Thus uloborid and araneoid orb webs were not thought to be convergent because they were adaptive; if anything they were thought to be adaptive because they were convergent. Convergence therefore is an a posteriori explanation of taxonomic discordance. If there is no discordance the hypothesis of common ancestry is always simpler.

But what of fabricational constraints? If, in the absence of discordance, homology is a preferable alternative to convergence, then fabricational constraint is preferable to homology for the reasons outlined earlier. Uloborid and araneoid orbs share the following distinctive architectural features: (1) a two-dimensional frame connected to the substrate at relatively few points, (2) a pattern of nonsticky radial lines converging on a point, (3) a NS spiral, later mostly removed, and (4) a permanent spiral of sticky silk. By "sticky silk" I simply mean silk to which things stick, without specifying the mechanism of adhesion (i.e., true viscid silk or cribellate silk). All of these features are potential synapomorphies of the taxa. None appears to be the consequence of fabricational constraints. For example, coincidence of a hub in the web and a two-dimensional frame bounding the lines exiting from the central point is only one architectural style, as demonstrated by the existence of diverse other designs; compare, e.g., the webs of *Eidmanella* (Fig. 12.6), *Synotaxus* (Fig. 12.4), *Ogulnius* (Fig. 12.12), *Mysmena* (Fig. 12.19), *Achaeranea* (Fig. 12.5), and *Wendilgarda* (Fig. 12.11). Point symmetry bounded by a two-dimensional frame clearly is not the only architecture possible. (That all these webs are probably derived from orb webs does not affect the argument that the "orb" architecture is not inevitable—on the contrary, the webs of *Ogulnius*, *Mysmena*, and *Wendilgarda* are certainly derived from orbs, thus showing that the orb is not an inescapable adaptive optimum.)

As mentioned previously, hub-reinforcing loops are nearly indistinguishable behaviorally from the temporary spiral. In orb weavers that spin an extensive temporary spiral such as araneids, nephilines, and uloborids, it is obviously of some use, but many orb weavers spin quite regular sticky lines (not necessarily spirals) without the help of a temporary spiral, e.g., *Epeirotypus* (Figs. 12.7, 12.8), *Theridiosoma* (Figs. 12.9, 12.10), *Anapis* sp. (Fig. 12.14), *Anapis heredia* (Fig. 12.16), *Poecilopachys* (Clyne 1973), *Cyrtarachne* (Stowe, this volume), and *Pasilobus* (Robinson and Robinson 1975). Thus a temporary spiral is not necessary for sticky spiral construction. The simplest interpretation seems to be that its occurrence in both uloborids and araneoids is a synapomorphy rather than a feature either so adaptive that its identical form in the two taxa is convergent or so necessary that its existence is inevitable.

The occurrence of a sticky spiral in both groups also does not seem to be an inevitable consequence of the design of the NS network. The webs of both *Synotaxus* and *Achaearanea* combine sticky and nonsticky silk without anything like a spiral. The recent discovery of an orblike web in the psechrid genus *Fecenia* (Robinson and Lubin 1979b) and Szlep's (1966) earlier work on *Titanoeca* show that when vaguely spirallike designs are present in other taxa, they are still quite different in form and mode of construction from those of true orb weavers. It seems simplest to accept the spirals spun by uloborids and araneoids as homologues.

The similarities in web form thus do not seem to pose any real difficulties for the monophyletic hypothesis for the origin of the orb web. The real difficulties, at least as they were articulated by Kullmann (1972a), have been the nature of the transition between a cribellate and an ecribellate orb. Kullmann listed four serious problems (p. 401):

(1) It is highly improbable that, while preserving the same method of capture, the highly developed and extremely efficient dry cribellate snare is merely exchanged for the ecribellate viscid capture-thread. Simultaneously, the behavior of weaving cribel-

late threads plus the typical cribellate organs must have been reduced while being replaced by new spinning-organs plus any necessary motor pattern.

(2) Since orb-webs without any capture threads are totally useless, in the time of transition either spiders which simultaneously produced cribellate and ecribellate capture-threads must have existed, or the change from one system to the other happened suddenly. Both ideas are fantastic.

(3) One absurdity more—the Uloboridae are the only spiders without poison glands. Since they are not primitive, they must have descended from other spiders; these organs obviously have been reduced. After becoming ecribellate, they must have regained them, as all Araneidae are poisonous.

(4) The typically structured hairs covering the body differ in both families. The Uloboridae have hairs like all cribellates—they are plumose. Those of Araneids are smooth or more or less serrated only; they are identical with those of other ecribellate families, especially theridiids and linyphiids. There is no doubt that these families are related, but it is difficult to imagine that cob-webs of theridiids and sheet-webs of linyphiids developed via orb-webs of Araneids from those of Uloborids.

Kullmann's first two points dwell on the fantastic nature of the necessary transition. I can at least suggest why the transition is less than fantastic, although the reasons are as hypothetical as Kullmann's objections. From the point of view of selection or adaptive value, the length of time required to build the scaffolding for either an uloborid or an araneid orb is about the same for similar-sized webs—about 20 minutes on the average. However, an uloborid requires far longer to construct a cribellate sticky spiral (1–2 hours?) than an araneid requires to construct an equivalent viscid spiral (as little as 10–20 minutes). The extra time is consumed in the tedious process of combing out the cribellate silk; in araneids, viscid silk is produced as a by-product of the spider's movement. Thus araneids (and araneoids) spend less time exposed to predators and parasites, while performing an activity that probably renders the spider less alert than when it is at the hub or in a retreat. The convergent occurrence of barrier webs, signal lines, and retreats among many orb weavers attests to the selection pressure exerted by predators and parasites.

Second, although both uloborid and araneoid orb weavers move from one radius to the next, uloborids go quite slowly, all the while frenetically combing out cribellate silk. The araneid moves quickly, with no such frenzied movements, presumably a metabolic economy.

Third, Tillinghast, Kavanaugh, and Kolbjornsen (1981) have shown that the viscid silk of *Argiope trifasciata* (Araneidae) contains substantial quantities of carbohydrates—much more than is contained in ampullate silk. Kovoor (1977b) showed that the aggregate silk of *Latrodectus* (Theridiidae) also contains mucosubstances. If it is presumed that the carbohydrate content of all types of uloborid silk is comparable to that of araneid nonsticky silk because they are all fibrous, the dramatic increase in the carbohydrate content of viscid silk (by length the largest component of any orb web) may represent a considerable economy in the total protein investment in the web.

Peakall and Witt (1976) argued that because *Araneus diadematus* recycles its web there is little or no "protein" cost associated with an orb web, but that argument may be invalid. Webs in nature are often damaged to the extent that the spider loses some or all of the silk. Therefore, while the silk is in a web, the spider is at risk for the material. On the other hand, spiders eat a high-protein diet (Foelix 1982); perhaps carbohydrates are for them more "costly" than proteins. Without data on that variable and on the amounts of protein in webs, we can neither deny nor confirm that web loss may be a substantial risk for orb weavers.

Fourth, authors have remarked on the persistent, almost paradoxical stickiness of cribellate silk (Foelix and Jung 1978), and also implied that cribellate silk might be more effective in retaining prey than viscid silk (Opell 1982a), but its superiority has not been shown in field studies. Eberhard (1980b) noted that the stickiness of cribellate silk in nature was often substantially reduced within a few days or less owing to rain or dust accumulation. Finally, the idea of an ancestral cribellate spider producing slightly sticky viscid silk from aciniform glands exiting on the posterior lateral spinnerets while still producing cribellate silk is not unimaginable.

The transition from a cribellate- to a viscid-silk orb is also not as improbable as Kullmann imagines. Second-instar uloborid spiderlings lack functional cribella and spin dry-silk orbs (Szlep 1961). The spiderlings catch some prey, and this fact confounds Kullmann's second point; it also suggests that any second-instar spiderling that could produce sticky silk might profit. Szlep (1961) states that the supernumerary radii characteristic of juvenile uloborid webs are made from a different silk, finer and softer than radial silk. Although she did not identify it, the logical guess is that it is aciniform silk. As noted, Kovoor (1977c) believed that aciniform silk may have been the precursor of aggregate silk, so that, if required, a plausible scenario exists for the origin of the viscid-silk orb web.

We know rather little about silk and its functions, and what we know is confined to only a few groups—there is virtually no evidence to settle these questions one way or the other. The points that I have made are merely speculations to render less objectionable implications that Kullmann felt to be fantastic. Certainly the monophyly of orb weavers, based on some evidence, should not be rejected because of preconceptions about hypothetical adaptive situations.

Kullmann's third and fourth points do not really concern the implausibility of the homology between uloborid and araneoid orb webs. The third point stems from the presumption that modern uloborids would have to have been ancestors of araneoids; all that the monophyletic hypothesis requires is that they share a common orb-weaving ancestor with poison glands. The fourth point merely states that serrate hairs are a derived feature of Araneoidea, an interpretation that does not conflict with the relationship between uloborids and araneoids suggested here. The idea of theridiids and linyphiids as orb weavers is unsettling, but when the orbs of other araneoid families are considered, the idea may become less objectionable.

In sum, the four similarities in web form given earlier are unique to orb weavers. There is no compelling evidence against them as synapomorphies. The evidence from behavior and silk glands is consonant with that form, and together they imply (1) that the orb web arose only once, (2) that orb weavers are a monophyletic group, (3) that uloborids (and/or dinopids) are the sister taxon of araneoids, and (4) that the orb web is primitive for all araneoids.

The total number of species descended from orb-weaving ancestors therefore approaches one-third of all described spider species (as estimated from Levi 1982). Far from confirming that the orb web is the adaptive pinnacle of spider-web architecture, achieved once or twice during the evolution of the order, the evidence suggests that the orb web is an ancient feature, discarded by many derived groups such as Linyphiidae, Nesticidae, and Theridiidae. The araneoid orb did not emerge from the chaos of the theridiid cobweb, as most previous authors have claimed (Pocock 1895; Comstock 1912; Kaston 1964; Kullmann 1972a)—it was the other way around.

Web Architecture Within the Araneoidea

I shall now briefly discuss the web architecture of the araneoids that still spin orb webs. For taxa such as linyphiids and theridiids no comparable information exists. The preceding discussion orients the transformation series in web architecture within the Araneoidea. The primitive web architecture, and how it was constructed, can be deduced. This information permits a reasonable first hypothesis for phylogeny within the araneoid superfamily.

In the preceding discussion it was argued that the primitive web form of the Araneoidea was a two-dimensional orb web, i.e., a point-symmetrical array of radial lines bearing a viscid-silk spiral, the whole bounded by a two-dimensional frame attached to the substrate at relatively few points. The construction of each frame line involved the simultaneous construction of a radial line, and those secondary radii met at a protohub that was reinforced by the construction of a nonsticky spiral outward from the protohub. Nonsticky spiral construction was interrupted frequently to construct more frame and radial lines. Radius construction ceased early in the spinning of the nonsticky spiral and, after completion of the nonsticky spiral, construction of the sticky spiral began. This was an uninterrupted sequence consisting of stereotyped movements, probably involving a sideways tap of the outer first leg to locate the innermost loop of sticky spiral. In addition, the outside third and fourth legs were used to grip the radius during SS attachment, and the inside fourth leg pulled more sticky silk from the spinnerets, just before the spinnerets attached the SS line to the radius. Completion of the sticky spiral was the last step in web construction, and the center of the hub was neither removed nor modified.

This description of the plesiomorphic web architecture and the method used to construct it is a generalization based on the distribution of architectural and behavioral characters among the orb weavers. Nearly all the available information on behavior concerns the uloborids, nephilines, and araneids. As the information presented below indicates, rather little is known about how the rest of the araneoid orb weavers construct their webs.

Metine-Tetragnathines

As Eberhard's (1982) data show, the building behavior of metines and tetragnathines is nearly identical to that of araneids. They differ primarily in using an iL_1 forward tap to locate the inner SS loop, as in theridiosomatids, symphytognathids, anapids, and mysmenids (Eberhard 1982, character A1).

Theridiosomatidae

Despite the small size of the family, theridiosomatid webs are astonishingly diverse. Only the webs of *Epeirotypus* (Fig. 12.7) and *Theridiosoma* (Fig. 12.9) resemble orbs; the webs of *Wendilgarda* (Fig. 12.11; see also Coddington and Valerio 1980) and *Ogulnius* (Fig. 12.12) are very different. Like all araneoids except the nephilines, *Epeirotypus* and *Theridiosoma* bite out the center of the hub after sticky spiral construction, but hub modification in these genera is much more extensive and results in a completely rebuilt hub (Eberhard 1981a; pers. obs.). Both *Epeirotypus* and *Theridiosoma* appear to have "nonsticky circles" instead of a nonsticky spiral (Figs. 12.8, 12.10). The building behavior of *Wendilgarda* and *Ogulnius* differs substantially from the araneoid sequence in Fig. 12.25 but is consonant with it (pers. obs).

Eberhard (1981a, 1982) has shown that theridiosomatids are similar to anapids in a number of features. During sticky spiral construction, *Epeirotypus* and *Theridiosoma* locate the innermost loop of the SS line with an iL_1 forward tap, their body axis parallel to the radius (Eberhard 1982, character A1). Eberhard (1981a) also observed that *Epeirotypus*, *Theridiosoma*, *Ogulnius*, and two anapid species hold the SS line away from the radius with their fourth legs as they move between SS attachments, presumably to prevent it from adhering to the radius. He did not mention if symphytognathids or mysmenids did the same.

A third similarity is the apparent anastomosis of radii before they reach the hub (Eberhard 1981a). In theridiosomatids, true radial anastomosis (i.e., of structural radii after sticky spiral construction) occurs only in the genus *Theridiosoma.* (As presently defined, the genus *Theridiosoma* is paraphyletic, and will have to be relimited.) Radial anastomosis in symphytognathids, anapids, and mysmenids has not been shown to involve other than the accessory radii added after sticky spiral construction. No *Epeirotypus* species joins its radii either before sticky spiral construction or after. The presence of the plesiomorphic condition in *Epeirotypus* (lack of radial anastomosis), combined with the incontestable monophyly of theridiosomatids, suggests that radial anastomosis in *Theridiosoma* must have arisen independently in that group.

One similarity unites theridiosomatids exclusively with symphytognathids, anapids, and mysmenids. All of these taxa add hub loops after the hub bite-out as part of hub modification (Fig. 12.25). Other araneoids may add lines to fill in the open center of the hub, but they do not add them to the periphery of the hub, nor do they add them in a stereotyped sequence reminiscent of nonsticky spiral construction.

Symphytognathidae

Symphytognathids spin strictly two-dimensional orbs. Few people have seen them build, but as Eberhard (1981a) points out, some details of construction behavior can be inferred from the appearance of the web. Symphytognathid webs are peculiar in two respects. First, after sticky spiral construction, the spider returns to the hub and constructs many accessory radial lines that are not specifically attached to the SS line (Fig. 12.23; see also Forster and Platnick 1977, Fig. 1). Figure 12.24 is a photograph of the web of a *Patu* sp. immature and shows the smaller number of structural radii. Accessory radii are also characteristic of second-instar uloborid webs (Szlep 1961), but Szlep stated that the accessory radii of uloborid webs were not made of the same silk as the structural radii. A similar comparison of the accessory and structural radii of symphytognathids has not been made. The use of a silk type other than ampullate-gland silk in a radial position in orb webs would be unique, but until Szlep's assertion is corroborated and the situation in symphytognathid webs evaluated, any suggestion that symphytognathids might be allied to uloborids by this character would be premature.

Second, like the webs of anapids and *Maymena* (Mysmenidae), symphytognathid webs exhibit "radial anastomosis." Eberhard (1982) states that in anapids this extensive modification of the hub and radii occurs after the SS line is complete, but he did not specifically say whether the structural radii were joined to each other (as in *Theridiosoma*; see above) or whether only the accessory radii were joined to the structural radii. My own observations of anapids and *Maymena* support the latter alternative, so that the character is more appropriately termed "accessory radius anastomosis." As a result of this process, fewer radial lines meet at the hub, and the tension in all radial lines is reduced (Eberhard 1981a: 194–95). Radial anastomosis of either type is unknown as a general feature in araneids, tetragnathines, nephilines, and uloborids. Eberhard (1981a) has seen radial lengthening in a few species in these groups (in two tetragnathine species and one *Gasteracantha* species), but this occurrence is by no means general and does not result in radial anastomosis.

Anapidae

All known webs of anapid species resemble one another quite closely, e.g., the webs of *Anapis heredia* (Fig. 12.16), *Anapis* sp. (Fig. 12.14), *Anapisona felida* (Platnick and Shadab 1978a, Fig. 1—printed upside down), *Anapisona simoni* (Fig. 12.15), and *Chasmocephalon shantzi* (Fig. 12.17). The webs share the usual two-dimensional frame and point symmetry, but during radius construction some lines are laid out of the plane of the orb and attached directly to the substrate above the plane of the web. (Note the straight, roughly vertical lines in Figs. 12.14–12.17, and see also Fig. 12.18, which shows a *C. shantzi* web before sticky spiral construction.) When the spider subsequently encounters one of these out-of-plane radii during sticky spiral construction, it may either skip it or treat it like a planar radius. The result is sporadic catenary SS lines connecting the out-of-plane radii with the rest of the web (Figs. 12.14–12.18).

Eberhard (1981a: 194) states that the anapid species that he has observed lengthen their radii in the same way as symphytognathids do. Anapids also lack extensive nonsticky spirals (Eberhard 1982, character H3) and place their egg sacs just above the hub (Figs. 12.14, 12.15). In addition, at least *C. shantzi* (pers. obs.) uses an iL_1 forward tap to locate the inner loop of the SS line during SS construction (Eberhard 1982, character A1). Like symphytognathids, anapids construct accessory radii after SS construction (Figs. 12.15–12.17).

Mysmenidae

At present the mysmenids include two genera that are known to have very dissimilar web architectures, *Maymena* and *Mysmena*. Thus far, known *Mysmenopsis* species do not spin webs (Platnick and Shadab 1978b), and the webs of other genera are unknown. Figure 12.22 is a photograph of the web of *Maymena ambita*, and Figs. 12.19–12.21, are a series of photographs of webs of various *Mysmena* species.

Maymena webs include the out-of-plane radii characteristic of anapid webs (see above), accessory radius construction, accessory radius anastomosis like that of symphytognathid webs but not like that of *Theridiosoma* webs (pers.

obs.), and radial lengthening (Eberhard 1981a). Like anapids, *Maymena* place their egg sacs just above the hub (pers. obs.). To my knowledge the webs of *Maymena* and those of anapids do not differ substantially, although detailed study of the building behavior of these spiders may disclose some contrasts.

Mysmena, on the other hand, spin an orb web that is among the most derived architectures known. Yet, as Figs. 12.19 and 12.20 show, this three-dimensional orb is produced mainly as a result of a proliferation of out-of-plane radii. The spiders do not spin a nonsticky spiral (pers. obs.); the sticky spiral is laid in a typical fashion, using the iL_1 forward tap to locate the innermost loop of sticky silk (Eberhard 1982, character A1). Apparently the spiders do not always move out on the exit radius far enough to contact the inner loop; i.e., some SS attachments are made without inner-loop contact (pers. obs.). As in *Maymena*, anapids, and the theridiosomatid *Ogulnius* (Fig. 12.12), the trajectory of the SS line is simply a fabricational consequence of radial lines in three dimensions. Close inspection of the hubs of completed webs reveals the presence of radial anastomosis, but it is not known whether radial lengthening occurs. Also like *Maymena* and anapids, *Mysmena* hang their egg sacs just above the hub (Fig. 12.21).

The Phylogeny of Orb Weavers

The implications of the preceding discussion are summarized in Fig. 12.26. The cladogram was produced using component analysis (Nelson and Platnick 1981). Often the occurrence of a feature in a few members of a taxon has been assumed to be enough to propose it as a general feature of the taxon. Some of these assumptions are critical, such as the presence of aggregate and flagelliform glands in all araneoid higher taxa, the use of a viscid-silk wrap attack by nesticids and theridiids, or the generalizations about anapids from the few species described in the literature or that I have observed.

Some characters mentioned in the text, but omitted in the cladogram, may turn out to be informative when better studied. These include SS-R cement (Kavanaugh and Tillinghast 1979; Work 1981), data on wrap attacks, and the various forms of araneoid paracymbia. Other characters not even considered here, such as the kind and number of spigots on spinnerets, may also turn out to be informative. The lack of a functional cribellum in araneoids is probably synapomorphic, but until homologies between coluli are better understood (Lehtinen 1967), the derived nature of the araneoid colulus cannot be defined.

The interpretation of some characters is problematic. For example, uloborids do not cut and reel radii, but they do cut and reel frames. Nephilines

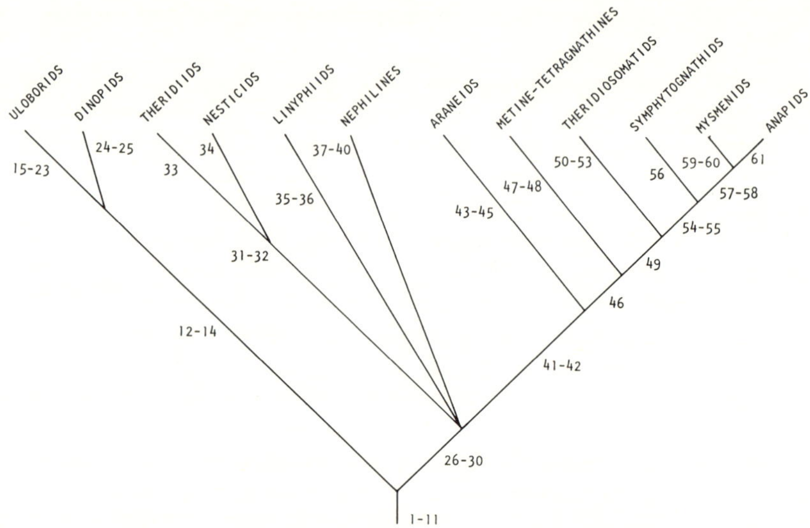

Fig. 12.26. Cladogram for orb-weaving spiders. Characters defining the branch points are listed below and explained in the text.

1–11: 1, specialized glands on posterior lateral spinnerets; 2, duct-levator muscle and control valve in anterior lateral spinnerets; 3, two-dimensional frame; 4, hub; 5, nonsticky spiral; 6, sticky spiral; 7, frame behavior (cut and reeled); 8, radius behavior (cut and reeled); 9, frame and radius behavior during nonsticky spiral construction; 10, oL_1 lateral tap to locate inner sticky-silk loop; 11, oL_3L_4 nonsticky-line grip.

12–14: 12, puffed cribellate silk; 13, pseudoserrate plumose hairs; 14, fourth-tarsal macrosetae.

15–23: 15, cymbial setae; 16, inverted minor ampullate glands; 17, pseudoflagelliform glands; 18, loss of poison glands; 19, second-instar web with accessory radii; 20, uloborid frame behavior; 21, protohub bite-out; 22, iL_4 radius grip during sticky spiral construction; 23, radii not cut and reeled.

24–25: 24, eyes in three rows; 25, web structure (Fig. 12.13).

26–30: 26, aggregate glands; 27, flagelliform glands; 28, serrate hairs; 29, paracymbium on male palp; 30, iL_4 push during sticky-silk attachment.

31–32: 31, sticky-silk wrap attack; 32, fourth-tarsal comb setae.

33: lobed aggregate glands.

34: basal plate of trichobothria.

35–36: 35, conformation of male palp; 36, route of ejaculatory duct.

37–40: 37, temporary spiral not removed; 38, radii doubly attached to frame; 39, oL_4 downward slide to locate inner sticky-silk loop; 40, frames and radii not cut and reeled.

41–42: 41, final hub bite-out; 42, hub modification.

43–45: 43, rotation of palpal bulb; 44, spin-wrap attack; 45, reduced tapeta in posterior median eyes.

46: iL_1 forward tap to locate inner sticky-silk loop.

47–48: 47, modified paracymbium?; 48, elongate chelicerae.

49: hub loops after sticky spiral construction.

50–53: 50, sternal pit organs; 51, palp conformation; 52, route of ejaculatory duct; 53, temporary spiral as circles, when present.

54–55: 54, accessory radii; 55, radii lengthened.

56: fused chelicerae.

57–58: 57, out-of-plane radii; 58, egg sac suspended above hub.

59–60: 59, male metatarsal spurs; 60, femoral tubercle.

61: labral spur between chelicerae.

cut and reel neither, and the remaining araneoids cut and reel both. The interpretation taken here is that absence of cutting and reeling is secondary, wherever it occurs within orb weavers. One might, for example, consider cut and reeled frames as a synapomorphy for orb weavers, nephilines autapomorphic in lacking them, and cut and reeled radii as a synapomorphy for the remaining araneoids. But the mere presence of cutting and reeling during movement via draglines in many spider groups suggests that most taxa are capable of it. Thus, its absence in nephilines is probably derived.

The method used by orb weavers to locate the inner loop of sticky silk during sticky spiral construction is a similar character. Uloborids and araneids use an oL_1 lateral tap, nephilines use an infrequent oL_2 lateral tap and a frequent oL_4 downward slide, and the remaining araneoids use an iL_1 forward tap. Depending on the situation in dinopids, the plesiomorphic technique is probably an oL_1 lateral tap, but other possibilities could be argued. A third case is the use by uloborids of an iL_4 grip to brace the current radius during SS attachment in addition to the oL_3L_4 grip. Eberhard (1982) calls this an oiL_4 grip because these legs are in fact closest to the site of attachment. In view of the prevalence of the oL_3L_4 grip elsewhere in uloborid building behavior and in orb-weaver building behavior generally, it seems simpler to interpret the iL_4 grip as an addition to the plesiomorphic orb-weaver behavior. A similar argument justifies uloborid frame behavior and the bite-out of the protohub as additional uloborid autapomorphies. Regardless of the alternative favored, the cladistic structure of Fig. 12.26 remains unaffected because of the numerous other characters supporting the monophyly of orb weavers on the one hand and of Araneoidea on the other.

The cladogram makes several implicit predictions, besides the validity of the generalizations just mentioned. For example, it predicts that Dinopidae will possess a pair of specialized glands serving the posterior lateral spinnerets (homologues of the pseudoflagelliform and flagelliform glands), a duct-levator muscle in the anterior lateral spinnerets, and an ampullate-duct control valve of the type characterizing uloborids and araneoids (Wilson 1969). In addition, the four architectural features and five behaviors listed as synapomorphies of all orb weavers are predicted to occur in Dinopidae.

The main point of this review has been to demonstrate the monophyly of all orb weavers, and, in the absence of any contrary evidence, the 11 characters documenting that monophyly seem adequate. Some of the branch points in Fig. 12.26, however, are not well established. For example, the monophyly of the uloborid-dinopid lineage is based on only three characters, primarily because so little is known about dinopids. Because araneoids are so autapomorphic, the characters unique to uloborids and dinopids may be retained plesiomorphies; thus "puffed" cribellate silk and pseudoserrate plumose hairs may actually be additional orb-weaver synapomorphies. In that case the lineage of orb weavers would exhibit a basal trichotomy. However, rather than postulating gain of those characters in the stem taxon of orb weavers and their subsequent loss or modification in araneoids, it is simpler to view them as single gains uniting uloborids and dinopids.

Several nodes of the cladogram are especially weak, e.g., that uniting the araneids with the rest of the araneoids, which is based solely on the occurrence of hub bite-out and hub modification. Although the generality of those features in all araneoids except nephilines is fairly well established (Eberhard 1982; pers. obs.), their absence in nephilines could be secondary. Nephilines appear to have lost or modified many features, e.g., cut and reeled frames and radii, an oL_1 lateral tap to locate the inner SS loop, and temporary spiral removal.

Similarly, the only character uniting the metine-tetragnathine lineage (if indeed that lineage is monophyletic) with the remaining araneoids is the iL_1 forward tap to locate the inner SS loop. This feature is, however, universal in the terminal taxa of the tree; i.e., all of the species observed in these taxa use an identical behavior. An oL_1 lateral tap does occur jointly with an iL_1 forward tap in at least one *Tetragnatha* species and in some *Chrysometa* species (Eberhard 1982).

The only really derived localization behavior among all orb weavers may be that of nephilines, which use the iL_4 in a downward slide. As Eberhard (1982) points out, that behavior is practical only in species that are large compared with the size of their webs, and all the nephilines are large spiders building closely meshed webs. It is significant that *Nephila clavipes* uses an infrequent oL_2 lateral tap to touch the SS line several radii in advance of the current SS-R attachment. Both uloborids and araneoids do a "leg over leg" behavior, using the first two outer legs to pull the current inner SS-R junction towards them (pers. obs.). This behavior supports the hypothesis both that the iL_1 forward tap is derived and that nephilines belong outside that group.

Multiple loss of the cribellum among orb weavers is a possibility. In that case the femoral trichobothria common to uloborids and tetragnathines might be synapomorphic, as might the accessory radii common to symphytognathids and juvenile uloborids. The similarity between uloborids and tetragnathines in posture at the hub may also be significant. (The first and second legs are appressed and extend straight out.) But these suggestions require a convergent development of all five characters supporting the monophyly of the Araneoidea, especially of the aggregate glands. On the other hand, the aggregate glands as a synapomorphy is based mainly on the morphological similarity of SS lines in araneoids, and not on any knowledge that aggregate glands are present in all araneoids.

The theridiid-nesticid and linyphiid lineages can be placed only at the basal node of the Araneoidea, on the basis of behavioral data. That placement implies that both groups certainly are araneoids, but may also turn out to be sister taxa of any of the araneoid lineages more precisely placed in Fig. 12.26. Since the data reviewed here are insufficient to allow either of the lineages to be placed with its sister taxon, they are placed at the most restrictive node possible, i.e., the basal node of the araneoids. The cladogram of Fig. 12.26 logically includes all the more fully resolved possibilities (Nelson and Platnick 1981).

Linyphiids and theridiids are among the most derived araneoid taxa behaviorally, and it may not be possible to discern any action patterns typical of orb weavers in their building behavior beyond a L_3L_4 NS-line grip and a L_4 pull. Few detailed studies exist of the building behavior of theridiids (but see Cutler 1972a; Freisling 1961; Szlep 1965). It is my impression that both *Latrodectus* and *Achaearanea* spiders touch the substrate with a L_1 forward tap just before SS attachment, but in doing so the spiders are hanging head down on a vertical line, and in this position few alternatives exist to touching the substrate with the first legs. Some theridiids do have sticky silk doubly connected in their webs, e.g., *Theridion impressum* (Wiehle 1933) and *Synotaxus* (Fig. 12.4), and it would be important to know which legs are used in attachment behavior. Szlep (1965) did speculate that the building behavior involved in the construction of the lower sheet of a *Latrodectus* web was vaguely similar to that of orb weavers, but her observations are too general to allow any conclusions.

It is unlikely that behavioral synapomorphies will be found uniting theridiids or linyphiids with any single taxon within the Araneoidea, since those two groups are very specialized. The webs of theridiids and linyphiids are probably not the result of elaboration of the orb-web algorithm, but rather of simplification—i.e., of prolongation and elaboration of the initial stages of web construction, so that the action patterns uniquely characteristic of orb weaving never appear. Exact placement of these taxa will probably be based on morphological evidence. In linyphiids more attention might be paid to the erigonine genera because sticky silk is said to be more abundant in their webs (Bristowe 1958). The signature, as it were, of theridiid webs is the use of segmental sticky silk, usually attached at one end to the substrate, rather than a continuous SS line attached only to silk, as in most orb weavers.

Ironically, this review has not really addressed the question of how the orb web first evolved. Its main point has been the demonstration that the orb webs of uloborids and araneoids are homologues. One effect of that argument is to remove the question of the origin of the orb web from the domain of people who work on araneoid spiders. As noted previously, the existence of a cribellate orb weaver within a monophyletic taxon of orb weavers makes it advisable to presume that the sister taxon of orb weavers was cribellate, and probably still is (excepting the possibility that the sister lineage independently lost the cribellum). Interestingly, nearly all the "proto-orbs" proposed by previous authors are spun by cribellate spiders, e.g., *Titanoeca* (Amaurobiidae) and *Stegodyphus* (Eresidae) (Szlep 1966), *Dictyna* (Dictynidae) (Kaston 1964), *Filistata* (Filistatidae) (Norgaard 1951), *Zoropsis* (Zoropsidae) and *Acanthoctenus* (Acanthoctenidae) (Wiehle 1931), *Fecenia* (Psechridae) (Robinson and Lubin 1979b). If indeed any of these webs is homologous to orbs, the definition of an orb web becomes a matter of taste depending on what degree of "orbness" merits the name. Eberhard (1972b) has shown that, in uloborids, frame, hub, and radius construction are difficult to distinguish. This basic similarity in behavioral pattern suggests that the NS network of an orb arose through a process of simplification and stereotypy of the most basic knot an orb weaver ties: the addition of one NS line to another to make a three-way junction, or trivalent node. This act might well be called "3-NS" behavior; one wonders how general a behavior it is in web-building spiders. As Denny (1976) has pointed out, orb webs actually consist of a network of nodes of which only one, the "hub," has a valence higher than 3. (In topology the valence of a node is the number of lines meeting at that node: Loeb 1976.) In fact, even the hub of an orb web consists of trivalent nodes, as hub construction in uloborids and nephilines clearly demonstrates. The hub of an orb web appears to have a high valence only because its trivalent nodes are closely spaced. The NS network qualitatively consists simply of the iteration of 3-NS behavior; quantitatively it consists of the distortion of sequential trivalent nodes so that short series of nodes are closely spaced (forming the hub), and the nodes alternate to those series are dispersed (forming the frames and radii). This pattern of 3-NS attachments in conjunction with a centripetal-motion pattern accounts for the point symmetry of orbs, and reveals how simple the algorithm resulting in an orb web may actually be. Of course, in some araneoid orbs the series of closely spaced trivalent hub nodes are often truly coalesced, at least during radius construction, into a single node of exceptional valence. We can therefore describe the architecture of an orb web as consisting of a number of trivalent nodes combined with one interior node that has a valence of approximately two-thirds the total number of nodes. Any network fitting this topological description will appear vaguely orblike in the architectural sense.

On the other hand, the diversity of orb webs within Araneoidea demonstrates that a purely architectural notion of an orb web is insufficient. For example, if our concept of an orb web is based only on the coincidence of point symmetry and a two-dimensional frame, we shall recognize orbs only when they look like orbs. The recognition of highly derived orb webs such as those of *Wendilgarda* (Fig. 12.11), *Miagrammopes* (Lubin, Eberhard, and Montgomery 1978), or *Mastophora* (Eberhard 1980a) would therefore be nearly impossible if the taxonomic affinities of the builder in question did not strongly suggest that the web must be a derived orb. Obviously a more heuristic way to view orb webs is as a behavioral process involving the use of silks from particular glands in a particular behavioral sequence. But is there any silk or action pattern so peculiar to orb webs that we can confidently identify an orb when we recognize that feature? At present we know very little about the fundamental behaviors and silks of webs, or even the basic features of web architecture, that might have led to the evolution of the orb web. It is therefore entirely fitting to end this review with a question: what, if anything, is an orb web?

Acknowledgments

I thank N. I. Platnick for the identification of anapid voucher specimens, and H. W. Levi and W. G. Eberhard for their support and generous sharing of ideas—in particular, for access to their manuscripts prior to publication. I also thank M. K. Stowe for the use of his photographs of *Synotaxus* and *Dinopis*, C. Lenk for editorial comments, and C. Villars for many long hours of help in preparing the manuscript. I thank the Organization for Tropical Studies, the Jessie Smith Noyes Foundation, and the Richmond, Anderson, Atkins, and Barbour Funds of Harvard University for funding several opportunities to study the building behavior of tropical spiders. Voucher specimens for web photographs (see figure legends) have been deposited in the Museum of Comparative Zoology at Harvard University.

13

The Evolution of Web-Building Behavior in Spiders: A Third Generation of Hypotheses

WILLIAM A. SHEAR

EVERY STUDENT of spider behavior in the past hundred years has surely given some consideration to the implications of the stereotyped, programmed behavior exhibited by web-building spiders. Most general accounts of spider web building remark on two features: that young spiders make their webs, even the most complicated ones, without significant experience, and that these webs are generally similar in design to those made by their parents, and other members of their species. (Although many species do show pronounced ontogenetic changes in web structure, these changes occur in an unvarying way in all members of a given species, so the web-building program may still be seen as closed.) The more evolutionarily inclined authors have noted also that there are constant and repeatable differences between the webs of two species, even closely related ones. The first two observations suggest that the kind of web that is built is largely inherited, and that the program for building it develops without the need for "practice." The third observation suggests that this behavior is species-specific and probably the result of natural selection to fit the circumstances at hand. Web building in spiders has evolved.

One of the earliest schemes proposed to describe the course taken by this evolutionary history was that of Pocock (1895). His ideas were taken up and elaborated by Comstock (1912). Bristowe followed suit in his papers of 1930, and in books published in 1941 and 1958. These earlier authors concentrated on hypotheses of the origin of silk use in spiders, the development of silk snares used in predation, and the sequence of changes that might have taken place in the course of finer and finer adaptation. Their reviews were based on limited knowledge of spider behavior, and drew from information about the habits of relatively few species in the Northern Hemisphere (although Bristowe especially knew a good deal about these species).

Kaston, writing in 1964, summarized past schemes and presented one of his own: a second generation of hypotheses about the origin of web building and the course of its evolutionary development. Kaston gave little direct argumentation, but described a great variety of web types and summarized his ideas about the course of evolution of web building in two diagrams. He proposed two schemes because he believed that the cribellate and ecribellate araneomorphs represented separate and coherent lines of evolution. Unfortunately, a study of these diagrams could lead to wrong impressions, for although some of the branchings correlate well with the evolution of spiders themselves as understood through morphological evidence, others certainly do not, and seem to suggest such absurdities as the development of the lycosid *Pirata* from the Atypidae, and the Dipluridae from a "lycosid." No doubt Kaston meant only to suggest that these taxonomic names be used for web types, not spiders, but the message is not clear. We are now well aware that cribellate and ecribellate spiders may occur even in the same genus, and that the cribellum, though probably a synapomorphy for all araneomorph spiders, has been lost many times independently. Writing today, Kaston would undoubtedly find a way to integrate his pair of diagrams.

In 1972, Kullmann published an article postulating, as Kaston had done, the convergent development of the orb web in both cribellate and ecribellate spiders. Kullmann wrote in response to a revival of Thorell's (1886) suggestion that the orb web arose but once, and that cribellate orb weavers gave rise to ecribellate ones; Thorell's hypothesis is given new life by Coddington (this volume). Like Kaston, Kullmann was not yet convinced of the multiple losses of the cribellum that we now believe to have taken place. Loss of the cribellum at the orb-weaving stage was, for Kullmann, "difficult to imagine."

Common to, and important in, the schemes of both Kaston and Kullmann was the concept of convergent evolution. Kaston proposed two entirely separate converging schemes for the cribellate and ecribellate spiders because he assumed that the cribellum represented a synapomorphy for a monophyletic group of spiders. It does, but the group is all araneomorph spiders, and not just some of them. Kullmann postulated two converging schemes (not at all as detailed as Kaston's, and so not as readily falsifiable) because he could not imagine that Uloboridae and Araneidae might be sister groups. And, in each scheme, a few important webs wound up misplaced—near the base of certain lines when they should have been near the top. Both suggested a pre-orb position for the web of *Polenecia* (Uloboridae; then called *Sybota*), and regarded the webs of *Cyrtophora* and *Mecynogea* as primitive rather than derived orbs.

Convergence and parallelism have both been important in the evolution of web building in spiders, but not in the form of the dichotomy established by Kaston and Kullmann. Below I shall suggest that convergence has in fact occurred many times, though the orb web seems to have originated at most twice, perhaps only once.

Before continuing, it may be a good idea to list and evaluate a number of assumptions that seem to have guided past writers on the subject and, indeed, have helped to form this essay.

1. That web building in individual spiders is strongly, almost entirely, dependent on a genetic program impervious to permanent environmental influence (learning). In other words, though details of form and size of a spider's web may be altered by changes in the spider's physiology, or by the course of ontogeny, changes of the first kind do not alter the basic architectonic of the web, and merely represent the function of the genetic program under physical constraints, and changes of the second kind are repeatable from individual to individual of the same species and so do not depend on experience. The daily changes in web form noted by Eberhard (this volume) and others still defy analysis, but are not likely to be due to anything but the interaction of a wide variety of physiological givens.

2. That the sequence of steps (polarity of the transformation series) in the evolution of web building has been from irregular to more regular webs. This assumption has been based on the general (but not exact) correlation of irregular webs with spider species acknowledged to have plesiomorphic anatomical characters.

3. That the sequence as established by this correlation has been general for spiders. In other words, the phylogeny of web-building behavior is close (or identical) to the phylogeny of spiders as organisms.

4. That the amount of silk used in webs, graphed against the course of evolution, follows first a rising, then a falling, curve. If one assumes that the earliest webs consisted of a few silken lines or a silken collar extending from a tunnel or retreat, then the transformation of the earliest web form into a retreat plus sheet, or into an aerial sheet with a maze, involves more silk. And if one assumes that the orb is derived from an aerial sheet (not a necessary assumption), the amount of silk used is reduced by this step, and may be reduced further by additional steps leading to orb derivatives, until finally a stage is reached by the most advanced species that is virtually identical to the starting point: a simple retreat (sometimes not even that). The end points are differentiated only by relative morphological change. (*Arcys* or *Pachygnatha* show morphology that clearly differentiates them from theraphosids or *Ariadna*.)

5. That the driving force behind the sequence of steps has been, first, the increased use of silk in prey capture and, second, the attempt to economize

in the relative amount of protein used. The validity of this assumption seems almost self-evident, though one must still account for the so-called primary hunters that are claimed to show no evidence of having had a web. The second part of this assumption is often explicitly stated but supported only by slight evidence (J. F. Anderson 1970; Peakall 1971; Peakall and Witt 1976; Ford 1977; Prestwich 1977; Greenstone and Bennett 1980).

If these crucial assumptions are evaluated, it becomes clear that there is as yet little hard evidence supporting them. Perhaps assumption 1 is the sturdiest, deriving its support from the long series of papers by Witt and his colleagues, particularly LeGuelte and Witt 1968, Witt, Reed, and Peakall 1968, LeGuelte 1969, Reed et al. 1970, and Witt, Rawlings, and Reed 1972. In these papers it was shown that although severe environmental restrictions (such as keeping the spider in a glass tube and not allowing it to spin an orb) produced minor perturbations in the first few webs made after confinement, the basic structure of the web remained that characteristic of the species. In other work, summarized in Witt 1971, the alterations produced in the webs were the result of substantial physiological disturbance by means of psychoactive drugs; in nearly all cases the web was nevertheless still identifiable as an orb. Risch (1977) showed that there are not only essential similarities in web design within species but also constant differences between species. References supporting this assumption further are scattered throughout the taxonomic literature, and these arguments are given more credence by the ability of experienced arachnologists to identify genera (and even species) by their webs. Though there is still a need for a detailed study of numerous webs made by numerous individuals in numerous populations of closely related species, I tend to accept assumption 1 on the basis of what we know so far.

Assumption 2 will hold up as long as it is restricted as stated. If the level of web complexity is correlated with independent morphological features along a transformation series of established polarity, the assumption seems reasonable. It cannot be stated a priori that webs that are simpler structurally are in fact more primitive than webs that are structurally complex. The definition of the terms in this statement must be clarified. Primitive webs are not necessarily less complex than advanced ones. A sheet web, consisting of a flattened tangle of threads, would be much more difficult to describe with precision than an orb; there are vastly more threads and vastly more intersections between them—and these connections are not predictable. The space-filling, apparently random tangle may be a derivative of the orb. "Primitive" suggests both early evolutionary origin and (often) simplicity. A primitive web may be defined as one that exhibits a preponderance of plesiomorphic characters relative to others. These characters may include simplicity of design. The plesiomorphy may be established by the correlation of the ethological (web-building) characters with plesiomorphic morphological characters. Derived webs show chiefly apomorphic characters, or are correlated with apomorphic morphological characters. Derived webs are usually regular, but may or may not be of complex design. It may be useful also to suggest that a spider constructing an irregular space web or an irregularly meshed sheet is behaving heuristically, whereas a spider weaving an orb is behaving according to an algorithm. A heuristic process involves exploratory methods—carrying out one step may suggest another, but not always the same one. The job is done "when it's done," when some unspecified final state is reached. Behavior controlled by an algorithm is orderly and proceeds to a final result by a set of steps that follow one another in a predictable way. The end point is achieved when the result corresponds to a predetermined set of parameters, coded, in this case, in the spider's nervous system.

The point is an important one because complexity alone does not necessarily indicate that a web type is derived. The several types of simplified webs derived from orbs provide good examples of this; these webs are thought to be derived, not primitive, because they are made by spiders that are members of advanced families and because the algorithm for making them resembles that for making the orb. Lubin (this volume) has used both of these lines of evidence in evaluating the webs of *Miagrammopes* (Uloboridae), which are composed of simple threads with cribellate silk laid on them, but which nevertheless seem to be the remnants of reduced orbs (see also Lubin, Eberhard, and Montgomery 1978). There is good evidence from detailed studies of web construction that the retitarius snare of *Dinopis* is also a modified orb (Theuer 1954; Coddington 1982).

The first part of assumption 4 is really an observation. If assumption 3 is granted, the amount of silk used does seem to increase at first, and then to decrease first to the level used in orbs, then to the level found in orb derivatives, and finally to the level employed by, e.g., adult *Pachygnatha*, which make scanty use of silk. But there is a time factor here: many araneid orb weavers take down their webs daily (Carico, this volume), whereas most sheet weavers leave the sheet up once it has been woven, and add to it only in the course of moving across it to reach prey. I know of no longitudinal study that would establish that, e.g., linyphiids and araneids of comparable size expend different amounts of silk during their whole lifetimes. Although it is clear that the araneids that replace their webs also recycle silk proteins, this requires verification from studies of a number of species. And, of course, some orb weavers do not replace their webs daily, but simply repair damage. Many uloborids even throw old webs away.

The economy of design, supposed efficiency as an insect trap (Eberhard, this volume), and tendency to recycling of the orb web have been used as arguments to support assumption 5, that economy in silk use is an important selective force, and it may well be, despite the lack of clear, detailed supporting evidence. Such a selective pressure must be balanced not only against the costs of muscular activity other than spinning (Prestwich 1977) but also against possible countervailing pressures having to do with the relation between mesh size and prey-capture success (Chacón and Eberhard 1980; Eberhard, this volume) as well as the need for protection, which can be greatly enhanced by using more silk to produce an impenetrable tangle.

Stowe (this volume) makes a good case for the correlation of web reduction with specialization on certain sorts of prey, such as moths or pedestrian insects. Perhaps it would be better to suggest that each spider maximizes its efficiency of silk use in its particular environmental situation, given the constraints of its spinning apparatus and web-building program, as well as the pressures exerted by predators and/or parasites and by prey availability.

With these assumptions as background, I intend to show in this essay that in many only distantly related lines of spiders a repeated sequence of web-building evolution has taken place, beginning with a silk-lined retreat and generally ending with the development of a horizontal sheet of silk. In a few cribellate species that operate their sheet webs from below, this sheet is exchanged in later life for a regular array of threads symmetrical about a point and often bounded by a two-dimensional frame—the orb web.

The origins of the ecribellate orb are not so clear, but older hypotheses seem less attractive since certain examples of "primitive" orbs are now thought to be derived relative to others. Thorell's 1886 hypothesis, propounded also by Coddington (this volume), suggests that the largest group of cribellate orb weavers (Uloboridae) is the sister group of the ecribellate ones, and some of their relatives, taken together (Araneoidea). (The orb-weaving araneoid families are Araneidae *sensu latu*, Theridiosomatidae, Anapidae, Symphytognathidae, and Mysmenidae; the non-orb-weaving families are Theridiidae, Linyphiidae, Nesticidae, Mimetidae, and Cyatholipidae (of dubious validity). The web structure and relationships of Micropholcommatidae (sheets?) and Textricellidae (minute orbs?) are as yet unclear.) A corollary to this hypothesis is that the orb has been lost in those araneoids that make other types of webs. However, an independent origin for the ecribellate orb cannot be entirely discarded because of doubts about the monophyly of Araneoidea, and a lack of really convincing morphological synapomorphies linking Araneoidea and Uloboridae. In order to support an independent origin, it must be shown that some group other than Uloboridae is a reasonable candidate for sister group of the araneoids. These two hypotheses will be discussed later.

Subsequent evolution of orb making in spiders may have been driven in part by prey specialization and the metabolic cost of silk, and has led to active spiders that manipulate webs reduced from orbs. Space-filling webs originate in two ways, from the reduction of sheets in sheet-plus-tangle webs and, in a few cases, from orbs in which out-of-plane radii proliferate.

What follows can be organized into four subjects: the origins of web building, the convergent development of sheet webs, the origins of the orb web, and developments from orb webs.

Origins of Silk Use and Web Building

The anatomy of spinnerets and of the internal silk-producing apparatus has been the subject of a number of recent papers (e.g., Kovoor 1972, 1977c, 1978; Mikulska 1966, 1967, 1969; Wąsowska 1969, 1970), although a great deal obviously remains to be learned. The actual origins of silk making are so far in the past as to be totally obscure. Silk glands are not a feature of the Amblypygida, the living sister group of the spiders, nor has any evidence for them been found in exceedingly well preserved fossils of the Trigonotarbi, an extinct order that is probably the sister group of the amblypygids and spiders taken together. The earliest fossil spiders with preserved spinnerets (Carboniferous) evidently have highly developed ones, but there is an urgent need to restudy these fossils. It seems unlikely that the habit of spinning silk developed before the protospider emerged onto dry land. (Trichopterous insects, however, spin silken webs underwater, as does the "water spider" *Argyroneta aquatica.*) The ability to make silk appears sporadically in other arachnid orders—here and there in the mites, and in the Pseudoscorpionida—but the silk is always produced from prosomal glands. These groups are unrelated to spiders.

Externally, the number of spinnerets varies from eight (or seven) in the Mesothelae, where they are found near the middle of the ventral surface of the segmented abdomen, to as few as two in *Lutica* (Zodariidae), where they occur in the more common terminal position. The reduction in number has taken place convergently in both araneomorph and mygalomorph lines. The spinnerets arise on the tenth and eleventh segments of the developing spider's abdomen and seem to be persistent abdominal appendages; the rudimentary appendage primordia on other abdominal segments quickly regress. The persistent ones, originally one pair on each segment, divide before hatching in all spiders but the Mesothelae, where the division takes place afterward. In cribellate spiders, the anterior median pair are expanded and fused into a cribellum, or spinning plate, densely set with spigots, from which the exceedingly fine cribellate silk is spun. Originally a feature of all araneomorph spiders, this spinning plate has been lost many times (Lehtinen 1967), and there may be no evidence of its having been present, or there may be a median structure (the colulus) ranging from a broad lobe, to a small, fingerlike projection, to a pair of setae. The spinnerets are operated by a combination of muscles and hemolymph pressure.

Internally, the silk glands range in number of kinds from one or two in mygalomorphs to as many as six in araneids. Each gland produces a silk with different properties, and each type is used differently by the spiders (S. O. Anderson 1970; Denny 1976). Silk is a multiple-use material for spiders; the variety of uses was enumerated by Kaston (1964).

Of particular interest in the present context is the use of silk to produce structures that protect the spider and are of use in detecting, restraining, and capturing prey. In at least two groups of spiders, the function of prey restraint and capture is enhanced by the production and use of "sticky" silk. This is of two kinds. In cribellate spiders, the sticky silk is dry, and probably is able to entangle and adhere because the fine fibers get caught on small irregularities, as well as by virtue of its own surface properties acting at the level of molecular adhesive forces (Kullmann et al. 1975). Cribellate silk does not lose its "stickiness" over long periods of time, unless it is wetted by rain or becomes dusty (Eberhard 1980b). The cribellate fibers are combed out by special curved setae on the fourth legs and laid over a core fiber, which comes from different glands and a different pair of spinnerets. The second type of "sticky" silk is evidently limited to the present superfamily Araneoidea, but neither the silk nor the glands that make it have been demonstrated in all the families included. This silk is wet. It consists of a core fiber produced by the flagelliform glands, which open on the posterior spinnerets, and a viscous cement, produced by the aggregate glands, that is laid on the core fiber. The viscous matter disperses in discrete droplets on the core fibers and eventually dries out, at which time it loses both its stickiness and its elasticity. I shall call this form of silk "viscous" silk, and use the term "sticky silk" only when I wish to refer to both viscous and cribellate silk together.

The trend seems to be that the silk glands become more specialized, and multiply in kind, as spiders themselves become more specialized. The embryology of the glands requires more study, and clues might be found thereby as to their evolutionary origin. The best current hypothesis suggests that they are derivatives of the coxal glands, and since these glands incorporate parts of coelomic pouches, careful attention to the origin of silk glands might allow the hypothesis to be tested. The silk glands have been observed to originate as ectodermal invaginations in some spiders.

As to the origin of silk use, Kaston (1964) has ably summarized the two major hypotheses. Pocock (1895), and later Comstock (1912) and Bristowe (1930a), believed that silk was used by early spiders only to protect their eggs. In this scenario, a silklike matter was first produced from the mouth region, probably from the present poison glands, and smeared over the egg mass. It is not indicated how the switch was made from prosomal to opisthosomal silk glands. Some spiders then attached their egg sacs to the substrate, mixing the gum with soil; others hung the sacs among vegetation. The first group constructed a retreat around the egg mass and eventually led to one lineage of web builders making sheets along the ground and operating them from above; the second group wandered around in the foliage and led to a line of spiders hanging upside down in aerial tangles.

One problem here is that the whole hypothesis derives from the mistaken impression that the sister group of the spiders, the Amblypygida, wrap their eggs in silk. In fact, these animals carry their egg masses attached to the gonopore, covered by a membrane formed from a hardened secretion of associated glands. They do not produce silk of any kind, nor do they use secretions from the mouth area, fecal material, excretions, or "mortar" made from soil to protect the eggs.

The part of the hypothesis suggesting the use by spiders of a cheliceral "presilk" was based by Bristowe (1932) on the observations of Millot (1930) on the bilobed poison gland of *Scytodes*, from one lobe of which sticky gum is squirted for use in predation. Bristowe suggested that this type of bilobed gland is primitive, and that the present, single-lobed poison glands of other spiders are derived from this "cheliceral spinning apparatus." Soil was supposedly incorporated into this material, to provide additional protection and camouflage. Secretions from the mouth are used in this way by some Clubionidae and Gnaphosidae today.

There is not a whit of evidence to support the notion that the poison glands of spiders were ever a "cheliceral spinning apparatus." The gum-squirting habit of *Scytodes* is obviously specialized, not primitive. *Scytodes* species lightly wrap their egg bundles with silk from the spinnerets and do not use their cheliceral gum to cover them. The secretions used by clubionids and gnaphosids to cover their eggs are added after a silken egg sac has been constructed, and come from the salivary glands, not the poison glands.

It is difficult to see how this egg-sac theory enlightens us as to the origin of web-building behavior. The transition from covering the eggs with a prosomally produced "presilk" from the chelicerae to the use of real silk fibers from opisthosomal glands is not explained.

The second hypothesis is that of McCook (1889) and Savory (1928), who suggested that, as in some living centipedes, the primitive arachnids trailed excretory material from coxal glands after them as they walked about. An immediate adaptive advantage of this habit would lie in being able to follow such a trail back to a safe retreat. It would also be of use in finding mates, via chemical senses. Gradually the prosomal coxal glands took over all coxal-gland excretory functions, and the opisthosomal glands of the tenth and eleventh segments became specialized to produce a continuous thread of durable, proteinaceous material. Accumulations of this thread around and within a retreat would have obvious protective advantages. Further, wandering protospiders would be able to explore aerial vegetation more fully by using their draglines to move from one level to another. Eventually a radiating group of lines from a retreat might so expand a spider's sensory area (Coyle, this volume) that leaving the retreat except for short dashes after prey would be unnecessary. The use of silk in making egg sacs would be a secondary development.

I find this hypothesis much more attractive than that involving the initial use of silk as a covering for eggs, because the transition from prosomal to opisthosomal glands is not required, and the major multiple functions of silk remain the same as they are today—protection, sensory enhancement, and predation.

Recently, Decae (1984) has argued that aquatic protospiders, if they existed, might have used silk even under water. Drawing analogies with marine littoral crustaceans and annelids, he suggests that silk might have been of use to keep the walls of a mud burrow from collapsing, or that silk caps would have prevented burrows from being filled by sediment. Decae missed two important analogical bits of evidence for his theory. The use of silk nets in prey capture and tube construction by caddisfly (Trichoptera) larvae demonstrates that silk can be spun under water. Likewise, the few known species of intertidal spiders (generally members of the family Desidae) make silk bags for themselves, or seal old barnacle shells with silk, to avoid submergence at high tide. Rovner (1984) has recently expanded on Decae's ideas, suggesting from experimental evidence that many terrestrial spiders can survive the flooding of their habitat because their silken retreats hold an air bubble and function as a physical gill. He speculates that one of the original functions of such a retreat might have been to allow early spiders in the intertidal zone or in low-lying habitats to survive periodic flooding.

Attempting to argue against Savory's "dragline" theory, Decae claims that silk draglines are not trailed out by other animals, and that some mygalomorph spiders do not now always lay draglines. The first of these claims ignores the common observation of lepidopterous larvae dropping from trees on silk lines, or as in the processionary caterpillars, laying silk on the substrate as they move. The second claim is not really relevant; simply because all living spiders do not trail draglines behind them at all times does not mean that a frequent occurrence of this habit in ancestral spiders could not have led to silk-lined burrows, as I have described above.

However, Decae's suggestion cannot be discarded as unreasonable. It still leads to the conclusion that the earliest type of silk structure made by spiders or their ancestors was a silk-lined burrow or a silk tube in a crevice.

Evolutionary Development of Web Types

My line of argument is this: Web evolution has proceded from the primitive lined retreat, through expanded collars, to sheets along the substrate or guyed in vegetation, to aerial sheets lacking a retreat, and finally, through the regularization of the arrangement of elements, to orb webs in cribellates. (The origin of the ecribellate orb will be discussed separately.) From there, prey specialization and substitution of muscular activity for silk synthesis have led to further and further reduced webs. Although my scheme resembles in outline that proposed by Kaston (1964), there are differences. I draw no distinction between cribellate and ecribellate groups of spiders; many modern lower-level taxa include both. Nor do I suggest only a single line leading to each of the major types of webs. Instead, I propose that the projected line of web development has recurred many times in many mygalomorph and araneomorph phyletic lines. To support my argument, I present first a brief survey of web types, with descriptions, and then a discussion of their distribution among taxonomic groups of spiders.

A Survey of Web Types

Silk-Lined Retreat

Many spiders modify crevices and holes in their environment by lining them with silk. These chambers are used as places to remain quiescent during periods of inactivity, frequently as molting retreats, or as places to put (and sometimes guard) egg sacs. The opening is sometimes closed. This habit occurs in many mygalomorph genera, many "haplogynes," and several families of "entelegynes." A modification involves building the retreat as a closed bag of silk not necessarily conforming to an environmental space, such as the retreats made by salticids, thomisids, and clubionoids. The original function of such a retreat was undoubtedly protection, so it is logical to assume that advantages were to be gained by extending this protection even further, by using the retreat as base from which to hunt. It has been shown that *Aphonopelma* and possibly other theraphosids do exactly this—spend most of their time in the retreat, and leap out at passing prey (Stradling 1981). Having seized prey, the spiders quickly return to the retreat. Modifications of this retreat may be associated with either the protective or the predatory function. Protection can be increased by attaching a movable lid or cover to the retreat entrance, in the manner of the Migidae (Forster and Wilton 1968), or by camouflaging the retreat itself with bits of debris from the environment. Since spiders pay out dragline silk as they move, each dash from the retreat entrance would leave behind a line, which eventually might serve as a tripline expanding the spider's sensory area. The importance of this function is illustrated by the fact that among certain ctenizids (Aganippinae: Main 1957) and antrodiaetids (*Aliatypus*: Coyle 1971, this volume) that rarely exit their secure burrows, and thus do not lay down draglines outside the entrance, leaves and twigs may be attached to the burrow margin in a radiating pattern, and serve the same function as lines of silk.

The ctenizids and scattered genera in other families, of course, build webs that represent a basic modification of this type, since they dig a tunnel, often of considerable complexity (Moggridge 1873), line it partly with silk, and close it with a hinged door. This habit is also present in certain araneomorph spiders (some Lycosidae).

The retreat with triplines (Fig. 10.1) is exemplified by the web of *Liphistius batuensis* (Bristowe 1976). Liphistiids may have more plesiomorphic characters than any other group of living spiders (but see Murphy and Platnick 1981). *Ariadna* (Dysderidae) are also commonly cited (Beatty 1970) as making this type of web: a tube in a deep crevice with a definite pattern of trip-lines extending from the lip. Less known are the ecribellate oecobiid species of *Uroctea*, which make a free-standing retreat (usually under a stone) with several entrances, and trip-lines rigged on stiff-silk pylons (Kullmann 1975).

The lip of the tube may become more important. At least two distinct atypoid groups, Atypidae and *Atypoides* (Antrodiaetidae), expand the lip of the burrow retreat into a long tube that may extend along the ground surface (European *Atypus* species), extend up through a litter layer (*Atypoides*: Coyle 1971), or be attached to a tree trunk (American *Sphodros* species: Coyle and Shear 1981, see Figs. 10.18, 10.19). At least in the atypids, the spider never has to leave the protection of its tube and attacks its prey by stabbing through the tube wall with enormously long fangs. Clearly this modification is an attempt to maximize the protective function.

On the other hand, the subaerial web of *Hypochilus* species eliminates the retreat (Gertsch 1958; Shear 1970b; Ferguson 1972). The spiders spin a filmy sheet against a rock wall and from it guy out a broad, lampshade-like tube. Seen from below, this web is surprisingly regular, suggesting features derived both from the trip-lines and from the extended collar of the burrow (Fig. 13.1). To support the web, and probably also to provide additional protection for the somewhat exposed spider, an extensive irregular tangle is built below the lampshade (Fig. 13.2).

Retreat plus Sheet

In retreat-plus-sheet webs, one side of the collar at the retreat entrance is continuous with a sheet of densely woven silk (Figs. 2.1, 10.4–10.7). Examples of this type of web are too numerous to list, but in the Northern

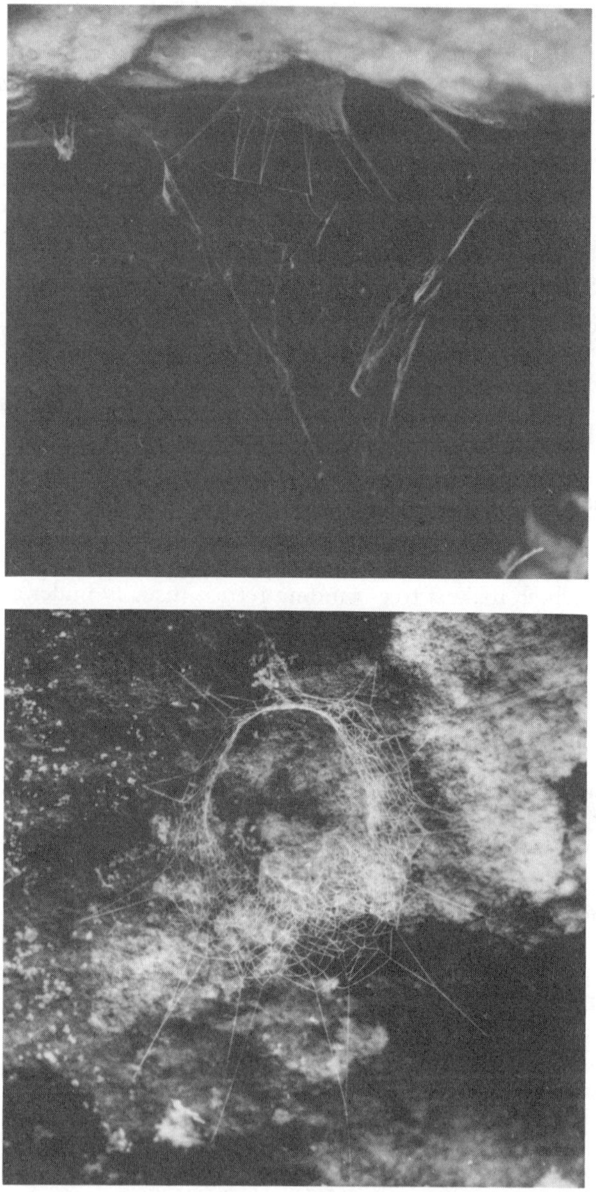

Fig. 13.1. *Top*, the web of *Hypochilus gertschi*, seen from the side. The large tangle extending below the "lampshade" serves for support and seems to have no function in predation. *Bottom*, the upper part of the web as seen from below. Note the regular arrangement of web elements. Photographs taken near Athens, West Virginia.

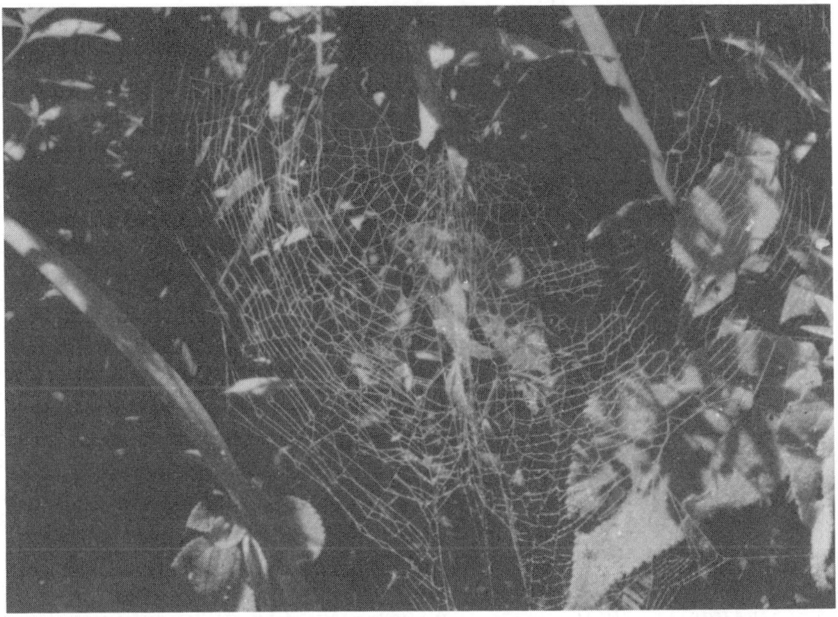

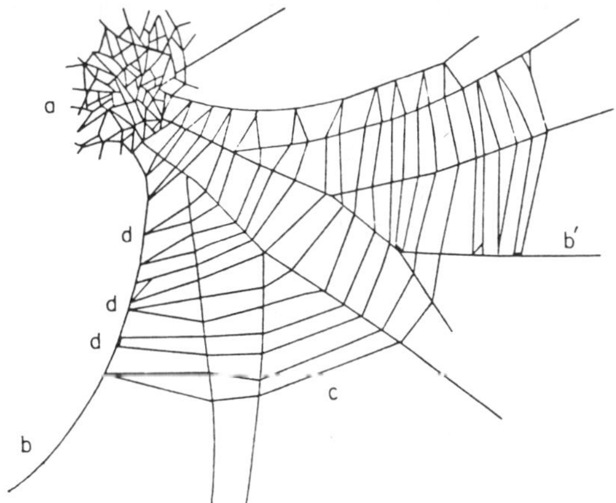

Fig. 13.2. "Pseudo-orb" webs made by two unrelated species of spiders. *Above*, web of *Fecenia ochracea*. The spider's retreat is in the curled leaf near the top of the photograph. Note the absence of clear radial elements and the irregular "hub" below the retreat. Photograph taken near Wau, Papua New Guinea, by H. W. Levi. *Below*, web of *Titanoeca albomaculata*. Note the presence in the web of several "hubs," and loops, spirals, and half circles of cribellate silk. *a*, the spider's retreat; *b*, primary "radius"; *b′*, secondary "radius"; *c*, cribellate silk; *d*, turning points in loops of cribellate silk. From Szlep 1966.

Hemisphere the most obvious examples are webs made by many members of the Agelenidae, particularly *Agelenopsis* species in North America. Kaston (1964) and Forster (1967) list numerous other families and genera. The threads of the sheet impede the movements of insects walking over it, and the vibrations that their struggles produce alert the spider. *Agelenopsis* run out over the upper surface of the web and attack the prey, often first by leaping over and running around it, paying out silk. Captured prey are taken to the retreat and fed upon. Each excursion of the spider makes the sheet more dense and presumably more effective in retarding insects. In many species, the sheet is simply laid out over the ground or another horizontal surface, but in some *Agelenopsis* species and many other spiders an extensive maze is constructed above the sheet. This maze can help in two ways: flying insects may be stopped and knocked down onto the sheet, and the increased tension on the sheet may make it more efficient in signaling the location of prey (Kaston 1964). Such a maze also makes it possible for some agelenids (and members of other families) to extend the sheet freely in space. The original single capture surface is thereby doubled, although the spider generally will continue to utilize either the top or the bottom surface alone for its movements, probably owing to the difficulty of switching sides. The members of the Theridiidae with this sort of web (especially *Latrodectus* and *Lithyphantes* species) use the lower surface, whereas nearly all agelenids use the upper. Prey may be attacked on either side of the aerial sheet, however, by biting through the web and cutting holes through which the prey can be pulled.

The sicarioid *Diguetia* produce a variant of this plan, resembling a *Hypochilus* web freed from the overhanging ledge and with the lampshade expanded into a broad but diffuse sheet. The retreat is a free-standing tube with the opening at the bottom; the spider emerges from the tube onto the lower surface of the sheet.

Protection for those spiders using the lower surface of the sheet is increased, particularly if the web is rather close to the ground, and many species have dispensed with the retreat altogether.

Aerial Sheets Without a Retreat

Aerial sheets without an associated retreat are found in a wide variety of families, most prominently in Linyphiidae (Fig. 12.5), Agelenidae, and Theridiidae; often additional sheets appear, providing further protection. Webs of *Frontinella* species (Linyphiidae) have a cup-like upper sheet on which the spider runs hanging from the undersurface; below this is a second, flatter sheet. *Calymmaria cavicola* is an agelenid often found in shallow caves in the Appalachians, inhabiting the roof, or rock overhangs. The spider runs on the upper surface of a sheet spun close to the ceiling, but under this is a funnel-shaped cone of silk with an opening at the apex and the broad end downward. The cone may guide upward-flying insects onto the catching sheet.

The webs of agelenids do not incorporate viscid silk, and the distribution of this type of silk in linyphiid webs is not well documented, though it has been detected by Nielsen (1932), Kullmann (1964), and Eberhard (1972b). Where the occurrence of viscid silk in linyphiid webs has been documented, it has been found in both the sheet and the maze, or tangle, above the sheet. The dense mesh of the linyphiid sheet may be sufficient to retain the smaller insects upon which they feed, and which they attack speedily hanging below the sheet.

Theridiid webs often have viscid elements, but only rarely or not at all on the sheet. Instead, viscid threads are frequently the ones that extend down from the sheet to the substrate, producing the characteristic "gum-footed" web, also found among the Nesticidae (Bristowe 1958; Gertsch 1979; Eberhard 1972a; see Figs. 12.4, 12.8). These viscid threads come from glands that are different in appearance from, but probably homologous to, the glands producing viscid silk in the araneids. A difference pointed out by Eberhard (1977b) is that theridiids (*Synotaxus*) often spin dry and viscid silk alternately, whereas araneids first spin all the dry elements of their webs and then lay in the viscid silk. Further, lines in "gum-footed" webs are dry with viscid segments.

Sheet with Regularly Arranged Threads

Many cribellate families and genera have members in which some elements of the sheet become regularly arranged. The regular arrangement is usually described as a series of roughly convergent "radii" in the same plane, across which zigzags of cribellate silk are laid. Such regular structures appear orblike to our eyes because they suggest the point symmetry of true orb webs (Coddington, this volume). Crude examples are found in the webs of *Eresus* and *Stegodyphus* and in webs of numerous species of *Dictyna* (Kullmann 1972a, pers. obs.). *Eresus* and *Stegodyphus* belong to the Eresidae, a highly specialized cribellate family not related to the Dictynidae, which family also contains both cribellate and ecribellate species.

The best-studied examples to date of these "pseudo-orbs" are those of *Titanoeca albomaculata* in Israel (Szlep 1966) and *Fecenia ochracea* in Papua New Guinea (Robinson and Lubin 1979b; Levi 1982). Lehtinen (1967) placed *Titanoeca* in the Titanoecidae, a family perhaps related to the Psechridae, which includes *Fecenia*. The web of *T. albomaculata* is spun near the ground, and the spider has a retreat under a rock (Fig. 13.2). No frame threads are spun, and radii are not constructed in any particular order; some may be laid during the spinning of the cribellate thread. The cribellate thread appears nearly continuous and is laid across the whole range of radii, the spider turning back at the last radius in each series, and, according to Szlep (1966), the measuring activities of the spider resemble those of the orb-weaving uloborids. Additional elements, including sectors and semicircles, may be added later, so that there appear to be several "hubs" in the same web.

The activity of *Fecenia ochracea* was observed in New Guinea by Robinson and Lubin (1979b), and I examined a few of the unusual and exciting webs of these spiders during my stay there in 1980 (Fig. 13.2). Young *Fecenia* make a sort of conical sheet centering on a retreat of densely woven silk. As in some species of the related *Psechrus*, the sheet is operated from the underside. By about the seventh instar, the spider changes its web to a structure rather like that of *T. albomaculata*. However, the "radii" are placed even more irregularly. The spider lays the cribellate band while facing the retreat, which now incorporates a rolled or folded leaf. The spider never makes a complete spiral, but zigzags after the fashion of *T. albomaculata*. The most interesting thing about the behavior of this species is the ontogenetic change from a tented sheet to a planar pseudo-orb. (The final result is remarkably similar to the web of the South American araneid *Spilasma tubulofaciens* (Quintero 1974).)

Such regular webs, built on polar coordinates, do not seem to occur in the ecribellate sheet weavers such as Agelenidae and Linyphiidae, where sticky silk is sparse or lacking. Bristowe (1958) has suggested that the tangled, messy look of theridiid webs may be misleading, and that substantial regularity may exist.

Point-Symmetrical Planar Orb

The variety of point-symmetrical planar orbs, made by the members of two distinct groups (Araneoidea and Uloboridae), is surveyed in chapters in this volume by Eberhard, Stowe, Lubin, and Coddington, and reference to papers describing still more of these webs is made within those chapters. Also valuable are the behavioral notes in the continuing series of revisions of genera of Araneidae by Levi (1973, 1977a, 1977b, 1977c, 1980a, 1981), and two important papers by Robinson and Lubin (1979a, 1979b).

The mode of construction of the orb has been described many times and need not be repeated here. The major features of the web are (1) a planar polygonal frame of strong, dry silk, (2) a set of radial lines convergent on a point (because of modifications of the hub carried out by many species this point often exists only in the observer's imagination), (3) a spiral or circular scaffold of dry silk that may be temporary, may persist, or may not be made at all, and (4) a spiral of cribellate or viscid silk that serves as the major prey-catching element, though in one group (*Cyrtophora* and *Mecynogea*) where the scaffold persists the viscid spiral does not appear (Fig. 7.1).

The orientation of the orb is variable, and all inclinations from horizontal to vertical occur. The webs may be placed among leaf litter on the ground or high in trees or on cliffs. (I have observed *Araneus cavaticus* webs with supporting lines to the frame more than 10 m long.) One specialized guild of unrelated species place their webs close to the surfaces of tree trunks (Robinson and Lubin 1979a). Webs may occur randomly or be closely grouped, even leading to forms of social organization (Uetz, this volume; see Figs. 6.9, 7.2).

The basic elements of the web listed here seem subject to almost endless modifications. The radii vary greatly in number from as few as three (*Epeirotypus gloriae*: Petrunkevitch 1930) to more than 100 (*Cyrtophora*), and may be functionally single lines, branched, or anastomosed. In nephilines, they are pulled out of line and appear "notched." In one genus of Uloboridae, *Polenecia*, cribellate snaring threads are laid on radii (Wiehle 1931; Opell 1979). The hub upon which the radii converge may be strongly modified (see the photographs in Eberhard, this volume), may be the site of a retreat (e.g., *Spilasma*: Quintero 1974; *Cyrtophora*; *Mecynogea*), or may be removed altogether. The dry scaffold may be elaborate and persistent, substituting entirely for the usual catching spiral, reduced to a single circle (Coddington, this volume), or not spun at all (*Eucta*: Crome 1954). When present, the viscid spiral varies greatly in its density and regularity. The mesh made by the intersecting radii and spirals has been evaluated in terms of its prey-catching efficiency by Chacón and Eberhard (1982) and Eberhard (this volume).

There may also be added elements, such as stabilimenta (Robinson and Robinson 1973b; Horton 1980), mazes (constructed independently or derived from additional orbs), retreats, or strings of egg sacs. Most of these elements seem to have to do with protection either of the spider or of its eggs.

As presently defined by most arachnologists, point-symmetrical planar orbs are restricted to one cribellate family (Uloboridae) and five families with viscid silk (Araneidae *sensu latu*, Anapidae, Symphytognathidae, Theridiosomatidae, and Mysmenidae). (J. E. Carico and R. R. Forster (pers. comm.) have discovered a sixth orb-weaving family—not named at this writing—in New Zealand, Australia, and New Guinea.)

Orb-Web Derivatives

No discussion here can improve on those given by Stowe and Coddington in this volume. The cribellate-orb derivatives are treated in Lubin, Eberhard, and Montgomery (1978) and in a thesis by Theuer (1954) extensively quoted in Kaestner (1968).

As Stowe ably points out, most of the modifications of the orb, ranging from simple reduction of a few of the basic elements to nearly complete transformation, have to do with prey specialization. His observations and conclusions serve to verify the insight of Quintero (1974) that each of the basic elements of the orb web may be subject to modification through selective pressures, and that these modifications can take place to a surprising degree independently of modifications of the other elements.

Eberhard (1977b) has remarked on the use of transposable modules of square mesh in the web of *Synotaxus* (Theridiidae), implying that this web can be adjusted thereby to suit a wide variety of situations and prey densities. Such modification of an orb web would be difficult, though *Zygiella* (Araneidae) leave out a wedge-shaped sector of the orb, *Hyptiotes* (Uloboridae) build only such a sector and *Dinopis* (Dinopidae) seem to construct a truncated sector using only the outer segment. Many of the orb weavers with adaptive reduced webs have treated the web elements (radii, scaffold, catching spiral, hub, etc.) as if they were analogues of the modules of *Synotaxus*.

Space-Filling Webs

Space-filling webs often appear highly irregular, though little systematic work has been done on their structure. They appear in several haplogyne families, most notably the Pholcidae, Leptonetidae, and Ochyroceratidae, a group of families probably closely related to one another, and the Diguetidae of North American deserts.

Theridiid spiders spin the quintessential space web, the prototype of the cobweb. Even webs of sheet-spinning theridiids incorporate large numbers of lines arranged irregularly in three dimensions. When an included sheet or horizontal orb is domed, as it is in *Cyrtophora* (Kullmann 1958) or *Frontinella* (Gertsch 1979), and in some *Achaearanea* (Eberhard 1972a), it becomes difficult to distinguish the sheet or orb from the space-filling web. For that reason, I would include in this category only those webs in which the volume that they occupy does not contain a clearly formed, densely woven sheet or orb. However, carefully viewing an apparent space-filling web as just defined often leads to the detection of a planar or domed region of more densely meshed silk.

Spiders using this sort of web always move through it hanging upside down. Sticky silk may be included by theridiids making these webs, usually on lines in the lower part of the tangle.

Some small araneids make space-filling webs that feature regularity to a surprising degree; Coddington (this volume) has pictured and discussed some of these. I observed a small red *Ogulnius*(?) in New Guinea that constructed a basketball-sized web that was obviously point-symmetrical, but not planar.

Some theridiids make webs that are highly regular, albeit reduced. *Episinus* have reduced the number of threads in their "gum-footed" web to a regularly arranged few (Stowe, this volume), *Phoroncidia studo* constructs single, horizontal sticky lines (Eberhard 1981c), and the most remarkable theridiid web yet studied, that of *Synotaxus*, is a regularized web that is built on rectangular, not polar, coordinates, and consists of variable numbers of modules (Eberhard 1977b; see Fig. 12.7). These regular webs are strikingly different from the sheets with regularly arranged threads described above because, instead of being derived from a sheet, the regular portions of the webs are undoubtedly homologous to lines of the tangle, or stopping, maze.

Webless Hunters

This survey may be closed by mentioning the many species of spiders that are obviously secondary hunters. Jackson (this volume) has argued that virtually every araneomorph spider regarded as a hunter (not using a premade silk construction in its predatory activity) is probably descended from web-making forebears. In all but a few of these spiders, where even the "weapon" or snaring function of silk has been taken over by other secretions, such as the cheliceral gum of *Scytodes thoracica*, the silk-spinning apparatus remains as highly specialized and developed as in related web builders.

Such a loss of the web-building habit is most striking in some species of two families, Theridiidae and Araneidae. Here we have species and whole genera that have abandoned webs for a sit-and-wait strategy, active foraging, or even the use of chemical lures (Stowe, this volume). In each case we can clearly connect the species with others making highly reduced webs, and these with still others producing webs typical of their families. The structure of the genitalia of the members of the family Mimetidae (Shear 1981) suggests a relationship with the Araneidae, and mimetids evidently make no capture threads of their own. Instead they invade the webs of other spiders, entice the web builders to draw near, and kill and eat them. They may occupy their victim's web for a brief time. Even in a whole family so obviously characterized by visual pursuit of prey as the Salticidae there is evidence of a former web-building habit (L. M. Forster 1982a; Jackson 1982d, this volume; Jackson and Blest 1982a).

Survey of the Taxonomic Distribution of Web Types

The following survey is not complete, and is taxonomic rather than phylogenetic, though groups of families are dealt with in what most arachnologists, basing their conclusions on comparative morphology, regard as a plesiomorphic to apomorphic order. The point is to show that many of the types of webs just described occur in several lines, which may or may not be related.

Mesothelae

The liphistiomorph spiders make a tube with triplines (Fig. 10.1). An important feature is the regularity in both number and spacing of the lines. Recent studies of these primitive spiders have revealed a surprising wealth of apomorphic characters in the genitalia (Murphy and Platnick 1981). It now seems clear that they are highly specialized, and not ancestral to mygalomorph or araneomorph spiders.

Mygalomorphae

Coyle (this volume) surveys the use of silk in the mygalomorph spiders. Much remains to be done in examining the web-making behavior of these spiders. There are highly specialized types such as the atypids (Coyle and Shear 1981), as well as groups (many theraphosids) that hunt from retreats, by dashing out at prey detected through substrate vibrations. The most evolved type of web present is the tube-and-sheet web built by diplurids, mecicobothriids, and a few other families (Coyle, this volume).

Hypochiloidea

Shear (1970b) and Ferguson (1972) have described the *Hypochilus* web and its use. Again, as in the liphistiids, an important feature is the regular arrangement of the radiating elements in the web, though cribellate silk is laid down as a loose mat without discernible regularity. The single species of the related family Thaididae is said to make a large, cone-shaped sheet in vegetation, which it operates from below (Zapfe and Gertsch 1955).

Forster and Gray (1979) recently described the web of a cribellate gradungulid, *Progradungula*. The web is a small, irregular patch of cribellate silk made near the floor of a cave, but pulled into the vertical plane by threads more than 2 m long. It seems adapted to catch pedestrian insects.

Filistatidae

Observations of *Kukulkania* indicate that filistatid spiders construct a silk-lined retreat, from which radiating lines extend; cribellate silk is laid over these lines. *Filistatoides* species observed in Mexico make a small sheet under rocks, and hang beneath it.

Dysderoidea

The dysderoid families have not been well studied, except for *Segestria* and *Ariadna*, of the Dysderidae (Segestriinae). Their webs somewhat resemble filistatid retreats with triplines, but all the known species are ecribellate, and the webs lack an adhesive element. Most dysderids and oonopids seem to use silk only to make retreat bags for periods of inactivity, and as a covering for eggs.

Scytodoidea

Many species of *Drymusa* (Scytodidae) make nondescript sheets frequently with a retreat at one side. Valerio (1974b) has described a more organized, tubelike web for some *Drymusa* species. *Scytodes thoracica*, a common house spider, no longer uses silk in predatory behavior, but relies on a gum squirted from the chelicerae (Millot 1930; Bristowe 1958). Plectreuridae construct a lining for a retreat, and sometimes also an irregular collar or small sheet. The sheet is usually under a stone and the spiders hang from the underside. The plectreurids are surprisingly like ecribellate filistatids. *Diguetia* is the only genus of Diguetidae; these spiders make an aerial sheet, slightly domed, with a tubular retreat suspended in the center. The exit from the retreat leads to the undersurface of the sheet. Eberhard (1967) has described the prey wrapping of one species, *Diguetia canites*. Lehtinen (1981) has recently reported that species of Tetrablemmidae make small sheet webs in litter.

Pholcoidea

Pholcid spiders hang inverted in tangled space webs, but in several species that I have observed, including the common *Pholcus phalangioides*, there is a planar region in the center of the tangle that, seen from the side, suggests a sheet. Gertsch (1982) reports a finely meshed sheet for some pholcid webs. Little is known of the webs of the litter-dwelling species of Leptonetidae and Ochyroceratidae, but Gertsch (1974) has indicated that some cave-dwelling leptonetids make webs like small pholcid webs.

Amaurobioidea and Dictynoidea

Many families, sometimes difficult to distinguish from one another (Lehtinen 1967; Forster 1967, 1970), fall into this general grouping. Although most that utilize webs make irregular sheets in the style of the well-known North American *Callobius* (Amaurobiidae), *Titanoeca* make webs that sometimes include one or more pseudo-orbs, which are also found in webs of dictynids, and in those of the desid *Matachia* (Forster 1970). Social habits are frequent in this group (Shear 1970a; Tietjen, this volume). Evidently the cribellum has been lost many times in different lines, and most of the families and some genera listed by Lehtinen (1967) include both cribellate and ecribellate members. The webs of many of them are not known.

Psechridae and Stiphidiidae

Psechrus species make a large sheet with a tangle above and below, and operate the web from a retreat, running out on the underside of the sheet (Robinson and Lubin 1979b). The interesting web of *Fecenia* (Psechridae) has already been described; it is a pseudo-orb (Robinson and Lubin 1979b).

According to Forster (1970), *Cambridgea* species (Stiphidiidae) also operate their sheet from below.

Hersilioidea

Uroctea species (Oecobiidae) weave one or several sheets, one above the other. The sheet web evidently serves mostly as a retreat, and the spiders rig guy lines on pylons of stiff silk extending out in all directions from the polygonal sheet (Kullmann 1975). They rush out and attack prey by combing swathing silk through special setae on the anal tubercle. The same habit is shared by the cribellate *Oecobius* (Glatz 1967). *Tama* species (Hersiliidae) have lost the web entirely, lack the greatly enlarged anal tubercle, and attack prey with bands of silk delivered from enormously long spinnerets, as they circle the prey facing away.

Araneoidea

The array of web types in the large family Araneidae and its satellite groups (Theridiosomatidae, Anapidae, Mysmenidae, and Symphytognathidae) has been surveyed in the chapters by Eberhard, Stowe, and Coddington in this volume. All are point-symmetrical planar orbs or their derivatives, with the possible exception of the horizontal, regularly woven sheets of *Mecynogea* and *Cyrtophora* (Araneidae), but evidence is strong that these sheets are derived from more typical orbs and are not their precursors (Lubin 1980b). Linyphiids spin aerial sheets (often multiple), small sheets near the ground, or retreats with sheets; the sheets are always operated from the underside. Some theridiids make webs similar to linyphiid webs, but the distribution of viscid silk is different, being mostly on the vertical lines either above or below the web. Other theridiids spin tangled, space-filling webs, "gum-footed" webs, or webs of only a few lines. At least one theridiid, *Synotaxus* (Eberhard 1977b), has developed a highly regular web with both dry and viscid elements. A number of members of Araneidae and Theridiidae have secondarily given up webs as predatory devices, though some still use silk as an offensive weapon (*Dipoena*: Carico 1978), or rely on chemical attractants (Stowe, this volume; *Phoroncidia*: Eberhard 1981c). All of these families produce a viscid type of sticky silk from aggregate glands, which are not found in other families.

Dinopoidea

The variety of uloborid web types is surveyed by Lubin (this volume). Web building and predatory behavior in some dinopids have been treated by Kaestner (1968) and Robinson and Robinson (1971). The dinopid snare is probably an orb derivative; all uloborid webs are point-symmetrical planar orbs or their derivatives.

Lycosoidea; Other Families

Members of the lycosid genus *Sosippus* make sheet webs with a silk-lined retreat at one end, after the fashion of the agelenids, and these sheets are operated from above as well (Brady 1962). Many species of *Pirata* make silken tubes and hunt from the entrances, whereas *Lycosa* and *Geolycosa* dig tunnels that may have large collars, or trap doors. In the Oxyopidae, juvenile *Peucetia* make small sheet webs, but do not use them to catch prey (Kaston 1972b). Griswold (in press) reports web-building behavior in adult oxyopids (*Tapinillus longipes*, Costa Rica). While adult *Inola* of three species and at least one species each of *Dendrolycosa* and *Architis* (Pisauridae) make horizontal sheet webs (Davies 1982; Nentwig 1985), other pisaurids (*Pisaurina*; Carico 1972, 1985) make webs as juveniles, apparently only as resting sites or for protection while molting. Incidentally, the assertion by Davies (1982) that *Inola* webs are "sticky" should be investigated further. These observations suggest (but do not prove) that the hunting habit so characteristic of this group of families is secondarily acquired. In this connection might also be mentioned such families as the Salticidae, Thomisidae, Heteropodidae, Aphantochilidae, Philodromidae, and Selenopidae, all of which have (probably secondarily) adopted hunting ways, frequently relying on vision.

Eresoidea

Eresids make sheets with retreats and extensive tangles, often with regular arrays of radiating elements at the margins covered with zigzag bands of cribellate silk. A number of species of *Stegodyphus* are highly social.

Gnaphosoidea and Clubionoidea

Members of these groups seem to use silk retreats only during periods of inactivity. Little is known of their hunting behavior, though clubionids use silk to immobilize prey before biting them. No web-building members are known to me.

Conclusions on Web Evolution

An assessment of the evidence just presented seems to me to lead to the following conclusions.

1. Among spider families that are acknowledged to be primitive, the most common kind of silk structure is the retreat, often lined with silk, and having guy lines or a collar. A number of spider groups, emphasizing the protective function of silk, have developed the ability to dig elaborate burrows and close them with trap doors, or have extended the silk collar of the burrow retreat into a long, closed tube. But closed burrows have appeared

independently in several lines, e.g., in Ctenizidae (Mygalomorphae) and Lycosidae (Araneomorphae).

2. In many spiders not closely related, the retreat and collar have been elaborated into sheets. In a few groups, the sheet has become aerial and the retreat more or less lost. This sequence of events has also happened convergently in several groups (see especially Coyle, this volume).

3. Although many spiders operate their sheet webs from above, some spiders hang from beneath the sheet, and evidently have done so from the start. Bristowe (1932), and later Kaston (1964), recognized the importance of this functional difference, but in my opinion interpreted it incorrectly. They claimed that the first group came from spiders that hid their egg sacs on the ground or in a burrow, and the second group from spiders that suspended their egg sacs in foliage. I think it is unlikely that the use of silk for protection and predation developed from the use of it as a covering for eggs. Instead I suggest that the ancestors of sheet-making spiders that operate their webs from above lived in retreats or burrows close to or on the surface of the ground, or beneath objects lying on the ground, so that emergence from the retreat was, perforce, in a dorsal-side-up position. The spiders that work their sheet webs from below came from ancestors with retreats in vertical surfaces, so that the most efficient way to emerge from the retreat to run out on silk lines was to hang from beneath them. Each of these habits has been acquired independently several times, e.g., in the Pholcidae, Stiphidiidae, and Theridiidae, three quite unrelated families that all operate sheets from beneath. Space-filling webs (including those few that are obviously orb-derived) seem all to be evolved from sheets worked from beneath.

4. The orb web has originated perhaps twice, perhaps only once. The cribellate orb weavers probably derived from cribellate sheet-weaving ancestors operating their webs from beneath, as do the Psechridae and cribellate members of Stiphidiidae, Amphinectidae, and Neolanidae today. In one genus of Psechridae (*Fecenia*), sheet webs made by juveniles and operated from beneath become highly regular and orblike in adults. If the retreat is eliminated, this web strongly resembles the uloborid orb. However, it has not been established which group of cribellate sheet weavers, if any, is the sister group of the uloborids. The psechrids are good candidates. Forster and Wilton (1973) have found that the male palp of New Zealand neolanids is similar to the dinopid palp. Dinopids may be related to uloborids (see below). Kaston (1964) and Kullmann (1972a) have erred in identifying *Polenecia* (= *Sybota*) as pre-orb-making uloborids. Opell (1979) has shown that *Polenecia* are specialized relative to other uloborids, and it is likely that their webs are, too.

The origin of the ecribellate orb is the subject of considerable debate, and will be given separate treatment.

Origins of Orb Webs

There appear to be few problems in accounting for the origins of the orb web in the Uloboridae. Not only do several cribellate groups (whose relationships to the uloborids are not clear) make highly regular webs that in some cases might even qualify as orbs, but also the orb-making algorithm in Uloboridae is evidently relatively simple (Eberhard 1969; Lubin, this volume; see Fig. 12.6). Second-instar uloborids, which lack cribellate silk, make a filmy, nearly sheetlike web, reminding one of the ontogenetic development of an orblike web in *Fecenia* (Psechridae).

What is needed is a thorough study of the cribellate groups that might be sister groups of uloborids. The answer well may come when ultrastructural studies now in progress on these groups, which are found mostly in the Southern Hemisphere, are completed (R. R. Forster, in lett.). As noted, the Psechridae are good candidates, but the Stiphidiidae and Neolanidae should also be examined. Each of these families has members that weave aerial or subaerial sheets and operate them from beneath (Forster 1970; Forster and Wilton 1973). The hypothesis put forward here is that the Uloboridae originated from a cribellate ancestor that made an aerial or subaerial sheet web, operated it from beneath, and added regularly designed sectors (as in *Matachia*, *Titanoeca*, or *Dictyna*) that gradually came to dominate the entire web (as in *Fecenia*). I would also hypothesize that this ancestor had already invented the wrap attack, and probably had reduced poison glands. These characteristics, therefore, might also be expected in the cribellate sister group of the uloborids.

Just here I would like to revive a proposal that I first put forward in 1970, in a paper on the predatory behavior of *Hypochilus*. The web of that unusual and primitive spider is surprisingly regular when seen from the proper angle. There is a frame, to which radiating double lines are connected. (The radiating lines converge on the frame threads and diverge widely at the filmy silk sheet spun against the substrate.) A haphazard mat of cribellate silk is combed out over these radiating lines. Few alterations would be required to produce an orb from such a web—one would only have to make the web two-dimensional, make the double radii single, and regularize the cribellate meshwork. The presence of radiating elements in *Hypochilus*'s web, as well as in those made by several genera of mygalomorphs, *Liphistius*, and *Ariadna*, led me to suggest the possible direct origin of the orb web from a web form like that of *Hypochilus*. No further evidence in support of this idea has been found, but it still holds out the tantalizing possibility that the orb web may be much older phylogenetically than we now think. Such an idea could be construed to support either the monophyletic or the convergent hypothesis of the origin of the orb web.

Position of the Dinopidae

Of particular interest in relation to the uloborids are the members of the cribellate "ogre-faced" family Dinopidae. Actually, the "ogre-face," with posterior median eyes vastly enlarged, occurs in only one genus, *Dinopis*. *Avella*, *Avellopsis*, and *Menneus* have normal eyes. In this family, a trapezoidal web incorporating cribellate silk is held free in the spider's anterior legs and is used to trap pedestrian insects against twigs, tree trunks, or leaves (Robinson and Robinson 1971). The algorithm for making this web strongly suggests that it may be an orb derivative (Kaestner 1968; Coddington 1982), but the case for this has been weakened somewhat by the discovery of *Fecenia*'s web.

Dinopids have long been regarded as the sister group of the uloborids (the two are often united in the superfamily Dinopoidea), but few convincing synapomorphies have been found. In *Avella* and *Dinopis* specimens that I dissected there were large poison glands extending into the cephalothorax. This would seem to exclude an uloborid ancestry for the dinopids, but not a sister-group relationship. In *Avella*, but not in *Dinopis*, the calamistrum is located in a curved depression on the fourth metatarsus, as it is in the uloborids, but in all dinopid genera the palpal organs of the males are very different from anything found in the uloborids. The dinopid embolus makes numerous spiral turns around the bulb, from the center of which arises a small sclerite of unknown function. *Avella* and *Dinopis* have trichobothria on the tarsi of the legs; uloborids do not. I could not find femoral trichobothria in *Avella* or *Dinopis*; all uloborids have them. The polarity of these trichobothrial characters will be discussed later.

The cribellate silk of the dinopids and uloborids is strikingly similar when examined under a scanning electron microscope (Kullmann et al. 1975; Kullmann 1975; Opell 1979), with regularly arranged "puffs" of cribellate fibers. The silk of relatively few cribellates has been examined closely, but in *Hypochilus* the puffs are absent (demonstrated to me by J. Palmer). The light micrographs published by Comstock (1948) show that the cribellate threads of *Callobius*, *Dictyna*, and *Filistata* also lack this "puffed" structure, but further investigation of a wide variety of cribellate silk is required. Kullmann (1975) showed that the cribellate silk of *Stegodyphus* (Eresidae) incorporates larger, curled threads and is not puffed. Both "curls" and "puffs" are probably derived.

The anatomy of the cribellum is also worthy of future study. Dinopids and uloborids both have undivided cribella, but undivided cribella are widespread in spiders. There is also a distinct median cribellar lobe in these two families, projecting between the anterior spinnerets, but this character has not been looked for in a range of cribellate families, and the polarity has not been determined.

The data are suggestive of a sister-group relationship; certainly, aside from factors associated with the general architectonic of the web, the dinopids have more morphological features in common with uloborids than do araneoids. Is the dinopid web really a reduced orb? The answer to this question is important, because if dinopids and uloborids are sister groups, and the dinopid web is not an orb derivative, then the ancestor of the two groups in all likelihood was not an orb weaver. A demonstration that the dinopids are the sister group of the uloborids and that their web is not an orb derivative could force adherents of the monophyletic hypothesis who propose sister-group status for the Dinopoidea and Araneoidea to postulate a third loss (in dinopids) of the orb. Such a demonstration would not, however, be inconsistent with a multiple origin of the orb.

If dinopids and uloborids are sister groups, we need to search for a sister group of Dinopoidea, not of Uloboridae alone.

The Ecribellate Orb

The origin of orb webs in ecribellates is problematic, and complicated by the lack of a clear statement of the possible alternatives. The two dominant ideas, broadly stated, are as follows. Kullmann (1972a) and Kaston (1964) have argued for the convergent development of orb webs in two unrelated groups: the Uloboridae and the Araneoidea *sensu strictu*. The convergent hypothesis is also favored by Eberhard (1982), who bases his argument on a broad survey of many genera and species of orb-weaving spiders. Following Thorell (1886), Coddington (this volume) has summarized the evidence for believing that the orb originated but once, in a cribellate common ancestor of the Uloboridae and the Araneoidea *sensu latu* (including some non-orb-weaving families). The resurgence of the monophyletic hypothesis has been due in part to the interest and support of Brignoli (1979), of Opell (1979), and, informally, of R. R. Forster (in lett.). Levi, an authority on the taxonomy of the Araneidae, has voiced some support for the monophyly of the orb web (1980b), but has difficulty accepting the sister-group relationship of Uloboridae and Araneoidea (1980a).

The supporters of both hypotheses accept the monophyly of the traditional Araneoidea (Araneidae and its satellite families plus Theridiidae and Linyphiidae and their satellite families). The convergent hypothesis denies a sister-group relationship between Uloboridae and Araneoidea, but no one favoring this hypothesis has yet suggested what the sister groups of the two taxa might otherwise be—aside from Dinopidae for Uloboridae. If it is assumed that the most important synapomorphies for Araneoidea are the presence of viscid silk produced from aggregate glands and the presence of a paracymbium on the palpus, the convergent hypothesis requires that this combination of characters appear but once, and that the orb be invented twice. (The palpal paracymbium is a lobe (which tends to become a separate sclerite) at the edge of the cymbium, or modified tarsus, of the male palpal organ. It has been found only in the traditional Araneoidea.)

The monophyletic hypothesis considers the Uloboridae (or possibly the traditional Dinopoidea) as the sister group of a monophyletic Araneoidea. Like the convergent hypothesis, this idea requires only one invention of viscid silk and the paracymbium (in Araneoidea), but differs in postulating that the orb was invented once, by a cribellate common ancestor of the uloborids and araneoids, then lost at least twice, in the linyphiids and their satellites and in the theridiids and their satellites. Some cribellate group, not required to be orb-weaving, would be the sister group of Uloboridae-Araneoidea.

Therefore both the convergent and the monophyletic hypothesis as presently understood accept (require?) the monophyly of the old Araneoidea, including the non-orb-weaving families. Araneoid monophyly, however, is not so clear as we may think. The Theridiidae, for example, differ in a range of characters from the Araneidae and Linyphiidae. The theridiids have different-looking viscid-silk glands, use viscid silk in a wrap attack, and spin viscid and dry silk intermittently. They have a comb of specialized bristles on the fourth tarsus, which they use to throw the viscid silk during the wrap attack. Their legs lack macrosetae, and the organization of palpal sclerites differs in never having postembolic elements (Levi 1961). All these features but the last, an obvious plesiomorphy, might be autapomorphies for Theridiidae, not necessarily disproving membership in the old Araneoidea.

A number of suggested synapomorphies for Araneoidea listed by Lehtinen (1967) are not very convincing, such as the presence of a globular abdomen and the lack of tarsal trichobothria. Many included species have elongate abdomens; Lehtinen (1980) has demonstrated the lack of tarsal trichobothria in Hypochilidae, Eresidae, Hersiliidae, Segestriidae, Stenochilidae, Nicodamidae, Mecysmaucheniidae, Atypidae, Filistatidae, and Dictynidae, as well as in the Araneoidea (which he calls Araneomorpha) and Uloboridae. Generally descriptions of this character refer to "lost" trichobothria, but Lehtinen argues convincingly that a lack of tarsal trichobothria represents a primitive pattern of sparse trichobothria on the legs. The more complicated patterns are derived. Absence of tarsal trichobothria cannot, therefore, be a good synapomorphy for uniting the the families in Araneoidea.

The best nonbehavioral synapomorphies, it seems to me, are the production of viscid silk from homologous (aggregate) glands and the presence of the palpal paracymbium. These two characters are absent from all other spiders, Uloboridae included.

For the present, I am willing to consider the Araneoidea monophyletic. Within the superfamily, the orb weavers seem a compact unit itself monophyletic; their behavioral and anatomical characters lead to a number of similar, testable schemes of relationship (Levi 1980a; Eberhard 1982; Coddington, this volume). This group of orb-weaving families is the sister group of the Linyphiidae (possibly including the doubtful family Cyatholipidae); the best synapomorphies are the tendency to develop postembolic palpal sclerites, the tendency for the paracymbium to separate from the cymbium, and the usual presence of an epigynal scape in females. The sister group of these two families is Theridiidae, whose abundant autapomorphies (many shared with the closely related Nesticidae) have already been described.

As I have stated, both the convergent and the monophyletic hypothesis accept the Araneoidea as outlined above. This presents no problems for the convergent hypothesis, but is a source of real difficulty for the monophyletic hypothesis. If the Uloboridae and Araneoidea share a common ancestor that was a cribellate orb weaver, then the orb must have been lost twice, in the Theridiidae and in the Linyphiidae.

Actually losing the orb is not the problem, because that has happened a number of times. Through the pressures of prey specialization, and perhaps for the sake of the more efficient use of silk (usually more active use, substituting relatively cheap muscular activity for relatively expensive protein synthesis), many members of Araneidae have reduced and even lost the orb web (Stowe, this volume). In all of these cases, however, the result has been the use of less silk, not more, and if webs are still built and used by these spiders, the algorithm for making the orb is frequently recognizable (Coddington 1982). On the other hand, the webs of most linyphiids and theridiids contain more silk than the webs of orb weavers of similar size, though recycling of silk by orb weavers might cause the differences to disappear over the lifetime of a spider. More important, the scattered work that has been done on web building in linyphiids and theridiids has revealed not a trace of the orb-weaving algorithm. This is especially true in theridiids, where *Synotaxus* make a highly regular web that bears no resemblance whatsoever to an orb, either in the details of construction or in final appearance (Fig. 12.7). Both of these families use viscid silk in ways quite different from the ways in which it is used by araneids.

Are there any convincing synapomorphies, other than those connected with making the orb web, uniting araneoids and uloborids? Only one has been suggested—the presence of femoral trichobothria on the legs of all uloborids and in a handful of tetragnathine and metine genera of araneids (Opell 1979). Femoral trichobothria are considered derived (Lehtinen 1980), and, if they were found only in the Uloboridae and Araneoidea, would constitute a synapomorphy. The case is weakened by the presence of femoral trichobothria in the Liphistiidae as well (Lehtinen 1980), a group that could not be related to either of the families in question. Liphistiid femoral trichobothria must be a convergent development. We can consider it convergent in Uloboridae and Araneoidea as well, unless other synapomorphies are found.

Coddington (this volume) has listed a number of similarities involving nonadaptive features of orb-web construction, but the difficulty here lies in making outgroup comparisons. Only the two groups under consideration make "real" orbs—two-dimensional, point-symmetrical, and bounded by frame lines—so we cannot really know which among these features might be set by the constraints (Eberhard 1982, this volume) of making such a web. And, of course, monophyly of the orb is what is in question. Naturally each group will have many autapomorphies. The lack of significant synapomorphies for the uloborids and the whole superfamily Araneoidea is a great difficulty for the monophyletic hypothesis. In his discussion, Coddington frequently lapses into an argument that assumes that the orb is the basic type of web for the Araneoidea, and that the demonstration of a sister-group relationship between the Uloboridae and the Araneidae (or at least part of one subfamily of araneids) suffices to bring in all araneoids. But in fact this is something that would follow from such a demonstration, and cannot be a first assumption.

If one accepts the monophyly of Araneoidea and the usual arrangement within the superfamily, then the sister groups of the orb-weaving families for at least two steps down in the cladogram are *not* orb weavers. The morphological synapomorphies for these groups are almost impossible to discount, and we cannot accept the relationship of the uloborids postulated by the monophyletic hypothesis on the basis of the femoral trichobothria and a few behavioral similarities. All the tantalizing suggested behavioral synapomorphies concerning web building are coupled with numerous strong differences; e.g., uloborids lay cribellate silk on preexisting dry radii and frame lines, have a different web-building algorithm, and uniquely carry prey in the palps held over the cephalothorax (Lubin, this volume). But, of course, simply showing masses of autapomorphies in each group does not per se lead to a rejection of a sister-group relationship. What is required is a convincing demonstration that there exists another cribellate family that does not weave real orbs and is more closely related to uloborids than are araneoids. Taken together with the convincing evidence of the monophyly of the Araneoidea, and a strong demonstration, based on studies of even more species, that linyphiid and theridiid web-building behavior is not orb-derived, such data would put the monophyletic hypothesis to rest permanently. On the other hand, if further synapomorphies in behavior and anatomy are found uniting the Uloboridae and the Araneoidea, and if some trace of the orb-weaving algorithm could be found in linyphiids and theridiids, the convergent hypothesis would be untenable.

I find the monophyletic hypothesis of the origin of the orb web very attractive. But, for all its shortcomings, the convergent hypothesis has the advantage of being more congruent with the facts that we have in hand. The weakness of the convergent hypothesis lies in the difficulty of making falsifiable predictions based on it—except that uloborids and araneoids should be unrelated. The monophyletic hypothesis is stronger scientifically because it contains more detail, makes more predictions, and is easier to falsify. But, as Coddington (this volume) has said, when everything else has been eliminated the remaining explanation, no matter how strange, must be right.

In summary, a comparison of the monophyletic and convergent hypotheses of orb-web origins, including an acceptance of a relationship between the uloborids and the dinopids, together with the generally accepted monophyly of the Araneoidea, leads to the following:

The monophyletic hypothesis is favored if (1) the dinopid web is an orb derivative, (2) Theridiidae and Linyphiidae show traces of the orb algorithm, and (3) no other convincing sister-group relationships except for those of Uloboridae (Dinopoidea?) and Araneoidea can be shown for the two groups.

The convergent hypothesis is favored if (1) the dinopid web is not an orb derivative, (2) Theridiidae and Linyphiidae show no trace of the orb algorithm, and (3) either Dinopoidea or Araneoidea can be shown to have closer, non-orb-weaving relatives than one another.

Beyond the Orb

Two types of webs have developed from orbs: reduced orb webs and orb webs with added structures.

Reduced Webs

Stowe (this volume) has provided an excellent survey of a number of highly specialized reduced webs derived from orbs. These webs are considered derived because they retain traces of the orb-weaving algorithm, or because the spiders involved (some of which make no webs at all) are, by morphological evidence, members of the Araneidae.

I have already suggested that the reduction in silk use in some of these spiders may have resulted from the substitution of relatively cheap muscular activity for relatively expensive protein synthesis, looking at spider energetics somewhat along the lines suggested by Ford (1977) and Prestwich (1977). Riechert and Gillespie (this volume) summarize work on *Agelenopsis aperta* that approaches the problem of energetics and strategy from an ecological viewpoint. However, the evidence for the relative metabolic costs of a variety of strategies has not yet emerged. Crucial to web architecture, and the degree of the spider's involvement in it, is a sort of cost-benefit ratio including, on the debit side, the spider's own metabolism (J. F. Anderson 1970), the costs of silk manufacture (Peakall and Witt 1976) and web construction (Ford 1977; Prestwich 1977), and both of these factors relative to muscular activity after the web has been made; on the payoff side is the profit in prey captured and eaten. There is increasing interest in problems relating web architecture to prey-capture efficiency (Chacón and Eberhard 1980; Uetz, Johnson, and Schemske 1978; Eberhard, this volume; Lubin, this volume; Uetz, this volume). Combining these studies with further data on spider metabolism, we may expect to get some answers to questions about the efficiency of web reduction.

Although many spiders making reduced webs based on orbs have increased their activity in connection with the web, by shaking or otherwise manipulating it, many have not. Certainly the moth-attracting spiders in the group Celaeniae are as sedentary as any orb weaver passively waiting at the hub of its orb. The economic argument will take us only so far. Of at least equivalent, and possibly greater, importance are other selective factors, such as protection from potentially dangerous prey, which, it may be argued, led to the development in spiders of wrapping behavior. Of course, wrapping behavior also prevents the escape of still-struggling prey while a second victim is attacked. The wrap attack probably developed from this behavior in response to the need to handle a variety of prey types, especially potentially dangerous ones, such as beetles with robust mandibles, or Hymenoptera with their potent stings (Eberhard 1967). The specializations of *Miagrammopes* (Lubin, Eberhard, and Montgomery 1978), *Hyptiotes*, and *Theridiosoma* seem to be mechanisms that disable the prey as much as possible before the spider comes in contact with it. Some web reduction may have evolved simply as a way of providing an efficient means of manipulating silk to accomplish this end. Of course, smaller webs are less visible to predators.

Stowe (this volume) has emphasized prey specialization, particularly on moths, as a major force in web reduction. He describes numerous suspected and documented cases of araneids utilizing moth sexual attractants, and constructs an appealing series from the building of a full orb to the behavior of *Celaenia*, which evidently attract moths with a chemical lure produced directly from glands on the legs or prosoma. Stowe emphasizes that the behaviors of all of these spiders are not necessarily homologous, but suggests how such predatory techniques might have evolved.

The prey-specialization argument does not hold, however, for those several araneids like *Pachygnatha*, the young of which make orb webs and the adults of which hunt actively (Levi 1980a), and perhaps not for *Archemorus* (Robinson 1980; pers. obs.), which sit on leaves and jump at passing insects and spiders. More research is needed, but there is no present indication that members of these genera are prey specialists.

Members of classic web-making families that produce highly reduced webs are found not only in araneids and uloborids but also in theridiids and linyphiids. The reduced webs of the latter families have been studied with far less thoroughness than those made by araneids. Eberhard (1981c) has reported on the single-line web of the theridiid *Phoroncidia studo*, which he believes may attract flies. *Dipoena* species, which evidently make no webs, prey on ants using a typical theridiid wrap attack with viscid silk (Carico 1978); they may have evolved from an ant-specialist web-making ancestor like *Steatoda fulva*, which makes its web over entrances to ant hills, and moves to other entrances when the ants plug the one that they have chosen, or stop using it (Hoelldobler 1970). These appear to be cases of prey specialization analogous to those in the araneids, and much could be learned by making detailed comparisons.

Of web reduction in the Linyphiidae even less is known. Kullmann (1964) has examined the behavior of *Drapetisca socialis*, which had usually been described as a hunter not making a web. He found that this spider makes a filmy web, tightly appressed to the bark of trees or the sides of buildings. Thus, this spider may be a part of a guild of arboricoles analogous to that studied in an entirely different environment by Robinson and Lubin (1979a). Other linyphiids are known that do not seem to spin webs (e.g., *Tapinopa*). A number of small linyphiids and theridiids are closely associated with ants and may be prey specialists.

Loss of web-building behavior and the substitution of active hunting have occurred independently in other lines. Species of Hersiliidae, a family related to Oecobiidae, are not known to make webs. Instead, they live on tree trunks and attack pedestrian prey by circling around them facing away, paying out silk from enormously elongated spinnerets. This behavior seems related to that described by Glatz (1967) for *Oecobius*, except that in that cribellate genus there are species making web retreats, and the swathing silk is combed through special hairs on the anal tubercle.

I have already mentioned the strong possibility that all of the hunting araneomorphs are descended from web-building ancestors (see also Jackson, this volume). The reduction and even loss of the web have taken place independently at various levels of web evolution, in nearly all the major lines of spiders.

Yet another path to web reduction appears to have passed through kleptoparasitism. The best-known spiders exhibiting this fascinating way of life are the theridiids of the genus *Argyrodes*. Only a little work has been done on the behavior of *Argyrodes* as it relates to web building (Vollrath 1979a), and most of the observations on the group are scattered about within the literature on other species and genera (see especially Robinson and Robinson 1973a). Species of *Argyrodes* are most frequently found in the webs of other spiders, especially the webs of tropical orb weavers, and other genera of

Theridiidae. They may be present in large numbers, and several species may occupy the web of a single large orb weaver—perhaps creating enough of a load to cause the host to move its web (Rypstra 1981). (It has even been speculated that the daily removal of webs by many orb weavers is in part adaptive because it prevents kleptoparasites from accumulating.) *Argyrodes* have been observed to feed on trapped prey that are too small for the host to notice, but they also "steal" prey either by feeding along with the host or by surreptitiously cutting wrapped prey out of the web and making off with them. Some species may cut out small sections of the host web and replace them with their own threads, but probably many species rely entirely on the host web for prey capture (Kullmann 1959; Vollrath 1979a). This habit appears to have led some *Argyrodes* to araneophagy, in which they attack the host (Trail 1980). *Argyrodes fictilium* may be a specialist on the linyphiid *Frontinella communis*. I have observed this bizarre, elongate theridiid systematically destroy a half dozen *F. communis* in as many days. The attack is direct, the predator moving easily through the web to seize its prey. Eberhard (1979a) published some observations on *Argyrodes attenuatus* in tropical America. This species makes single lines that are not sticky, but that spiderlings and males of other species explore; these are caught by *A. attenuatus* by means of thrown silk. *A. attenuatus* also catches flies that habitually perch on spider silk. A highly unusual species of *Argyrodes* from Papua New Guinea, as yet undescribed, lives in the nurseries of a social theridiid species and attacks and eats the spiderlings (Y. D. Lubin, in lett.; Lubin and Robinson 1982).

The family Mimetidae seem closely related to orb weavers on the basis of the anatomy of the male palpal organ (Shear 1981), and the presence on the anterior legs of a rakelike arrangement of spines also found on araneids of the *Arcys* group. (These araneids probably do not make a web.) Recently Forster and Platnick (1984) have challenged this traditional wisdom and claimed a relationship with Palpimanidae, Archaeidae, and Mecysmaucheniidae for the mimetids. However, the case is weak and rests on two synapomorphies of the chelicerae for which most other spider families have not been examined, and which admittedly occur in other families clearly not in Palpimanoidea. The structure of the mimetid palpus and the resemblance of the leg armature to that found in several araneid genera is not addressed in their paper (Forster and Platnick 1984). As far as is known, all of the species of the Mimetidae are araneophagous. (Cutler (1972), however, has induced *Mimetus puritanus* to take insect prey while in captivity.) The few available published data suggest that mimetids may lure their prey closer by simulating the movements in the web of captured insects or courting males; the rakelike spines are used to immobilize the legs of the prey while a quick bite, evidently delivering a fast-acting venom, is given (Bristowe 1958). Nothing is known of prey specialization or of possible web-building activity in this family. I have observed *Mimetus* species to occupy the web of a recent victim for as long as 1 day.

Complete loss of web-building behavior also appears to have taken place in *Curimagua bayano* (Vollrath 1978), a tiny symphytognathid. Forster and Platnick (1977) have noted that several other species of *Curimagua* have, like *C. bayano*, dropped the usual orb-making habits of the family and associated with species of *Diplura*, large, sheet-weaving mygalomorphs. The details of the behavior of these other species is not known, but *C. bayano* is definitely a parasite—its tiny chelicerae could not be used to subdue even the smallest prey. Instead, it walks about on its relatively huge host and sucks up the mixture of enzymes and partially digested prey found around the host's chelicerae as it feeds. Because it deprives its host of valuable protein and evidently cannot feed itself, Vollrath (1978), following the somewhat whimsical terminology of Robinson and Robinson (1977), dubbed it a "dipsoparasite."

Jackson (this volume) has suggested that the large family Salticidae are derived from web-building ancestors through the route of kleptoparasitism as well. See his detailed discussion.

Webs with Added Structures

Web specialization frequently has other results. Studies of the behavior of some orb weavers have shown that many of them have modified the basic orb by adding new elements, or by multiplying standard elements disproportionally. Coddington (this volume) refers to several of these, and they require only a brief review here. Nephilines, for example, leave in the dry scaffold, and they and some other forms, such as *Mecynogea* and *Cyrtophora* (Cyrtophorinae), produce branched or supernumerary radii. These two genera use dry silk to make the catching spiral in their webs. In recent years the usual assumption (Kaston 1964; Kullmann 1972a) that the web of *Cyrtophora* is primitive among the orb weavers and represents a transition from the horizontal sheets of the linyphiids has been reversed. Lubin (1980b) carried out a detailed study of the predatory behavior of *Cyrtophora* species and found them to be specialized, probably derived from an *Argiope*-like ancestor. Levi (1980a) expressed similar conclusions based on the male palpus and other anatomical considerations. Although closely meshed, and having viscid silk only in the tangles immediately above and below, the web of *Cyrtophora* is still an orb. There is no evidence that it represents a transition from linyphiid webs, or that it represents a transition *to* linyphiid webs, as those supporting the monophyletic hypothesis of orb-web origin may argue.

In addition to elaborating or emphasizing certain standard web elements, spiders may introduce new structures. The most interesting of these are stabilimenta: ribbons, spirals, or disks of silk placed in orb webs. As the name suggests, the original hypothesis concerning the function of these structures was that they stabilized the webs that had them, or somehow served as reinforcement; this idea was persuasively argued by Robinson and Robinson (1973b). However, Eberhard (1973) used evidence from construction behavior to suggest that stabilimenta were antipredatory devices. Horton (1980) provided experimental evidence that birds normally preying on spiders learn to avoid webs with stabilimenta, and thus avoid the unpleasant experience of being tangled in silk. Similar evidence is lacking at this time for the functional significance of the irregular tangles of silk that some orb weavers add to their webs (Uetz, this volume), but presumably these are protective also.

Conclusion

I think that the data show the extreme plasticity of web-building behavior under natural selection, particularly in tropical regions where prey specialization, spider-eating predators and parasites, and opportunities for kleptoparasitism, araneophagy, and even dipsoparasitism abound. Data presented earlier in this chapter have demonstrated the potential, if not the actual, existence of extensive convergent behavioral evolution in many spider families. Taken together, these observations should make us cautious about drawing firm evolutionary conclusions later to be translated into a taxonomic scheme. Levi (1980a) voiced a warning repeated by Eberhard (1982) and Coddington (this volume): we must seek characteristics shielded as much as possible from selection in order to find those ethological features of webs that will lead us to a better understanding of spider evolution.

Taxonomic Glossary

Taxonomic Glossary

THE TAXONOMIC GLOSSARY is preceded by a list of family-level or higher taxa. The list is intended simply as a guide to the Glossary, and is not meant to be a new classification of spiders. Many spider families are not mentioned, because the list is restricted to taxa discussed in this volume. However, the arrangement of the families shown probably represents as close to a consensus of spider taxonomists as is now possible, spider classification being an area of a great deal of ferment and argumentation, especially since the publication of Lehtinen's seminal paper of 1967, and the ongoing series of papers by R. Forster and his collaborators. Generally speaking, the older classifications that made distinctions between major groups based on the presence or absence of such features as the cribellum and calamistrum, entelegyne genitalia, claw tufts and scopulorae (or simply the number of claws), and a host of other characters now known to be distributed in a disorderly way among spider families, or to be simply shared primitive features, have been rejected. Subfamilies are not always given, because for some families the subfamilial classification is either poorly worked out or entirely absent. Working on the principle that misinformation is worse than no information at all, I have not tried to impose subfamily names in these groups.

The Taxonomic Glossary proper is alphabetically arranged. Only generic-level or higher taxa are listed; important species are handled under the appropriate genera. For most names listed, the text contains additional information that can be found by using the Index. The purpose of the Glossary is to provide basic orientation only. It would be impossible in this context to provide complete information on spider genera and families, so wherever possible a reference to more detail, usually taxonomic in nature, has been given. A simple citation (e.g., "Levi 1981") indicates that full bibliographic information may be found in the Bibliography; a citation giving the journal title, volume, year, and page numbers has been used for references cited only in the Glossary. Some information for the Glossary was provided by H. W. Levi, N. Platnick, and F. Wanless, but the editor assumes full responsibility for any errors.

Order ARANEAE

Suborder MESOTHELAE

Family Liphistiidae

Suborder MYGALOMORPHAE

Family Atypidae
Family Antrodiaetidae
Family Mecicobothriidae
Family Actinopodidae
Family Ctenizidae
Family Dipluridae
Family Hexathelidae
Family Migidae
Family Theraphosidae

Suborder ARANEOMORPHAE

Superfamily Hypochiloidea
Family Hypochilidae
Family Thaididae
Family Gradungulidae
Superfamily Scytodoidea
Family Scytodidae
Family Diguetidae
Family Plectreuridae
Family Tetrablemmidae
Family Loxoscelidae
Family Sicariidae
Family Filistatidae
Superfamily Pholcoidea
Family Pholcidae
Family Leptonetidae
Family Ochyroceratidae
Superfamily Dysderoidea
Family Dysderidae
Subfamily Segestriinae
Family Oonopidae
Family Telemidae
Superfamily Palpimanoidea
Family Zodariidae
Family Mecysmaucheniidae
Superfamily Clubionoidea
Family Clubionidae
Family Gnaphosidae
Superfamily Philodromoidea
Family Philodromidae

[Superfamily Philodromoidea, *cont'd*]
Family Heteropodidae
Family Selenopidae
Family Cycloctenidae
Superfamily Hersilioidea
Family Hersiliidae
Family Oecobiidae
Family Hadrotarsidae
Superfamily Eresoidea
Family Eresidae
Superfamily Dictynoidea
Family Dictynidae
Family Desidae
Family Anyphaenidae
Family Amaurobioididae
Family Argyronetidae
Family Cybaeidae
Family Hahniidae
Superfamily Agelenoidea
Family Agelenidae
Family Amaurobiidae
Family Neolanidae
Family Amphinectidae
Family Tengellidae
Superfamily Lycosoidea
Family Lycosidae
Family Acanthoctenidae
Family Ctenidae
Family Oxyopidae
Family Pisauridae
Family Psechridae
Family Zoropsidae
Family Stiphidiidae
Family Senoculidae
Family Zoridae
Superfamily Thomisoidea
Family Thomisidae
Subfamily Aphantochilinae
Superfamily Salticoidea
Family Salticidae
Subfamily Spartaeinae
Subfamily Lyssomaninae
Superfamily Dinopoidea
Family Dinopidae
Family Uloboridae
Subfamily Miagrammopinae

Superfamily Araneoidea
  Family Araneidae
    Subfamily Araneinae
    Subfamily Cyrtophorinae
    Subfamily Mastophorinae
      Tribe Exechocentreae
      Tribe Cyrtarachneae
      Tribe Celaenieae
    Subfamily Nephilinae
    Subfamily Gasteracanthinae
    Subfamily Metinae
    Subfamily Tetragnathinae
  Family Archaeidae

[Superfamily Araneoidea, *cont'd*]
  Family Mimetidae
  Family Textricellidae
  Family Theridiidae
  Family Theridiosomatidae
  Family Symphytognathidae
  Family Nesticidae
  Family Micropholcommatidae
  Family Mysmenidae
  Family Linyphiidae
  Family Cyatholipidae
  Family Anapidae
  Family Nicodamidae

Acacesia. Araneidae, Araneinae. A genus of a few species of medium-sized araneids, of which one, *A. hamata*, is relatively common in the eastern United States. *A. hamata* adult females weave orb webs about a fourth of a meter in diameter, with close spacing between turns of the viscid spiral. They are able to capture moths by rapidly attacking prey held momentarily in the web. Species are described by Levi (1977c).

Acantharachne. Araneidae, Mastophorinae. A genus found in Africa and Madagascar. Their anatomy has led Emerit (1978) to suggest that they are bolas spiders; both cephalothorax and abdomen are set with peculiar arrays of spines.

Acanthoctenidae. A family of cribellate spiders, most genera of which occur in the Southern Hemisphere, though a few species of *Acanthoctenus* occur in Central America and the Caribbean. Those that weave webs make sheets in low vegetation. The family seems to be related to the ecribellate lycosid-ctenid complex.

Acanthoctenus. Acanthoctenidae, q.v.

Achaearanea. Theridiidae. A large genus of worldwide distribution, of which one species, *A. tepidariorum*, is virtually ubiquitous, especially around buildings and in settled areas. Species of *Achaearanea* weave the typical irregular "cobweb" of folklore (Fig. 12.5). *A. mundula* is a species of tropical America. *A. wau* is a highly social species occurring in Papua New Guinea, where its large colonies may make sheets more than 1 m across (Lubin and Robinson 1982). See H. W. Levi, Bull. Mus. Comp. Zool. Harv. Univ. 121 (1959): 57–163.

Actinopodidae. A family of mygalomorph spiders found in South America and Australia. The species for which behavior is known dig burrows closed with a trap door. See Coyle, this volume.

Actinoxia. Ctenizidae. *A. versicolor* is a trap-door spider of the western United States.

Aganippe. Ctenizidae, Aganippinae, q.v.

Aganippinae. Ctenizidae. A subfamily found in Africa and Australia, where there are several genera. They all make burrows of varying forms with trap doors. The eyes are usually broadly distributed across the front of the carapace. See Main 1957.

Agatostichus. Araneidae, Mastophorinae. A genus of bolas spiders found in the American tropics, one species occurring northward into the United States. See Gertsch 1955.

Agelena. Agelenidae. A genus of sheet-web-weaving spiders found in Eurasia and Africa. The sheets of solitary species are generally less than 0.5 m^2 in area, and have a silk-lined retreat tube at one side. Two social species are known, from Africa: *A. consociata* and *A. republicana*. Their large social webs may have a volume of several cubic meters, and house more than 1,000 spiders (Darchen 1965).

Agelenidae. A family of both cribellate and ecribellate genera of true spiders, worldwide in occurrence. In North America and Europe, the most common representatives are the "grass spiders" or "funnel spiders," which weave funnel-shaped webs in grass or low vegetation.

Agelenopsis. Agelenidae. A genus of about a dozen species, limited to North America. These spiders are weavers of expansive sheets (those of *A. pennsylvanica* can be nearly 1 m^2 in area) with tubular retreats at one end. *A. aperta* is a much-studied species of the southwestern United States and northwestern Mexico; for a detailed account of its life and habitat, see Riechert and Gillespie, this volume.

Aliatypus. Antrodiaetidae. A genus of atypical tarantulas found in California and Arizona. These medium-sized mygalomorphs dig tunnels that they close with hinged trap doors (Figs. 10.15–10.17). Information on the taxonomy and natural history of the genus may be found in Coyle 1971. *A. aquilonius*, *A. californicus*, *A. isolatus*, *A. janus*, and *A. thompsoni* are mentioned in Coyle, this volume.

Alpaida. Araneidae, Araneinae. *A. calix* is a small orb weaver found in the eastern United States, from Maryland to Illinois and south to western Florida. The genus is predominantly tropical, with as many as 100 species (H. W. Levi, Bull. Mus. Comp. Zool. Harv. Univ. 147 (1976): 351–91).

Amaurobiidae. A family of cribellate and ecribellate genera of true spiders, once called Ciniflonidae, worldwide in distribution. Many of the web-building species make irregular, messy, sheetlike tangles on the ground or under objects. A few social species are known. For information on North American species, see R. Leech, Mem. Entomol. Soc. Can. 84 (1972): 5–182; for information on New Zealand species, see Forster and Wilton 1973.

Amaurobioididae. A family of ecribellate genera of true spiders known only from the Southern Hemisphere, and including some species that inhabit rocky intertidal zones, withstanding immersion in saltwater in a tightly woven silk bag (Forster 1970).

Amaurobius. Amaurobiidae. *Amaurobius* species (including *A. ferox*, *A. similis*, and *A. terrestris*) are found in Europe and temperate Asia. Some species live in burrows with silk collars (Bristowe 1958; 198); others make tangles of matted silk under stones on the ground. *A. ferox* has been imported accidentally to America and has established itself in several areas; I have observed it on limestone cliffs along the shorelines of islands in Lake Erie.

Amphinectidae. A family of cribellate and ecribellate genera of true spiders, endemic to New Zealand and eastern Australia. Forster and Wilton (1973) describe 16 genera from New Zealand; species are either vagrant hunters or make small sheet webs extending from a silk-lined retreat.

Anapidae. A family of tiny, heavily armored araneoids. The female palps are reduced or absent, and both sexes have a peculiar spur on the labrum. Many are found in tropical or subtropical areas. Some of the known species make tiny orb webs near the ground in rain forests.

Anapis. Anapidae. Twenty-one species are known from tropical America (Platnick and Shadab 1978a). All are tiny, and where webs are known, they are small orbs (Fig. 12.14). The photograph of the web of *A. felidia* in Platnick and Shadab's (1978a) revision of the genus is upside down. *A. heredia* occurs in leaf litter in Costa Rican forests; females are about 1.2 mm long. The web is shown in Fig. 12.16.

Anapisona. Anapidae. A genus of 11 known species, all from tropical America (Platnick and Shadab 1978a). Details of eye arrangement and genitalia separate them from other anapids. Probably most species make small orb webs featuring out-of-plane radii (Fig. 12.15). *A. simoni* is from Panama.

Anelosimus. Theridiidae. A cosmopolitan genus making typical irregular space webs. A few species are (facultatively?) social, e.g., *A. studiosus* of North America, *A. eximius*, and *A. jucundus*. See H. W. Levi, Trans. Am. Microsc. Soc. 75 (1956): 407–21.

Antrodiaetidae. A family of mygalomorph spiders almost endemic to North America. (There are two species in Japan). All are burrowers, but treat the burrow entrance in various ways: by building a trap door, a folding silk collar, or a long silk tube incorporating debris. See Coyle 1971.

Antrodiaetus. Antrodiaetidae. Thirteen species are known, 11 from North America and two from Japan. All are burrowing mygalomorphs that close their burrows with a folding silk collar (Figs. 10.8, 10.9). *A. unicolor* is common on banks in moist forests from New Jersey to Mississippi. The genus was revised by Coyle (1971).

Anyphaenidae. A dictynoid family of ecribellate, medium-sized to small spiders, usually with pale coloration. None makes webs, most run on foliage. The courtship behavior of some of the species involves buzzing sounds made by rapidly vibrating the abdomen against leaves. The American genera have been revised by N. I. Platnick (Bull. Mus. Comp. Zool. Harv. Univ. 146 (1974): 205–66).

Aparua. Dipluridae. Eight species are recorded from New Zealand (Forster and Wilton 1968). These mygalomorphs construct burrows (not closed with trap doors) or make silk tubes under stones or logs.

Aphantochilidae. A family not consistently recognized by arachnologists, and usually included in Thomisidae, q.v. They make no webs. *Aphantochilus* species are ant mimics.

Aphonopelma. Theraphosidae. A large genus that has not been well studied by taxonomists, *Aphonopelma* includes many of the famous tarantulas of the American West (Baerg 1958). Most species live in burrows with a diaphanous silk collar (Fig. 10.20). The males wander in late summer and create substantial comment because of their large size. *A. chalcodes* has been studied by Minch (1979a).

Aporoptychus. Ctenizidae. An African genus of trap-door spiders (Blandin and Célérier 1977).

Arachnura. Araneidae, Araneinae. A genus of orb weavers from tropical Asia that, in their resemblance to fallen flower petals, seem adapted to concealment. The New Guinea species, *A. melanura*, has a long, tailed abdomen and is yellow or brown. The web is a normal orb, but lacks spiral elements in two sectors (Robinson and Lubin 1979a).

Araneidae. The major family of orb-weaving spiders, cosmopolitan in distribution. Perhaps 3,000 species are known. Most make orb webs or webs derived from orbs; some make no webs at all. The family is too diverse to characterize briefly; see the series of revisions by Levi (cited in the Bibliography) for impressions of the variation in body form and size within the family. The monophyly of the family has recently been questioned, and there is as yet no reliable distribution of the genera into subfamilies and tribes, though Cyrtophorinae, Mastophorinae, Celaeniae, Araneinae, Nephilinae, Argiopinae, Cyrtarachneae, Exechocentreae, Metinae, Gasteracanthinae, and Tetragnathinae (which is sometimes ranked as a family) are mentioned in this volume.

Araneinae. Araneidae. A subfamily including orb weavers large and small. The definition is based on the structure of the male palp (Levi 1980a). Typical genera: *Araneus*, *Nuctenea*, *Cyclosa*, *Metepeira*, *Acacesia*, *Eriophora*, *Neoscona*, and *Verrucosa*.

Araneoidea. A difficult-to-define superfamily containing the families Araneidae, Archaeidae, Theridiidae, Linyphiidae, Anapidae, Theridiosomatidae, Mysmenidae, Textricellidae, Micropholcommatidae, Symphytognathidae, Nicodamidae, Nesticidae, Cyatholipidae, and Mimetidae. All have serrate hairs and aggregate and (probably) flagelliform silk glands, are ecribellate, and have certain similarities of eye structure. Habits are diverse. See Coddington, this volume.

Araneomorphae. A suborder of spiders (sometimes called Labidognatha) usually taken to include all "true" spiders, i.e., all with diaxial chelicerae and respiratory tracheae.

Araneus. Araneidae, Araneinae. Typical orb weavers—the spiders that biologists think of when orb weavers are mentioned (H. W. Levi, Bull. Mus. Comp. Zool. Harv. Univ. 141 (1971): 131–79). Usually they are large or medium-sized spiders, with distinctively marked globular abdomens. One group of small-bodied species evidently inhabits the forest canopy (Levi 1973). The webs may be operated from the hub, or the spider may hide in a nearby retreat, linked to the hub by a signal line. *A. bandelieri*, a subsocial species of northern South America, probably does not belong to the genus. *A. diadematus* may be the most intensively studied spider; see the references to Witt and co-workers in the Bibliography. *A. trifolium* is a large, brightly colored species of eastern North American meadows, where it makes an orb a little less than 1 m in diameter. The spider hides in a nearby "tent" of leaves tied together with silk.

Araniella. Araneidae, Araneinae. A small genus of small spiders. *A. displicata*, the most common North American species, makes an orb on the upper surfaces of cupped leaves or on flowers. The orb may be only a few centimeters across. See H. W. Levi, Bull. Mus. Comp. Zool. Harv. Univ. 146 (1974) 291–316.

Archaea. Archaeidae. A genus of spiders from the Southern Hemisphere of uncertain taxonomic position; they may be araneoids, though Platnick and Forster (1984) divide the group into four families and place them in Palpimanoidea. Reportedly they feed only on other spiders, which they spear with enormously long chelicerae. Web building, if present, is unknown.

Archaeidae. An araneoid family of ecribellate spiders including the genus *Archaea*, q.v.

Archemorus. Araneidae. A genus of small, hard-bodied spiders of the tropics of the western Pacific. In New Guinea, they sit on leaves and jump at passing insects and spiders; some are camouflaged by their resemblance to a spot of bird dropping. A further account is given in Stowe, this volume.

Arcys. Araneidae. These brilliantly colored araneids of Australia make no webs (Fig. 5.3b). They may attract moths through the production of analogues of the moths' sex pheromones. *A. clavatus*, *A. lancearius*, and *A. nimdol* are mentioned in the text; more detail can be found in Stowe, this volume.

Argiope. Araneidae, Argiopinae. Large orb weavers usually of open habitats: fields, meadows, roadsides, marshes, and forest edges (H. W. Levi, Bull. Mus. Comp. Zool. Harv. Univ. 136 (1968): 319–52). Their typical orbs are placed low in annual or shrubby vegetation, and usually include a ribbon of silk, the stabilimentum, across the hub. The spiders are brightly colored, often with metallic-looking spots. *A. aurantia*, the garden spider, is vivid yellow and black with a silver carapace; it is a favorite of North American ecologists. *A. argentata* is found in tropical America and is a large, silver species. *A. trifasciata* shares its range with *A. aurantia* but is distinct in having the abdomen marked with crossways stripes of silver, yellow, and black.

Argiopinae. Araneidae. See *Argiope.*

Argyrodes. Theridiidae. A fascinating genus, most species of which seem to live as kleptoparasites in other spiders' webs, or to prey on other spiders by invading their webs and killing them. Both *A. attenuatus* (Eberhard 1979a) and *A. fictilium* have very elongated abdomens, and look almost wormlike. A taxonomic revision was published by Exline and Levi (1962); kleptoparasitic habits were studied by Vollrath (1979a).

Argyroneta. Argyronetidae. *A. aquatica*, the water spider of Europe, makes a dome-shaped web underwater, which it fills with air carried from the surface. It preys on freshwater invertebrates, small fish, and tadpoles. See Bristowe 1958, pp. 201–7, Pl. XXII.

Argyronetidae. See *Argyroneta.*

Ariadna. Dysderidae, Segestriinae. Species of *Ariadna* line holes and crevices in vertical or near-vertical surfaces with tough silk, which flares out to form a small collar. They rush out to attack passing insects and spiders. The American species were revised by Beatty (1970).

Ariamnes. Theridiidae. A generic name still sometimes applied to a few of the more bizarre species of *Argyrodes*—those with extremely long abdomens—though the name was synonymized by Exline and Levi (1962).

Ariston. Uloboridae. Tiny uloborids, four species of which are found from Mexico to Honduras (Opell 1979).

Asemonea. Salticidae. See Wanless 1980a.

Atrax. Hexathelidae. The venom of *A. formidabilis*, the funnel-web spider of Australia, is highly toxic to man. The spiders have aposematic coloration in bright red and black. See McKeown 1952 and Mascord 1980.

Atypidae. A mygalomorph family found in Europe, Africa, North America, Japan, and the Asian tropics. Typical webs are tubes that the spiders never leave, except when the males mature and go on mating excursions. The family was revised for North America by Gertsch and Platnick (1980).

Atypoides. Antrodiaetidae. Mygalomorphs found only in Oregon, California, Missouri, and Illinois. There are three species. *A. gertschi* and *A. hadros* close their burrows with a short, collapsible collar (Fig. 10.11); *A. riversi* extends the collar into a long tube held erect by incorporated debris (Figs. 10.12–10.14). See Coyle 1971 for a detailed account of their natural history.

Atypus. Atypidae. An Old World mygalomorph genus of atypical tarantulas. European species live in burrows lined with silk; the silk lining is extended as a tough, closed tube that in most species lies flat on the ground under leaf litter or grass. The spider attacks insects crossing the tube by stabbing through it with enormous cheliceral fangs. One species, *A. snetsingeri*, has been established in Pennsylvania, coming from parts unknown. The European fauna was revised by O. Kraus and Baur (Abh. Verh. Naturwiss. Ver. Hamburg 17 (1974): 85–116).

Aulonia. Lycosidae. A genus of web-building wolf spiders most diverse in Africa (Job 1974).

Australothele. Dipluridae. Seven species of medium-sized to large Australian ischnotheline spiders living in tubular, often branched silk retreats with very little (if any) exposed capture web (Raven 1984).

Avella. Dinopidae. A genus found in the Asian tropics and Australia. *Avella* share the web-throwing habits of *Dinopis* (pers. obs.) but do not have their enormously enlarged eyes.

Avellopsis. Dinopidae. From South Africa, these dinopids, like *Avella* species, have normal-sized eyes. Their web-building habits are unknown.

Avicularia. Theraphosidae. *A. avicularia*, a large tarantula of the southern Caribbean and northern South America, makes its web, a flattened silk tube, on tree trunks, and rushes out at night to attack large insects and small lizards (Stradling 1981). This species supposedly also attacks birds on their nests; hence its name.

Azilia. Araneidae, Tetragnathinae. A genus including four or five species of weavers of typical orb webs, from South and Central America. One species, *A. affinis*, is also found in the southern United States (Levi 1980).

Badumna. Amaurobiidae? Desidae? An Australian genus of cribellate spiders, including at least one social species, *B. insignis* of Queensland. Lehtinen (1967) places the genus in Amaurobiidae, whereas Forster (1970) implies membership in Desidae by including in Desidae (Desinae) the related *Matachia*.

Barychelidae. A mygalomorph family from South America, Africa, Australia, and India, species of which either dig burrows or make tubes on tree bark. Many make two entrances, each closed with a trap door. See Coyle, this volume.

Boethus. Salticidae, Boethinae. See Wanless 1978.

Bothriocyrtum. Ctenizidae. A common genus of trap-door spiders in California. See Coyle, this volume, for information on *B. californicum*.

Brettus. Salticidae. See Wanless 1979.

Bymainiella. Hexathelidae. Species of this Australian mygalomorph genus live in silk tubes under rocks. *B. boycei* is arboreal; *B. lugubris* flares its burrow lining into a large funnel at the entrance. See Coyle, this volume, and Raven 1978.

Calathotarsus. Migidae. *C. coronatus* is found in Chile, and lives in a burrow with a large silk funnel at the entrance (Buecherl 1962).

Calisoga. Ctenizidae. A North American genus of trap-door spiders; a number of species occur in the western part of the continent (R. V. Chamberlin, Bull. Univ. Utah Biol. Ser. 3 (1937): 1–11).

Callobius. Amaurobiidae. A genus of cribellate spiders common in forest regions of the United States, living in irregular, matted webs under rocks, in the cracks of vertical cliffs, or under tree bark. The American species were formerly listed under *Amaurobius*. For a revision see R. Leech, Mem. Entomol. Soc. Can. 84 (1972): 5–182.

Calommata. Atypidae. An Asian and African genus. Little is known of the behavior of *Calommata* species. Probably they live in closed silk tubes like those of *Atypus* and *Sphodros*.

Calymmaria. Agelenidae. A group of small to medium-sized spiders. A few species are found in the Appalachians and many more in the Pacific Northwest of the United States. The small species make filmy sheets under stones or logs. *C. cavicola* is a larger Appalachian species that builds a complicated funnel-and-sheet web under rock ledges.

Cambridgea. Stiphidiidae. More than a dozen ecribellate species are known in New Zealand. The spiders resemble agelenids, but operate their sheet webs from below (Forster and Forster 1973).

Cantuaria. Ctenizidae. A genus of trap-door spiders from New Zealand, including about 45 species. Some do not close their burrows with trap doors, e.g., *C. huttoni*; others make doors, e.g., *C. vellosa*. See Forster and Wilton 1968 for descriptions of species.

Carrai. Dipluridae. A monotypic genus of medium-sized Australian ischnotheline spiders living in tubular silk retreats with multiple entrances (Raven 1984).

Cataxia. Ctenizidae. An Australian genus of trap-door spiders (Main 1969).

Celaeneia. Araneidae, Mastophorinae, Celaenieae. An Australian and New Zealand genus of webless araneids thought to attract moths by mimicking the female sex attractants of target species. *C. kinbergi* (Fig. 5.7a) is discussed by Stowe (this volume); see also McKeown 1952 and Mascord 1980.

Celaenieae. Araneidae, Mastophorinae. See *Celaeneia*.

Cethegus. Dipluridae. Australian diplurids weaving the most aerial webs of any mygalomorphs. See Coyle, this volume, and Main 1960, 1976.

Chasmocephalon. Anapidae. A genus of tiny, armored orb weavers. The web of the Californian *C. shantzi* has anastomosing radii and out-of-plane radii (Fig. 12.17).

Chilobrachys. Theraphosidae. A little-known mygalomorph genus from tropical Asia.

Chinoscopus. Salticidae, Lyssomaninae. A genus of supposedly primitive jumping spiders.

Chorizopes. Araneidae. Little is known about this tropical Asian genus of orb weavers; they may be related to the bolas spiders.

Chrosiothes. Theridiidae. A genus of more than a dozen species of small spiders from the Americas (Colorado to Ecuador) weaving typical theridiid cobwebs under ledges in forested regions. See H. W. Levi, Psyche 71 (1964): 73–92.

Chrysometa. Araneidae, Metinae. The orb of a tropical species is shown in Fig. 4.2c.

Chrysso. Theridiidae. A genus of small cobweb weavers from Asia and North and South America. The species are often brightly colored in red, yellow, black, and white. For descriptions of American species see H. W. Levi, J. N.Y. Entomol. Soc. 63 (1955): 59–81.

Ciniflo. Amaurobiidae. Synonym of *Amaurobius*.

Ciniflonidae. Synonym of Amaurobiidae.

Cladomelea. Araneidae, Mastophorinae. A genus of African bolas spiders. See Stowe, this volume.

Clubiona. Clubionidae. A large, cosmopolitan genus of vagrant, nocturnal, hunting spiders, small to medium-sized, usually white, yellow, or brown in color. *C. cambridgei* is an Australian species that eats the eggs of other spiders. For a review of United States species, see Edwards, Bull. Mus. Comp. Zool. Harv. Univ. 118 (1958): 365–436.

Clubionidae. A widely distributed family of ecribellate spiders characterized by generalized genitalia, claw tufts, and usually a soft, light-colored body. Webs are not constructed, only silk bags used as retreats.

Clubionoidea. A grouping of ecribellate families with two claws and claw tufts. All are hunters, making silk constructs only as retreats. Includes Clubionidae and Gnaphosidae.

Cocalodes. Salticidae. See Wanless 1983.

Cocalus. Salticidae. See Wanless 1981a.

Codeta. Salticidae. F. R. Wanless (pers. comm.) considers this genus a synonym of *Gelotia*. The spiders are medium-sized, inconspicuous forest inhabitants found in southeastern Asia, Borneo, and Sumatra.

Coelossia. Araneidae, Mastophorinae, Exechocentreae. An African genus of peculiarly modified spiders. Habits are unknown, but their morphology suggests similarities to other genera of Mastophorinae. See Stowe, this volume, and Emerit 1978.

Coleosoma. Theridiidae. Species of cobweb weavers are included; *C. floridanum* is cosmotropical. The females look like typical theridiids; the males are ant mimics (H. W. Levi, Breviora 110 (1959): 1–11).

Conifaber. Uloboridae. *C. parvus* makes a unique orb-plus-cone web. See Lubin, this volume.

Conothele. Ctenizidae. *C. arboricola* of Australia is one of a few arboreal trap-door spiders, living in a silk tube camouflaged with bits of bark (Main 1957).

Corasoides. Amaurobiidae? Desidae? An Australian genus of cribellate sheet-and-tangle weavers. Lehtinen (1967) places the genus in Amaurobiidae (Desinae), whereas Forster (1970) recognizes Desidae as a separate family without mentioning *Corasoides*.

Cribellatae. An obsolete taxon including all cribellate spiders. Since we now understand that the cribellum and calamistrum are synapomorphies for all araneomorph spiders, the term has only informal meaning.

Ctenidae. A lycosoid family defined by eye placement and structure. It may be only a subfamily of Lycosidae.

Ctenizа. Ctenizidae. A European genus of trap-door spiders. *C. sauvagesi* runs trip-lines from its burrow entrance (Buchli 1969).

Ctenizidae. The classic trap-door spider family, cosmopolitan in distribution. See Coyle, this volume, for a review of their predatory behavior.

Cupiennius. Ctenidae. A South American genus of large, cursorial, hunting spiders commonly found on foliage. *C. salei* has been the subject of intensive study concentrating on behavior; see the references in the Bibliography to the work of Barth.

Curimagua. Symphytognathidae. *C. bayano*, one of the two tiny species of this genus, makes no web, but lives in the nests of large diplurids in Central America. The spiders cannot feed themselves (their chelicerae are fused and reduced) and share the meals of their hosts as "dipsoparasites." Vollrath (1978) provides behavioral information; Forster and Platnick (1977) describe the genus.

Cyatholipidae. A family of Araneoidea that make sheet webs and operate them from beneath. Affinities are with linyphiids and theridiids. V. T. Davies (Symp. Zool. Soc. London 42 (1978): 293–302) originally recognized the family, but called it Teemenaariidae.

Cybaeidae. A dictynoid family of both cribellate and ecribellate spiders, most of which were formerly included in Agelenidae. Typically ground-dwelling, the spiders build sheet webs and operate them from the upper surface. The family is predominantly Holarctic.

Cyclocosmia. Ctenizidae. A genus of trap-door spiders (Fig. 10.22) of the southern United States, Mexico, and Asia, including species with a posteriorly truncated abdomen ending in a hard, round, ridged plate used in emergencies to block a narrow section of the burrow (W. J. Gertsch and N. I. Platnick, Am. Mus. Novit. 2580 (1975): 1–20).

Cycloctenidae. A small philodromoid family of ecribellates from Australia and New Zealand placed by Lehtinen (1967) in Lycosoidea. Webs are not constructed.

Cyclosa. Araneidae, Araneinae. A cosmopolitan genus of typical orb weavers with modified abdomens; the abdomens are soft but drawn out into "tails" of various lengths (Levi 1977a). Together with a stabilimentum of debris and egg sacs in the center of the orb, this modification may help in camouflage. Six species (*C. bifurca*, *C. caroli*, *C. conica*, *C. insulana*, *C. trifida*, and *C. turbinata*) are mentioned in the text. For the web-takedown behavior of *C. turbinata*, see Carico, this volume.

Cyphonesia. Barychelidae. *C. obesa* digs a shallow, short tube in the soil with two trap doors only millimeters apart facing in exactly opposite directions (Blandin and Célérier 1977).

Cyrba. Salticidae. A small genus of brightly colored jumping spiders from the Mediterranean region. *C. algerina* is found under stones in North Africa and southern France.

Cyrtarachne. Araneidae, Mastophorinae, Cyrtarachneae. An Old World genus of medium-sized, brightly colored weavers of extremely sticky, horizontal, reduced orb webs (Fig. 5.4b). *C. bufo* and *C. nagasakiensis* have been studied in Japan. See Stowe, this volume.

Cyrtarachneae. Araneidae, Mastophorinae. A tribe of orb weavers that seem to have invented an especially sticky glue for their webs. This has allowed for substantial web reduction, as in *Pasilobus*. See Stowe, this volume.

Cyrtocarenum. Ctenizidae. A North African genus of trap-door spiders. See Coyle, this volume.

Cyrtophora. Araneidae, Cyrtophorinae. Including many species of tropical orb weavers, *Cyrtophora* occur the world around in the tropics and subtropics. See Cyrtophorinae for a description of the web. Several species are social, e.g., *C. citricola* (Fig. 7.1) and *C. molluccensis*; others are solitary, e.g., *C. cylindroides* and *C. monulfi*. A few species, especially *C. citricola*, have been intensively studied; see the references to the work of Lubin and of Kullmann in the Bibliography.

Cyrtophorinae. Araneidae. A subfamily of orb weavers building horizontal, closely meshed orbs suspended in dense tangles of silk (Fig. 7.2); the spiral is not viscid.

Damarchus. Ctenizidae. *D. cavernicolus* is a cave-dwelling Malaysian mygalomorph that lives in an open burrow (Bristowe 1952).

Daramuliana. Uloboridae. Two species from the western Pacific are included; their web-building behavior is unknown (Opell 1979).

Dekana. Dipluridae. Australian mygalomorphs that excavate burrows and line them with silk (Main 1976).

Desidae. A dictynoid family of cribellate and ecribellate spiders most diverse in the Southern Hemisphere, and including both web-building species and wandering hunters. Many of the genera are associated with water or are intertidal. Forster (1970) revises the New Zealand species.

Dicrostichus. Araneidae, Mastophorinae. *D. magnificus*, the "magnificent spider," is an Australian bolas spider (McKeown 1952; Mascord 1980).

Dictyna. Dictynidae. One of the largest spider genera in number of species, *Dictyna* occur in both the North and the South Temperate Zone. Species are small, cribellate spiders found in a variety of habitats, making webs on leaves, in crevices, between twigs, etc. A few produce web sectors that have a regular arrangement of cribellate threads, and suggest parts of orb webs. *D. calcarata* is common in the eastern United States; *D. follicola* is found in Japan. American species are described by Chamberlin and Gertsch (1958).

Dictynidae. A large, cosmopolitan family of cribellate spiders, the concept of which has been expanded greatly by Lehtinen (1967) to include some ecribellates formerly placed in Argyronetidae and Agelenidae; not everyone agrees with him. *Dictyna* is a typical genus.

Dictynoidea. A superfamily of cribellate and ecribellate true spiders, including the families Dictynidae, Hahniidae, Desidae, Cybaeidae, Argyronetidae, Amaurobioididae, and Anyphaenidae (Forster 1970). Other groups may also belong here.

Diguetia. Diguetidae, q.v.

Diguetidae. A small family of six-eyed haplogyne spiders including the genus *Diguetia*, with species in the southwestern United States, Mexico, and Argentina (W. J. Gertsch, Am. Mus. Novit. 1904 (1958): 1–24). The web is a horizontal sheet suspended in a tangle, with a vertical retreat tube in the center.

Dinopidae. A family of cribellate spiders, perhaps related to Uloboridae, and distributed throughout the warm regions of the world. The genera for which behavior is known are characterized by a unique web, described under *Dinopis*.

Dinopis. Dinopidae. A cosmotropical genus of spiders with enormous posterior median eyes at the front of the carapace. They weave small, rectangular webs that they carry with the front legs and throw at passing prey (Fig. 5.1b). The web may be a highly modified orb (Fig. 12.13). *D. spinosus* is present in the southernmost United States, *D. longipes* in Panama. Robinson and Robinson (1971) give a typical account of the predatory behavior; see also Coddington, this volume.

Dinopoidea. A superfamily of cribellate spiders including the families Uloboridae and Dinopidae. See Coddington, this volume.

Diplura. Dipluridae. Many species of *Diplura* are found in the American tropics, where they weave sheet webs. See Coyle, this volume.

Dipluridae. A mygalomorph family containing both sheet-web-building and burrow-dwelling species. The spinnerets are often extraordinarily long. See Coyle, this volume.

Dipoena. Theridiidae. Small spiders with triangular abdomens, often strikingly marked with silver and black. Males have unusually boxlike carapaces.

Dolichognatha. Araneidae, Tetragnathinae. The several species of this genus make horizontal to vertical orbs between the buttress roots of trees (Fig. 4.21), and hang their egg sacs in a line above or behind the hub, where they also sit, camouflaged (Levi 1981).

Drapetisca. Linyphiidae. *D. socialis* of Europe was thought to be a cursorial hunter (nearly all linyphiids make webs) until the web was discovered on tree trunks by Kullmann (1964).

Drymusa. Scytodidae. Six-eyed haplogyne spiders of tropical America, making tubular webs open at either end (Valerio 1974b).

Dugesiella. Theraphosidae. Many species of North American tarantulas belong here. *D. hentzi* is said to occur in Texas, but the name has been misapplied by laboratory biologists to virtually any large, black to brown, North American theraphosid. There are many undescribed species.

Dyarcyops. Ctenizidae. An Australian genus of trap-door spiders (*D. fuscipes*, *D. pulchellus*, and *D. robertsi*) that extend their burrow linings into large tubes or funnels (Main 1976).

Dysderidae. A family of six-eyed haplogyne spiders with great diversity in the Mediterranean region, where they live mostly under rocks or bark or in caves. Species of subfamily Segestriinae make silk tubes in holes in vertical surfaces.

Dysderoidea. A poorly defined superfamily, usually including Dysderidae and Oonopidae, which seem to be just smaller dysderids. Oonopids are common in tropical leaf litter. Some arachnologists would include the lungless Telemidae in Dysderoidea.

Eidmanella. Nesticidae. *E. pallida* makes an irregular, space-filling web in caves and the basements of old houses. Most species are cave dwellers, but their usual habitat might be small, deep crevices or the spaces under deeply buried stones (W. J. Gertsch, Bull. Texas Mem. Mus. 31 (1984): 1–91).

Enoplognatha. Theridiidae. A large genus of ground-living theridiids of the Northern Hemisphere. Webs are usually made under stones or logs. North American species are treated by H. W. Levi (Bull. Am. Mus. Nat. Hist. 112 (1957): 1–123).

Epeiriformes. An obsolete name for a taxon roughly corresponding to Araneoidea.

Epeirotypus. Theridiosomatidae. *E. gloriae*, a tiny spider of Puerto Rico, makes a much-reduced orb web, with just three radii and a sticky "spiral" in the form of two concentric triangles (Petrunkevitch 1930). The web of an undetermined species is shown in Fig. 12.7.

Episinus. Theridiidae. This genus is cosmopolitan, but species are most common in the tropics and subtropics. The abdomen is broadly triangular. The H-shaped construction of threads on which the spiders are often found (Bristowe 1958, Pl. XXVI) seems to be a resting web rather than a capture web (Stowe, this volume). American species are treated by H. W. Levi (J. N.Y. Entomol. Soc. 62 (1955): 65–90).

Eresidae. A family of cribellate spiders found in Africa, southwestern Asia, India, and southern Europe. Species have compact bodies and a square-fronted carapace; males are often brilliantly colored. Some species of *Stegodyphus* and *Eresus* are subsocial to eusocial. For beautiful color pictures, see E. Kullmann, M. St. Nawabi, and W. Zimmerman (Z. Kölner Zoo 14 (1971): 87–108).

Eresoidea. A superfamily coextensive with Eresidae.

Eresus. Eresidae. A genus of cribellate spiders (see Eresidae). Most of the species are solitary. *E. niger* occurs in southern Europe and is infrequently found in England.

Erigone. Linyphiidae. A genus that includes many species of small spiders ranging from the arctic tundra to the northern limits of the tropics. Small tangles or sheet webs are made among low plants, in grass, or under stones. *E. arctica* occurs in Europe and North America (Van Wingerden 1978).

Eriophora. Araneidae, Araneinae. Large tropical orb weavers (Fig. 4.2d). Some social species are included, e.g., *E. bistriata*. Species of the *E. ravilla* group were revised by H. W. Levi (Psyche 77 (1970): 280–302; Fig. 1a shows a large *Eriophora* eating a bat).

Euagrus. Dipluridae. Sheet-web-weaving mygalomorphs of the southwestern United States, Mexico, and Central America (Figs. 10.2–10.5). *E. mexicanus* occurs in Arizona and Mexico. The genus is being revised, and the behavior of several species studied, by F. A. Coyle (in prep.).

Eucta. Araneidae, Tetragnathinae. Synonym of *Tetragnatha*. The behavior of *E. kaestneri*, a European species, was studied by Crome (1954).

Euglyptila. Araneidae, Exechocentreae. Peculiarly modified mastophorines(?) from Asia. See Emerit 1978.

Euophrys. Salticidae. A common genus of jumping spiders from the tropics and Old World temperate zone. The many South American species are brightly colored, actively hopping over foliage during the day.

Euryopis. Theridiidae. A widespread genus, often with striking color patterns of black and metallic silver. Most species seem to specialize in preying on ants. Carico (1978) gives an account of the predatory behavior of *E. funebris*, and S. Porter and D. Eastmond (J. Arachnol. 10 (1982): 275–77) describe attacks on ants by *E. coki*.

Eurypelma. Theraphosidae. A poorly defined genus of New World tarantulas. The name is often used as a synonym for "tarantula," especially for North American species; but most real *Eurypelma* species are South American.

Eustala. Araneidae, Araneinae. About a dozen species of these typical orb weavers occur in the United States. They are related to *Cyclosa* species. The web-takedown behavior is described by Carico (this volume); the American species have been studied by Levi (1977a), who believes that the widespread *E. anastera* may be a complex of sibling species.

Exechocentreae. Araneidae, Mastophorinae(?). The anatomical modification of this tribe of African and Asian spiders suggests that they may be bolas spiders (Emerit 1978).

Exechocentrus. Araneidae, Exechocentreae, q.v.

Fecenia. Psechridae. *F. ochracea* (=*F. angustata*) occurs in Papua New Guinea. Young spiders make an inclined sheet web, adults a "pseudo-orb." See Robinson and Lubin 1979b, and Shear, this volume.

Filistata. Filistatidae. A genus of primitive cribellate spiders—the only cribellates with fully haplogyne male and female genitalia. European species line holes in walls with silk and extend a large, messy collar of cribellate silk out in all directions.

Filistatidae. See *Filistata*.

Filistatoides. Filistatidae. Several species of this small genus occur in the southern United States and Mexico. They make small webs under stones in the desert.

Floronia. Linyphiidae. *F. bucculenta* is a small, European linyphiid making little sheet webs in grass tufts. The abdomen can change color (Bristowe 1958).

Frontinella. Linyphiidae. *F. communis* is the bowl-and-doily spider of the eastern United States; the web consists of two sheets, the lower one flat and the upper one bowl-shaped (Gertsch 1979: 160).

Gasteracantha. Araneidae, Gasteracanthinae. Gasteracanthine orb weavers have brightly colored, spiny abdomens, and make outsized orbs in openings in tropical forests (Fig. 4.2j). *G. cancriformis* is found in Florida, and elsewhere in tropical America (but not in Puerto Rico).

Gasteracanthinae. Araneidae. Characterized by brightly colored, bizarrely spiny abdomens, these weavers of large orb webs are common sights in the tropics. Their spinnerets are surrounded by a sclerotized ring. *G. cancriformis* and several species of *Micrathena* are found in the United States (H. W. Levi, Bull. Mus. Comp. Zool. Harv. Univ. 148 (1978): 417–42).

Geolycosa. Lycosidae. A genus of wolf spiders from North America, living in burrows, usually in sandy soil. Some of the southern species were revised by J. Mc-Crone (Am. Midl. Nat. 70 (1963): 47–73).

Gnaphosidae. A family of wandering, hunting spiders common everywhere. Most species are black or dark brown, but a few are colorful, and many are ant mimics. N. I. Platnick has published a series of generic revisions in the *Bulletin of the American Museum of Natural History*.

Gnaphosoidea. In some publications, the family Gnaphosidae is placed in its own superfamily, together with a few small satellite families probably better included in Gnaphosidae itself (but see especially N. I. Platnick, Amer. Mus. Novitates 2801 (1984): 1–17, and 2808 (1985): 1–17).

Goleba. Salticidae. Species of this genus of jumping spiders are found in East Africa and Madagascar; they seem adapted to life on foliage, and some of the species are green (Wanless 1980a).

Gradungulidae. See *Progradungula*.

Hadrotarsidae. A family of ecribellate spiders of uncertain affinities. Most species occur in the southwestern Pacific. They look a little like oecobiids; the natural history is unknown.

Hahniidae. A dictynoid family of ecribellate spiders, found in the temperate regions of both the Northern and the Southern Hemisphere. The spinnerets are arranged in a characteristic transverse row. The webs are sheets operated from the upper surface.

Hebestatis. Ctenizidae. An Old World genus of trap-door spiders. See Coyle, this volume.

Heptathela. Liphistiidae. *H. nishihirai* occurs in Japan. The genus is distinguished from all other spider genera by having seven spinnerets.

Herennia. Araneidae, Nephilinae. *H. ornatissima* of southeastern Asia to New Guinea weaves a peculiar orb web that curves around tree trunks close to the bark (Robinson and Lubin 1979a).

Hermachastes. Ctenizidae. Another African ctenizid genus that extends the lip of the burrow into a tube held rigid by incorporated debris (Purcell 1903).

Hersiliidae. A hersilioid family of ecribellate spiders not known to make webs; they circle their prey facing away, paying out entangling silk from enormously long spinnerets. Most species live on tree trunks.

Hersilioidea. A superfamily including both cribellate and ecribellate spiders: Oecobiidae, Hersiliidae, and Hadrotarsidae(?).

Heteropodidae. A philodromoid family of large, ecribellate spiders with flattened bodies and crablike legs. Most species are tropical, and the large spiders sometimes found in banana shipments belong to this family.

Hexathele. Hexathelidae. The web-building behavior of members of this mygalomorph genus varies; Coyle (this volume) describes the behavior of *H. hochstetteri*, *H. huka*, *H. petriei*, and *H. waipa*.

Hexathelidae. A mygalomorph family of Australia and Africa including both web-building and burrowing species (Raven 1980b).

Hexura. Mecicobothriidae. A genus of small to medium-sized mygalomorphs making sheet-and-funnel webs in leaf litter in the northwestern United States.

Hexurella. Mecicobothriidae. Web building in this genus resembles that of *Hexura*, but the spiders are much smaller, and are found in the southwestern United States.

Homogona. Migidae. An Australian genus of mygalomorphs building tubes with both turrets and doors (Main 1976).

Hypochilidae. Spiders in this family (which has both cribellate and ecribellate members) are sometimes described as "missing links," combining as they do characteristics of both mygalomorphs and araneomorphs. Gertsch (1958) revised the family.

Hypochiloidea. A phylogenetically important superfamily of spiders with characters almost exactly intermediate between Mygalomorphae and Araneomorphae: the chelicerae are angled in some species so that they are neither orthognathous nor labidognathous, and most have two pairs of book lungs. Included are Hypochilidae, Gradungulidae, Hickmaniidae, and Thaididae, the latter two families comprising just one species each. See Platnick 1977.

Hypochilus. Hypochilidae. A genus of four or five North American cribellate species making peculiar "lampshade" webs under rocky ledges. The behavior has been described by Ferguson (1972) and Shear (1970b).

Hypophthalma. Araneidae, Araneinae. A genus of tropical orb weavers. Little is known of their behavior or of the number of species that might be included.

Hyptiotes. Uloboridae. *H. paradoxus* and its relatives make triangle webs representing a sector of an orb web. Opell (1982a, 1982b) provides information on the behavior and development of *H. cavatus*.

Idiops. Ctenizidae. *I. colletti* incorporates linear debris in its trap door or tunnel rim, probably to expand its sensory area (Coyle, this volume).

Inola. Pisauridae. A genus of spiders endemic to northern Australia, and one of two pisaurid genera known to weave webs as adults. As described by Davies (1982), the webs are "sticky" horizontal sheets with a silk-lined retreat at one side, but it is doubtful that the silk is actually viscid.

Ischnothele. Dipluridae. A South American and African genus of web-building mygalomorphs (Figs. 10.6, 10.7) (Petrunkevitch 1929).

Ixamatus. Dipluridae. Australian web-building mygalomorphs (Main 1968).

Kaira. Araneidae, Araneinae(?). An unusual genus of American orb weavers making extremely reduced webs (Fig. 5.7b) and attracting moths by chemical mimicry (Stowe, this volume).

Kiama. Dipluridae. The single species, *K. lachrymoides*, was described from Australia (Main and Mascord 1969).

Kukulkania. Filistatidae. The numerous American species once in *Filistata* are now placed here. See *Filistata*.

Labidognatha. Synonym of Araneomorphae.

Lagnus. Salticidae. *L. kochi* is a frequent prey item of *Portia fimbriata* in Australia (Jackson, this volume).

Lampona. Gnaphosidae. An Australian genus containing hunting spiders frequently found in houses; one species, *L. cylindrata*, is reputed to be dangerously venomous to humans, and has been introduced into New Zealand (Main 1976: 107–8).

Lampropodus. Barychelidae. *L. iridescens* of Australia attaches debris to its tunnel entrance (Main 1976).

Latrodectus. Theridiidae. The widow spiders, including the notorious black widow (*L. mactans*), are best known for their potent venom, responsible for occasional human fatalities (see Z. Maretic and J. Lebez, *Araneism* (Pula, Ljubljana 1979). Some make smallish tangled webs in debris; others make large sheets that have tubular retreats to one side and are operated from beneath. All have unusually strong silk. Kaston (1970) has written about the biology of the American species.

Leptonetidae. A family of tiny, long-legged, six-eyed spiders inhabiting moist leaf litter or caves. The American species were revised by Gertsch (1974).

Leptopelma. Barychelidae. *L. elongata* of Africa extends its tunnel mouth into a densely woven white silk funnel, which may attract insects by its resemblance to a flower (Simon 1873).

Leucauge. Araneidae, Metinae. A genus of weavers of rather open orb webs slightly inclined from the horizontal. More common in warm regions, species are often bright silver and marked with green and vivid red (Levi 1980a).

Linyphia. Linyphiidae. A widespread genus of sheet-weaving spiders from the Holarctic region. Usually of medium size, these spiders place their filmy sheets in low vegetation or in grass. See Gertsch 1979: 157–61.

Linyphiidae. A family in the superfamily Araneoidea, the species of which are all ecribellate. A few make no webs, but most construct dense sheets that they operate from beneath. Millidge (1977, 1980) suggests that though no nonsexual characters clearly define the family, features of the male palp set them apart from other araneoids.

Liphistiidae. The only family in the suborder Mesothelae, a line of spiders not clearly related to the Mygalomorphae or Araneomorphae, and distinguished mainly by having a segmented abdomen and carrying the spinnerets near the middle of the ventral surface. Like mygalomorphs, they have paraxial chelicerae and two pairs of book lungs. For anatomical, evolutionary, and behavioral details see Bristowe 1932, 1952, 1976, Murphy and Platnick 1981, and Coyle, this volume.

Liphistiomorphae. Synonym of Mesothelae.

Liphistius. Liphistiidae. *Liphistius* species (e.g., *L. batuensis* and *L. malayanus*) are found in southeastern Asia. They make silk tubes or burrows closed with trap doors; radiating silk lines extend from the burrow entrances (Fig. 10.1). See Coyle, this volume.

Lithyphantes. Theridiidae. An Old World genus of ground-dwelling cobweb weavers. Kullmann (1964) describes some aspects of the behavior. Now considered a synonym of *Steatoda*.

Loxoscelidae. A scytodoid family closely related to Scytodidae, and once included with them as a subfamily. It includes the notoriously poisonous members of the genus *Loxosceles*, all of which make tangled, irregular sheet webs (W. J. Gertsch and F. Ennik, Bull. Amer. Mus. Nat. Hist. 175 (1983): 264–360).

Lubinella. Uloboridae. A New Guinea genus containing only the species *L. morobensis*. See Lubin, this volume, and Opell 1984a.

Lutica. Zodariidae. A genus distinguished by having (apparently) only two spinnerets.

Lycosa. Lycosidae. One of the largest genera of spiders, containing medium-sized to very large species of wolf spiders. Many live in burrows, which are sometimes closed with trap doors. Most species are nocturnal hunters. See Gertsch (1979: 184–92).

Lycosidae. The wolf spiders. The family is difficult to define, but includes hunters, web builders, and burrow dwellers. See Gertsch 1979 and Bristowe 1958.

Lycosoidea. A superfamily of ecribellate spiders defined by Lehtinen (1967) as including Lycosidae, Psechridae, Zoridae, Zoropsidae, Ctenidae, Cycloctenidae, Stiphidiidae, Selenopidae, and part of the classical Pisauridae. Others include all of Pisauridae, plus Oxyopidae and Senoculidae.

Lyssomanes. Salticidae, Lyssomaninae. A genus of tropical jumping spiders thought to be primitive members of the family; some use silk quite extensively. *L. viridis* is found as far north as North Carolina; it is translucent apple green, with startling black retinas visible through the semitransparent carapace. See Jackson, this volume.

Lyssomanidae. Now classified as a subfamily of Salticidae. See *Lyssomanes*.

Lyssomaninae. Salticidae. See *Lyssomanes*.

Macopaeus. Salticidae. See Wanless 1980b.

Macrothele. Hexathelidae. *M. darcheni*, of Africa, is the only communal mygalomorph known (Darchen 1967).

Mallos. Dictynidae. A genus of cribellate spiders of western North America and Mexico, distinguished from *Dictyna* by larger size and details of the genitalia. Solitary and communal species are included. The solitary species make irregular tangles among twigs. *M. gregalis* is communal (Tietjen, this volume).

Mangora. Araneidae, Araneinae(?). A genus of small to medium-sized orb weavers distinguished by a peculiar arrangement of trichobothria on the third-leg tibiae. The orb webs are made in grass or herbaceous vegetation. See H. W. Levi, Bull. Mus. Comp. Zool. Harv. Univ. 147 (1975): 101–35).

Marpissa. Salticidae. The American species have been reviewed by R. D. Barnes (Am. Mus. Novit. 1867 (1958): 1–50).

Mastophora. Araneidae, Mastophorinae. The classic bolas spiders of the Americas. *M. dizzydeani* has been thoroughly studied by Eberhard (1980a). Gertsch (1955) revised the genus for North America. Stowe (this volume) mentions *M. bisaccata*, *M. cornigera* (Fig. 5.8), and *M. phrynosoma*.

Mastophorinae. Araneidae. The subfamily of the bolas spiders.

Matachia. Desidae. A genus of cribellates found in the Southern Hemisphere (Forster 1970). Some species make webs that look (in part) like orb sectors.

Maymena. Mysmenidae. A genus of tiny orb weavers. Species are more common in warm regions, but a few are found in the temperate United States. The web of *M. ambita* is shown in Fig. 12.22.

Mecicobothriidae. A family of mygalomorph spiders making small sheet webs, silk tubes, or burrows. Gertsch and Platnick (1979) revised the family.

Mecynogea. Araneidae, Cyrtophorinae. *M. lemniscata*, the basilica spider, is a temperate-zone species that makes a web similar to that of solitary *Cyrtophora* species. Carico (this volume) describes the web-takedown behavior.

Mecysmaucheniidae. A family of haplogyne spiders of the Southern Hemisphere, once included in Archaeidae, which implies that they prey on other spiders. The group has been revised by Forster and Platnick (1984).

Megadictyna. Nicodamidae. The New Zealand species of this cribellate genus have been treated by Forster (1970).

Megahexura. Mecicobothriidae. *M. fulva* of California weaves a sheet-and-funnel web (Coyle, this volume).

Meioneta. Linyphiidae. Found throughout the North Temperate Zone, these tiny spiders make small sheet webs in leaf litter or grass.

Menneus. Dinopidae. *M. camelus* of Australia is a typical web-throwing dinopid (Akerman 1926).

Mesothelae. A suborder of spiders including only the family Liphistiidae, and possibly some extinct forms. See Liphistiidae.

Meta. Araneidae, Metinae. A genus of weavers of rather open orbs in shady, moist places. *M. menardi* is common in cave entrances in Europe and the eastern United States. For a revision see Levi (1980a).

Metabus. Araneidae, Metinae. The only species, *M. gravidus*, is a communal orb weaver of Central America (Buskirk 1975a, 1975b). The genus may be a synonym of *Leucauge*.

Metazygia. Araneidae, Araneinae. A mainly tropical genus of undistinguished orb weavers with three species in the United States (Levi 1977a). They are superficially similar to species of the mainly European genus *Nuctenea*.

Metepeira. Araneidae, Araneinae. An exclusively American genus making orb webs often incomplete in the upper sectors, and with an extensive barrier web above, which incorporates a thimblelike retreat (Levi 1977b). *M. labyrinthea*, a solitary species, is common in the eastern United States. *M. spinipes*, a communal species, is found in Mexico, where its immense colonies may persist for years (Uetz, this volume).

Metinae. Araneidae. A subfamily thought to be primitive and perhaps close to the basic stock of Araneidae, Theridiidae, and Linyphiidae. Levi (1980a) revises the American members of the subfamily and discusses its phylogenetic position. It may have to be raised to full family status.

Miagrammopes. Uloboridae, Miagrammopinae. A genus of curiously modified cribellate spiders, usually with long, sticklike bodies. There are many species in the world tropics; their webs consist of only a few lines or a single line (Lubin, Eberhard, and Montgomery 1978; Opell 1984b).

Miagrammopidae. Synonym of Uloboridae, sometimes recognized as a subfamily. See *Miagrammopes*.

Micrathena. Araneidae, Gasteracanthinae. A large, mostly tropical American genus of orb weavers, which, like *Gasteracantha*, make rather open orbs in forest openings (Fig. 12.1). They have hard, spiny, colorful abdomens. *M. duodecimspinosa* is tropical; *M. gracilis* and *M. sagittata* are common in the eastern United States. See H. W. Levi, Bull. Mus. Comp. Zool. Harv. Univ. 148 (1978): 417–22, and A. M. Chickering, Bull. Mus. Comp. Zool. Harv. Univ. 125 (1961): 391–470.

Microhexura. Dipluridae. A genus of two species of minute mygalomorph spiders, one from the northwestern United States and the other from the highest mountain tops of Tennessee and North Carolina. They make small, tubular webs in mats of moss and debris (Coyle 1981a).

Micropholcommatidae. A family of tiny, armored araneoids (most are about 1.0 mm long) from Australia, where they are found in mosses. The web of only one species has been described by V. V. Hickman (Pap. Proc. R. Soc. Tasmania 1943 (1944): 179–95), who says that it is an irregular tangle resembling a theridiid web.

Migas. Migidae. A mygalomorph genus from New Zealand; most species make tubes on tree trunks, but a few burrow (Forster and Wilton 1968).

Migidae. A mygalomorph family of the Southern Hemisphere, including a number of arboreal species that make camouflaged tubes on tree trunks (Coyle, this volume). New Zealand species were revised by Forster and Wilton (1968).

Mimetidae. See *Mimetus*.

Mimetus. Mimetidae. An araneoid genus of worldwide distribution not known to make webs of their own, but invading the webs of other spiders and killing them (Gertsch 1979: 163–64). *M. puritanus* is a common American species. The family seems close to the Araneidae (Shear 1981), but Forster and Platnick (1984) place it in Palpimanoidea.

Missulena. Actinopodidae. A genus of mygalomorphs with a complex burrow entrance structure (Coyle, this volume). Species are found in Australia.

Misumena. Thomisidae. A genus of crab spiders frequently found on flowers, ready to ambush insects. Some species are evidently capable of a limited degree of color change, matching either a yellow or a white background. This seems directed more at potential predators than at potential prey.

Misumenops. Thomisidae. Members of this genus of crab spiders are commonly found on flowers, where they wait, front legs spread wide, for the visits of bees and flies.

Moggridgea. Ctenizidae. In Africa, species of this mygalomorph genus either burrow (e.g., *M. mordax*) or make tubes on tree trunks (e.g., *M. abrahami*, *M. dyeri*, and *M. microps*). See Coyle, this volume.

Mopsus. Salticidae. An Australian genus of jumping spiders with large, strikingly colored males. R. Jackson (Austr. J. Zool. 31 (1983): 39–53) describes the display behavior of *M. mormon*, from Queensland.

Mygalomorphae. A suborder of spiders characterized by orthognathous chelicerae and two pairs of book lungs (tracheae are lacking). Tarantulas (American usage), trap door spiders, and the like are included. See Araneomorphae and Mesothelae.

Mysmena. Mysmenidae. The orbs of *Mysmena* species are three-dimensional—point-symmetrical but not planar—and enclosed in an "envelope" of viscid silk (Figs. 12.19, 12.20).

Mysmenidae. An araneoid family of the world tropics, with a few species in the temperate zones. The webs are modified orbs, some with out-of-plane radii, some with an "envelope" of viscid silk. Most species are very small. See Coddington, this volume.

Mysmenopsis. Mysmenidae, q.v.

Namirea. Dipluridae. Five species of small Australian ischnotheline spiders living in thin-walled tubular silk retreats, which, in at least one species, open out into conspicuous space webs (Raven 1984).

Nemesia. Ctenizidae. A genus of trap-door spiders (*N. caementaria*, *N. dubia*, and *N. meridionalis*) of the Mediterranean region.

Nemoscolus. Araneidae, Metinae. A little-known genus that may be related to *Leucauge* (H. W. Levi, pers. comm.).

Neoarchemorus. Araneidae, Mastophorinae(?). See *Archemorus*, and Fig. 5.3a.

Neolana. Neolanidae, q.v.

Neolanidae. A family including only the New Zealand genus *Neolana*, the webs of which are pictured by Forster and Wilton (1973, Fig. 971). The genitalia suggest a relationship with Dinopidae.

Neoscona. Araneidae, Araneinae. A genus of typical orb weavers, found in North and South America. *N. domiciliorum* and *N. hentzii* occur in the eastern United States (Berman and Levi 1971).

Nephila. Araneidae, Nephilinae. These large spiders are sometimes called the golden-silk weavers, for their large webs (often more than 1 m in diameter) of tough yellow silk. Other distinctive features of the orbs include a persistent scaffold and branched radii. The species are all fairly similar; all are found in the tropics and subtropics. *N. edulis* and *N. maculata* are from the Asian and Pacific tropics; *N. clavipes*, with tufts of hairs on the legs, is American. The Robinsons (1973a, 1973b, 1976b) have studied the ecology and behavior of several species.

Nephilengys. Araneidae, Nephilinae. Tropical orb weavers related to *Nephila*.

Nephilinae. Araneidae. A subfamily including relatively few genera of large orb weavers, all from the tropics. See *Nephila* and *Herennia*.

Neriene. Linyphiidae. *N. radiata*, a Holarctic species, is the common filmy-dome spider, named for its distinctive dome-shaped web in low shrubs and herbs. Many American references have this species as *Linyphia marginata* (van Helsdingen 1969).

Nesticidae. An araneoid family closely related to Theridiidae, and usually separated by details of the genitalia. However, the families share the distinctive comb of setae on the fourth tarsi, and probably both use sticky silk as a weapon to attack prey. *Nesticus cellulanus* is generally found in buildings and caves. The web of *Eidmanella pallida* is shown in Fig. 12.6. The family has been revised for North America by W. J. Gertsch (Bull. Texas Mem. Mus. 31 (1984): 1–91.

Nicodamidae. An obscure family of spiders from Australia (the ecribellate genus *Nicodamus*) and New Zealand (the cribellate genus *Megadictyna*). Australian species are brightly colored in red and black; probably all make a sheet web operated from below (Forster 1970).

Nuctenea. Araneidae, Araneinae. A Holarctic genus of typical orb weavers, related to *Araneus* and *Metazygia*. The orbs of *N. cornutus* are frequently over water; the common name of *N. sclopetaria* (the bridge spider) reflects similar habits. During the day, the spiders hide in a retreat near the web. For descriptions of species, see H. W. Levi, Bull. Mus. Comp. Zool. Harv. Univ. 146 (1974): 291–316.

Ochyroceratidae. A pholcoid family of delicate, six-eyed haplogyne spiders, nearly all from tropical caves or leaf litter. A number of tiny, parthenogenetic species are known.

Octonoba. Uloboridae. One species of this genus is found in most of the eastern United States; it makes a typical horizontal uloborid orb web. Opell (1979) mentions several additional species from northeastern Asia.

Oecobiidae. A family of cribellate and ecribellate spiders making baglike webs from which they rush to seize prey. Victims are attacked as the spiders circle around them, combing out silk through special hairs on the anal tubercle. The family was reviewed for North America by W. A. Shear (Bull. Mus. Comp. Zool. Harv. Univ. 140 (1970): 129–64).

Oecobius. Oecobiidae. Small, cribellate spiders, of which several species are common on buildings. At least one species (from Mexico) is communal. See Oecobiidae.

Oedothorax. Linyphiidae. Tiny, dark-colored spiders making small sheets or tangles in leaf litter, under tree bark, or in the bases of grass clumps.

Ogulnius. Theridiosomatidae. A genus of small, tropical orb weavers. The web of one species is shown in Fig. 12.12; frame lines are lacking, and the radii attach directly to the substrate.

Onomastus. Salticidae. See Wanless 1980c.

Oonopidae. A dysderoid family of ecribellate, haplogyne spiders ranging in size from small to minute. Most species inhabit tropical soils or leaf litter; a few are found in temperate desert regions. At least three species live in buildings. Little is known of their web-building behavior; probably most are hunters or sit-and-wait predators.

Orbiculariae, Orbitelae, Orbitelariae, Orbitèles. Obsolete names for various groups of orb-weaving spiders. See Coddington, this volume.

Ordgarius. Theridiosomatidae. A genus of small orb weavers.

Orinomana. Uloboridae. A genus of unknown habits, found in the high Andes (Opell 1979).

Orsinome. Araneidae, Metinae. A little-known genus probably close to *Leucauge* (H. W. Levi, pers. comm.).

Oxyopes. See Oxyopidae.

Oxyopidae. The lynx spiders. A family of ecribellate, wandering, daytime hunters with a distinctive hexagonal eye arrangement. They are variously said to be related to Lycosidae or Salticidae. The American species are treated by A. R. Brady (Bull. Mus. Comp. Zool. Harv. Univ. 131 (1964): 429–518).

Pachygnatha. Araneidae, Tetragnathinae. A Holarctic genus of spiders in which the young make small, horizontal orb webs, but later instars are wandering hunters. See Levi 1980.

Palpimanoidea. A superfamily of ecribellate spiders extensively redefined by Forster and Platnick (1984); at present it seems to be composed of a rather heterogeneous group of families.

Pandisus. Salticidae. These small, often green jumping spiders from Madagascar have (in the males) peculiarly modified palpal femurs that may produce a sex pheromone (Wanless 1980a).

Paracenobiopelma. Barychelidae. *P. gerecormophila* may be the spider species with the longest name. It is an arboreal tube builder of Brazil (Feio 1952).

Paraneus. Araneidae, Araneinae. African *Paraneus* species make webs with incomplete frames, often pulled into a cone by lines attached to the hub (Edmunds 1978).

Paraplectana. Araneidae, Cyrtarachneae. The web of the Japanese species *P. tsushumensis* has been described by Chigira (1978). See Fig. 5.6b.

Paraplectanoides. Araneidae, Cyrtarachneae. This Tasmanian spider (*P. crassipes*) makes a peculiar web, consisting of a few radii enclosed in a tough silk bag with a hole at the bottom (Fig. 5.2); insects enter the hole and are captured (Hickman 1975).

Pardosa. Lycosidae. A genus containing many species of small, diurnal wolf spiders, found all over the world.

Pasilobus. Araneidae, Cyrtarachneae. Species of this Australian genus make a much reduced, horizontal, triangular web (Fig. 5.6c). See Stowe, this volume.

Patu. Symphytognathidae. *P. digua*, 0.3 mm long, has the distinction of being the smallest known spider. The web of an unknown species of *Patu* is shown in Fig. 12.24. See also Forster and Platnick 1977.

Pellenes. Salticidae. *P. nigrociliata* sleeps at night in old snail shells suspended by a thread.

Peucetia. Oxyopidae. Some young *Peucetia* make webs, but the adults are wandering hunters (Kaston 1972). American species are large and apple green with red spots. See Oxyopidae.

Phaecius. Salticidae. See Wanless 1981c.

Phiale. Salticidae. Considerable confusion has existed in the classification of this South American genus. Recently M. E. Galiano (Bull. Br. Arachnol. Soc. 5 (1981): 205–16) discovered that in at least two of the species multiple color forms exist among the females, whereas the males are monomorphic.

Phidippus. Salticidae. Species in this genus are generally large, bulky, brightly colored jumping spiders. There are about 20 species in North America. Jackson (1977b) describes the courtship behavior of *P. johnsoni*.

Philodromidae. A family of crablike spiders with long legs. All are hunters; none is known to make a web. The family was formerly included in Thomisidae, but the eye structure and hair type are fundamentally different. The crablike body form has evidently arisen independently in spiders several times.

Philodromoidea. A superfamily including ecribellate families having a special type of eye pigmentation: a ring of black around and above the cup of each eye. The best-known families, Philodromidae, Selenopidae, and Heteropodidae, have a crablike appearance.

Philoponella. Uloboridae. Several species occur in South and Central America (and as far north as the southern United States) and in tropical Asia and the western Pacific. Eight species are treated by Lubin (this volume). Many of the known species are communal (Opell 1979).

Pholcidae. A cosmopolitan family of haplogyne spiders, more common in warm regions. Species are usually pale in color and have very long legs. The webs are tangles or vague sheets; there are unconfirmed reports that they contain viscid silk. For descriptions of some typical species, see Gertsch 1982.

Pholcoidea. A superfamily usually including Pholcidae, Ochyroceratidae, and Leptonetidae.

Pholcus. Pholcidae. *P. phalangioides*, found in cellars and buildings everywhere in the United States, makes large, tangled cobwebs. When disturbed, the spiders vibrate in the web so rapidly that they virtually disappear from view.

Phoroncidia. Theridiidae. *P. studo* of the American tropics makes a single-line web (Eberhard 1981c). The spiders have peculiar lumpy or spiny abdomens.

Phrurolithus. Clubionidae. These small spiders run rapidly over the ground, often in the company of ants. No webs are built.

Phryganoporus. Amaurobiidae. A genus of cribellate spiders found in Australia. *P. candidus* is communal.

Pirata. Lycosidae. A genus of many species of small wolf spiders distinguished by a Y-shaped mark on the cephalothorax. Most are associated with wet regions. A few species make tubular webs in saturated mosses. The American species were revised by H. K. Wallace and H. Exline (J. Arachnol. 5 (1978): 1–112).

Pisaura. Pisauridae, q.v.

Pisauridae. A family of medium-sized to large hunting ecribellates, though a few web-building species were found recently in Australia (Davies 1982) and South

America (Nentwig 1985). Many live at the water's edge, and are capable of skating over the surface film to catch aquatic insects, small fish, and tadpoles. Females make an extensive bag of silk around the eggs and guard this "nursery" until the young disperse. The American species of *Dolomedes*, the largest genus, were revised by J. E. Carico (Bull. Mus. Comp. Zool. Harv. Univ. 144 (1973): 435–88).

Pisaurina. Pisauridae. Typical Pisauridae, q.v. Young *Pisaurina* have been observed to make webs (Carico 1985), and adults make nursery webs. Carico (1972) describes the American species.

Plectreuridae. A family of eight-eyed haplogyne spiders endemic to the western United States, where they make tangled webs or silk tubes under rocks. They look like ecribellate filistatids (W. J. Gertsch, Am. Mus. Novit. 1920 (1958): 1–53).

Plexippus. Salticidae. This genus contains many species of colorful jumping spiders from warm regions. *P. paykulli* is found in the southernmost parts of the United States, and is often quite abundant.

Poecilopachys. Araneidae, Cyrtarachneae. In Australia, *P. australasia* makes a horizontal, much-reduced orb with a few loops of viscid silk. See Stowe, this volume, especially Fig. 5.6a, and Clyne 1973.

Polenecia. Uloboridae. *P. producta* (= *Sybota producta*) is found in the Mediterranean region, and makes a reduced orb without a capture spiral; cribellate silk is laid directly on radii (Wiehle 1931). The genus could be an evolutionary step to *Miagrammopes*.

Poltys. Araneidae, Araneinae. These Australian spiders make a densely meshed orb (Mascord 1980). Perhaps they are related to *Neoscona* or *Parawixia* (H. W. Levi, pers. comm.).

Ponella. Uloboridae. A single species has been described, from southern South America; habits are unknown (Opell 1979).

Porrhothele. Hexathelidae. Laing (1973 et seq.) has studied *P. antipodiana* of New Zealand in detail. This mygalomorph makes a tube-and-sheet web.

Portia. Salticidae. The only salticids presently known that make a web for prey capture; they also invade the webs of other spiders to loot and kill. Jackson (this volume) reviews their behavior and its evolutionary implications.

Progradungula. Gradungulidae. A cribellate genus. The family may belong in Hypochiloidea; like the hypochilids proper, it seems to combine primitive and advanced characteristics. The web of one species is described by Forster and Gray (1979).

Psalistops. Barychelidae. *P. melanopygia* of Venezuela makes a burrow without a door; the entry funnel incorporates debris (Coyle, this volume).

Psechridae. A family of cribellate web weavers from tropical Asia and Australia. Typically, the webs are large sheets with a tube at one side and below, and are operated from beneath (Robinson and Lubin 1979b). The eye structure is lycosoid.

Psechrus. Psechridae, q.v.

Pseudidiops. Ctenizidae. *P. opifex* is one of a small number of arboreal ctenizids, making a tube on tree trunks (Coyle, this volume).

Purumitra. Uloboridae. The only species, *P. grammicus*, has been found in the Philippines and Palau; nothing is known of its web-building behavior (Opell 1979).

Salassina. Araneidae, Araneinae. A little-known group of species identified in the literature as *Edricus*. The genus is probably related to *Wixia* (H. W. Levi, pers. comm.).

Salticidae. Perhaps the largest spider family (in number of species), characterized by compact bodies and enormously enlarged main eyes, which give them acute daytime vision. The vast majority of species are active hunters; only a few (*Portia*: Jackson, this volume) make webs.

Sason. Barychelidae. A genus of arboreal barychelids from Sri Lanka making short double-doored tubes on tree trunks (Berland 1932).

Savignia. Linyphiidae. A genus of sheet weavers known from the temperate regions of Europe and Asia.

Schizocosa. Lycosidae. A genus of medium-sized wolf spiders from North America. The males often have tufts of black hairs on the front legs. The genus was revised by C. D. Dondale and J. Redner (Can. Entomol. 110 (1978): 143–81).

Scoloderus. Araneidae, Araneinae. A genus of orb weavers with elongate, bumpy abdomens. The species for which webs have been described make inverted "ladder" webs (Fig. 5.5b), with a long catching area arranged perfectly vertically and the hub at the bottom (Eberhard 1975; Levi 1977b).

Scytodes. Scytodidae. A genus of six-eyed haplogyne spiders from the world tropics; a few make tangled webs under bark or stones, but the best-known species, *S. thoracica*, is found in and around houses, where it captures prey by squirting a sticky gum from its chelicerae (Bristowe 1958: 103–9).

Scytodidae. See *Scytodes*.

Scytodoidea. A superfamily of haplogyne spiders including Filistatidae, Scytodidae, Sicariidae, Loxoscelidae, Plectreuridae, Diguetidae, and Tetrablemmidae. All have chelicerae that are joined at the base by a membrane and have a large, drawn-out tooth opposing the fang.

Segestria. Dysderidae, Segestriinae. A genus of haplogyne spiders living in silk tubes in vertical surfaces with triplines extending from the entrance (Bristowe 1958, Pl. IX). Most species are found in dry regions.

Segestriidae. See Segestriinae.

Segestriinae. Dysderidae. See *Segestria* and *Ariadna*.

Selenocosmia. Theraphosidae. An Australian genus of large mygalomorphs living in open burrows (Main 1976).

Selenopidae. A family of ecribellate hunting spiders with flattened, crablike bodies. No webs are made; the spiders hide during the day in narrow crevices. For descriptions of American species see M. H. Muma, Am. Mus. Novit. 1619 (1953): 1–55.

Senoculidae. A lycosoid family including only the South American ecribellate genus *Senoculus*. Members of the genus all seem to be hunters; webs are not constructed.

Sericopelma. Theraphosidae. A genus of large, arboreal mygalomorphs. The sensory physiology was studied by Den Otter (1974).

Sicariidae. At present, only the genus *Sicarius* can be definitively included in this scytodoid family. These large, ecribellate, haplogyne spiders are found in the deserts of South America; rather than making webs, they bury themselves in the sand and await prey (J. Reiskind, Am. Zool. 9 (1969) 195–200).

Siratoba. Uloboridae. Two species are found in the southwestern United States and northwestern Mexico; nothing is known of their behavior (Opell 1979).

Sosippus. Lycosidae. A genus of large wolf spiders of warm regions, making extensive sheet webs with tubular retreats. Brady (1962) describes American species.

Spartaeinae. Salticidae. A newly described subfamily of jumping spiders generally with large posterior median eyes, and thought for this reason to be primitive (Wanless 1984).

Sphodros. Atypidae. All but one species of North American atypids belong here. The webs are closed, tubular extensions of burrow linings, and are made against tree trunks (Figs. 10.18, 10.19). Coyle and Shear (1981) describe the natural history of two common species.

Spilasma. Araneidae, Araneinae. *S. tubulofaciens* of South America makes a "half orb," with a vertical silk retreat occupying the hub (Quintero 1974).

Spintharus. Theridiidae. The web of *S. flavidus*, an American species, is shown in Fig. 5.1c. *Spintharus* seems related to *Argyrodes* (H. W. Levi, J. N.Y. Entomol. Soc. 62 (1954): 65–90).

Stanwellia. Dipluridae. An Australian genus of burrowing mygalomorphs (Main 1972).

Steatoda. Theridiidae. A large cosmopolitan genus of ground-living theridiids, making tangled webs under stones, or sheets with tubular retreats. *S. triangulosa* (= *Teutana triangulosa*) is a house spider found throughout the United States; *S. fulva* is an ant specialist in the southwestern United States (Hoelldobler 1970). The genus was revised by H. W. Levi (Bull. Mus. Comp. Zool. Harv. Univ. 117 (1957): 367–424).

Stegodyphus. Eresidae. A genus of cribellate spiders from Asia and Africa. Several species are communal and make large, matted webs over vegetation (Figs. 7.3, 7.4). *S. sarasinorum* of southwestern Asia has been studied by Kullmann and Zimmerman (1971).

Stemonyphantes. Linyphiidae. A genus of medium-sized spiders that weave irregular sheets under stones and other objects on the ground. A single species, *S. blauvelti*, occurs infrequently in the northern United States (W. J. Gertsch, Am. Mus. Novit. 1514 (1951): 1–11).

Stenogrocercus. Dipluridae. Synonym of *Cethegus*, q.v.

Stiphidiidae. A lycosoid family of cribellate and ecribellate spiders from Australia and New Zealand. Species make sheet webs or large, inverted cone webs, working them from beneath (Forster and Wilton 1973).

Stoidis. Salticidae. Species of this jumping spider genus are specialists preying on ants (Edwards, Carroll, and Whitcomb 1974).

Stothis. Barychelidae. *S. astuta* digs a U-shaped burrow and closes each end with a trap door (McCook 1890; Coyle, this volume).

Sybota. Uloboridae. Synonym of *Polenecia*.

Symphytognathidae. A family of minute araneoids distinguished by reduced or absent pedipalps in the females and fused chelicerae in both sexes. Where webs are known, they are small, horizontal orbs (Fig. 12.24), but some tropical species are web parasites in the nests of much larger spiders. Forster and Platnick (1977) have reviewed the family.

Synotaxus. Theridiidae. A genus of tropical spiders. The webs are unique—a rectangular array of dry and viscid threads (Fig. 12.4). See Eberhard 1977b.

Taczanowskia. Araneidae, Mastophorinae. A South American genus of bolas spiders (Eberhard 1981b).

Taieria. Gnaphosidae. A genus of ground-dwelling spiders found only in New Zealand, inhabiting forests, river beds, beaches, and swamps (Forster and Blest 1979).

Tama. Hersiliidae, q.v. Species occur on tree trunks and buildings in Mexico and Central America, one of them reaching southern Texas.

Tangaroa. Uloboridae. A small genus of cribellate orb weavers from the western Pacific; webs are unknown, but probably are standard orbs (Opell 1979).

Tapinillus. Oxyopidae. Members of this neotropical genus of lynx spiders are reported by Griswold (in press) to weave webs.

Tapinocyba. Linyphiidae. Small spiders of the North Temperate Zone, typically living in leaf litter or under stones on the ground, where they make small, irregular sheet webs.

Tapinopa. Linyphiidae. A genus of cursorial, hunting linyphiids that probably make no webs.

Teemenaariidae. Synonym of Cyatholipidae.

Telemidae. Tiny, pale, lungless spiders are included in this dysderoid family. They are found in scattered places around the world, usually in caves or moist litter; American species (all in the genus *Usofila*) have been recorded from Oregon and California. Nothing is known of their behavior.

Tengella. Tengellidae. A small genus of cribellate spiders found in Central and South America. W. Wolff (J. Arachnol. 5 (1977): 139–44) has revised the genus.

Tengellidae. See *Tengella*.

Tetrablemmidae. A family of minute, bizarrely modified, heavily armored scytodoid spiders with eyes reduced from six, to four, to two, to even one. They make small, generalized webs in tropical leaf litter (Lehtinen 1981).

Tetragnatha. Araneidae, Tetragnathinae. Species of *Tetragnatha* are a conspicuous component of the American spider fauna. Their open-meshed orbs are commonly suspended over water. The spiders are long-bodied and sticklike, and have extraordinary elongated jaws. The North American species have been revised by Levi (1980b).

Tetragnathidae. See Tetragnathinae. It is uncertain at present if the group should have family or subfamily status.

Tetragnathinae. Araneidae. A subfamily close to Metinae and containing species with simple genitalia and enlarged chelicerae. A few make no webs as adults (*Pachygnatha*), but most make loose, open orbs over water (e.g., *Tetragnatha*). Levi (1980a, 1980b) has revised the North American genera.

Teutana. Theridiidae. See *Steatoda*.

Textricellidae. A family of minute spiders, usually about 1 mm long, with heavily armored bodies, and with lungs and posterior tracheae absent. About 20 species are known from mosses in the Australian region and Chile, where they make small sheet webs. Only one of the species is Chilean; it is described, and references are given, by R. R. Forster and N. I. Platnick (Bull. Am. Mus. Nat. Hist. 170 (1981): 263–70).

Teyl. Dipluridae. An Australian genus of burrowing mygalomorphs (Main 1975).

Thaididae. A South American family of hypochiloid spiders. The web is a large sheet in vegetation (Zapfe and Gertsch 1955).

Theraphosidae. A family of large mygalomorph spiders found in tropical forests and deserts; in North America they occur as far north as Utah and Arkansas. Habits are diverse; some live in silk-lined burrows, others are entirely arboreal. Recently some of the more colorful species have been offered for sale in pet shops.

Theridiidae. The cobweb weavers—though many species make sheet webs, and a few make arrays of regularly arranged threads (see *Episinus* and *Synotaxus*). The genera in the family are allied by the comb of special setae on the fourth tarsus and their ability to spin viscid and dry silk alternately. The family occurs everywhere and may include more than 2,000 species. Levi and Levi (1962) diagnose most of the valid genera.

Theridion. Theridiidae. One of the largest spider genera, typical for the family Theridiidae. Species make three-dimensional, irregular webs, often in low vegetation or among the leaves of trees. For descriptions of American species see H. W. Levi, Bull. Am. Mus. Nat. Hist. 112 (1957): 1–123; Bull. Mus. Comp. Zool. Harv. Univ. 121 (1959): 57–163.

Theridiosoma. Theridiosomatidae. Small orb weavers of damp places, especially tropical forests, though one species is common in the northeastern United States. The web is a small orb with anastomosing radii; the spider often holds it under tension by pulling it up into a cone (Figs. 12.9, 12.10).

Theridiosomatidae. A family of small orb weavers, most of which are tropical species. See *Theridiosoma*. J. Coddington is working on a revision of the group. The family is united by the presence of a peculiar sternal organ of unknown function (Wunderlich 1980).

Thomisidae. The crab spiders, classic sit-and-wait predators. Species are not known to make webs, but they do weave baglike retreats in which to spend their resting hours. See Gertsch 1979 and Bristowe 1958.

Thomisoidea. See Thomisidae.

Titanoeca. Amaurobiidae (Titanoecidae?). A genus of web-weaving cribellates. *T. albomaculata* adds orblike sectors to its web (Szlep 1966).

Titanoecidae. Usually considered a synonym of Amaurobiidae, q.v. Lehtinen (1967) argues (not very convincingly) for the separate status of this family, placing it near Psechridae. This idea has gained little or no support.

Trogloneta. Mysmenidae. A genus of tiny orb weavers. *T. paradoxum* occurs in Europe.

Tylorida. Araneidae, Metinae. The species reported as makers of long "ladder" web in New Guinea (Robinson and Robinson 1972) belongs to an as-yet-unnamed genus close to *Tylorida*.

Uloboridae. The family of the cribellate orb weavers, worldwide in occurrence, but much more diverse in the tropics. See Lubin, this volume, and Opell 1979.

Uloborus. Uloboridae. Typical orb-weaving uloborids, though some have modified webs (e.g., *U. conus*). Fifteen of the many known species are discussed by Lubin (this volume); the American species are revised by Opell (1979).

Ummidia. Ctenizidae. North American and European trap-door spiders that close their burrows with a corklike lid. The special construction of the third leg, which has a deeply bowed tibia, allows them to hold the lid down with surprising tenacity (Coyle 1981b).

Uroctea. Oecobiidae. Ecribellate spiders of the Mediterranean region, North Africa, and southern Asia. They make sheet webs under stones, but evidently rely on triplines extending out in all directions to alert them to prey; prey-capture behavior is as described under Oecobiidae. See Kullmann et al. 1975.

Verrucosa. Araneidae, Araneinae. One species, *V. arenata*, is found in the United States. The triangular, lumpy abdomen is brightly colored in yellow or white. The orb is typical, but the spider rests in it head up, the reverse of the usual position (H. W. Levi, Bull. Mus. Comp. Zool. Harv. Univ. 147 (1976): 351–91).

Wagneriana. Araneidae, Araneinae. A genus of little orb weavers with lumpy abdomens. Most species are Neotropical, but a single species, *W. tauricornis*, reaches the southern United States (H. W. Levi, Bull. Mus. Comp. Zool. Harv. Univ. 147 (1976): 351–91).

Waitkera. Uloboridae. A single species of cribellate orb weavers from New Zealand belongs here; the orb is small and made among grasses (Opell 1979).

Wendilgarda. Theridiosomatidae. Some members of this genus make what is certainly one of the most weirdly modified orb webs (Fig. 12.11)—a single line strung over tropical streams, from which viscid lines extend to the moving water surface (Coddington and Valerio 1980).

Wixia. Araneidae, Araneinae. *W. ectypa* of the eastern United States weaves a reduced "asterisk" web (Fig. 5.1a) and preys on pedestrian insects. See Stowe 1978, this volume.

Xamiatus. Dipluridae. Five species of large, burrowing Australian diplurine spiders, which use very little silk (Raven 1981, 1982).

Zodariidae. An ill-defined family that has long served as a taxonomic dumping ground for undistinguished groups of spiders. See Lehtinen 1967 for attempts to redefine the family.

Zoridae. A poorly defined lycosoid family considered by Lehtinen (1967) to be close to Ctenidae. A few species assigned to *Zora* occur in the United States, one of them being commonly collected from leaf litter in Oregon. No webs are constructed by any of the known species.

Zoropsidae. See *Zoropsis.*

Zoropsis. Zoropsidae. An Old World genus of cribellate sheet-web weavers. Their affinities are unclear, but they may be related to Acanthoctenidae.

Zosis. Uloboridae. *Z. geniculatus* is a large orb weaver often found around buildings in tropical regions. See Lubin, this volume, and Opell 1979, 1981.

Zygiella. Araneidae, Metinae. These orb weavers leave an open sector in their webs (usually in the 12 to 2 o'clock position) and use the bare radius as a signal line running to their tubular retreat. H. W. Levi (Bull. Mus. Comp. Zool. Harv. Univ. 146 (1974): 267–90) has revised the genus.

Bibliography

Akerman, C. 1923. A comparison of the habits of a South African spider, *Cladomelea*, with those of an Australian *Dicrostichus*. Ann. Natal Mus. 5: 83–88.

———. 1926. On the spider *Menneus camelus* Pocock, which constructs a moth-catching, expanding snare. Ann. Natal Mus. 5: 411–22.

———. 1932. On the spider *Miagrammopes* sp., which constructs a single-line snare. Ann. Natal Mus. 7: 137–43.

Almquist, S. 1970. Thermal tolerances and preferences of some dune-living spiders. Oikos 21: 230–36.

———. 1973a. Habitat selection by spiders on coastal sand dunes in Scanie, Sweden. Entomol. Scand. 4: 134–54.

———. 1973b. Spider associations in coastal sand dunes. Oikos 24: 444–57.

Anatasiu-Dumitrescu, M. 1941. Contributions à l'étude anatomique et cytologique de l'appareil séricigène des araignées, I. An. Acad. Rom. Mem. Sect. Stiint. 16: 733–840.

———. 1942. Contributions à l'étude anatomique et cytologique de l'appareil séricigène des araignées, II. An. Acad. Rom. Mem. Sect. Stiint. 17: 263–349.

Anderson, J. F. 1970. Metabolic rates of spiders. Comp. Biochem. Physiol. 33: 51–72.

Anderson, J. F., and K. Prestwich. 1982. Respiratory gas exchange in spiders. Physiol. Zool. 55: 72–90.

Anderson, S. O. 1970. Amino acid composition of spider silks. Comp. Biochem. Physiol. 35: 705–13.

Apstein, C. 1889. Bau und Function der Spinndrüsen der Araneida. Arch. Naturgesch. 55: 29–74.

Archer, A. F. 1942. Alabama spiders of the family Mimetidae. Pap. Mich. Acad. Sci. 27: 183–93.

———. 1947. The Theridiidae or comb-footed spiders of Alabama. Pap. Ala. Mus. Nat. Hist. 22: 5–67.

———. 1948. A modified classification of spiders based in part on the evolution of the male palpus. J. Ala. Acad. Sci. 20: 3–9.

———. 1953. Studies in the orbweaving spiders (Argiopidae), 3. Am. Mus. Novit. 1622: 1–27.

Arita, T. 1963. Web construction by *Cyrtarachne nagasakiensis* (in Japanese). Atypus 28: 10–12.

Aspey, W. P. 1979. Wolf spider sociobiology, II: Density parameters influencing agonistic behavior in *Schizocosa crassipes*. Behavior 62: 142–63.

Austin, A. D., and D. T. Anderson. 1978. Reproduction and development of the spider *Nephila edulis* (Koch) (Araneae: Araneidae). Aust. J. Zool. 26: 501–18.

Austin, A. D., and A. D. Blest. 1979. The biology of two Australian species of dinopid spiders. J. Zool. London 189: 145–56.

Bacelar, A. 1933. Sur les moeurs des *Nemesia* et des *Pachylomerus*. Bull. Soc. Port. Sci. Nat. 11: 291–95.

Baerg, W. J. 1958. *The Tarantula*. Univ. Kansas Press, Lawrence.

Balogh, J. 1934. Vorläufige Mitteilung über radnetzbauende Pachygnathen. Folia Zool. Hydrobiol. 6: 94–96.

Barnes, R. D. 1953. The ecological distribution of spiders in nonforest maritime communities at Beaufort, North Carolina. Ecol. Monogr. 23: 315–37.

Barrows, W. M. 1915. The reactions of an orb-weaving spider, *Epeira sclopetaria* Clerck, to rhythmic vibrations of its web. Biol. Bull. 29: 316–33.

Barth, F. G. 1967. Ein einzelnes Spaltsinnesorgan auf dem Spinnentarsus: Seine Erregung in Abhängigkeit von den Parametern des Luftshallreizes. Z. Vgl. Physiol. 55: 407–49.

———. 1972. Die Physiologie der Spaltsinnesorgane, II: Funktionelle Morphologie eines Mechanoreceptors. J. Comp. Physiol. 81: 159–86.

———. 1978. Slit sense organs: "Strain gauges" in the arachnid exoskeleton. Symp. Zool. Soc. London 42: 439–48.

———. 1981. Strain detection in the arthropod exoskeleton. *In* M. S. Laverack and D. J. Cosens, eds. *Sense Organs*. Blackie, Glasgow, pp. 112–41.

———. 1982. Spiders and vibratory signals: Sensory reception and behavioral significance. *In* P. N. Witt and J. S. Rovner, eds. *Spider Communication: Mechanisms and Ecological Significance*. Princeton Univ. Press, Princeton, N.J., pp. 67–122.

Barth, F. G., and J. Bohnenberger. 1978. Lyriform slit sense organ: Thresholds and stimulus amplitude ranges in a multi-unit mechanoreceptor. J. Comp. Physiol. 125: 37–43.

Barth, F. G., and E. A. Seyfarth. 1979. *Cupiennius salei* Keys. (Araneae) in the highlands of central Guatemala. J. Arachnol. 7: 255–63.

Bates, H. W. 1863. *The Naturalist on the River Amazons*, 2 vols. John Murray, London.

Baum, J. 1938. On the habits of the Australian spider *Dinopis subrufus* L. Koch. Vestn. Cesk. Spol. Zool. 5: 28–33.

Bays, S. M. 1962. A study of the training possibilities of *Araneus diadematus* Cl. Experientia 18: 423.

Beatty, J. A. 1970. The spider genus *Ariadna* in the Americas. Bull. Mus. Comp. Zool. Harv. Univ. 139: 433–528.

Beaver, D., and P. Baldwin. 1975. Ecological overlap and the problem of competition and sympatry in the Western and Hammond's Flycatchers. Condor 77: 1–13.

Beer, C. G. 1977. What is a display? Am. Zool. 17: 155–65.

Benforado, J., and K. H. Kistler. 1973. Growth of the orb weaver *Araneus diadematus* and correlation with web measurements. Psyche 80: 90–100.

Bentzien, M. M. 1973. Biology of the spider *Diguetia imperiosa* (Araneida: Diguetidae). Pan-Pacific Entomol. 2: 110–23.

———. 1975. Biosystematics of the spider genus *Brachythele* Ausserer (Araneida: Dipluridae). Ph.D. thesis, Univ. Calif., Berkeley.

Berland, L. 1932. *Les arachnides*. Vol. 16 of *Encyclopédie entomologique*. Lechevalier, Paris.

Berman, J. E., and H. W. Levi. 1971. The orb-weaver genus *Neoscona* in North America (Araneae: Araneidae). Bull. Mus. Comp. Zool. Harv. Univ. 141: 465–500.

Bertkau, P. 1878. Versuch einer natürlichen Anordnung der Spinnen, nebst Bemerkung zu einzelnen Gattungen. Arch. Naturgesch. 44: 351–410.

———. 1882. Über das Cribellum und Calamistrum: Ein Beitrag zur Histologie, Biologie, und Systematik der Spinnen. Arch. Naturgesch. 48: 316–62.

Bhattacharya, G. C. 1936. Observations on some peculiar habits of the spider *Marpissa melanognathus*. J. Bombay Nat. Hist. Soc. 39: 142–44.

Biere, J. M., and G. W. Uetz. 1981. Web orientation in the spider *Micrathena gracilis* (Araneae: Araneidae). Ecology 62: 336–44.

Blackwall, J. 1833. Notice of several recent discoveries in the structure and economy of spiders. Trans. Linn. Soc. London 16: 471–85.

———. 1841. The difference in the number of eyes with which spiders are provided, proposed as the basis of their distributions into tribes. . . . Trans. Linn. Soc. London 18: 601–70.

———. 1877. A list of spiders captured in the Seychelles Islands by Professor E. Percival Wright, M.D., F.L.S., with descriptions of species supposed to be new to arachnologists. Proc. R. Ir. Acad. 3: 1–22.

Blandin, P. 1971. Découverte en Côte d'Ivoire de *Calommata simoni* Pocock. Bull. Inst. Fondam. Afr. Noire, Sér. A, 1: 48–52.

Blandin, P., and M. Célérier. 1977. Observations sur les mygales terricoles récoltées à la Station d'Ecologie Tropicale de Lamto (Côte d'Ivoire). Rev. Arachnol. 1: 75–83.

Blanke, R. 1972. Untersuchungen zur Ökethologie von *Cyrtophora citricola* Forskål (Araneae, Araneidae) in Andalusien. Forma Functio 5: 125–206.

———. 1973. Nachweis von Pheromonen bei Netzspinnen. Naturwissenschaften 60: 481.

———. 1975a. Das Sexualverhalten der Gattung *Cyrtophora* als Hilfsmittel für phylogenetische Aussagen. Proc. Int. Arachnol. Congr. 6: 116–19.

———. 1975b. Untersuchungen zum Sexualverhalten von *Cyrtophora cicatrosa* (Stoliczka) (Araneae, Araneidae). Z. Tierpsychol. 37: 62–74.

Blest, A. D. 1978. The rapid synthesis and destruction of photoreceptor membrane by a dinopid spider: A daily cycle. Proc. R. Soc. London, Ser. B, 200: 463–83.

———. 1983. Ultrastructure of secondary retinae of primitive and advanced jumping spiders (Salticidae, Araneae). Zoomorphology 102: 125–42.

Blest, A. D., R. C. Hardie, P. McIntyre, and D. S. Williams. 1981. The spectral sensitivities of identified receptors and the function of retinal tiering in the principal eyes of a jumping spider. J. Comp. Physiol. A 145: 227–39.

Blest, A. D., and M. F. Land. 1977. The physiological optics of *Dinopis subrufus* L. Koch: A fish lens in a spider. Proc. R. Soc. London, Ser. B, 196: 197–222.

Blest, A. D., D. S. Williams, and L. Kao. 1980. The posterior median eyes of the dinopid spider *Menneus*. Cell Tissue Res. 211: 391–403.

Blum, M. S., and J. M. Brand. 1972. Social insect pheromones: Their chemistry and function. Am. Zool. 12: 553–76.

Bohnenberger, J. 1981. Matched transfer characteristics of a single unit in a compound slit sense organ. J. Comp. Physiol. 142: 391–402.

Bonnet, P. 1959. *Bibliographia Araneorum*, Vol. 2, Pt. 5, T–Z. Published by the author, Toulouse.

Boys, C. V. 1880. The influence of a tuning fork on the garden spider. Nature (London) 23: 149–50.

Brach, V. 1975. The biology of the social spider *Anelosimus eximius* (Araneae: Theridiidae). Bull. South. Calif. Acad. Sci. 74: 37–41.

———. 1976. Subsocial behavior in the funnel-web wolf spider *Sosippus floridanus* (Araneae: Lycosidae). Fla. Entomol. 59: 225–29.

———. 1977. *Anelosimus studiosus* (Araneae: Theridiidae) and the evolution of quasi-sociality in theridiid spiders. Evolution 31: 154–61.

Bradoo, B. L. 1972a. The cocoon-spinning behavior and fecundity of *Stegodyphus sarasinorum* Karsch (Araneae: Eresidae) from India. J. Bombay Nat. Hist. Soc. 2: 394–400.

———. 1972b. Some observations on the ecology of the social spider *Stegodyphus sarasinorum* Karsch (Araneae, Eresidae) from India. Orient. Insects 6: 193–203.

———. 1979. *Uloborus ferokus* n. sp. (Araneae: Uloboridae), a commensal of *Stegodyphus sarasinorum* Karsch. Bull. Br. Arachnol. Soc. 4: 353–55.

———. 1980. Feeding behavior and recruitment display in the social spider *Stegodyphus sarasinorum* Karsch (Araneae: Eresidae). Tijdschr. Entomol. 123:89–104.

Brady, A. R. 1962. The spider genus *Sosippus* in North America, Mexico, and Central America (Araneae, Lycosidae). Psyche 69: 129–64.

Braendegaard, J. 1937. Observations on spiders starting off on "ballooning excursions." Vidensk. Medd. Dan. Naturhist. Foren. Khobenhavn 101: 115–17.

Breed, A. L., V. D. Levine, D. B. Peakall, and P. N. Witt. 1964. The fate of the intact orb of the spider *Araneus diadematus* Cl. Behaviour 23: 43–60.

Brignoli, P. M. 1966. La societa eterotipiche degli araneidi. Rend. Accad. Naz. Quaranta 16: 219–46.

———. 1979. Contribution à la connaissance des Uloboridae paléoarctiques (Araneae). Rev. Arachnol. 2: 275–82.

Bristowe, W. S. 1930a. Notes on the biology of spiders, I: The evolution of spiders' snares. Ann. Mag. Nat. Hist., Ser. 10, 6: 334–42.

———. 1930b. Notes on the biology of spiders, III: Miscellaneous. Ann. Mag. Nat. Hist., Ser. 10, 6: 347–53.

———. 1931. Spitting as a means of capturing prey by spiders. Ann. Mag. Nat. Hist., Ser. 10, 8: 469–71.

———. 1932. The liphistiid spiders. Proc. Zool. Soc. London 103: 1015–45.

———. 1938. The classification of spiders. Proc. Zool. Soc. London, Ser. B, 108: 285–322.

———. 1941. *The Comity of Spiders*, Vol. 2. The Ray Society, London.

———. 1952. The arachnid fauna of the Batu Caves in Malaysia. Ann. Mag. Nat. Hist., Ser. 12, 5: 699–706.

———. 1954. The chelicerae of spiders. Endeavour 13: 42–49.

———. 1958. *The World of Spiders*. Collins, London.

———. 1975. A family of living fossil spiders. Endeavour 34: 115–17.

———. 1976. A contribution to the knowledge of liphistiid spiders. J. Zool. London 178: 1–6.

Brock, V. E., and R. H. Riffenburgh. 1960. Fish schooling: A possible factor in reducing predation. J. Cons. Int. Explor. Mer 25: 307–17.

Broen, B., and M. Moritz. 1964. Zur Biologie und Verbreitung der deutschen *Atypus*-Arten (Araneae, Atypidae). Zool. Anz. 172: 147–51.

Brown, J. L., and G. H. Orians. 1970. Spacing patterns in mobile animals. Ann. Rev. Ecol. Syst. 1: 239–62.

Brown, S. G., E. M. Hill, K. E. Goist, P. A. Wenzl, and T. E. Christenson. 1985. Ecological and seasonal variations in a free-moving population of the golden-web spider, *Nephila clavipes*. Bull. Br. Arachnol. Soc. 6: 313–19.

Brownell, P. H. 1977. Compressional and surface waves in sand: Use by desert scorpions to locate prey. Science 197: 479–82.

Buchhave, P. 1975. Laser Doppler vibration measurements using variable frequency shift. DISA Inf. (DISA Elektronik A/S, DK-2740, Skovlunde, Denmark) 18: 15–20.

Buchli, H. 1969. Hunting behavior in the Ctenizidae. Am. Zool. 9: 175–93.

Buecherl, W. 1962. *Sudamerikanische Vogelspinnen.* A. Zimsen Verlag, Wittenburg.

Burch, T. L. 1979. The importance of communal experience to survival for spiderlings of *Araneus diadematus.* J. Arachnol. 7: 1–18.

Burgess, J. W. 1976. Social spiders. Sci. Am. 234 (March): 100–106.

———. 1978. Social behavior in group-living spider species. Symp. Zool. Soc. London 42: 69–78.

———. 1979a. The spider species *Mallos gregalis* Simon: Behavioral adaptations for living in Mexico. Ph.D. diss., North Carolina State Univ., Raleigh.

———. 1979b. Web-signal processing for tolerance and group predation in the social spider *Mallos gregalis* Simon. Anim. Behav. 27: 157–64.

Burgess, J. W., and G. W. Uetz. 1982. Social spacing strategies in spiders. *In* P. N. Witt and J. S. Rovner, eds. *Spider Communication: Mechanisms and Ecological Significance.* Princeton Univ. Press, Princeton, N.J., pp. 317–52.

Burgess, J. W., and P. N. Witt. 1976. Spider webs: Design and engineering. Interdiscip. Sci. Rev. 1: 322–35.

Burlacu, F. 1972. Investigations bearing on spatial distribution in spiders. Rev. Roum. Biol. Sér. Zool. 17: 3.

Buskirk, R. E. 1975a. Aggressive display and orb defence in a colonial spider, *Metabus gravidus.* Anim. Behav. 23: 560–67.

———. 1975b. Coloniality, activity patterns, and feeding in a tropical orb-weaving spider. Ecology 56: 1314–28.

———. 1981. Sociality in the Arachnida. *In* H. R. Hermann, ed. *Social Insects,* Vol. 2. Academic, London, pp. 282–367.

Cady, A. B. 1978. Microhabitat selection and locomotor activity of *Schizocosa ocreata* (Araneae: Lycosidae) as viewed at a woodland site. M.S. thesis, Ohio Univ., Athens.

Caporiacco, L. 1938. Il sistèma degli araneidi. Arch. Zool. Ital. 25: 35–155.

Carico, J. E. 1972. The Nearctic spider genus *Pisaurina* (Pisauridae). Psyche 79: 295–310.

———. 1973. The Nearctic species of the genus *Dolomedes* (Araneae: Pisauridae). Bull. Mus. Comp. Zool. Harv. Univ. 144: 435–88.

———. 1977. A simple device for coating orb webs for field photography. Bull. Br. Arachnol. Soc. 4: 100.

———. 1978. Predatory behavior in *Euryopis funebris* (Hentz) (Araneae: Theridiidae) and the evolutionary significance of web reduction. Symp. Zool. Soc. London 42: 51–58.

———. 1985. Description and significance of the juvenile web of *Pisaurina mira* (Walck.) (Araneae: Pisauridae). Bull. Br. Arachnol. Soc. 6: 295–96.

Carrel, J. E. 1978. Behavioral thermoregulation during winter in an orb-weaving spider. Symp. Zool. Soc. London 42: 41–50.

Carroll, D. 1977. Nocturnal behavior in day-wandering arboreal spiders. Peckhamia 1: 30–31.

Castillo, J., and W. G. Eberhard. In press. Use of artificial webs to determine prey available to orb-weaving spiders. Ecology.

Cazier, M. A., and M. A. Mortenson. 1964. Bionomical observations on the tarantula hawks and their prey (Hymenoptera: Pompilidae: *Pepsis*). Ann. Entomol. Soc. Am. 57: 533–41.

Chacón, P., and W. G. Eberhard. 1980. Factors affecting numbers and kinds of prey caught in artificial spider webs, with consideration of how orb webs trap prey. Bull. Br. Arachnol. Soc. 5: 29–38.

Chamberlin, R. V. 1924. The spider fauna of the shores and islands of the Gulf of California. Proc. Calif. Acad. Sci. 12: 561–694.

Chamberlin, R. V., and W. J. Gertsch. 1958. The spider family Dictynidae in America north of Mexico. Bull. Am. Mus. Nat. Hist. 116: 1–152.

Chamberlin, R. V., and W. Ivie. 1945. On some Nearctic mygalomorph spiders. Ann. Entomol. Soc. Am. 38: 549–58.

Charnov, E. L. 1976. Optimal foraging: The marginal-value theorem. Theor. Popul. Biol. 9: 129–36.

Charnov, E. L., G. H. Orians, and K. Hyatt. 1976. The ecological implications of resource depression. Am. Nat. 109: 247–59.

Cherrett, J. M. 1964. The distribution of spiders on the Moor House National Nature Reserve, Westmoreland. J. Anim. Ecol. 33: 27–48.

Chew, R. M. 1961. Ecology of the spiders of a desert community. J. N.Y. Entomol. Soc. 69: 5–41.

Chigira, Y. 1978. A note on the web of *Paraplectana tsushimensis* (in Japanese). Atypus 72: 19–24.

Christiansen, A., R. Baum, and P. N. Witt. 1962. Changes in spider webs brought about by mescaline, psilocybin, and an increase in body weight. J. Pharmacol. Exp. Ther. 136: 31–37.

Christensen, T. E., and K. C. Goist, Jr. 1979. Costs and benefits of male-male competition in the orb-weaving spider, *Nephila clavipes*. Behav. Ecol. Sociobiol. 5: 87–92.

Clerck, C. 1757. Araneae Suecici. Stockholmiae : 1–154.

Cloudsley-Thompson, J. L. 1957. Studies in diurnal rhythms, V: Nocturnal ecology and water relations of the British cribellate spiders of the genus *Ciniflo*. J. Linn. Soc. London Zool. 43: 134–52.

———. 1978. Biological clocks in Arachnida. Bull. Br. Arachnol. Soc. 4: 184–91.

———. 1981. Comments on the nature of deception. Biol. J. Linn. Soc. 16: 11–14.

Clyne, D. 1967. Note on the construction of the net and sperm-web of the cribellate spider *Dinopis subrufus* (Koch) (Araneida: Dinopidae). Aust. Zool. 14: 189–97.

———. 1969. *A Guide to Australian Spiders*. Thomas Nelson, Melbourne.

———. 1973. Notes on the web of *Poecilopachys australasia* (Griffith and Pidgeon) (Araneida: Argiopidae). Aust. Entomol. Mag. 1: 23–29.

———. 1979. *The Garden Jungle*. Collins, Sydney.

Coddington, J. 1982. Orb-web construction in *Dinopis longipes* (abstract). Am. Arachnol. 26: 8.

Coddington, J., and C. E. Valerio. 1980. Observations on the web and behavior of *Wendilgarda* spiders (Araneae: Theridiosomatidae). Psyche 87: 93–106.

Colebourne, P. H. 1974. The influence of habitat structure on the distribution of *Araneus diadematus* (Clerck). J. Anim. Ecol. 43: 401–9.

Coleman, C. 1978. A spider aberration—*Linus fimbriata*. North. Queensl. Nat. 43: 5.

———. 1980. Web construction by *Portia*—A spider of the family Salticidae, commonly known as jumping spiders. North. Queensl. Nat. 45: 7–8.

Comstock, J. H. 1912. The evolution of the webs of spiders. Ann. Entomol. Soc. Am. 5: 1–10.

———. 1948. *The Spider Book*. Revised and edited by W. J. Gertsch. Comstock Publishing, Ithaca, N.Y.

Conner, W. E., T. Eisner, R. K. Van der Meer, A. Guerrero, D. Ghiringelli, and J. Meinwald. 1980. Sex attractant of an arctiid moth (*Utatheisa ornatrix*): A pulsed chemical signal. Behav. Ecol. Sociobiol. 7: 55–63.

Conner, W. E., and W. M. Masters. 1978. Infrared video viewing. Science 199: 1004.

Cook, R. M., and S. F. Hubbard. 1977. Adaptive searching strategies in insect parasites. J. Anim. Ecol. 46: 115–25.

Cooke, J. A. L. 1972. Stinging hairs: A tarantula's defense. Fauna 4: 3–8.

Coventry, S. J. 1967. Web orientation of the banded garden spider *Argiope trifasciata*. M.S. thesis, Univ. Wisc., Madison.

Cowie, R. J., and J. R. Krebs. 1979. Optimal foraging in patchy environments. Symp. Br. Ecol. Soc. 20: 183–205.

Coyle, F. A. 1971. Systematics and natural history of the mygalomorph spider genus *Antrodiaetus* and related genera (Araneae: Antrodiaetidae). Bull. Mus. Comp. Zool. Harv. Univ. 141: 269–402.

———. 1981a. The mygalomorph spider genus *Microhexura* (Araneae: Dipluridae). Bull. Am. Mus. Nat. Hist. 170: 64–75.

———. 1981b. Notes on the behaviour of *Ummidia* trapdoor spiders (Araneae: Ctenizidae): Burrow construction, prey capture, and functional morphology of the peculiar third tibia. Bull. Br. Arachnol. Soc. 5: 159–64.

Coyle, F. A., and W. A. Shear. 1981. Observations on the natural history of *Sphodros abboti* and *Sphodros rufipes* (Araneae, Atypidae), with evidence for a contact sex pheromone. J. Arachnol. 9: 317–26.

Crane, J. 1949. Comparative biology of salticid spiders at Rancho Grande, Venezuela, IV: An analysis of display. Zoologica (N.Y.) 34: 159–215.

Crome, W. 1954. Beschreibung, Morphologie, und Lebensweise der *Eucta kaestneri* sp. n. Zool. Jahrb. Abt. Syst. Ökol. Geogr. Tiere 82: 425–52.

———. 1955. Die Beziehungen zwischen den dorsalen Zeichnungmeister und die Metamerie des Spinnen Abdomens, II. Zool. Jahrb. Abt. Syst. Ökol. Geogr. Tiere 83: 541–638.

———. 1957. Bau und Funktion der Spinnapparates und Analhügels, Ernährungsbiologie, und allgemeine Bemerkungen zur Lebensweise von *Uroctea durandi* (Latreille). Zool. Jahrb. Abt. Syst. Ökol. Geogr. Tiere 85: 571–606.

———. 1962. Die Wohnröhre von *Conothele arboricola* Pocock nebst Bemerkungen über die Lebensweise dieser Falltürspinne. Zool. Anz. 168: 450–59.

Curio, E. 1976. *The Ethology of Predation*. Springer-Verlag, Berlin.

Curtis, D. J., and E. Morton. 1974. Notes on spiders of tree trunks with different bark textures: Indices of diversity and overlap. Bull. Br. Arachnol. Soc. 3: 1–5.

Cutler, B. 1972a. Notes on the behavior of *Coleosoma floridanum* Banks. J. Kans. Entomol. Soc. 45: 275–81.

———. 1972b. Notes on the biology of *Mimetus puritanus* Chamberlin (Araneae: Mimetidae). Am. Midl. Nat. 87: 554–55.

Cutler, B., D. T. Jennings, and M. J. Moodie. 1977. Biology and habitats of the lynx spider *Oxyopes scalaris* Hentz (Araneae: Oxyopidae). Entomol. News 88: 87–97.

Czajka, M. 1963. Unknown facts of the biology of the spider *Ero furcata* (Villers) (Mimetidae, Araneae). Pol. Pismo Entomol. 33: 229–31.

Dabelow, S. 1958. Zur Biologie der Leimschleuderspinne *Scytodes thoracica* (Latreille). Zool. Jahrb. Abt. Syst. Ökol. Geogr. Tiere 86: 85–126.

Dabrowska-Prot, E., J. Luczak, and Z. Wojcik. 1973. Ecological analysis of two invertebrate groups in the wet alder wood and meadow ecotone. Ekol. Pol. 21: 753–812.

Dahl, F. 1883. Analytische Bearbeitung der Spinnen Norddeutschlands, mit einer anatomisch-biologischen Einleitung. Schr. Naturwiss. Ver. Schleswig-Holstein 5: 13–88.

———. 1901. Über den Wert des Cribellums und Calamistrums für das System der Spinnen und eine Übersicht der Zoropsiden. Sitzungsber. Ges. Naturforsch. Fren. Berlin: 177–99.

———. 1904. Über das System der Spinnen (Araneae). Sitzungsber. Ges. Naturforsch. Fren. Berlin: 93–120.

Darchen, R. 1965. Ethologie d'une araignée sociale, *Agelena consociata*. Biol. Gabonica 1: 117–46.

———. 1967. Biologie d'une mygale gabonnaise nouvelle, *Macrothele darcheni* Benoit (Araneida, Dipluridae). Biol. Gabonica 4: 253–57.

———. 1979. Relation entre colonies d'agélénides sociaux du Gabon, II: Précisions sur les essaimages. Rev. Arachnol. 2: 156.

Darwin, C. 1845. *Voyage of the Beagle*. Annotated by L. Engel, 1962. Doubleday and Co., Garden City, N.Y.

———. 1876. *Journal of Researches into the Natural History and Geology of the Countries Visited During the Voyage of H.M.S. Beagle Round the World*. John Murray, London.

Davies, V. T. 1982. *Inola* nov. gen., a web-building pisaurid (Araneae: Pisauridae) from northern Australia, with descriptions of three species. Mem. Queensl. Mus. 20: 479–87.

Decae, Arthur E. 1984. A theory on the origin of spiders and the primitive function of spider silk. J. Arachnol. 12: 21–28.

Decae, A. E., G. Caranhac, and G. Thomas. 1982. The supposedly unique case of *Cyrtocarenum cunicularium* (Oliver, 1811) (Araneae, Ctenizidae). Bull. Br. Arachnol. Soc. 5: 410–19.

Denny, M. 1976. The physical properties of spiders' silk and their role in the design of orb-webs. J. Exp. Biol. 65: 483–506.

Den Otter, C. J. 1974. Setiform sensilla and prey detection in the bird-spider, *Sericopelma rubronitens* Ausserer (Araneae, Theraphosidae). Neth. J. Zool. 24: 219–35.

DeVoe, R. D. 1972. Dual sensitivities of the cells in wolf spider eyes at ultraviolet and visible wavelengths of light. J. Gen. Physiol. 59: 247–69.

DeVoe, R. D., and J. E. Zvargulis. 1967. Spectral sensitivities of vision in wolf spiders and jumping spiders. Fed. Proc. Fed. Am. Soc. Exp. Biol. 26: 655.

DeWilde, J. 1943. Some physical properties of the spanning threads of *Aranea diademata*. Arch. Néerl. Physiol. 28: 118–31.

Dice, L. R. 1947. Effectiveness of selection by owls of deer mice (*Peromyscus maniculatus*) which contrast in color with their background. Contrib. Lab. Vertebr. Biol. Univ. Mich. 34: 1–20.

Diguet, L. 1909a. Le mosquero: Nid d'araignée employé à Méxique comme piège à mouches. Naturaliste 31: 283–85.

———. 1909b. Sur l'araignée mosquero. C. R. Acad. Sci. Paris 148: 735–36.

———. 1915. Nouvelles observations sur le mosquero ou nid d'araignées comme piège à mouches. Naturaliste 62: 240–49.

Dill, L. M. 1975. Predatory behavior of the zebra spider *Salticus scenicus* (Araneae, Salticidae). Can. J. Zool. 53: 1284–89.

Dondale, C. D. 1977. Life histories and distribution patterns of hunting spiders (Araneae) in an Ontario meadow. J. Arachnol. 4: 73–93.

Drees, O. 1952. Untersuchungen über die angeborenen Verhaltensweisen bei Springspinnen (Salticidae). Z. Tierpsychol. 9: 169–207.

Drews, C. D., and R. A. Bernard. 1976. Electrophysiological responses of chemosensitive sensilla in the wolf spider. J. Exp. Zool. 198: 423–28.

Duffey, E. 1956. Aerial dispersal in a known spider population. J. Anim. Ecol. 25: 85–111.

———. 1962a. A population study of spiders in limestone grassland. J. Anim. Ecol. 31: 571–99.

———. 1962b. A population study of spiders in limestone grassland: The field-layer fauna. Oikos 13: 15–33.

———. 1963. A mass dispersal of spiders. Trans. Norfolk Norwich Nat. Soc. 20: 38–43.

———. 1966. Spider ecology and habitat structure. Senckenbergiana Biol. 47: 45–49.

———. 1968. An ecological analysis of the spider fauna of sand dunes. J. Anim. Ecol. 37: 641–74.

Dugdale, B. E. 1969. The weaving of an engineering masterpiece, a spider's orb web, done at Fayson Lakes, N.J., August 8, 1942, as observed and reported by B. E. Dugdale, Structural Engineer. Nat. Hist. 78: 36–41.

Dumpert, K. 1978. Spider odor receptor: Electrophysiological proof. Experientia 34: 754–55.

Dzimirski, I. 1959. Untersuchungen über Bewegungssehen und Optomotorik bei Springspinnen (Salticidae). Z. Tierpsychol. 16: 385–402.

Eakin, R. M., and J. L. Brandenburger. 1971. Fine structure of the eyes of jumping spiders. J. Ultrastruct. Res. 37: 618–63.

Eason, R. R., W. B. Peck, and W. H. Whitcomb. 1967. Notes on spider parasites, including a reference list. J. Kans. Entomol. Soc. 40: 422–34.

Eberhard, M. J. W. 1975. The evolution of social behavior by kin selection. Q. Rev. Biol. 50: 1–33.

Eberhard, W. G. 1967. Attack behavior of diguetid spiders and the origin of prey wrapping in spiders. Psyche 74: 173–81.

———. 1969. Computer simulation of orb-web construction. Am. Zool. 9: 229–38.

———. 1971a. The ecology of the web of *Uloborus diversus* (Araneae: Uloboridae). Oecologia (Berlin) 6: 328–42.

———. 1971b. Senile web patterns in *Uloborus diversus* (Araneae: Uloboridae). Dev. Psychobiol. 4:249–54.

———. 1972a. Observations on the biology of *Achaearanea tesselata* (Araneae: Theridiidae). Psyche 78: 209–12.

———. 1972b. The web of *Uloborus diversus* (Araneae: Uloboridae). J. Zool. London 166:417–65.

———. 1973. Stabilimenta on the webs of *Uloborus diversus* (Araneae: Uloboridae) and other spiders. J. Zool. London 171: 367–84.

———. 1974. Maternal behavior in a South American *Lyssomanes*. Bull. Br. Arachnol. Soc. 3: 51.

———. 1975. The "inverted ladder" orb web of *Scoloderus* sp. and the intermediate orb of *Eustala*(?) sp. (Araneae: Araneidae). J. Nat. Hist. 9: 93–106.

———. 1976. Physical properties of sticky spirals and their connections: sliding connections in orb webs. J. Nat. Hist. 10: 481–88.

———. 1977a. Aggressive chemical mimicry by a bolas spider. Science 198: 1173–75.

———. 1977b. "Rectangular orb" webs of *Synotaxus* (Araneae: Theridiidae). J. Nat. Hist. 11:501–7.

———. 1977c. The webs of newly emerged *Uloborus diversus* and of a male *Uloborus* sp. (Araneae, Uloboridae). J. Arachnol. 4: 201–6.
———. 1979a. *Argyrodes attenuatus* (Theridiidae): A web that is not a snare. Psyche 86: 407–13.
———. 1979b. Rates of egg production by tropical spiders in the field. Biotropica 11: 292–300.
———. 1980a. The natural history and behavior of the bolas spider *Mastophora dizzydeani* sp. n. (Araneae). Psyche 87: 143–69.
———. 1980b. Persistent stickiness of cribellum silk. J. Arachnol. 8: 283.
———. 1981a. Construction behavior and the distributions of tensions in orb webs. Bull. Br. Arachnol. Soc. 5: 189–204.
———. 1981b. Notes on the natural history of *Tackzanowskia* sp. (Araneae: Araneidae). Bull. Br. Arachnol. Soc. 5: 175–76.
———. 1981c. The single-line web of *Phoroncidia studo* Levi (Araneae, Theridiidae): A prey attractant? J. Arachnol. 9: 229–32.
———. 1982. Behavioral characteristics for the higher classification of orb-weaving spiders. Evolution 36: 1067–95.
———. In press. Predatory behavior of an assassin spider, *Chorizopes* sp. (Araneidae), and the defensive behavior of its prey. J. Bombay Nat. Hist. Soc.
Edgar, W. 1969. Prey and predators of the wolf spider *Pardosa lugubris*. J. Zool. Soc. London 159: 405–11.
Edgar, W., and M. Loenen. 1974. Aspects of the overwintering habitat of the wolf spider *Pardosa lugubris*. J. Zool. Soc. London 172: 383–88.
Edmunds, J. 1978. The web of *Paraneus cyrtoscapus* (Pocock, 1899) (Araneae: Araneidae) in Ghana. Bull. Br. Arachnol. Soc. 4: 191–96.
Edmunds, M. 1974. *Defence in Animals*. Longmans, Essex.
Edwards, G. B. 1981. Sound productions by courting males of *Phidippus mystaceus* (Araneae: Salticidae). Psyche 88: 199–214.
Edwards, G. B., J. F. Carroll, and W. H. Whitcomb. 1974. *Stoidis aurata* (Araneae: Salticidae), a spider predator of ants. Fla. Entomol. 57: 337–46.
Eisner, T. 1980. Chemistry, defense, and survival: Case studies and selected topics. *In* M. Locke and D. S. Smith, eds. *Insect Biology in the Future "VBW 80."* Academic, New York, pp. 847–78.
Eisner, T., R. Alsop, and G. Ettershank. 1964. Adhesiveness of spider silk. Science 146: 1058–61.
Eisner, T., and J. Dean. 1976. Ploy and counter-ploy in predator-prey interactions: Orb-weaving spiders versus bombardier beetles. Proc. Natl. Acad. Sci. U.S.A. 73: 1365–67.
Eisner, T., and S. Nowicki. 1983. Spider-web protection through visual advertisement: Role of the stabilimentum. Science 219: 185–87.
Emerit, M. 1978. Deux sous-familles d'Araneidae (les Cyrtarachninae et Mastophorinae) nouvelles pour Madagascar, à propos de prétendues lacunes faunistiques de la région malgache. Symp. Zool. Soc. London 42: 358–65.
Enders, F. 1973. Selection of habitat by the spider *Argiope aurantia* Lucas (Araneidae). Am. Midl. Nat. 90: 47–55.
———. 1974. Vertical stratification in orb-web spiders (Araneidae: Araneae) and a consideration of other methods of coexistence. Ecology 55: 317–28.
———. 1975a. Airborne pheromone probable in the orb-web spider *Argiope aurantia* (Araneidae). Br. Arachnol. Soc. News Lett. 13: 5–6.
———. 1975b. Change of web site in *Argiope* spiders (Araneidae). Am. Midl. Nat. 94: 484–90.

———. 1975c. The influence of hunting manner on prey size, particularly in spiders with long attack distances (Araneidae, Linyphiidae, and Salticidae). Am. Nat. 109: 737–63.

———. 1977. Web-site selection by orb-web weavers, particularly *Argiope aurantia* Lucas. Anim. Behav. 25: 694–712.

Enock, F. 1885. The life history of *Atypus piceus* Sulz. Trans. Entomol. Soc. London: 389–420.

Erber, J. 1868. Bericht über eine Reise nach Rhodus. Verh. Zool.-Bot. Ges. Wien 18: 903–8.

Esteban, C. J. de la S. de. 1976. Presencia de tráqueas en el prosoma de algunas especies de Salticidae (Araneae). Physis (Buenos Aires) 35: 251–56.

Ewer, R. F. 1972. The devices in the web of the West African spider *Argiope flavipalpus*. J. Nat. Hist. 6: 159–67.

Exline, H. 1945. Spiders of the genus *Conopistha* from northwestern Peru and Ecuador. Ann. Entomol. Soc. Am. 38: 505–28.

Exline, H., and H. W. Levi. 1962. American spiders of the genus *Argyrodes* (Araneae, Theridiidae). Bull. Mus. Comp. Zool. Harv. Univ. 127: 75–204.

Farr, J. A. 1976. Social behavior of the golden-silk spider *Nephila clavipes*. J. Arachnol. 4: 137–44.

Feio, J. L. A. 1952. Uma curiosa migalomorpha arboricola—*Paracenobiopelma gerecormophila* g. n., sp. n. (Araneae, Barychelidae). Bol. Mus. Nac. Rio de Janeiro Zool. 113: 1–23.

Ferguson, I. C. 1972. Natural history of the spider *Hypochilus thorelli* Marx (Hypochilidae). Psyche 78: 179–99.

Finck, A. 1972. Vibration sensitivity in an orb weaver. Am. Zool. 12: 539–43.

Finck, A., and C. F. Reed. 1979. Behavioral response to whole body vibration in the orb-weaver *Araneus sericatus* Clerck (Araneae, Araneidae). J. Arachnol. 7: 139–47.

Finck, A., G. Stewart, and C. F. Reed. 1975. The orb web as an acoustic detector. J. Acoust. Soc. Am. 57: 753–54.

Fincke, O. M. 1981. The association between two Neotropical spiders (Araneae: Uloboridae and Tengellidae). Biotropica 13: 301–7.

Foelix, R. F. 1970. Chemosensitive hairs in spiders. J. Morphol. 132: 313–34.

———. 1982. *The Biology of Spiders*. Harvard Univ. Press, Cambridge, Mass.

Foelix, R. F., and I. Chu-Wang. 1973a. The morphology of spider sensilla, I: Mechanoreceptors. Tissue Cell 5: 451–60.

———. 1973b. The morphology of spider sensilla, II: Chemoreceptors. Tissue Cell 5: 461–78.

Foelix, R. F., and H. Jung. 1978. Some anatomical aspects of *Hypochilus thorelli*, with special reference to the calamistrum and cribellum. Symp. Zool. Soc. London 42: 417–22.

Ford, M. J. 1977. Energy costs of the predation strategy of the web-spinning spider *Lepthyphantes zimmermanni* Bertkau (Linyphiidae). Oecologia (Berlin) 28: 341–49.

Forster, L. M. 1977a. A qualitative analysis of hunting behaviour in jumping spiders (Araneae: Salticidae). N. Z. J. Zool. 4: 51–62.

———. 1977b. Some factors affecting feeding behaviour in young *Trite auricoma* spiderlings (Araneae: Salticidae). N. Z. J. Zool. 4: 435–43.

———. 1979. Visual mechanisms of hunting behaviour in *Trite planiceps*, a jumping spider. N. Z. J. Zool. 6: 79–93.

———. 1982a. Vision and prey-catching strategies in jumping spiders. Am. Sci. 70: 165–75.

———. 1982b. Visual communication in jumping spiders (Salticidae). *In* P. N. Witt and J. S. Rovner, eds. *Spider Communication: Mechanisms and Ecological Significance.* Princeton Univ. Press, Princeton, N.J., pp. 161–212.

Forster, R. R. 1967. The spiders of New Zealand, I. Otago Mus. Zool. Bull. 1: 1–124.

———. 1970. The spiders of New Zealand, III. Otago Mus. Zool. Bull. 3: 1–184.

Forster, R. R., and A. D. Blest. 1979. The spiders of New Zealand, V. Otago Mus. Zool. Bull. 5: 1–173.

Forster, R. R., and L. M. Forster. 1973. *New Zealand Spiders: An Introduction.* Collins, Auckland.

Forster, R. R., and M. R. Gray. 1979. *Progradungula*, a new cribellate genus of the spider family Gradungulidae. Aust. J. Zool. 27: 1051–71.

Forster, R. R., and N. I. Platnick. 1977. A review of the spider family Symphytognathidae (Arachnida: Araneae). Am. Mus. Novit. 2619: 1–29.

———. 1984. A review of the archaeid spiders and their relatives, with notes on the limits of the Superfamily Palpimanoidea. Bull. Am. Mus. Nat. Hist 178(1): 1–106.

Forster, R. R., and C. L. Wilton. 1968. The spiders of New Zealand, II. Otago Mus. Zool. Bull. 2: 1–180.

———. 1973. The spiders of New Zealand, IV. Otago Mus. Zool. Bull. 4: 1–309.

Fowler, H. G., and J. Diehl. 1978. Biology of a Paraguayan colonial orb weaver, *Eriophora bistriata* (Rengger). Bull. Br. Arachnol. Soc. 4: 241–50.

Freisling, J. 1961. Netz und Netzbauinstinkte bei *Theridion saxatile* Koch. Z. Wiss. Zool. 165: 396–421.

Fretwell, S. D. 1972. Populations in a seasonal environment. Monogr. Popul. Biol. 5: 1–218.

Fretwell, S. D., and H. L. Lucas. 1969. On territorial behavior and other factors influencing habitat distribution in birds, I: Theoretical development. Acta Biotheor. 19: 16–36.

Friedrich, V. L., and M. Langer. 1969. Fine structure of cribellate spider silk. Am. Zool. 9: 91–96.

Frings, H., and M. Frings. 1966. Reactions of orb-weaving spiders (Argiopidae) to airborne sounds. Ecology 47: 578–88.

Frohlich, C., and R. E. Buskirk. 1982. Transmission and attenuation of vibration in orb spider webs. J. Theor. Biol. 95: 13–36.

Gabel, J. R. 1972. Further observations on tarantula burrows. Pan-Pac. Entomol. 48: 72–73.

Galiano, M. E. 1976. Comentarios sobre la categoría sistemática del taxón Lyssomanidae (Araneae). Rev. Mus. Argent. Cienc. Nat. Bernardino Rivadavia Inst. Nac. Invest. Cienc. Nat. 5: 59–70.

———. 1980. Revisión del género *Lyssomanes* Hentz, 1845 (Araneae, Salticidae). Opera Lilloana 30: 1–104.

Galton, F. 1871. Gregariousness in cattle and men. MacMillan's Mag. 23: 353.

Gardner, B. T. 1964. Hunger and sequential responses in the hunting behavior of salticid spiders. J. Comp. Physiol. Psychol. 58: 167–73.

———. 1965. Observations on three species of *Phidippus* jumping spiders (Araneae: Salticidae). Psyche 72: 133–47.

Gerhardt, U. 1924. Weitere Studien über die Biologie der Spinnen. Arch. Naturgesch. 90: 85–192.

———. 1927. Neue biologische Untersuchungen an einheimischen und ausländischen Spinnen. Z. Morphol. Ökol. Tiere 8: 96–186.

———. 1933. Neue Untersuchungen zur Sexualbiologie der Spinnen, insbesondere an Arten der Mittelmeerländer und der Tropen. Z. Morphol. Ökol. Tiere 27: 1–75.

Gerhardt, U., and A. Kaestner. 1938. Araneae. *In* W. G. Kuekenthal and T. Krumbach, eds. *Handbuch der Zoologie*, Vol. 3. De Gruyter, Berlin, pp. 497–656.

Gering, R. L. 1953. Structure and function of the genitalia in some American agelenid spiders. Smithson. Misc. Collect. 121: 1–84.

Gertsch, W. J. 1947. Spiders that lasso their prey. Nat. Hist. 56: 152–59.

———. 1949. *American Spiders*. D. Van Nostrand, New York.

———. 1955. The North American bolas spiders of the genera *Mastophora* and *Agatostichus*. Bull. Am. Mus. Nat. Hist. 106: 221–54.

———. 1958. The spider family Hypochilidae. Am. Mus. Novit. 1912: 1–28.

———. 1971. A report on some Mexican cave spiders. Assoc. Mex. Cave Stud. Bull. 4: 47–111.

———. 1973. A report on cave spiders from Mexico and Central America. Assoc. Mex. Cave Stud. Bull. 5: 141–63.

———. 1974. The spider family Leptonetidae in North America. J. Arachnol. 1: 145–203.

———. 1979. *American Spiders*, 2d ed. Van Nostrand Reinhold, New York.

———. 1982. The spider genera *Pholcophora* and *Anopsicus* in North America, Central America and the West Indies. Tex. Mem. Mus. Bull. 28: 95–144.

———. 1984. The spider family Nesticidae (Araneae) in North America, Central America, and the West Indies. Texas Mem. Mus. Bull. 31: 1–91.

Gertsch, W. J., and N. Platnick. 1979. A revision of the spider family Mecicobothriidae (Araneae, Mygalomorphae). Am. Mus. Novit. 2687: 1–32.

———. 1980. A revision of the American spiders of the family Atypidae (Araneae, Mygalomorphae). Am. Mus. Novit. 2704: 1–39.

Gillespie, R. G. 1981. The quest for prey by the web-building spider *Amaurobius similis* (Blackwell). Anim. Behav. 29: 953–54.

Giulio, L. 1980. Analysis of the optomotor reaction of *Heliophanus muscorum* Walck. (Araneae: Salticidae) induced by rotating spirals. Monit. Zool. Ital., n.s., 14: 211–20.

Glatz, L. 1967. Zur Biologie und Morphologie von *Oecobius annulipes* Lucas (Araneae, Oecobiidae). Z. Morphol. Ökol. Tiere 61: 185–214.

Goerner, P. 1965. A proposed transducing mechanism for a multiple-innervated mechanoreceptor (trichobothrium) in spiders. Cold Spring Harbor Symp. Quant. Biol. 30: 69–73.

Goerner, P., and P. Andrews. 1969. Trichobothrien, ein Ferntastsinnesorgan bei Webspinnen (Araneen). Z. Vgl. Physiol. 64: 301–17.

Graeser, K. 1973. Die Übertragungseigenschaften des Netzes von *Zygiella x-notata* (Clerck) für transversale Sinusschwingungen im niederen Frequenzbereich und Frequenzanalyse beutetiererregter Netzvibrationen. Thesis, Univ. Frankfurt, Frankfurt.

Grasshoff, M. 1974. Zur Evolution der Kopulationsorgane der Spinnen. Nat. Mus. 104: 55–60.

———. 1975. Die Evolution komplizierter Kopulationsorgane—Ein separater Adaptationsverlauf. Nat. Mus. 105: 61–68.

Gravely, F. H. 1921. The spiders and scorpions of Barkuda Island. Rec. Indian Mus. 22: 399–459.

Green, L. A. 1970. Setal structure and spider phylogeny. M.S. thesis, Southern Illinois Univ., Carbondale.

Greenquist, E. A., and J. S. Rovner. 1976. Lycosid spiders on artificial foliage: Stratum choice, orientation preferences, and prey wrapping. Psyche 83: 196–209.

Greenstone, M. H. 1978. The numerical response to prey availability of *Pardosa ramulosa* (Araneae: Lycosidae) and its relationship to the role of spiders in the balance of nature. Symp. Zool. Soc. London 42: 183–93.

———. 1981. Ballooning frequency and habitat predictability in two wolf spider species (Lycosidae: *Pardosa*). Fla. Entomol. 65: 83–89.

Greenstone, M. H., and A. F. Bennett. 1980. Foraging strategy and metabolic rate in spiders. Ecology 61: 1255–59.

Griswold, C. E. In press. *Tapinillus longipes* (Taczanowski), a web-building lynx spider from the American tropics (Araneae: Oxyopidae). J. Nat. Hist.

Guenée, M. A. 1852. *Histoire naturelle des insectes species générales des lépidoptères*, Vol. 7. Libraire Encyclopédique de Roret, Paris.

Guppy, M., and W. Davison. 1982. The hare and the tortoise: Metabolic strategies in cardiac and skeletal muscles of the skink and the chameleon. J. Exp. Zool. 220: 289–95.

Hallander, H. 1967. Courtship display and habitat selection in the wolf spider *Pardosa chelata* (O. F. Mueller). Oikos 18: 145–50.

———. 1970. Courtship display and habitat selection in the wolf spiders *Pardosa chelata* (O. F. Mueller) and *P. pullata* (Clerck). Oikos 21: 337–40.

Hamilton, W. D. 1964. The genetical theory of social behavior, I and II. J. Theoret. Biol. 7: 1–52.

———. 1971. Geometry for the selfish herd. J. Theor. Biol. 31: 295–311.

Harcourt, A. H. 1978. Activity periods and patterns of social interaction: A neglected problem. Behaviour 66: 121–35.

Harris, D. J., and P. J. Mill. 1977a. Observations on the leg receptors of *Ciniflo* (Araneae: Dictynidae), I: External mechanoreceptors. J. Comp. Physiol. 119: 37–54.

———. 1977b. Observations on the leg receptors of *Ciniflo* (Araneae: Dictynidae), II: Chemoreceptors. J. Comp. Physiol. 119: 55–62.

Hartsock, S. P. 1983. The influence of web structure and web placement on prey capture in an orb-weaving spider. M.S. thesis, Univ. of Cincinnati.

Harwood, R. H. 1974. Predatory behavior of *Argiope aurantia* (Lucas). Am. Midl. Nat. 91: 130–39.

Hashimoto, R. 1963. Prey-capture method of *Miagrammopes orientalis* and *Argyrodes saganus* (in Japanese). Atypus 30: 1–3.

Haupt, J. 1977. Preliminary report on the mating behavior of the primitive spider *Heptathela kimurai* (Kishida) (Araneae, Liphistiomorphae). Z. Naturforsch. Teil C, 32: 312–14.

———. 1979. Lebensweise und Sexualverhalten der mesothelen Spinne *Heptathela nishihirai* n. sp. (Araneae, Liphistiidae). Zool. Anz. 202: 348–74.

Heil, K. H. 1936. Beiträge zur Physiologie und Psychologie der Springspinnen. Z. Vgl. Physiol. 23: 1–25.

Heimer, S. 1982. Interne Arretierungsmechanismen an den Kopulationsorganen männlichen Spinnen (Arachnida: Araneae): Ein Beitrag zur Phylogenie der Araneoidea. Entomol. Abh. Mus. Tierk. Dresden 45: 35–64.

Hensley, S. D. 1971. Management of sugarcane borer populations in Louisiana: A decade of change. Entomophaga 16: 133–46.

Hensley, S. D., W. H. Long, L. R. Roddy, W. J. McCormick, and E. J. Concienne. 1961. Effects of insecticides on predaceous arthropod fauna of Louisiana sugarcane fields. J. Econ. Entomol. 54: 146–49.

Hickman, V. V. 1964. On *Atrax infensus* sp. n. (Araneida: Dipluridae), its habits and a method of trapping the males. Pap. Proc. R. Soc. Tasmania 98: 107–12.

———. 1975. On *Paraplectanoides crassipes* Keyserling (Araneae: Araneidae). Bull. Br. Arachnol. Soc. 3: 166–74.

Hieber, C. S. 1979. Web orientations to wind and light in the laboratory for the spiders *Araneus diadematus* and *Araneus gemmoides*. M.S. thesis, Univ. N. Dak., Grand Forks.

Hildrew, A., and C. Townsend. 1980. Aggregation, interference, and foraging by larvae of *Plectrocnemia conspersa* (Trichoptera: Polycentropodidae). Anim. Behav. 28: 253–560.

Hill, D. E. 1977. The pretarsus of salticid spiders. Zool. J. Linn. Soc. 60: 319–38.

———. 1979. Orientation by jumping spiders of the genus *Phidippus* (Araneae: Salticidae) during the pursuit of prey. Behav. Ecol. Sociobiol. 5: 301–22.

Hill, E. M., and T. E. Christenson. 1981. Effects of prey characteristics and web structure on feeding and predatory responses of *Nephila clavipes* spiderlings. Behav. Ecol. Sociobiol. 8: 1–5.

Hingston, R. W. G. 1920. *A Naturalist in the Himalaya*. Small, Maynard, Boston.

———. 1922. The snare of the giant wood spider *Nephila maculata*, Part III. J. Bombay Nat. Hist. Soc. 28: 917–23.

———. 1927. Protective devices in spiders' snares, with a description of seven new species of orb-weaving spiders. Proc. Zool. Soc. London 18: 259–93.

Hoelldobler, B. 1970. *Steatoda fulva* (Theridiidae), a spider that feeds on harvester ants. Psyche 77: 202–7.

Hollander, J. Den, and H. Lof. 1972. Differential use of the habitat by *Pardosa pullata* (Clerck) and *Pardosa prativaga* (C. L. Koch) in a mixed population (Araneae: Lycosidae). Tijdschr. Entomol. 115: 205–15.

Hollar, W. D., and P. N. Witt. 1979. Living together with minimal coordination. Neurosci. Abstr. 5:117.

Holling, C. S. 1966. The functional response of invertebrate predators to prey density. Mem. Entomol. Soc. Can. 48: 1–86.

Holm, A. 1939. Beiträge zur Biologie der Theridiiden. Festschr. Embrik Strand 5: 56–67.

Homann, H. 1928. Beiträge zur Physiologie der Spinnenaugen. I. Untersuchungsmethoden. II. Das Schvermögen der Salticidae. Z. Vgl. Physiol. 7: 201–68.

———. 1950. Die Nebenaugen der Araneen. Zool. Jahrb. Abt. Anat. Ontog. Tiere 71: 1–144.

———. 1971. Die Augen der Araneen: Anatomie, Ontogenie, und Bedeutung für die Systematik (Chelicerata, Arachnida). Z. Morphol. Ökol. Tiere 69: 201–72.

Honjo, S. 1977. Social behavior in *Dictyna follicola* (Araneae: Dictynidae). Acta Arachnol. 27: 213–19.

Hopfmann, W. 1935. Bau und Leistung des Spinnapparates einiger Netzspinnen. Jena. Z. Naturwiss. 70: 65–112.

Horton, C. C. 1979. Apparent attraction of moths by webs of araneid spiders. J. Arachnol. 7: 88.

———. 1980. A defensive function for the stabilimenta of two orb-weaving spiders (Araneae, Araneidae). Psyche 87: 13–20.

Horton, C. C., and D. H. Wise. 1981. Effects of environment versus competition in two orb-weaving spiders (Araneae, Araneidae) (abstract). Am. Arachnol. 24: 17.

———. In press. The experimental analysis of competition among two synoptic species of orb-web spiders (Araneae: Araneidae). Ecology.

Hutchinson, C. E. 1903. A bolas-throwing spider. Sci. Am. 10 (October): 172.

Jackson, R. R. 1971. Fine structure of thread connections in the orb web of *Araneus diadematus*. Psyche 78: 12–31.

———. 1976. Predation as a selection factor in the mating strategy of the jumping spider *Phidippus johnsoni* (Salticidae, Araneae). Psyche 83: 243–55.

———. 1977a. Courtship versatility in the jumping spider *Phidippus johnsoni* (Araneae: Salticidae). Anim. Behav. 25: 953–57.

———. 1977b. Prey of the jumping spider *Phidippus johnsoni* (Araneae: Salticidae). J. Arachnol. 5: 145–49.

———. 1977c. Web sharing by males and females of dictynid spiders. Bull. Br. Arachnol. Soc. 4: 109–12.

———. 1978a. Comparative studies of *Dictyna* and *Mallos* (Araneae, Dictynidae), I: Social organization and web characteristics. Rev. Arachnol. 1: 133–64.

———. 1978b. Male mating strategies of dictynid spiders with differing types of social organization. Symp. Zool. Soc. London 42: 79–88.

———. 1978c. The mating strategy of *Phidippus johnsoni* (Araneae, Salticidae), I: Pursuit time and persistence. Behav. Ecol. Sociobiol. 4: 123–32.

———. 1978d. Predatory behavior of the social spider *Mallos gregalis*: Is it cooperative? Insectes Soc. 26: 300–312.

———. 1979a. Comparative studies of *Dictyna* and *Mallos* (Araneae, Dictynidae), II: The relationship between courtship, mating, aggression, and cannibalism in species with differing types of social organization. Rev. Arachnol. 2: 103–32.

———. 1979b. Nests of *Phidippus johnsoni* (Araneae, Salticidae): Characteristics, pattern of occupation, and function. J. Arachnol. 7: 47–58.

———. 1980a. Does the web of the social spider *Mallos gregalis* (Araneae: Dictynidae) attract flies? Bull. Br. Arachnol. Soc. 5: 91–94.

———. 1980b. Notes on the copulatory postures of salticid spiders. Peckhamia 2: 4–6.

———. 1981a. Comparative studies of *Dictyna* and *Mallos* (Araneae, Dictynidae), IV: Silk-mediated interattraction. Insectes Soc. 29: 15–24.

———. 1981b. Comparative studies of *Dictyna* and *Mallos* (Araneae, Dictynidae), V: Tolerance and resistance to starvation. Psyche 87: 211–20.

———. 1981c. Nest-mediated sexual discrimination by a jumping spider (*Phidippus johnsoni*). J. Arachnol. 9: 87–92.

———. 1981d. Notes on the reproductive biology of a web-building jumping spider, *Portia fimbriata*. Peckhamia 2: 23–27.

———. 1981e. Particularités des filières de l'araignée cribellate *Mallos gregalis* (Dictynidae). Rev. Arachnol. 3: 211–20.

———. 1982a. The behavior of communicating in jumping spiders (Salticidae). *In* P. N. Witt and J. S. Rovner, eds. *Spider Communication: Mechanisms and Ecological Significance*. Princeton Univ. Press, Princeton, N.J., pp. 213–47.

———. 1982b. The biology of ant-like jumping spiders: Intraspecific interactions of *Myrmarachne lupata* (Araneae, Salticidae). Zool. J. Linn. Soc. 76: 293–319.

———. 1982c. The biology of *Mopsus mormon*, a jumping spider (Araneae, Salticidae) from Queensland: Intraspecific interactions. Aust. J. Zool. 31: 39–53.

———. 1982d. The biology of *Portia fimbriata* (Araneae, Salticidae), a web-building jumping spider from Queensland: Intraspecific interactions. J. Zool. London 196: 295–305.

———. 1982e. The courtship behavior of *Phidippus femoratus* (Araneae, Salticidae). Southwest. Nat. 27: 187–96.

———. In press. The biology of primitive web-building jumping spiders (Salticidae). Natl. Geogr. Soc. Res. Rep.

Jackson, R. R., and A. D. Blest. 1982a. The biology of *Portia fimbriata*, a web-building jumping spider (Araneae, Salticidae) from Queensland: Utilization of webs and predatory versatility. J. Zool. London 196: 255–93.

———. 1982b. The distances at which a primitive jumping spider, *Portia fimbriata*, makes visual discriminations. J. Exp. Biol. 97: 441–45.

Jackson, R. R., and D. P. Harding. 1982. Intraspecific interactions of *Holoplatys* sp. indet., a New Zealand jumping spider (Araneae: Salticidae). N. Z. J. Zool. 9: 487–510.

Jackson, R. R., and S. E. Smith. 1978. Aggregations of *Mallos* and *Dictyna* (Araneae, Dictynidae): Population characteristics. Psyche 85: 65–80.

Jacobi-Kleemann, M. 1953. Über die Lokomotion der Kreuzspinne *Aranea diadema* beim Netzbau (nach Filmanalysen). Z. Vgl. Physiol. 34: 606–54.

Jacson, C. C., and K. J. Joseph. 1973. Life history, bionomics and behavior of the social spider *Stegodyphus sarasinorum* Karsch. Insects Soc. 20: 189–204.

Jaeger, R. G. 1974. Competitive exclusion: Comments on survival and extinction of species. BioScience 24: 33–39.

Janetos, A. C. 1980a. Foraging tactics of two guilds of web-spinning spiders. Ph.D. diss., Princeton Univ., Princeton, N.J.

———. 1980b. Strategies of female mate choice: A theoretical analysis. Behav. Ecol. Sociobiol. 7: 110–14.

———. 1982a. Active foragers vs. sit-and-wait predators: A simple model. J. Theor. Biol. 95: 381–85.

———. 1982b. Foraging tactics of two guilds of web-spinning spiders. Behav. Ecol. Sociobiol. 10: 19–27.

Janetos, A. C., and B. J. Cole. 1981. Imperfectly optimal animals. Behav. Ecol. Sociobiol. 9: 203–9.

Jennings, D. T. 1971. Plant associations of *Misumenops coloradensis* Gertsch (Araneae: Thomisidae) in central New Mexico. Southwest. Nat. 16: 201–7.

Jennings, D. T., and M. W. Houseweart. 1978. Spider preys on spruce budworm egg mass. Entomol. News 89: 183–86.

Job, W. 1974. Beiträge zur Biologie der Fangnetzbauen den Wolfspinnen *Aulonia albimana* (Walckenaer 1805) (Arachnida, Araneae, Lycosidae, Hippasinae). Zool. Jahrb. Abt. Syst. Ökol. Geogr. Tiere 101: 560–608.

Johnson, C. G. 1969. *Migration and Dispersal of Insects by Flight.* Methuen, London.

Kaestner, A. 1968. *Invertebrate Zoology*, Vol. 2. Translated and adapted by H. W. and L. R. Levi. John Wiley, New York.

———. 1980. *Invertebrate Zoology*, Vol. 2. Translated and adapted by H. W. and L. R. Levi. Robert E. Krieger, New York.

Kajak, A. 1965. An analysis of food relations between spiders—*Araneus cornutus* Clerck and *Araneus quadratus* Clerck—and their prey in meadows. Ekol. Pol. A 13: 717–64.

Kaston, B. J. 1936. The senses involved in the courtship of some vagabond spiders. Entomol. Am. 16: 97–167.

———. 1948. Spiders of Connecticut. Conn. State Geol. Nat. Hist. Surv. Bull. 70: 1–874.

———. 1964. The evolution of spider webs. Am. Zool. 4: 191–207.

———. 1965. Some little known aspects of spider behavior. Am. Midl. Nat. 73: 336–56.

———. 1966. Evolution of the web. Nat. Hist. 75: 26–33.

———. 1970. Comparative biology of American black widow spiders. Trans. San Diego Soc. Nat. Hist. 16: 33–82.

———. 1972a. *How to Know the Spiders*, 3d ed. William C. Brown, Dubuque, Iowa.

———. 1972b. Web making by young *Peucetia*. Notes Arachnol. Southwest 3: 6–7.

Kavanaugh, E. J., and E. K. Tillinghast. 1979. Fibrous and adhesive components of the orb webs of *Araneus trifolium* and *Argiope trifasciata*. J. Morphol. 160: 17–25.

Kayashima, I. 1952. Note on the ray-form orb web of *Cyrtarachne bufo* (in Japanese). Arachnol. News 1: 5–7.

Keyserling, E. 1882. Neue Spinnen aus Amerika, III. Verh. Zool.-Bot. Ges. Wien 31: 269–314.

King, J. A. 1973. The ecology of aggressive behavior. Ann. Rev. Ecol. Syst. 4: 117–38.

Kirchner, W., and E. Kullmann. 1972. Ökologische Untersuchungen an einer Freilandpopulation von *Nesticus cellulanus* im Siebengebirg unter besonder Berücksticktigung der Kältresistant (Araneae: Nesticidae). Decheniana 125: 219–27.

Kirkendall, L. 1979. *Mysmenopsis* in diplurid webs. Am. Arachnol. 20: 13.

Klingel, H. 1966. Beobachtungen an *Liphistius batuensis* Abr. (Araneae, Mesothelae). Verh. Dtsch. Zool. Ges. Göttingen 30: 246–53.

Knost, S. J., and J. S. Rovner. 1975. Scavenging by wolf spiders (Araneae: Lycosidae). Am. Midl. Nat. 93: 239–44.

Koenig, M. 1951. Beiträge zur Kenntnis des Netzbaues orbiteiler Spinnen. Z. Tierpsychol. 8: 462–92.

Kovoor, J. 1972. Etude histochimique et cytologique des glandes séricigènes de quelques Argiopidae. Ann. Sci. Nat. Zool. Biol. Anim. 14: 1–40.

———. 1977a. L'appareil séricigène dans le genre *Uloborus* Latr. (Araneae: Uloboridae), I: Anatomic. Rev. Arachnol. 1: 89–102.

———. 1977b. Données histochimiques sur les glandes séricigènes de la veuve noire *Latrodectus mactans* Fabr. (Araneae: Theridiidae). Ann. Sci. Nat. Zool. Biol. Anim. 19: 63–87.

———. 1977c. La soie et les glandes séricigènes des arachnides. Ann. Biol. 16: 97–171.

———. 1978. L'appareil séricigène dans le genre *Uloborus* Latreille (Araneae: Uloboridae), II: Données histochimiques. Ann. Sci. Nat. Zool. Biol. Anim. 20: 3–25.

Kovoor, J., and A. Lopez. 1980. Variation de l'appareil séricigène dans la famille des Araneidae: Cas des genres *Cyclosa* Menge et *Nemoscolus* Simon. C. R. 5th Colloq. Arachnol. 9: 119–27.

Kovoor, J., and L. Zylberberg. 1979. Ultrastructure du canal des glandes agregées et flagelliformes d'*Araneus diadematus* Clerck (Araneae: Araneidae). Zoomorphologie 92: 217–32.

Krafft, B. 1969. Various aspects of the biology of *Agelena consociata* Denis when bred in the laboratory. Am. Zool. 9: 201–10.

———. 1970a. Contribution à la biologie et à l'éthologie d'*Agelena consociata* Denis (araignée sociale du Gabon), I. Biol. Gabonica 6: 197–301.

———. 1970b. Contribution à la biologie et à l'éthologie d'*Agelena consociata* Denis (araignée sociale du Gabon), II. Biol. Gabonica 4: 308–69.

———. 1970c. Les rhythmes d'activité d'*Agelena consociata* Denis: Activité de tissage et activité locomotrice. Biol. Gabonica 4: 99–130.

———. 1971. Contribution à la biologie et à l'éthologie d'*Agelena consociata* Denis (araignée sociale du Gabon), III: Etude expérimentale de certains phénomènes sociaux (suite). Biol. Gabonica 7: 3–56.

———. 1975. La tolérance réciproque chez l'araignée sociale *Agelena consociata* Denis. Proc. Int. Arachnol. Congr. 6: 107–12.

———. 1978. The recording of vibratory signals performed by spiders during courtship. Symp. Zool. Soc. London 42: 59–67.

———. 1979. Organisation et évolution des sociétés d'araignées. J. Psychol. 1: 23–51.

———. 1982. The significance and complexity of communication in spiders. *In* P. N. Witt and J. S. Rovner, eds. *Spider Communication: Mechanisms and Ecological Significance*. Princeton Univ. Press, Princeton, N.J., pp. 15–66.

Krafft, B., and R. Leborgne. 1979. Perception sensorielle et importance des phénomènes vibratoires chez les araignées. J. Psychol. 3: 299–334.

Krafft, B., and C. Roland. 1979. Un labyrinthe appliqué à l'étude des attractions sociales et sexuelles et leur spécificité chez les araignées. Rev. Arachnol. 2: 165–71.

Krakauer, T. 1972. Thermal responses of the orb-weaving spider *Nephila clavipes* (Araneae: Argiopidae). Am. Midl. Nat. 88: 245–50.

Kraus, O. 1955. Spinnen aus El Salvador. Abh. Senckenb. Naturforsch. Ges. 493: 1–112.

Krebs, J. R., J. C. Ryan, and E. L. Charnov. 1974. Hunting by expectation or optimal foraging? A study of patch use by chickadees. Anim. Behav. 22: 953–64.

Kronk, A. E., and S. E. Riechert. 1979. Parameters affecting the habitat choice of a desert wolf spider, *Lycosa santarita* Chamberlin and Ivie. J. Arachnol. 7: 155–66.

Kullmann, E. 1958. Beobachtung des Netzbau und Beiträge zur Biologie von *Cyrtophora citricola* Forskål (Araneae, Araneidae). Zool. Jahrb. Abt. Syst. Ökol. Geogr. Tiere 86: 181–216.

———. 1959. Beobachtungen und Betrachtungen zum Verhalten der Theridiide *Conopistha argyrodes* Walckenaer (Araneae). Mitt. Zool. Mus. Berlin 35: 275–92.

———. 1960a. Beobachtungen an *Theridium tepidariorum* C. L. Koch als Mitbewohner von *Cyrtophora*-Netzen. Dtsch. Entomol. Z. 7: 146–63.

———. 1960b. Über parasitäres Verhalten der Spinne *Theridium tepidariorum* C. L. Koch. Verh. Dtsch. Zool. Ges. Münster 1959: 332–42.

———. 1961. Über das bisher unbekannte Netz und das Webeverhalten von *Drapetisca socialis* (Sundevall) (Araneae, Linyphiidae). Decheniana 114: 99–104.

———. 1964. Neue Ergebnisse uber den Netzbau und das Sexualverhalten einiger Spinnenarten (*Cresmatoneta mutinensis*, *Drapetisca socialis*, *Lithyphantes paykullianus*, *Cyrtophora citricola*). Z. Zool. Syst. Evolutionsforsch. 2: 41–122.

———. 1972a. The convergent development of orb webs in cribellate and ecribellate spiders. Am. Zool. 12: 395–405.

———. 1972b. Evolution of social behavior in spiders (Araneae: Eresidae and Theridiidae). Am. Zool. 12: 419–26.

———. 1975. The production and function of spider threads and spider webs. *In* IL 8, *Netze in Natur und Technik*. Mitt. Inst. Leichte Flächentragwerke Stuttgart, pp. 319–78.

Kullmann, E., F. Otto, T. Braun, and R. Raccanello. 1975. Fundamentals and classification—A survey of spider net constructions. *In* IL 8, *Netze in Natur und Technik*. Mitt. Inst. Leichte Flächentragwerke Stuttgart, pp. 304–16.

Kullmann, E., and W. Zimmerman. 1971. Versuche zur Toleranz bei der permanent-sozialen Spinnenart *Stegodyphus sarasinorum* Karsch (Eresidae). Proc. Int. Arachnol. Congr. 5: 175–82.

———. 1976. Ein neuer Beitrag zur Cribellaten-Ecribellaten-Problem: Beschreibung von *Uroecobius ecribellatus* n. gen., n. sp. und Diskussion seiner phylogenetischen Stellung (Arachnida: Araneae: Oecobiidae). Entomol. Ger. 3: 29–40.

Lahman, E. J., and W. G. Eberhard. 1979. Factores selectivos que afectan la tendencia a agruparse en la araña colonial *Philoponella semiplumosa* (Araneae: Uloboridae). Rev. Biol. Trop. 27: 231–40.

Laing, D. J. 1973. Prey and prey capture in the tunnel web spider *Porrhothele antipodiana*. Tuatara 20: 57–64.

———. 1975. The postures of the tunnel web spider *Porrhothele antipodiana*: A behavioral study. Tuatara 21: 108–20.

———. 1978. Studies on populations of the tunnel web spider *Porrhothele antipodiana*. Tuatara 23: 67–81.

———. 1979. Studies on populations of the tunnel web spider *Porrhothele antipodiana* (Mygalomorphae: Dipluridae). Tuatara 24: 1–21.

———. In press. Snail-eating behavior of the tunnel web spider *Porrhothele antipodiana* (Mygalomorphae: Dipluridae). Tuatara.

Lamore, D. H. 1957. The spider *Conopistha trigona* as a commensal of *Allepeira lemniscata* in Maryland. Proc. Entomol. Soc. Wash. 59: 79.

———. 1958. The jumping spider *Phidippus audax* Hentz and the spider *Conopistha trigona* Hentz, as predators of the basilica spider, *Allepeira lemniscata* Walckenaer, in Maryland. Proc. Entomol. Soc. Wash. 60: 286.

Land, M. F. 1969a. Structure of the retinae of the principal eyes of jumping spiders (Salticidae: Dendryphantinae) in relation to visual optics. J. Exp. Biol. 51: 443–70.

———. 1969b. Movements of the retinae of jumping spiders (Salticidae: Dendryphantinae) in response to visual stimuli. J. Exp. Biol. 51: 471–93.

———. 1971. Orientation by jumping spiders in the absence of visual feedback. J. Exp. Biol. 54: 15–40.

———. 1972a. Mechanisms of orientation and pattern recognition by jumping spiders (Salticidae). *In* R. Wehner, ed. *Information Processing in the Visual Systems of Arthropods*. Springer-Verlag, Berlin, pp. 231–47.

———. 1972b. Stepping movements made by jumping spiders during turns mediated by the lateral eyes. J. Exp. Biol. 57: 133–51.

———. 1974. A comparison of the visual behavior of a predatory arthropod with that of a mammal. *In* C. A. G. Wiersma, ed. *Invertebrate Neurons and Behavior*. MIT Press, Boston, pp. 411–18.

Langer, R. 1969. Elementary physics and spider webs. Am. Zool. 9: 81–89.

Latreille, P. A. 1825. *Familles naturelles du règne animal*. Paris.

Laughlin, S., A. D. Blest, and S. Stowe. 1980. The sensitivity of receptors in the posterior median eye of the nocturnal spider *Dinopis*. J. Comp. Physiol. A 141: 53–65.

Lawler, N. 1972. Notes on the biology and behavior of *Mimetus eutypus* Chamberlin and Ivie (Araneae: Mimetidae). Notes Arachnol. Southwest 3: 7–10.

Lawrence, R. F. 1981. The cannibal spiders (Mimetidae). Naturalist 25: 4–9.

Leborgne, R., and B. Krafft. 1979. Technique d'enregistrement et d'analyse des signaux vibratoires intervenant dans les comportements des araignées sédentaires. Rev. Arachnol. 2: 173–82.

Legendre, R. 1960. Quelques remarques sur le comportement des *Argyrodes* malgaches. Ann. Sci. Nat. Zool. Biol. Anim. 2: 507–12.
———. 1961. Etudes sur les *Archaea*: La capture des proies et la prise de nourriture. Bull. Soc. Zool. Fr. 86: 316–19.
Legendre, R., and D. Llinares. 1970. L'accouplement de l'araignée Salticidae *Cyrba algerina* (Lucas 1846). Ann. Soc. Hort. Hist. Nat. Heravit. 110: 169–74.
LeGuelte, L. 1966. Structure de toile de *Zygiella x-notata* Cl. (Argiopidae) et facteurs qui régissent le comportement de l'araignée pendant la construction de la toile. Thèse, Publ. Univ. Nancy, Nancy.
———. 1969. Learning in spiders. Am. Zool. 9: 145–52.
LeGuelte, L., and P. N. Witt. 1968. Données morphologiques permettant de localiser, sur l'animal vivant, les différentes régions du système nerveux central de l'araignée argiopide *Araneus diadematus* Cl. soumises à des lésions par action du laser. Bull. Mus. Nat. Hist. Natur., 2e Sér., 40: 742–44.
Lehtinen, P. T. 1967. Classification of the cribellate spiders and some allied families, with notes on the evolution of the suborder Araneomorpha. Ann. Zool. Fenn. 4: 199–468.
———. 1975. Notes on the phylogenetic classification of Araneae. Proc. Int. Arachnol. Congr. 6: 26–29.
———. 1978. Definition and limitation of supraspecific taxa in spiders. Symp. Zool. Soc. London 42: 255–71.
———. 1979. Evolutionary trends as taxonomic characters (abstract). Zool. Scr. 8: 315.
———. 1980. Trichobothrial patterns in high level taxonomy of spiders. Proc. Int. Arachnol. Congr. 8: 493–98.
———. 1981. Spiders of the Oriental-Australian region, III: Tetrablemmidae, with a world revision. Acta Zool. Fenn. 162: 1–151.
Lehtinen, P. T., and M. I. Saaristo. 1980. Spiders of the Oriental-Australian region, II: Nesticidae. Ann. Zool. Fenn. 17: 47–66.
Lenler-Eriksen, P. 1969. The hunting web of the young spider *Pisaurina mirabilis* J. Zool. London 157: 391–98.
Lesar, C. D., and J. D. Unzicker. 1978. Life history, habits, and prey preferences of *Tetragnatha laboriosa*. Environ. Entomol. 7: 879–84.
Levi, H. W. 1961. Evolutionary trends in the development of palpal sclerites in the spider family Theridiidae. J. Morphol. 108: 1–9.
———. 1973. Small orb-weavers of the genus *Araneus* north of Mexico (Araneae: Araneidae). Bull. Mus. Comp. Zool. Harv. Univ. 145: 473–552.
———. 1977a. The American orb-web genera *Cyclosa*, *Metazygia*, and *Eustala* north of Mexico (Araneae: Araneidae). Bull. Mus. Comp. Zool. Harv. Univ. 148: 61–127.
———. 1977b. The orbweaver genera *Metepeira*, *Kaira*, and *Aculepeira* in America north of Mexico (Araneae: Araneidae). Bull. Mus. Comp. Zool. Harv. Univ. 148: 185–238.
———. 1977c. The orbweaver genera *Verrucosa*, *Acanthepeira*, *Wagneriana*, *Acacesia*, *Wixia*, *Scoloderus*, and *Alpaida* north of Mexico (Araneae: Araneidae). Bull. Mus. Comp. Zool. Harv. Univ. 147: 351–90.
———. 1978a. Orb-weaving spiders and their webs. Am. Sci. 66: 734–42.
———. 1978b. Orb webs and phylogeny of orb weavers. Symp. Zool. Soc. London 42: 1–15.
———. 1980a. The orb-weaver genus *Mecynogea*, the subfamily Metinae, and the

genera *Pachygnatha*, *Glenognatha*, and *Azilia* of the subfamily Tetragnathinae north of Mexico (Araneae: Araneidae). Bull. Mus. Comp. Zool. Harv. Univ. 149: 1–74.

———. 1980b. Orb webs: Primitive or specialized. Proc. Int. Arachnol. Congr. 8: 367–70.

———. 1981. The American orb-weaver genera *Dolichognatha* and *Tetragnatha* north of Mexico (Araneae: Araneidae, Tetragnathinae). Bull. Mus. Comp. Zool. Harv. Univ. 149: 271–318.

———. 1982. Araneae. *In* S. B. Parker, ed. *Synopsis and Classification of Living Organisms*, Vol. 2. McGraw-Hill, New York, pp. 77–95.

———. 1982a. The spider genera *Psechrus* and *Fecenia* (Araneae: Psechridae). Pacific Insects 24: 114–38.

———. 1983. The orb-weaver genera *Argiope*, *Gea*, and *Neogea* from the western Pacific region (Araneae: Araneidae: Argiopinae). Bull. Mus. Comp. Zool. Harv. Univ. 150: 247–338.

Levi, H. W., and L. R. Levi. 1962. The genera of the spider family Theridiidae. Bull. Mus. Comp. Zool. Harv. Univ. 127: 1–71.

———. 1968. *Spiders and Their Kin*. Golden Press, New York.

Levins, R. 1968. *Evolution in Changing Environments*. Princeton Univ. Press, Princeton, N.J.

Levitt, V. 1961. The funnel-web spider in captivity. Proc. R. Zool. Soc. N.S.W. 1958–59: 80–84.

Liesenfeld, F. J. 1956. Untersuchungen am Netz und über den Erschütterungssinn von *Zygiella x-notata* (Cl.) (Araneidae). Z. Vgl. Physiol. 38: 563–92.

———. 1961. Über Leistung und Sitz des Erschütterungssinnes von Netzspinnen. Biol. Zentralbl. 80: 465–75.

Lin, N., and C. D. Michener. 1972. Evolution of sociality in insects. Q. Rev. Biol. 47: 131–59.

Lister, M. 1678. *Historiae animalium Angliae Tres Tratratus*. Unus de Araneis Londini, London.

Lloyd, J. E. 1975. Aggressive mimicry in *Photuris* fireflies: Signal repertoires of femmes fatales. Science 187: 452–53.

Loeb, A. L. 1976. *Space Structures*. Addison-Wesley, Reading, Mass.

Longman, H. A. 1922. The magnificent spider: *Dicrostichus magnificus* Rainbow: Notes on cocoon spinning and methods of catching prey. Proc. R. Soc. Queensl. 33: 91–98.

Lowrie, D. C. 1948. The ecological succession of spiders of the Chicago area. Ecology 29: 334–51.

———. 1971. Effects of time of day and weather on spider catches with a sweep net. Ecology 52: 348–51.

Lubin, Y. D. 1973. Web structure and function: The non-adhesive orb web of *Cyrtophora moluccensis* (Doleschall) (Araneae: Araneidae). Forma Functio 6: 337–58.

———. 1974. Adaptive advantages and the evolution of colony formation in *Cyrtophora* (Araneae: Araneidae). Zool. J. Linn. Soc. 54: 321–39.

———. 1975. Stabilimenta and barrier webs in the orb webs of *Argiope argentata* (Araneae, Araneidae) on Daphne and Santa Cruz Islands, Galápagos. J. Arachnol. 2: 119–26.

———. 1978. Seasonal abundance and diversity of web-building spiders in relation to habitat structure on Barro Colorado Island, Panama. J. Arachnol. 6: 31–51.

———. 1980a. Population studies of two colonial orb-weaving spiders. Zool. J. Linn. Soc. 70: 265–87.

———. 1980b. The predatory behavior of *Cyrtophora* (Araneae: Araneidae). J. Arachnol. 8: 159–85.

Lubin, Y. D., and S. Dorugl. 1982. The effectiveness of single-thread webs as insect traps: Sticky trap models. Bull. Br. Arachnol. Soc. 5: 399–407.

Lubin, Y. D., W. G. Eberhard, and G. G. Montgomery. 1978. Webs of *Miagrammopes* (Araneae: Uloboridae) in the Neotropics. Psyche 85: 1–23.

Lubin, Y. D., B. D. Opell, W. G. Eberhard, and H. W. Levi. 1982. Orb plus cone webs in Uloboridae (Araneae), with a description of a new genus and four new species. Psyche 89: 29–64.

Lubin, Y. D., and M. H. Robinson. 1982. Dispersal by swarming in a social spider. Science 216: 319–21.

Lucas, F., J. B. T. Shaw, and S. G. Smith. 1960. Comparative studies of fibroins, I: The amino acid composition of various fibroins and its significance in relation to their crystal structure and taxonomy. J. Mo. Biol. 2: 339–49.

Luczak, J. 1970. Behavior of a spider population in the presence of mosquitoes. Ekol. Pol. 18: 625–34.

McAlister, W. H. 1961. The spitting habit in the spider *Scytodes intricata* Banks. Texas J. Sci. 12: 17–20.

MacArthur, R. H. 1972. *Geographical Ecology*. Harper and Row, New York.

MacArthur, R. H., and E. R. Pianka. 1966. On optimal use of patchy environments. Am. Nat. 100: 603–9.

MacArthur, R. H., and E. O. Wilson. 1967. *The Theory of Island Biogeography*. Princeton Univ. Press, Princeton, N.J.

McBride, G. 1976. The study of social organisations. Behaviour 59: 96–115.

McCook, H. C. 1887. Prolonged life of invertebrates: Notes on the age and habits of the American tarantula. Proc. Acad. Nat. Sci. Philadelphia: 369–86.

———. 1888. Nesting habits of the American purseweb spider. Proc. Acad. Nat. Sci. Philadelphia: 203–20.

———. 1889. *American Spiders and Their Spinningwork*, Vol. 1. Philadelphia.

———. 1890. *American Spiders and Their Spinningwork*, Vol. 2. Philadelphia.

McDaniel, S. G., and W. L. Sterling. 1982. Predation of *Heliothis virescens* (F.) eggs on cotton in east Texas. Environ. Entomol. 11: 60–66.

McDaniel, S. G., W. L. Sterling, and D. A. Dean. 1981. Predators of tobacco budworm larvae in Texas cotton. Southwest. Entomol. 6: 102–8.

MacKay, W. P. 1982. The effect of predation of Western widow spiders (Araneae: Theridiidae) on harvester ants (Hymenoptera: Formicidae). Oecologia (Berlin) 53: 406–11.

McKeown, K. C. 1952. *Australian Spiders*. Angus and Robertson, Sydney.

———. 1963. *Australian Spiders*. Ed. N. Roberts. Angus and Robertson, Sydney.

Main, B. Y. 1956. Observations on the burrow and natural history of the trapdoor spider *Missulena* (Ctenizidae). West. Aust. Nat. 5: 73–80.

———. 1957. Biology of aganippine trapdoor spiders (Mygalomorphae, Ctenizidae). Aust. J. Zool. 5: 402–73.

———. 1960. The genus *Cethegus* Thorell (Mygalomorphae: Macrothelinae). J. R. Soc. West. Aust. 43: 30–34.

———. 1968. Distribution and adaptive diversity of trapdoor spiders. Aust. Nat. Hist. 16: 49–53.

———. 1969. The trapdoor spider genus *Cataxia* Rainbow (Mygalomorphae: Ctenizidae)—Taxonomy and natural history. J. Aust. Entomol. Soc. 8: 192–209.

———. 1972. The mygalomorph genus *Stanwellia* Rainbow and Pulleine (Dipluridae), and its relationship to *Aname* Koch and certain other diplurine genera. J. R. Soc. West. Aust. 55: 100–114.

———. 1975. The citrine spider: A new genus of trapdoor spider (Mygalomorphae, Dipluridae). West. Aust. Nat. 13: 73–78.

———. 1976. *Spiders*. Australian Naturalist Library, Collins, Sydney.

———. 1978. Biology of the arid-adapted trapdoor spider *Anidiops villosus* (Rainbow). Bull. Br. Arachnol. Soc. 4: 161–75.

———. 1982a. Notes on the reduced web, behaviour, and prey of *Arcys nitidiceps* Simon (Araneidae) in southwestern Australia. Bull. Br. Arachnol. Soc. 5: 425–32.

———. 1982b. Notes on the revised taxonomic position of the black wishbone spider *Dekana diversicolor* Hogg (Mygalomorphae: Dipluridae). J. R. Soc. West. Aust. 65: 25–29.

Main, B. Y., and R. Mascord. 1969. A new genus of diplurid spider (Araneae: Mygalomorphae) from New South Wales. J. Entomol. Soc. Aust. 6: 24–30.

———. 1974. Description and natural history of a "tube-building" species of *Dyarcyops* from New South Wales and Queensland (Mygalomorphae: Ctenizidae). J. Entomol. Soc. Aust. 8: 15–21.

Marler, P. 1968. Visual system. *In* T. A. Seboek, ed. *Animal Communication*. Indiana Univ. Press, Bloomington, pp. 103–26.

Marples, B. J. 1949. An unusual type of web constructed by a Samoan spider of the family Argiopidae. Trans. R. Soc. N.Z. 77: 232–33.

———. 1955. A new type of web spun by spiders of the genus *Ulesanus*, with the description of two new species. Proc. Zool. Soc. London 125: 751–60.

———. 1962. Notes on spiders of the family Uloboridae. Ann. Akad. Zool. Agr. India 4: 1–11.

Marples, B. J., and M. J. Marples. 1972. Observations on *Cantuaria toddi* and other trapdoor spiders in central Otago, New Zealand. J. R. Soc. N.Z. 2: 179–85.

Marples, M. J., and B. J. Marples. 1937. Notes on the spiders *Hyptiotes paradoxus* and *Cyclosa conica*. Proc. Zool. Soc. London 107: 213–21.

Marson, J. 1974. Some observations on the variation in the camouflage used by *Cyclosa insulana* (Costa), an Asiatic spider, in its web. Proc. Zool. Soc. London 11: 598–605.

Martin, D. 1978. Zum Radnetzbau der Gattung *Pachygnatha* Sund. (Araneae: Tetragnathidae). Mitt. Zool. Mus. Berlin 54: 83.

Mascord, R. 1970. *Australian Spiders in Colour*. A. H. and A. W. Reed, Sydney.

———. 1980. *Spiders of Australia*. A. H. and A. W. Reed, Sydney.

Masters, W. M., and H. Markl. 1981. Vibration signal transmission in spider orbwebs. Science 213: 363–65.

Matsumoto, S., E. Shinkai, and H. Ono. 1976. *Spiders*. Gakken, Tokyo.

Mayer, G. 1952. Untersuchungen über Herstellung und Struktur des Radnetzes von *Aranea diadema* and *Zilla x-notata*, mit besonderer Berücksichtigung des Unterschiedes von Jugend- und Altersnetzen. Z. Tierpsychol. 9: 337–62.

Mayer, M. S., and J. R. McLaughlin. 1975. An annotated compendium of insect sex pheromones. Fla. Agric. Exp. Stn. Monogr. Ser. 6: 1–133.

Maynard-Smith, J. 1976. Evolution and the theory of games. Am. Sci. 64: 41–45.

Meijer, J. 1977. The immigration of spiders (Araneida) into a new polder. Ecol. Entomol. 2: 81–90.

Menge, A. 1866. Preussische Spinnen, I. Schr. Naturforsch. Ges. Danzig, N.F. 1: 1–152.

Merrett, P. 1963. The palpus of male spiders of the family Linyphiidae. Proc. Zool. Soc. London 140: 347–67.

Meyer, E. 1928. Neue sinnesbiologische Beobachtungen an Spinnen. Z. Morphol. Ökol. Tiere 12: 1–69.

Michelsen, A., and O. Larsen. 1978. Biophysics of the ensiferan ear, 1. Tympanal vibrations in bushcrickets (Tettigoniidae) studied with laser vibrometry. J. Comp. Physiol. 123: 193–203.

Michener, C. D. 1969. Comparative social behavior of bees. Ann. Rev. Entomol. 14: 299–342.

Mikulska, I. 1961. Parental care in a rare spider, *Pellenes nigrociliata* (L. Koch) var. *bilunulata* Simon. Nature (London) 190: 365–66.

———. 1966. The spinning structures on the spinnerets (thelae) of *Nephila clavipes* L. Zool. Polon. 16: 209–22.

———. 1967. The external spinning structures on the thelae of the spider *Argiope aurantia* Lucas. Zool. Pol. 17: 357–65.

———. 1969. Variability in the number of external spinning structures in female spiders *Clubiona phragmitis* C. L. Koch in populations to varying degrees isolated. Zool. Pol. 19: 279–91.

Millidge, A. F. 1977. The conformation of the male palpal organs of linyphiid spiders, and its application to the taxonomic and phylogenetic analysis of the family (Araneae: Linyphiidae). Bull. Br. Arachnol. Soc. 4: 1–60.

———. 1980. The erigonine spiders of North America, 1: Introduction and taxonomic background. J. Arachnol. 8: 97–107.

Millot, J. 1930. Glandes venimeuses et glandes séricigènes chez les sicariides. Bull. Soc. Zool. Fr. 55: 150–75.

———. 1948. Faits nouveaux concernant les *Archaea* (aranéides). Mém. Inst. Sci. Madagascar, Sér. A, 1: 3–14.

———. 1949. Aranéides. *In* P.-P. Grassé, ed. *Traité de zoologie*, Vol. 6. Masson, Paris, pp. 589–743.

Millot, J., and R. Legendre. 1964. Les *Archaea* (aranéides) et leur cocoon. C. R. Acad. Sci. Paris 258: 4835–38.

Minch, E. W. 1978a. Daily activity patterns in the tarantula *Aphonopelma chalcodes* Chamb. Bull. Br. Arachnol. Soc. 4: 231–37.

———. 1978b. The effectiveness of chemical defenses against predation by the tarantula *Aphonopelma chalcodes* Chamb. Bull. Br. Arachnol. Soc. 4:275–78.

———. 1979a. Burrow-entrance plugging behaviour in the tarantula *Aphonopelma chalcodes* Chamb. (Araneae: Theraphosidae). Bull. Br. Arachnol. Soc. 4: 414–15.

———. 1979b. Reproductive behaviour of the tarantula *Aphonopelma chalcodes* Chamb. (Araneae: Theraphosidae). Bull. Br. Arachnol. Soc. 4: 416–20.

Moggridge, J. T. 1873. *Harvesting Ants and Trapdoor Spiders*. L. Reeve, London.

———. 1874. *A Supplement to Harvesting Ants and Trapdoor Spiders*. L. Reeve, London.

Monterosso, B. 1927. Osservazioni preliminari sulla biologia del gènere *Scytodes*. Rend. R. Accad. Lincei 6: 171–74.

Montgomery, T. H. 1909. On the spinnerets, cribellum, colulus, tracheae, and lung books of araneads. Proc. Acad. Nat. Sci. Philadelphia 61: 299–320.

Morel, G. 1978. Les maladies microbiennes des arachnides (acariens exceptés). Symp. Zool. Soc. London 42: 477–81.

Morse, D. H. 1980. *Behavioral Mechanisms in Ecology*. Harvard Univ. Press, Cambridge, Mass.

Morse, D. H., and R. S. Fritz. 1982. Experimental and observational studies of patch choice at different scales by the crab spider *Misumena vatia*. Ecology 63: 172–82.

Morse, R. A. 1972. Environmental control in the beehive. Sci. Am. 226 (June): 92–98.

Muma, M. H., and W. J. Gertsch. 1964. The spider family Uloboridae in North America north of Mexico. Am. Mus. Novit. 2196: 1–43.

Murphy, J., and F. Murphy. 1983. More about *Portia* (Araneae: Salticidae). Bull. Br. Arachnol. Soc. 6: 37–45.

Murphy, J., and N. I. Platnick. 1981. On *Liphistius desultor* Schiodte (Araneae, Liphistiidae). Bull. Am. Mus. Nat. Hist. 170: 46–56.

Myers, J. G. 1927. Ethological notes on some New Zealand spiders. N. Z. J. Sci. Technol. 9: 129–36.

Nachtigall, W. 1974. *Insects in Flight*. McGraw-Hill, New York.

Nelson, G. 1978. Ontogeny, phylogeny and the biogenetic law. Syst. Zool. 27: 324–45.

Nelson, G., and N. I. Platnick. 1981. *Systematics and Biogeography*. Columbia Univ. Press, New York.

Nentwig, W. 1982. Why do certain insects escape from a spider's web? Oecologia (Berlin) 53: 412–17.

———. 1985. *Architis nitidopilosa*, a neotropical pisaurid with a permanent catching web (Araneae, Pisauridae). Bull. Br. Arachnol. Soc. 6: 297–303.

Nielsen, E. 1932. *The Biology of Spiders*, 2 vols. Levin and Munksgaard, Copenhagen.

Norgaard, E. 1951. Notes on the biology of *Filistata insidiatrix* (Forsk.). Entomol. Medd. 26: 170–84.

Nyffeler, M., and G. Benz. 1978. Die Beutespektren der Netzspinnen *Argiope bruennichi* (Scop.), *Araneus quadratus* Cl. und *Agelena labyrinthica* (Cl.) in Ödlandwiesen bei Zürich. Rev. Suisse Zool. 85: 747–57.

———. 1979. Zur ökologischen Bedeutung der Spinnen der Vegetationsschicht von Getreide- und Rapsfeldern bei Zürich (Schweig). Z. Angew. Entomol. 87: 348–76.

———. 1980. Kleptoparasıtismus von juvenilen Kreuzspinnen und Skorpionsfliegen in den Netzen adulter Spinnen. Rev. Suisse Zool. 87: 907–18.

———. 1981. Einige Beobachtungen über die Flucht von adulten Lepidopteren aus den Netzen orbiteler Spinnen. Anz. Schädlingskd., Pflanz. Umweltschutz 54: 113–14.

Oaten, A. 1977. Optimal foraging in patches: a case for stochasticity. Theor. Popul. Biol. 12: 263–85.

Okuma, C. 1957. Web construction by *Cyrtarachne bufo* (in Japanese). Atypus 13: 3–4.

Olive, C. W. 1980. Foraging specializations in orb-weaving spiders. Ecology 61: 1133–44.

———. 1981a. Co-adapted foraging traits in a guild of orb-weaving spiders. Oecologia (Berlin) 49: 88–91.

———. 1981b. Optimal phenology and body size of orb-weaving spiders: Foraging constraints. Oecologia (Berlin) 49: 83–87.

———. 1982a. Behavioral response of a sit-and-wait predator to spatial variation in foraging gain. Ecology 63: 912–20.

———. 1982b. Sex pheromones in two orbweaving spiders (Araneae, Araneidae): An experimental field study. J. Arachnol. 10: 241–45.

Olivier, G. A. 1789. Araignée, Aranea. *In Encyclopédie méthodique, Histoire naturelle,* Vol. 4. Paris, pp. 173–240.

Ono, T., and T. Yaginuma. 1963. In search of the web of *Cyrtarachne* (in Japanese). Atypus 28: 7–9.

Opell, B. D. 1979. Revision of the genera and tropical American species of the spider family Uloboridae. Bull. Mus. Comp. Zool. Harv. Univ. 148: 445–549.

———. 1981. New Central and South American Uloboridae (Arachnida, Araneae). Bull. Am. Mus. Nat. Hist. 170: 219–28.

———. 1982a. Cribellum, calamistrum, and ventral-comb ontogeny in *Hyptiotes cavatus* (Hentz) (Araneae: Uloboridae). Bull. Br. Arachnol. Soc. 5: 338–43.

———. 1982b. Post-hatching development and web production of *Hyptiotes cavatus* (Hentz) (Araneae: Uloboridae). J. Arachnol. 10: 185–91.

———. 1983. A review of the genus *Tangaroa* (Araneae, Uloboridae). J. Arachnol. 11: 287–95.

———. 1984a. *Lubinella,* a new genus of Uloboridae (Arachnida, Araneae). J. Arachnol. 11: 441–46.

———. 1984b. Phylogenetic review of the genus *Miagrammopes* (Araneae, Uloboridae). J. Arachnol. 12: 229–40.

Opell, B. D., and W. G. Eberhard. 1983. Resting postures of orb-weaving uloborid spiders. J. Arachnol. 11: 369–76.

Oster, G. F., and E. O. Wilson. 1978. *Caste and Evolution in the Social Insects.* Princeton Univ. Press, Princeton, N.J.

Pack, Y. M. 1983. Masking behaviour and related biology of *Notomithrax ursus* (Oxyrhyncha: Majidae). Ph.D. thesis, Univ. Canterbury, Christchurch.

Parry, D. A. 1965. The signal generated by an insect in a spider's web. J. Exp. Biol. 43: 185–92.

Parry, D. A., and R. H. J. Brown. 1959. The jumping mechanism of salticid spiders. J. Exp. Biol. 36: 654–64.

Peakall, D. B. 1971. Conservation of web proteins in the spider *Araneus diadematus.* J. Exp. Zool. 176: 257–64.

Peakall, D. B., and P. N. Witt. 1976. The energy budget of an orb-building spider. Comp. Biochem. Physiol. A 54: 187–90.

Peaslee, J. E., and W. B. Peck. 1983. The biology of *Octonoba octonarius* (Muma) (Araneae: Uloboridae). J. Arachnol. 11: 51–67.

Peck, W. B., and W. H. Whitcomb. 1970. Studies on the biology of a spider, *Chiracanthium inclusum* (Hentz). Arkansas Agric. Exp. Stn. Bull. 753: 1–76.

Peckham, G. W., and E. G. Peckham. 1889. Observations on sexual selection in spiders of the family Attidae. Occas. Pap. Wis. Nat. Hist. Soc. 1: 3–60.

———. 1890. Additional observations on sexual selection in spiders of the family Attidae. Occas. Pap. Wis. Nat. Hist. Soc. 1: 117–51.

Peters, H. M. 1931. Die Fanghandlung der Kreuzspinne (*Epeira diademata* Cl.): Experimentelle Analysen der Verhaltens. Z. Vgl. Physiol. 15: 693–748.

———. 1932. Experimente über die Orientierung der Kreuzspinne *Epeira diademata* Cl. im Netz. Zool. Jahrb. Abt. Allg. Zool. Physiol. Tiere 51: 239–88.

———. 1933. Kleine Beiträge zur Biologie der Kreuzspinne *Epeira diademata* Cl. Z. Morphol. Ökol. Tiere 26: 447–68.

———. 1937. Studien am Netz der Kreuzspinne (*Epeira diademata* Cl.), II: Uber die Herstellung des Rahmens, der Radialfäden, und der Hilfspiral. Z. Morphol. Ökol. Tiere 33: 128–50.

———. 1938. Über das Netz der Dreieckspinne, *Hyptiotes paradoxus*. Zool. Anz. 121: 49–59.

———. 1939. Probleme des Kreuzspinnen-Netzes. Z. Morphol. Ökol. Tiere 36: 180–226.

———. 1953. Beiträge zur vergleichenden Ethologie und Ökologie tropischer Webespinnen. Z. Morphol. Ökol. Tiere 42: 278–306.

———. 1954. Estudios adicionales sobre la estructura de la red concéntrica de las arañas. Comun. Inst. Trop. Invest. Científ. Univ. El Salvador 3: 1–18.

———. 1955a. Contribuciones sobre la etología y ecología comparada de las arañas tejedoras tropicales. Comun. Inst. Trop. Invest. Científ. Univ. El Salvador 1/2: 37–46.

———. 1955b. Uber der Spinnapparat von *Nephila madagascariensis* (Radnetzspinnen; Fam. Argiopidae). Z. Naturforsch. Teil B, 10: 395–464.

———. 1982. Wie Spinnen der Familie Uloboridae ihre Beute einspinnen und verzehren. Verh. Naturwiss. Ver. Hamburg 25: 147–67.

Peters, P. J. 1970. Orb-web construction: Interaction of spider (*Araneus diadematus* Cl.) and thread configuration. Anim. Behav. 18: 478–84.

Petrunkevitch, A. 1911. Sense of sight, courtship, and mating in *Dugesiella hentzi*, a theraphosid spider from Texas. Zool. Jahrb. Abt. Syst. Ökol. Geogr. Tiere 31: 355–71.

———. 1926. The value of instinct as a taxonomic character in spiders. Biol. Bull. 50: 427–32.

———. 1928. Systema aranearum. Trans. Conn. Acad. Arts Sci. 29: 1–270.

———. 1929. The spiders of Porto Rico, I. Trans. Conn. Acad. Arts Sci. 30: 1–158.

———. 1930. The spiders of Porto Rico, II. Trans. Conn. Acad. Arts Sci. 30: 159–355.

———. 1933. An inquiry into the natural classification of spiders. Trans. Conn. Acad. Arts Sci. 31: 303–89.

———. 1952. The spider and the wasp. Sci. Am. 187 (April): 20–23.

Pickard-Cambridge, O. 1889. On some new species and a new genus of Araneida. Proc. Zool. Soc. London : 34–36.

———. 1897–1905. Arachnida: Araneida and Opiliones. Biol. Cent. Am. Zool. 2: 1–610.

Platnick, N. I. 1971. The evolution of courtship behavior in spiders. Bull. Br. Arachnol. Soc. 2: 40–47.

———. 1975. On the validity of Haplogynae as a taxonomic grouping in spiders. Proc. Int. Arachnol. Congr. 6: 30–32.

———. 1977. The hypochiloid spiders: A cladistic analysis, with notes on the Atypoidea (Arachnida, Araneae). Am. Mus. Novit. 2627: 1–23.

———. 1979. The Mecicobothriidae: Is convergence convincing? Am. Arachnol. 20: 12.

———. 1984. Studies on Malagasy spiders. I. The Family Galieniellidae (Araneae, Gnaphosoidea). Am. Mus. Novitates 2801: 1–17.

———. 1985. Studies on Malagasy spiders. II. The Family Trochanteriidae (Araneae, Gnaphosoidea), with a revision of the genus *Platyoides*. Am. Mus. Novitates 2808: 1–17.

Platnick, N. I., and M. U. Shadab. 1976. A revision of the mygalomorph spider genus *Neocteniza* (Araneae, Actinopodidae). Am. Mus. Novit. 2603: 1–19.

———. 1978a. A review of the spider genus *Anapis* (Araneae: Anapidae), with a dual cladistic analysis. Am. Mus. Novit. 2663: 1–23.

———. 1978b. A review of the spider genus *Mysmenopsis* (Araneae, Mysmenidae). Am. Mus. Novit. 2661: 1–22.

Plett, A. 1962a. Beobachtungen und Versuche zum Revier- und Sexualverhalten von *Epiblemum scenicum* Cl. und *Evarcha blancardi* Scop. (Salticidae). Zool. Anz. 169: 292–98.

———. 1962b. Untersuchungen zum Appetenzverhalten der Springspinne *Epiblemum scenicum* Cl. (Salticidae) und des Ameiselöwen *Euroleon nostras* Fourer. (Myrmeleonidae). Zool. Anz. 169: 280–91.

Pocock, R. I. 1895. Some suggestions on the origin and evolution of web spinning in spiders. Nature (London) 51: 417–20.

———. 1900. *The Fauna of British India, Including Ceylon and Burma: Arachnida.* London.

Pointing, P. J. 1965. Some factors influencing the orientation of the spider *Frontinella communis* (Hentz) in its web (Araneae: Linyphiidae). Can. Entomol. 97: 69–78.

Polis, G. A. 1981. The evolution and dynamics of intraspecific predation. Ann. Rev. Ecol. Syst. 12: 225–51.

Pollard, S. D. In press. Egg guarding by *Clubiona cambridgei* (Araneae: Clubionidae) against conspecific predators. J. Arachnol.

Pollard, S. D., and R. R. Jackson. 1982. The biology of *Clubiona cambridgei* (Araneae, Clubionidae): Intraspecific interactions. N. Z. J. Ecol. 198: 197–214.

Porter, S. D., and D. A. Eastmond. 1982. *Euryopis coki* (Theridiidae), a spider that preys on *Pogonomyrmex* ants. J. Arachnol. 10: 275–77.

Poteat, W. L. 1890. A tube-building spider. J. Elisha Mitchell Sci. Soc. 6: 134–47.

Poulton, E. B. 1890. *The Colours of Animals.* Kegan Paul, Trench, Truebner, London.

Precht, H. 1952. Über das angeborene Verhalten von Tieren: Versuche an Springspinnen (Salticidae). Z. Tierpsychol. 9: 207–30.

Precht, H., and G. Freytag. 1958. Über Ermüdung und Hemmung angeborener Verhaltensweisen bei Springspinnen zugleich ein Beitrag zum Triebproblem (Salticidae). Behaviour 13: 143–211.

Prestwich, K. N. 1977. The energetics of web building in spiders. Comp. Biochem. Physiol. A 57: 321–26.

Prószyński, J. 1968. Redescriptions of the type-species of genera of Salticidae (Araneida), VI–VII. Ann. Zool. Warsaw 26: 239–47.

———. 1971. Notes on systematics of Salticidae (Arachnida, Aranei), I–V. Ann. Zool. Warsaw 28: 227–55.

Purcell, W. F. 1903. New South African spiders of the families Migidae, Ctenizidae, Barychelidae, Dipluridae, and Lycosidae. Ann. S. Afr. Mus. 3: 69–142.

Pyke, G. H., H. R. Pulliam, and E. L. Charnov. 1977. Optimal foraging: A selective review of theory and tests. Q. Rev. Biol. 52: 137–54.

Quintero, D. 1974. The unusual web of *Spilasma tubulofaciens*, with taxonomic notes on the species (Araneae: Araneidae). Psyche 81: 307–14.

Ramousse, R. 1973. Body, web-building, and feeding characteristics of males of the spider *Araneus diadematus* (Araneae: Araneidae). Psyche 80: 22–47.

———. 1980. Temporal patterns in web-building in *Araneus diadematus* Clerck. Proc. Int. Arachnol. Congr. 8: 257–60.

Ramousse, R., and F. Davis, III. 1976. Web-building time for a spider: Preliminary application of ultrasonic detection. Physiol. Behav. 17: 997-1000.

Ramousse, R., and L. LeGuelte. 1979. Relations spatio- temporelles dans le comportement constructeur chez l'*Epeire diademe*. Rev. Arachnol. 2: 183–92.

Raven, R. J. 1976. A new spider of the genus *Hexathele* Ausserer (Dipluridae: Mygalomorphae) from Australia. Proc. R. Soc. Queensl. 87: 53–61.

———. 1978. Systematics of the spider subfamily Hexathelinae (Dipluridae: Mygalomorphae). Aust. J. Zool. Suppl. Ser. 65: 1–75.

———. 1980a. The Australian mygalomorph spider genus *Ixamatus* Simon (Dipluridae: Diplurinae) and its affinities. Bull. Br. Arachnol. Soc. 5: 43–49.

———. 1980b. The evolution and biogeography of the mygalomorph spider family Hexathelidae (Araneae, Chelicerata). J. Arachnol. 8: 251–66.

———. 1981. A review of the Australian genera of the mygalomorph subfamily Diplurinae (Dipluridae: Chelicerata). Aust. J. Zool. 29: 321–63.

———. 1982. On the mygalomorph spider genus *Xamiatus* Raven (Diplurinae: Dipluridae) with the description of a new species. Mem. Queensl. Mus. 20: 473–78.

———. 1984. Systematics of the Australian curtain-web spiders (Ischnothelinae: Dipluridae). Aust. J. Zool., Suppl. Sec. 93: 1–102.

Rayment, T. 1942. The onion net of a spider. Walkabout 8: 10–15.

Readshaw, J. L. 1973. The numerical response of predators to prey density. *In* J. Hughes, ed. *Quantitative Evaluation of Natural Enemy Effectiveness.* J. Appl. Biol. 10: 342–51.

Reed, C. F. 1968. Construction of the web. *In* P. N. Witt, C. F. Reed, and D. B. Peakall. *A Spider's Web: Problems in Regulatory Biology*. Springer-Verlag, New York, pp. 74–93.

———. 1969. Cues in the web-building process. Am. Zool. 9: 211–21.

Reed, C. F., P. N. Witt, M. B. Scarboro, and D. B. Peakall. 1970. Experience and the orb web. Dev. Psychobiol. 3: 251–65.

Richman, D. B. 1982. Epigamic display in jumping spiders (Araneae, Salticidae) and its uses in systematics. J. Arachnol. 10: 47–67.

Richman, D. B., and W. H. Whitcomb. 1981. The ontogeny of *Lyssomanes viridis* (Walckenaer) (Araneae: Salticidae) on *Magnolia grandiflora* L. Psyche 88: 127–33.

Richter, C. J. J. 1970a. Aerial dispersion in relation to habitat in eight wolf spider species (*Pardosa*, Araneae, Lycosidae). Oecologia (Berlin) 5: 200–214.

———. 1970b. Morphology and function of the spinning apparatus of the wolf spider *Pardosa amentata* (Cl.) (Araneae, Lycosidae). Z. Morphol. Tiere 68: 37–68.

———. 1970c. Relation between habitat structure and development of the glandulae ampullaceae in eight wolf spider species (*Pardosa*, Araneae, Lycosidae). Oecologia 5: 185–99.

Riechert, S. E. 1974a. The pattern of local web distribution in a desert spider: Mechanisms and seasonal variation. J. Anim. Ecol. 43: 733–46.

———. 1974b. Thoughts on the ecological significance of spiders. BioScience 24: 352–56.

———. 1976. Web-site selection in the desert spider *Agelenopsis aperta*. Oikos 27: 311–15.

———. 1978a. Energy-based territoriality in populations of the desert spider *Agelenopsis aperta*. Symp. Zool. Soc. London 42: 211–22.

———. 1978b. Games spiders play, I: Behavioral variability in territorial disputes. Behav. Ecol. Sociobiol. 3: 135–62.

———. 1979. Games spiders play, II: Resource assessment strategy. Behav. Ecol. Sociobiol. 6: 121–28.

———. 1981. The consequences of being territorial: Spiders, a case study. Am. Nat. 117: 871–92.

———. 1982. Spider interaction strategies: Communication vs. coercion. *In* P. N. Witt and J. S. Rovner, eds. *Spider Communication: Mechanisms and Ecological Significance.* Princeton Univ. Press, Princeton, N.J., pp. 281–316.

Riechert, S. E., and A. Cady. 1983. Patterns of resource use and tests for competitive release in a spider community. Ecology 64: 899–913.

Riechert, S. E., and J. Luczak. 1982. Spider foraging: Behavioral responses to prey. *In* P. N. Witt and J. S. Rovner, eds. *Spider Communication: Mechanisms and Ecological Significance.* Princeton Univ. Press, Princeton, N.J., pp. 353–86.

Riechert, S. E., and W. G. Reeder. 1972. Effects of fire on spider distribution on southwestern Wisconsin prairies. Proc. 2d Midwest Prairie Conf. 13: 1–19.

Riechert, S. E., W. G. Reeder, and T. A. Allen. 1973. Patterns of spider distribution (*Agelenopsis aperta* (Gertsch)) in desert grassland and recent lava-bed habitats, south central New Mexico. J. Anim. Ecol. 42: 19–35.

Riechert, S. E., and C. R. Tracy. 1975. Thermal balance and prey availability: Basis for a model relating web-site characteristics to spider reproductive success. Ecology 56: 265–84.

Riedl, R. 1979. *Order in Living Organisms.* John Wiley, New York.

Risch, P. 1977. Quantitative analysis of orb-web patterns in four species of spiders. Behav. Genet. 7: 199–238.

Roberts, N. L. 1955. The Australian netting spider *Deinopis subrufus*. Proc. R. Soc. Zool. N.S.W. 54: 24–33.

Robinson, J. V. 1981. The effect of architectural variation in habitat on a spider community: A field study. Ecology 62: 73–80.

Robinson, M. H. 1969a. Defenses against visually hunting predators. *In* T. Dobzhansky, M. K. Hecht, and W. C. Steere, eds. *Evolutionary Biology*, Vol. 3. Meredith, New York, pp. 225–59.

———. 1969b. Predatory behavior of *Argiope argentata* (Fabricius). Am. Zool. 9: 161–73.

———. 1973. Insect anti-predator adaptations and the behavior of predatory primates. Acta IV Congr. Latinoam. Zool. 2: 811–36.

———. 1975. The evolution of predatory behaviour in araneid spiders. *In* G. Baerends, C. Beer, and A. Manning, Eds. *Function and Evolution in Behaviour.* Clarendon, Oxford, pp. 292–312.

———. 1977a. Symbioses between insects and spiders: An association between lepidopteran larvae and the social spider *Anelosimus eximius* (Araneae: Theridiidae). Psyche 84: 225–32.

———. 1977b. Tropical spinners. New Sci. 76: 552–54.

———. 1980. The ecology and behavior of tropical spiders. Proc. Int. Arachnol. Congr. 8: 13–32.

———. 1982a. Courtship and mating behavior in spiders. Ann. Rev. Entomol. 27: 1–20.

———. 1982b. The ecology and biogeography of spiders in Papua New Guinea. Monogr. Biol. 42: 557–81.

Robinson, M. H., and Y. D. Lubin. 1979a. Specialists and generalists: The ecology and behavior of some web-building spiders from Papua New Guinea. I. *Herennia ornatissima*, *Argiope ocyaloides*, and *Arachnura melanura* (Araneae: Araneidae). Pac. Insects 21: 97–132.

———. 1979b. Specialists and generalists: The ecology and behavior of some web-building spiders from Papua New Guinea. II. *Psechrus argentatus* and *Fecenia* sp. (Araneae: Psechridae). Pac. Insects 21: 133–64.

Robinson, M. H., and H. Mirick. 1971. The predatory behavior of the golden-web spider *Nephila clavipes* (Araneae: Araneidae). Psyche 78: 123–39.

Robinson, M. H., H. Mirick, and O. Turner. 1969. The predatory behavior of some araneid spiders and the origin of immobilization wrapping. Psyche 76: 487–501.

Robinson, M. H., and J. Olazarri. 1971. Units of behavior and complex sequences in the predatory behavior of *Argiope argentata* (Fabricius) (Araneae: Araneidae). Smithson. Contrib. Zool. 65: 1–36.

Robinson, M. H., and B. Robinson. 1970a. Prey caught by a sample population of the spider *Argiope argentata* (Araneae: Araneidae) in Panama: A year's census data. Zool. J. Linn. Soc. 49: 345–57.

———. 1970b. The stabilimentum of the orb-web spider *Argiope argentata*: An improbable defense against predators. Can. Entomol. 102: 641–55.

———. 1971. The predatory behavior of the ogre-faced spider *Dinopis longipes* F. Cambridge (Araneae: Dinopidae). Am. Midl. Nat. 85: 85–96.

———. 1972. The structure, possible function, and origin of the remarkable ladder web built by a New Guinea orb-web spider. J. Nat. Hist. 6: 687–94.

———. 1973a. The ecology and behavior of the giant wood spider *Nephila maculata* (Fabricius) in New Guinea. Smithson. Contrib. Zool. 149: 1–76.

———. 1973b. The stabilimenta of *Nephila clavipes* and the origins of stabilimentum building in araneids. Psyche 80: 277–88.

———. 1974. The biology of some *Argiope* species from New Guinea: Predatory behaviour and stabilimentum construction (Araneae: Araneidae). Zool. J. Linn. Soc. 54: 145–59.

———. 1975. Evolution beyond the orb web: The web of the araneid spider *Pasilobus* sp., its structure, operation and construction. Zool. J. Linn. Soc. 56: 301–14.

———. 1976a. Discrimination between prey types: An innate component of the predatory behaviour of araneid spiders. Z. Tierpsychol. 41: 266–76.

———. 1976b. The ecology and behavior of *Nephila maculata*: A supplement. Smithson. Contrib. Zool. 218: 1–22.

———. 1977. Associations between flies and spiders: Bibiocommensalism and dipsoparasitism? Psyche 84: 150–57.

———. 1978. Thermoregulation in orb weavers: New descriptions of thermoregulatory postures and experiments on the effects of posture and coloration. Zool. J. Linn. Soc. 64: 87–102.

———. 1980. Comparative studies of the courtship and mating behavior of tropical araneid spiders. Pac. Insects Monogr. 36: 1–218.

Robinson, M. H., B. Robinson, and W. Graney. 1972. The predatory behavior of the nocturnal orb web spider *Eriophora fuliginea* (C. L. Koch) (Araneae: Araneidae). Rev. Peru. Entomol. 14: 304–15.

Robinson, M. H., and C. E. Valerio. 1977. Attack on large or heavily defended prey by tropical salticid spiders. Psyche 84: 1–10.

Roelofs, W. L., and R. T. Carde. 1974. Sex pheromones in the reproductive isolation of lepidopterous species. *In* M. C. Birch, ed. *Pheromones*. North-Holland, Amsterdam, pp. 96–114.

Ross, K., and R. L. Smith. 1979. Aspects of the courtship behavior of the black widow spider *Latrodectus hesperus* (Araneae, Theridiidae), with evidence for the existence of a contact sex pheromone. J. Arachnol. 7: 69–77.

Ross, S. M. 1970. *Applied Probability Models with Optimization Applications*. Holden-Day, San Francisco.

Rovner, J. S. 1968. Territoriality in the sheet-web spider *Linyphia triangularis*. Z. Tierspsychol. 25: 232–42.

———. 1980a. Adaptations for prey capture in oxyopid spiders: Phylogenetic implications. Proc. Int. Arachnol. Congr. 8: 233–37.

———. 1980b. Did oxyopid spiders evolve from an aerial web-building ancestor? Am. Arachnol. 22: 16.

———. 1984. Spider nests maintain a physical gill: flooding and the evolutionary origin of silk. Am. Arachnol. 30: 14.

Rovner, J. S., and F. G. Barth. 1981. Vibratory communication through living plants by a tropical wandering spider. Science 214: 464–66.

Rubenstein, D. I., and M. A. R. Koehl. 1977. The mechanisms of filter feeding: Some theoretical considerations. Am. Nat. 111: 951–94.

Rypstra, A. L. 1979. Foraging flocks of spiders: A study of aggregate behavior in *Cyrtophora citricola* (Forskål) (Araneae: Araneidae) in West Africa. Behav. Ecol. Sociobiol. 5: 291–300.

———. 1981. Building a better insect trap: An experimental investigation of prey capture in a variety of spider webs. Oecologia (Berlin) 52: 1–6.

———. 1983. The importance of food and space in limiting spider web densities; a test using field enclosures. Oecologia (Berlin) 59: 312–16.

———. 1985. Aggregations of *Nephila clavipes* (L.) (Araneae: Araneidae) in relation to prey availability. J. Arachnol. 13: 71–78.

Saaristo, M. I. 1975. On the evolution of the secondary genital organs of Lepthyphantinae (Araneae, Linyphiidae). Proc. Int. Arachnol. Congr. 6: 21–25.

Salzmann, L., and P. N. Witt. 1973. *Life on a Thread.* Blue Flower Films, Philadelphia.

Savory, T. H. 1928. *The Biology of Spiders.* Sidgwick and Jackson, London.

———. 1952. *The Spider's Web.* Frederick Warne, London.

———. 1960a. Notes on the biology of Arachnida. IV. J. Quekett Microsc. Club, Ser. 4, 5: 247–50.

———. 1960b. Spider webs. Sci. Am. 202 (October): 114–24.

———. 1971. *Evolution in the Arachnida.* Merrow Monographs, Zoology Series, London.

Schaefer, M. 1978. Some experiments in the regulation of population density in the spider *Floronia bucculenta* (Araneae: Linyphiidae). Symp. Zool. Soc. London 42: 203–10.

Scheuring, L. 1914. Die Angen der Arachnoideen, II. Zool. Jahrb. Abt. Anat. Ontog. Tiere 37: 369–464.

Schick, R. X. 1972. The early instars, larval feeding, and the significance of larval feeding in the crab spider genus *Misumenops* (Araneida: Thomisidae). Notes Arachnol. Southwest 3: 12–19.

Schoener, T. W. 1971. Theory of feeding strategies. Ann. Rev. Ecol. Syst. 2: 369–404.

Schoener, T. W., and C. A. Toft. 1983. Dispersion of a small island population of the spider *Metepeira datona* (Araneae: Araneidae) in relation to web site availability. Behav. Ecol. Sociobiol. 12: 121–28.

Seilacher, A. 1973. Fabricational noise in adaptive morphology. Syst. Zool. 22: 451–65.

Sekiguchi, K. 1952. On a new spinning gland found in geometric spiders and its function. Annot. Zool. Jpn. 25: 394–99.

Shear, W. A. 1967. Expanding the palpi of male spiders. Breviora Mus. Comp. Zool. Harv. Univ. 259: 1–27.

———. 1970a. The evolution of social phenomena in spiders. Bull. Br. Arachnol. Soc. 1: 65–76.

———. 1970b. Observations on the predatory behavior of the spider *Hypochilus gertschi* Hoffman (Hypochilidae). Psyche 76: 407–17.

———. 1981. Structure of the male palpal organ in *Mimetus*, *Ero*, and *Gelanor* (Araneoidea, Mimetidae). Bull. Am. Mus. Nat. Hist. 170: 257–62.

Sherriffs, W. R. 1931. South Indian arachnology, 5. Ann. Mag. Nat. Hist., Ser. 10, 7: 537–46.

Shimojana, M. 1965. Studies on morphology, habitats, and geographical distribution of *Heptathela kimurai* (Kishida). Biol. Mag. Okinawa 2: 1–8.

Shinkai, E. 1981. Observations of Asia's lasso spider (in Japanese). Atypus 79: 32.

Sholes, O. D., and J. E. Rawlins. 1979. Distribution of orb weavers (Araneidae: Araneae) in homogeneous old-field vegetation. Proc. Entomol. Soc. Wash. 81: 234–47.

Shorey, H. H. 1974. Environmental and physical control of insect sex pheromones. *In* M. C. Birch, ed. *Pheromones*. North-Holland, Amsterdam, pp. 62–80.

———. 1976. *Animal Communication by Pheromones*. Academic, New York.

Siegel, S. 1956. *Nonparametric Statistics for the Behavioral Sciences*. McGraw-Hill, New York.

Simon, E. 1864. *Histoire naturelle des araignées*. Libraire Encyclopédique de Roret, Paris.

———. 1873. Araignées nouvelles ou peu connues du midi de l'Europe. Mém. Soc. Liège 5: 187–351.

———. 1889. Arachnides. *In Voyage de M. E. Simon au Venezuela*. Ann. Soc. Entomol. Fr. 9: 169–220.

———. 1892. *Histoire naturelle des araignées*, Vol. 1. Paris.

———. 1901. *Histoire naturelle des araignées*, Vol. 2. Libraire Encyclopédique de Roret, Paris.

Simpson, G. G. 1953. *The Major Features of Evolution*. Columbia Univ. Press, New York.

Sivinski, J., and M. K. Stowe. 1980. A kleptoparasitic cecidomyiid and other flies associated with spiders. Psyche 87: 337–48.

Smith, C. P. 1908. A preliminary study of the Araneae Theraphosae of California. Ann. Entomol. Soc. Am. 1: 207–36.

Smith, D. T. 1982a. Evolution of sociality in spiders—The costs and benefits of communal behavior in *Philoponella oweni* (Araneae: Uloboridae). Ph.D. thesis, Cornell Univ., Ithaca, N.Y.

Smith, D. T. 1982b. Reproductive success of solitary and communal *Philoponella oweni* (Araneae: Uloboridae). Behav. Ecol. Sociobiol. 11: 149–54.

Smith, W. J. 1977. *The Behavior of Communicating*. Harvard Univ. Press, Cambridge, Mass.

Snyder, A. W., and W. H. Miller. 1978. Telephoto lens system of falconiform eyes. Nature (London) 275: 127–29.

Sokal, R., and F. J. Rohlf. 1973. *Introduction to Biostatistics*. W. H. Freeman, San Francisco.

Stearns, S. C. 1977. The evolution of life-history traits. Ann. Rev. Ecol. Syst. 8: 145–72.

Stern, H., and E. Kullmann. 1981. *Leben am seidenen Faden: Die rätselvolle Welt der Spinnen*. Kindler Verlag, Munich.

Stevenson, B. G., and D. L. Dindal. 1982. Effect of leaf shape on forest-litter spiders: Community organization and microhabitat selection of immature *Enoplognatha ovata* (Clerck) (Theridiidae). J. Arachnol. 10: 165–78.

Stevenson, C. W. 1908. On the spinning organs and architecture of *Euagrus*, a theraphosid araneid. Biol. Bull. 15: 105–10.

Stowe, M. K. 1978. Observations of two nocturnal orbweavers that build specialized webs: *Scoloderus cordatus* and *Wixia ectypa* (Araneae: Araneidae). J. Arachnol. 6: 141–46.

Stradling, D. J. 1981. The ecology of the tropical arboreal tarantula *Avicularia avicularia*. Paper presented at American Arachnological Society Meeting, Knoxville, Tenn.

Struhsaker, T. T. 1969. Notes on the spiders *Uloborus mundior* (Chamberlin and Ivie) and *Nephila clavipes* (Linnaeus) in Panama. Am. Midl. Nat. 82: 611–13.

Suginaga, A. 1963. The method of web construction by *Cyrtarachne bufo* and *Cyrtarachne inaequalis* (in Japanese). Atypus 31: 13–15.

Sundevall, J. C. 1833. Conspectus arachnidum. Londini Gothorum: 1–39.

Suter, R. B. 1978. *Cyclosa turbinata* (Araneae: Araneidae): Prey discrimination via web-borne vibration. Behav. Ecol. Sociobiol. 3: 283–96.

Szlep, R. 1961. Developmental changes in the web-spinning instinct of Uloboridae: Construction of the primary web. Behaviour 27: 60–70.

———. 1965. The web-spinning process and web-structure of *Latrodectus tridecimguttatus*, *L. pallidus*, and *L. revivensis*. Proc. Zool. Soc. London 145: 75–89.

———. 1966. Evolution of the web-spinning activities: The web spinning in *Titanoeca albomaculata* Luc. (Araneae, Amaurobiidae). Isr. J. Zool. 15: 83–88.

Theuer, B. 1954. Contributions to the life history of *Deinopis spinosus*. M.S. thesis, Univ. Fl., Gainesville.

Thompson, H. R. 1956. Distribution of distance to nth neighbor in a population of randomly distributed individuals. Ecology 37: 391–94.

Thompson, M. E. 1972. Notes on the biology of the genus *Mimetus*. Notes Arachnol. Southwest 3: 19.

Thorell, T. 1869. On European spiders. Nova Acta Regiae Soc. Sci. Ups. 7: 1–108.

———. 1886. On Dr. Bertkau's classification of the order Araneae, or spiders. Ann. Mag. Nat. Hist. 17: 301–26.

Thornhill, R. 1975. Scorpionflies as kleptoparasites of web-building spiders. Nature (London) 258: 709–11.

Tietjen, W. J. 1980. Sanitary behavior by the social spider *Mallos gregalis* (Dictynidae): Distribution of excreta as related to web density and animal movements. Psyche 87: 59–73.

———. 1981. An inexpensive computer-controlled image digitizer. Behav. Res. Methods Instrum. 13: 347–50.

———. 1982. Influence of activity patterns on social organization of *Mallos gregalis* (Araneae: Dictynidae). J. Arachnol. 10: 75–84.

Tietjen, W. J., and J. S. Rovner. 1982. Chemical communication in lycosids and other spiders. *In* P. N. Witt and J. S. Rovner, eds. *Spider Communication: Mechanisms and Ecological Significance*. Princeton Univ. Press, Princeton, N.J., pp. 249–80.

Tillinghast, E. K., E. J. Kavanaugh, and P. H. Kolbjornsen. 1981. Carbohydrates in the webs of *Argiope* spiders. J. Morphol. 169: 141–48.

Tilquin, A. 1942. *La toile géométrique des araignées*. Presses Universitaires, Paris.

Todd, V. 1945. Systematic and biological account of the New Zealand Mygalomorphae (Arachnida). Trans. R. Soc. N.Z. 74: 375–407.

Toft, S. 1976. Life histories of spiders in a Danish beechwood. Nat. Jutl. 19: 5–40.

Tolbert, W. W. 1975. Predator avoidance behaviors and web defensive structures in the orb weavers *Argiope aurantia* and *Argiope trifasciata* (Araneae, Araneidae). Psyche 82: 29–52.

———. 1976. Population dynamics of the orb-weaving species *Argiope trifasciata* and *Argiope aurantia* (Araneae: Araneidae): Density changes associated with mortality, natality, and migrations. Ph.D. diss., Univ. Tenn., Knoxville.

———. 1977. Aerial dispersal of two orb-weaving spiders. Psyche 84: 13–27.

———. 1979. Thermal stress of the orb-weaving spider *Argiope trifasciata* (Araneae). Oikos 32: 386–92.

Trail, D. S. 1980. Predation by *Argyrodes* (Theridiidae) on solitary and communal spiders. Psyche 87: 349–55.

Tretzel, E. 1961. Biologie, Ökologie, und Brutpflege von *Coelotes terrestris* (Wider) (Araneida, Agelenidae), II: Brütpflege. Z. Morphol. Ökol. Tiere 50: 375–542.

Trivers, R. L., and H. Hare. 1976. Haplodiploidy and the evolution of the social insects. Science 191: 249–63.

Turnbull, A. L. 1960. The prey of the spider *Linyphia triangularis* (Clerck) (Araneae, Linyphiidae). Can. J. Zool. 38: 859–73.

———. 1962. Quantitative studies of the food of *Linyphia triangularis*. Can. Entomol. 94: 1233–49.

———. 1964. The search for prey by a web-building spider, *Achaearanea tepidariorum* (C. L. Koch) (Araneae: Theridiidae). Can. Entomol. 96: 568–79.

———. 1965. The effects of prey abundance on the development of the spider *Agelenopsis potteri* (Araneae: Agelenidae). Can. Entomol. 97: 141–47.

———. 1966. A population of spiders and their prey in an overgrazed pasture in eastern Ontario. Can. J. Zool. 44: 557–83.

———. 1973. Ecology of the true spiders (Araneomorphae). Ann. Rev. Entomol. 18: 305–48.

Uetz, G. W., and J. Biere. 1980. Prey of *Micrathena gracilis* (Walckenaer) in comparison with artificial webs and other trapping devices. Bull. Br. Arachnol. Soc. 5: 101–7.

Uetz, G. W., and J. W. Burgess. 1979. Habitat structure and colonial behavior in *Metepeira spinipes* (Araneae: Araneidae), an orb-weaving spider from Mexico. Psyche 86: 79–89.

Uetz, G. W., and K. R. Cangialosi. In Press. Genetic differences in social behavior and spacing in populations of *Metepeira spinipes* Pickard-Cambridge (Araneae: Araneidae), a communal territorial orb weaver. J. Arachnol.

Uetz, G. W., A. D. Johnson, and D. W. Schemske. 1978. Web placement, web structure, and prey capture in orb-weaving spiders. Bull. Br. Arachnol. Soc. 4: 141–48.

Uetz, G. W., T. C. Kane, and G. E. Stratton. 1982. Variation in the tendency for social grouping of a communal web-building spider. Science 217: 547–49.

Uetz, G. W., T. C. Kane, G. E. Stratton, and M. J. Benton. In Press. Environmental and genetic influences on the social grouping tendency of a communal spider. *In* M. A. Huettel, ed. *Evolutionary Genetics of Invertebrate Behavior*. New York.

Uyemura, T. 1966. On the spiders called tarantula in Europe and in America. J. Jonan High School 3: 19–25.

———. 1967. On the theraphosid spider *Chilobrachys andersoni* from Thailand. Acta Arachnol. 20: 37–45.

Valerio, C. E. 1974a. Feeding on eggs by spiderlings of *Achaearanea tepidariorum* (Araneae: Theridiidae), and the significance of the quiescent instar in spiders. J. Arachnol. 2: 57–63.

———. 1974b. Prey capture by *Drymusa dinora* (Araneae: Scytodidae). Psyche 81: 284–87.

———. 1975. Population structure in the spider *Achaearanea tepidariorum* (Araneae, Theridiidae). J. Arachnol. 3: 185–90.

Van Dam, G., and A. Roberts. 1917. Notes on nests of some trapdoor spiders and the nest of *Calommata transvaalicus* Hwtt. Ann. Transvaal Mus. 5: 218–33.

Vane-Wright, R. I. 1980. On the definition of mimicry. Biol. J. Linn. Soc. 13: 1–6.

van Helsdingen, P. J. 1965. Sexual behaviour of *Lepthyphantes leprosus* (Ohlert) (Araneida, Linyphiidae), with notes on the function of the genital organs. Zool. Meded. Rijks Mus. Nat. Hist. Leiden 4: 15–42.

———. 1969. A reclassification of the species of *Linyphia* Latreille based on the functioning of the genitalia (Araneida: Linyphiidae), I. Zool. Verh. Rijksmus. Nat. Hist. Leiden 105: 1–303.

Van Wingerden, W. K. R. E. 1978. Population dynamics of *Erigone arctica* (White) (Araneae: Linyphiidae), II. Symp. Zool. Soc. London 42: 195–202.

Van Wingerden, W. K. R. E., and H. F. Vughts. 1974. Factors influencing aeronautic behaviour in spiders. Bull. Br. Arachnol. Soc. 3: 6–10.

Vincent, L. S. 1980. The population biology of *Atypoides riversi* O. P. Cambridge, a fossorial mygalomorph spider (Araneae: Antrodiaetidae). Ph.D. thesis, Univ. Calif., Berkeley.

Visscher, P. K. 1980. Adaptations of honeybees (*Apis mellifera*) to the problems of nest hygiene. Sociobiology 5: 249–60.

Vollrath, F. 1976. Konkurrenzvermeidung bei tropischen kleptoparasitischen Haubennetzspinnen der Gattung *Argyrodes* (Arachnida: Araneae: Theridiidae). Entomol. Ger. 3: 104–8.

———. 1978. A close relationship between two spiders (Arachnida, Araneidae): *Curimagua bayano* synoecious on a *Diplura* species. Psyche 85: 347–53.

———. 1979a. Behaviour of the kleptoparasitic spider *Argyrodes elevatus* (Araneae, Theridiidae). Anim. Behav. 27: 515–21.

———. 1979b. Vibrations: Their signal function for a spider kleptoparasite. Science 205: 1149–51.

Vughts, H. F., and W. K. R. E. Van Wingerden. 1976. Meteorological aspects of aeronautic behavior in spiders. Oikos 27: 433–44.

Walckenaer, C. A. 1802. Insectes, ou Histoire abrégée des insectes des environs de Paris. *In Faune parisienne*, Vol. 2. Paris, pp. 187–250.

———. 1837–47. *Histoire naturelle des insectes aptères*, 4 vols. Paris.

Walcott, C. 1963. The effect of the web on vibration sensitivity in the spider *Achaearanea tepidariorum* (Koch). J. Exp. Biol. 40: 595–611.

———. 1969. A spider's vibration receptor: Its anatomy and physiology. Am. Zool. 9: 133–44.

Walcott, C., and W. G. Van der Kloot. 1959. The physiology of the spider vibration receptor. J. Exp. Zool. 141: 191–244.

Waldorf, E. S. 1976. Spider size, microhabitat selection, and use of food. Am. Midl. Nat. 96: 76–87.

Wanless, F. R. 1978. A revision of the spider genus *Portia* (Araneae: Salticidae). Bull. Br. Mus. Nat. Hist. Zool. 34: 83–124.

———. 1979. A revision of the spider genus *Brettus* (Araneae: Salticidae). Bull. Br. Mus. Nat. Hist. Zool. 35: 183–90.

———. 1980a. A revision of the spider genera *Asemonea* and *Pandisus* (Araneae: Salticidae). Bull. Br. Mus. Nat. Hist. Zool. 39: 213–57.

———. 1980b. A revision of the spider genus *Macopaeus* (Araneae: Salticidae). Bull. Br. Mus. Nat. Hist. Zool. 35: 219–23.

———. 1980c. A revision of the spider genus *Onomastus* (Araneae: Salticidae). Bull. Br. Mus. Nat. Hist. Zool. 39: 179–88.

———. 1980d. A revision of the spider genus *Orthrus* (Araneae: Salticidae). Bull. Br. Mus. Nat. Hist. Zool. 38: 225–32.

———. 1981a. A revision of the spider genus *Cocalus* (Araneae: Salticidae). Bull. Br. Mus. Nat. Hist. Zool. 41: 253–61.

———. 1981b. A revision of the spider genus *Hispo* (Araneae: Salticidae). Bull. Br. Mus. Nat. Hist. Zool. 41: 179–98.

———. 1981c. A revision of the spider genus *Phaeacius* (Araneae: Salticidae). Bull. Br. Mus. Nat. Hist. Zool. 41: 199–212.

———. 1983. A revision of the spider genus *Cocalodes* with a description of a new related genus (Araneae: Salticidae). Bull. Br. Mus. Nat. Hist. Zool.

———. 1984. A review of the spider subfamily Spartaeinae nom. n. (Araneae: Salticidae), with descriptions of six new genera. Bull. Br. Mus. Nat. Hist. Zool. 46 (2): 135–205.

Wąsowska, S. 1969. Particularités de l'appareil séricigène des mâles *Clubiona phragmitis* C. L. Koch. Zool. Pol. 19: 505–15.

———. 1970. Structures fileuses extérieures sur les filières (thelae) de l'araignée *Argiope bruennichi* (Scopoli). Zool. Pol. 20: 257–68.

Wellington, W. G. 1957. Individual differences as a factor in population dynamics: The development of a problem. Can. J. Zool. 35: 293–323.

Westring, N. 1861. Araneae Suecici. Göteborg K. Vet. Handel. 7: 1–615.

Whitcomb, W. H. 1962. Egg-sac construction and oviposition of the green lynx spider *Peucetia viridans* (Oxyopidae). Southwest. Nat. 7: 198–201.

Whitcomb, W. H., and K. Bell. 1964. Predaceous insects, spiders, and mites of Arkansas cotton fields. Bull. Univ. Arkansas Agric. Exp. Stn. 690: 1–84.

Whitcomb, W. H., and R. Eason. 1965. The mating behavior of *Peucetia viridans* (Araneida: Oxyopidae). Fla. Entomol. 48: 163–67.

———. 1967. Life history and predatory importance of the striped lynx spider (Araneida: Oxyopidae). Proc. Arkansas Acad. Sci. 21: 54–58.

Whitcomb, W. H., and M. Tadic. 1963. Araneida as predators of the fall webworm. J. Kans. Entomol. Soc. 36: 186–90.

Whitham, T. G. 1980. The theory of habitat selection: Examined and extended using *Pemphigus* aphids. Am. Nat. 115: 449–66.

Wickler, W. 1968. *Mimicry in Plants and Animals*. Weidenfeld and Nicholson, London.

Wiehle, H. 1927. Beiträge zur Kenntnis des Radnetzbaues der Epeiriden, Tetragnathiden und Uloboriden. Z. Morphol. Ökol. Tiere 8: 468–537.

———. 1928. Beiträge zur Biologie der Araneen, insbesondere zur Kenntnis des Radnetzbaues. Z. Morphol. Ökol. Tiere 11: 115–51.

———. 1931. Neue Beiträge zur Kenntnis des Fanggewebes der Spinnen aus den Familien Argiopidae, Uloboridae und Theridiidae. Z. Morphol. Ökol. Tiere 23: 349–400.

———. 1933. Theridiidae oder Haubennetzspinnen (Kugelspinnen). *In* F. Dahl, ed. *Die Tierwelt Deutschlands*, Vol. 26. Gustav Fischer Verlag, Jena, pp. 119–222.

Williams, D. S. 1979. The feeding behaviour of New Zealand *Dolomedes* species (Araneae: Pisauridae). N. Z. J. Zool. 6: 95–105.

———. 1981. A jumping spider's view. Hemisphere 26: 178–83.

Williams, D. S., and P. McIntyre. 1980a. A deep fovea enhances visual acuity in animals with small heads. Proc. Aust. Physiol. Pharmacol. Soc. 11: 182.

———. 1980b. The principal eyes of a jumping spider have a telephoto component. Nature (London) 288: 578–80.
Williams, G. C. 1964. Measurement of consociation among fishes: Comments on the evolution of schooling. Publ. Mich. State Biol. Serv. 2: 351–83.
———. 1966. *Adaptation and Natural Selection*. Princeton Univ. Press, Princeton, N.J.
Williams, R. 1981. Environmental-buffering effects of the web of *Mallos gregalis* (Araneae, Dictynidae). M.S. thesis, Georgia College, Milledgeville.
Wilson, E. O. 1965. Chemical communication in social insects. Science 149: 1064–71.
———. 1971. *The Insect Societies*. Belknap, Cambridge, Mass.
———. 1975. *Sociobiology*. Belknap, Cambridge, Mass.
Wilson, R. S. 1962a. The control of dragline spinning in the garden spider. Q. J. Microsc. Sci. 104: 557–71.
———. 1962b. The structure of the dragline control valves in the garden spider. Q. J. Microsc. Soc. 103: 549–55.
———. 1969. Control of dragline spinning in certain spiders. Am. Zool. 9: 103–11.
Wise, D. H. 1975. Food limitation of the spider *Linyphia marginata*: Experimental field studies. Ecology 56: 637–46.
———. 1979. Effects of an experimental increase in prey abundance upon the reproductive rates of two orb-weaving spider species (Araneae: Araneidae). Oecologia (Berlin) 41: 289–300.
———. 1982. Predation by a commensal spider, *Argyrodes trigonum*, upon its host: An experimental study. J. Arachnol. 10: 111–16.
Witt, P. N. 1965. Do we live in the best of all possible worlds? Spider webs suggest an answer. Perspect. Biol. Med. 8: 475–87.
———. 1971. Drugs alter web building of spiders. Behav. Sci. 16: 98–113.
———. 1975. The web as a means of communication. BioSci. Commun. 1: 7–23.
———. 1978. Patterns of behavior in solitary and communal spiders. Soc. Neurosci. Abstr. 4.
Witt, P. N., J. O. Rawlings, and C. F. Reed. 1972. Ontogeny of web-building behavior in two orb-weaving spiders. Am. Zool. 12: 445–54.
Witt, P. N., and C. F. Reed. 1965. Spider web-building. Science 149: 1190–97.
Witt, P. N., C. F. Reed, and D. B. Peakall. 1968. *A Spider's Web: Problems in Regulatory Biology*. Springer-Verlag, Berlin.
Witt, P. N., and J. S. Rovner, eds. 1982. *Spider Communication: Mechanisms and Ecological Significance*. Princeton Univ. Press, Princeton, N.J.
Witt, P. N., M. B. Scarboro, and D. B. Peakall. 1978. Comparative feeding data in three spider species of different sociality: *Araneus diadematus* Cl., *Mallos trivittatus* (Banks), and *Mallos gregalis* (Simon). Symp. Zool. Soc. London 42: 89–97.
Work, R. W. 1976. The force-elongation behavior of web fibers and silks forcibly obtained from orb-web-spinning spiders. Text. Res. J. 46: 485–92.
———. 1977. Dimensions, birefringences, and force-elongation behavior of major and minor ampullate silk fibers from orb-web-spinning spiders—The effects of wetting on these properties. Text. Res. J. 47: 650–62.
———. 1981. Web components associated with the major ampullate silk fibers of orb-web spiders. Trans. Am. Microsc. Soc. 100: 1–20.
Workman, T. 1896. *Malaysian Spiders*, Vol. 1. Published by the author, Belfast.
Wunderlich, J. 1980. Sternal-Organe der Theridiosomatidae—Eine bisher übersehene Autapomorphie (Arachnida: Araneae). Verh. Naturwiss. Ver. Hamburg, N.F., 23: 255–57.

Yates, J. H. 1968. *Spiders of Southern Africa*. Books of Africa, Cape Town.

Yoshida, H. 1981. Seven new species of the genus *Octonoba* (Araneae: Uloboridae) from the Ryukyus, Japan. Acta Arachnol. 30: 21–32.

Yoshida, H., and Y. Suzuki. 1981. Silk as a cue for mate location in the jumping spider. *Carrhotus xanthogramma* (Latreille) (Araneae: Salticidae). Appl. Entomol. Zool. 16: 315–17.

Yoshida, M. 1977. Web site selection of 4 spider species at an inhabited stream site, and analysis of the factors of coexistence in relation to prey abundance (in Japanese). Acta Arachnol. 27: 261–82.

Zapfe, H., and W. J. Gertsch. 1955. Filogenia y función en *Austrochilus manni*. Trab. Lab. Zool. Univ. Chile 2: 5–52.

Index

Library of Congress Cataloging-in-Publication Data
Main entry under title:

Spiders—webs, behavior, and evolution.

Papers presented at the 1981 meeting of the American Arachnological Society at the University of Tennessee.
Bibliography: p.
Includes index.
1. Spiders—Congresses. I. Shear, William A. II. American Arachnological Society. III. Title: Spiders.
QL458.4.S65 1986 595.4'4 83-42833
ISBN 0-8047-1203-4 (alk. paper)